Progress in Mathematics
Volume 245

Gregor Fels
Alan Huckleberry
Joseph A. Wolf

Cycle Spaces of Flag Domains

A Complex Geometric Viewpoint

Birkhäuser
Boston • Basel • Berlin

Gregor Fels
Universität Tübingen
Fakultät für Mathematik und Physik
D-72076 Tübingen
Germany

Alan Huckleberry
Ruhr-Universität Bochum
Institut für Mathematik
D-44780 Bochum
Germany

Joseph A. Wolf
University of California
Department of Mathematics
Berkeley, CA 94720
USA

AMS Subject Classifications: 22E46, 32L25, 32N10, 32Q28, 53C30 (primary); 22F30, 32F10, 32M10, 81R25, 81S10 (secondary)

Library of Congress Control Number: 2005936610

ISBN-10 0-8176-4391-5   e-ISBN: 0-8176-4479-2
ISBN-13 978-0-8176-4391-1

Printed on acid-free paper.

Printed in the United States of America.   (JLS/EB)

9 8 7 6 5 4 3 2 1

www.birkhauser.com

Dedicated to our parents, our partners, and our children

Acknowledgments

The authors wish to acknowledge their gratitude for support of this work from the following institutions:

For Gregor Fels: Research was partially supported by a *Habilitationsstipendium* of the DFG.

For Alan Huckleberry: Research was partially supported by the DFG Schwerpunkt "Global methods in complex geometry," by the SFB/Tr 12 of the DFG, and by the JSPS.

For Joseph Wolf: Research was partially supported by NSF Grants DMS 99-88643 and DMS 04-00420, and by the DFG Schwerpunkt "Global methods in complex geometry."

Contents

Introduction

This research monograph is a systematic exposition of the background, methods, and recent results in the theory of cycle spaces of flag domains. Some of the methods are now standard, but many are new. The exposition is carried out from the viewpoint of complex algebraic and differential geometry. Except for certain foundational material, which is readily available from standard texts, it is essentially self-contained; at points where this is not the case we give extensive references.

After developing the background material on complex flag manifolds and representation theory, we give an exposition (with a number of new results) of the complex geometric methods that lead to our characterizations of (group theoretically defined) cycle spaces and to a number of consequences. Then we give a brief indication of just how those results are related to the representation theory of semisimple Lie groups through, for example, the theory of double fibration transforms, and we indicate the connection to the variation of Hodge structure. Finally, we work out detailed local descriptions of the relevant full Barlet cycle spaces.

Cycle space theory is a basic chapter in complex analysis. Since the 1960s its importance has been underlined by its role in the geometry of flag domains, and by applications in the representation theory of semisimple Lie groups. This developed very slowly until a few of years ago when methods of complex geometry, in particular those involving Schubert slices, Schubert domains, Iwasawa domains and supporting hypersurfaces, were introduced. In the late 1990s, and continuing through early 2002, we developed those methods and used them to give a precise description of cycle spaces for flag domains. This effectively enabled the use of double fibration transforms in all flag domain situations. That has very interesting consequences for the geometric construction of representations of semisimple Lie groups, especially for the construction of singular representations. It also has many potential interesting consequences for automorphic cohomology, other aspects of variation of Hodge structure, and moduli of compact complex manifolds. In this book we prove these recent results, filling in the background as necessary, and present new results that complete the picture.

Let us now roughly outline the contents of the book. Detailed references are, of course, contained throughout. Further notes and comments on a larger scope of the theory are contained at the end of each chapter.

We begin with a real linear semisimple group G_0 which for all practical purposes can be assumed to be simple. It is embedded in its complexification G and acts naturally on every G-homogeneous manifold $Z = G/Q$. Here we restrict to the case of flag manifolds; in other words Z is assumed to be compact Kähler, or equivalently Z is G-equivariantly projective algebraic. Also equivalently, Q is a parabolic subgroup.

Part I, "Introduction to Flag Domain Theory," is primarily devoted to preliminary and foundational material, and also to older results concerning flag manifolds and flag domains. Chapter 1 starts with a review of structure and finite-dimensional representation theory for semisimple Lie groups and algebras, introduces the structure theory for parabolic subalgebras and parabolic subgroups and ends with a discussion of homogeneous vector bundles and the Bott–Borel–Weil Theorem.

Chapter 2 contains the first combinatorial results for the G_0-action on Z. In particular, it is shown that the set $\mathrm{Orb}_Z(G_0)$ of G_0-orbits in Z is finite. Hence, by dimension at least one G_0-orbit is open in Z. This research monograph is devoted to the study of the cycle spaces of G_0-orbits on Z.

In Chapter 3 we give a complete description of the G_0-orbit structure for the case where G_0 is the group of a bounded symmetric domain and Z is the dual compact hermitian symmetric space, for example where G_0 corresponds to the open unit ball in $\mathbb{C}^n$ and $Z = \mathbb{CP}_n$.

We then turn our attention, in Chapter 4, to a discussion of the first results for the open G_0-orbits D in Z. We discuss their structure, compact subvarieties, and holomorphic functions. The compact subvarieties we discuss are of fundamental importance; they will be the base cycles in our treatment of cycle spaces.

A q-dimensional cycle in a complex manifold D is a linear combination $n_1C_1 + \cdots + n_kC_k$ where the coefficients are positive integers and the C_j are q-dimensional, irreducible compact subvarieties of D. One should think of a generic cycle as a compact subvariety $C = 1.C$. However, in general multiplicities cannot be avoided. For example, a family of quadric curves C in $\mathbb{P}_2(\mathbb{C})$ can degenerate to a sum $L_1 + L_2$ of two lines, and these can further degenerate to a double line. On the other hand, in the case of interest in this monograph the *base cycles* of our considerations are of the simple form $C = 1.C$.

We make serious use of certain recently developed methods in the theory of cycle spaces. These are explained in some detail in Section 7.4. In particular the complex structure on the space $\mathcal{C}_q(X)$ of q-dimensional cycles in a complex space X is explained. Then, also in Section 7.4, we explain the basics of a certain incidence geometry that leads to the construction of divisors and their associated meromorphic functions on the cycle space. This is of fundamental importance for our approach.

Cycles appear as follows in the study of open G_0-orbits D in Z: *Every maximal compact subgroup K_0 in G_0 possesses exactly one orbit C_0 in D that is a complex submanifold of Z.* This result is proved in Section 4.3. In this case the relevant cycle space would seem to be the connected component containing the q-dimensional cycle C_0 in $\mathcal{C}_q(D)$.

As a first step one considers a possibly smaller space $\mathcal{M}_D$ which is more easily defined by the group theory at hand. For this it is first appropriate to note that $\mathcal{C}_q(Z)$ is a locally finite-dimensional complex space whose irreducible components are compact projective varieties. The induced action of G is algebraic; in particular, the orbit $G.C_0$ is Zariski open in its closure X. Let $\mathcal{M}_D$ be the connected component of the intersection $G.C_0 \cap \mathcal{C}_q(D)$ which contains C_0. Therefore $\mathcal{M}_D$ is just the set of cycles C in $\mathcal{C}_q(D)$ for which there is a curve $\{g_t\}$ in G starting at the identity such that $\{g_t(C_0)\}$ is a curve in $\mathcal{C}_q(D)$ from C_0 to C.

Recent developments have shown that the component $\mathcal{C}$ of $\mathcal{C}_q(D)$ which contains C_0 may be much bigger than $\mathcal{M}_D$. A complete representation-theoretic local description of $\mathcal{C}$, i.e., of its tangent space at C_0, is given in Part IV. In fact global properties of $\mathcal{C}$ can often be derived from this description. However, with the exception of this local description, most of the present work is devoted to the global properties of cycle spaces of the type $\mathcal{M}_D$.

Also in Chapter 4 we look at the important case where the open orbit D carries an invariant measure, and, finally, in that case we construct a q-convex exhaustion function on D. This fact, which is proved in Section 4.7, implies that $\mathcal{C}$, and a fortiori $\mathcal{M}_D$, is a Stein space. A sketch of the complex analysis background required for this and related matters which appear throughout the monograph is presented in Section 4.6.

The q-convex exhaustion function on D gives us a very natural plurisubharmonic exhaustion function on $\mathcal{M}_D$. In some sense that Stein property was one of the first main goals of the theory, because it allows one to transfer a number of analytic considerations from the domain D, where more often than not there are no nonconstant holomorphic functions and cohomology can be unwieldy, to the cycle space $\mathcal{M}_D$ or to $\mathcal{C}$, which have optimal properties from the view of complex analysis. This reflects the philosophy of the *double fibration transformation* discussed in Chapter 14. In fact, in Part II we prove in general, by very different methods, that the cycle space is Stein.

At this point it should be mentioned that it is possible that $q = 0$. In this case $\mathcal{C} = \mathcal{M}_D = D$. Furthermore this only happens if D is a bounded symmetric domain $\mathcal{B}$ embedded as a G_0-orbit in its compact dual Z. This is a situation of great classical interest, but from the point of view of cycle spaces it can be regarded as well understood.

In the context of cycle spaces, the cases where G_0 is a group of hermitian type, in other words where K_0 has a positive-dimensional center, or, equivalently, where G_0/K_0 is a bounded symmetric domain, must occasionally be handled with special methods. This is primarily due to the fact that in the hermitian case the complexification $\mathfrak{k}$ of the Lie algebra $\mathfrak{k}_0$ of K_0 is not a maximal subalgebra of $\mathfrak{g}_0$.

Certain foundational results for the hermitian case are proved in Chapter 5. For example, the product $\mathcal{B} \times \overline{\mathcal{B}}$ of the bounded domain with its complex conjugate is identified as an open domain in G/K. Assuming that $\mathcal{M}_D$ is neither $\mathcal{B}$ nor $\overline{\mathcal{B}}$, in special cases that are easily identified, it is shown that $\mathcal{M}_D \subset \mathcal{B} \times \overline{\mathcal{B}}$. Later, in Section 9.1C, we use Schubert incidence geometry to show that $\mathcal{M}_D = \mathcal{B} \times \overline{\mathcal{B}}$ in this situation. In Chapter 5 this is shown for the classical groups of hermitian type by means of elementary flag geometry.

Part II, "Cycle Spaces as Universal Domains," is a systematic presentation of our recent work which describes in a precise way the cycle spaces associated to the

G_0-orbits in $Z = G/Q$. Let $\mathrm{Orb}_Z(G_0)$ denote the set of all G_0-orbits on Z. While much of the work is involved with the case of an open orbit D, any $\gamma \in \mathrm{Orb}_Z(G_0)$ has a naturally associated cycle space $\mathcal{C}(\gamma)$ that can be realized as an open set in a G-orbit in the appropriate cycle space $\mathcal{C}_q(Z)$. If γ is an open orbit D, it follows by definition that $\mathcal{C}(\gamma) = \mathcal{M}_D$.

The main result for the cycle domains $\mathcal{C}(\gamma)$ can be stated as follows: *For every real form G_0 there exists a precisely computable universal domain $\mathcal{U}$ so that, with a few well-understood exceptions which only occur in the hermitian cases, $\mathcal{C}(\gamma) = \mathcal{U}$ for all $\gamma \in \mathrm{Orb}_Z(G_0)$.*

The full cycle space $\mathcal{C}$ of an open orbit is considered later in Chapter 13. There, using somewhat delicate calculations of cohomology groups of certain homogeneous vector bundles on the base cycle C_0, it is shown that $\mathcal{C}$ is smooth, and the tangent space $T_{C_0}(\mathcal{C})$ is computed as a K-module.

For certain families of orbits of real forms G_0 in certain $Z = G/Q$ we show that this tangent space is just the tangent space of the orbit $G.C_0$. In such a case it follows that $\mathcal{C} = \mathcal{M}_D = \mathcal{U}$, where $\mathcal{U}$ is a certain universal domain. So in those cases there is nothing new.

For the remaining families it is shown that this tangent space is a K-direct sum of the tangent space of the G-orbit and a nontrivial concretely computed irreducible representation space. These cycle spaces $\mathcal{C}$ are therefore bigger than $\mathcal{M}_D$ and contain much more information than the universal domain $\mathcal{U}$.

We refer to the domain $\mathcal{U}$ as *universal*, because it appears as the basic G_0-manifold in a number of important contexts in algebraic and differential geometry, cycle space theory, and harmonic analysis. Its importance in representation theory was recognized early on by S. Gindikin, and it was described from the point of view of proper G_0-actions in 1990 by Akhiezer and Gindikin [AkG]. In this research monograph we look at it from a number of perspectives.

The definition of the universal domain can be motivated by considering possibilities for an appropriate complexification of the symmetric space G_0/K_0.

Let x_0 be the base point in the affine homogeneous space $\Omega := G/K$, i.e., the point $1K$ where G has isotropy subgroup K. Note the $\Omega_0 := G_0.x_0 \cong G_0/K_0$ is the symmetric space of noncompact type for G_0 and is embedded as a closed, totally real submanifold of Ω. Thus in a certain sense $\Omega = G/K$ can be regarded as a complexification of $\Omega_0 = G_0/K_0$. However, if one is looking for a G_0-invariant complexification where, e.g., invariant metrics or measures are available, then Ω is too large.

An appropriate smaller domain is defined as follows. Let $G_0 = K_0A_0N_0$ be an Iwasawa decomposition. Define the polyhedron ω_0 in $\mathfrak{a}_0$ to be the connected component containing $0 \in \mathfrak{a}_0$ of the intersection of the half-spaces

$$\left\{\xi \in \mathfrak{a}_0 \mid \alpha(\xi) \leq \frac{\pi}{2}\right\}$$

as α runs over the set of all roots. Then $\mathcal{U} := G_0.\exp(i\omega_0)(x_0)$.

In Part II we begin by describing results of Burns, Halverscheid and Hind [BHH] which show in particular that $\mathcal{U}$ can be naturally identified with the maximal domain of existence Ω_{adpt} of the *adapted complex structure* in the tangent bundle $T\Omega_0$. This underlines the complex differential geometric importance of $\mathcal{U}$.

Another basic result from [BHH] is proved in Section 6.3: *A G_0-invariant function $\rho : \mathcal{U} \to \mathbb{R}$ is plurisubharmonic if and only if its pullback to* $\mathfrak{u}$, *by the map* $\xi \mapsto \exp(i\xi)(x_0)$, *is convex.* Our proof, which is somewhat different from the original one, involves a description of the induced partial complex structure on the G_0-orbits in $\mathcal{U}$.

In Chapter 7 we introduce basic Schubert incidence geometry which is used for our description of $\mathcal{M}_D$, or more generally for $\mathcal{C}(\gamma)$ for any $\gamma \in \text{Orb}_Z(G_0)$.

A Schubert variety S is by definition the closure in Z of an orbit of a Borel subgroup B. The Borel groups that are appropriate for our considerations, i.e., for incidence geometry involving cycles, are those which contain the solvable group A_0N_0 of an Iwasawa decomposition $G_0 = K_0A_0N_0$. We refer to these as Iwasawa–Borel subgroups. These are just the Borel subgroups whose fixed points in Z lie on the unique closed G_0-orbit.

If $\mathcal{O}$ is an orbit of an Iwasawa–Borel group and $S = \mathcal{O}\dot{\cup}Y$ is the associated Schubert variety, then we consider the intersection of $\mathcal{O}$ with the base cycle C_0 in an open G_0-orbit D. For topological reasons there must be q-codimensional Schubert varieties with $S \cap C_0 \neq \emptyset$. For such an S one checks that this intersection is finite and is contained in $\mathcal{O}$. In fact, this holds not just for C_0 but for *every* $C \in \mathcal{M}_D$.

Thus there is a naturally associated B-invariant incidence variety,

$$I_Y := \{C \in \mathcal{C}_q(Z) \mid C \cap Y \neq \emptyset\},$$

which lies in the complement of $\mathcal{M}_D$ in $\mathcal{C}_q(Z)$. By intersecting it with the orbit $G.C_0$ and observing that K is contained in the G-stabilizer of C_0 we uniquely lift I_Y to a B-invariant variety H in the space $\Omega = G/K$ in which $\mathcal{U}$ lives. Doing the same for $\mathcal{M}_D$, we can directly compare $I_Y, \mathcal{U}$ and the cycle space.

Define $\mathcal{E}_H(\mathcal{M}_D)$ to be the connected component containing the neutral point x_0 of the complement in G/K of the closed set

$$\bigcup_{g \in G_0} g(H) = \bigcup_{k \in K_0} k(H).$$

This set is viewed as an envelope around $\mathcal{M}_D$ which is defined by the compact family $\{k(H)\}_{k \in K_0}$ of analytic hypersurfaces. Of course it depends on the Schubert cycle S, and therefore it is appropriate to define the Schubert domain $\mathcal{S}_D$ as the intersection of all of theses envelopes as S ranges over all q-codimensional Schubert varieties which have nonempty intersection with C_0. Obviously $\mathcal{M}_D \subset \mathcal{S}_D$.

As a first step in understanding $\mathcal{S}_D$ we consider an a priori smaller set, namely the intersection of the envelopes $\mathcal{E}_H$, where H ranges over *all* B-invariant hypersurfaces in G/K. Using properties of G_0-invariant plurisubharmonic function on $\mathcal{U}$, we show in Section 7.2B that this contains the universal domain $\mathcal{U}$. Since it is defined by the Iwasawa decomposition, we therefore refer to it as the Iwasawa envelope $\mathcal{E}_\mathcal{I}(\mathcal{U})$. The opposite inclusion $\mathcal{E}_\mathcal{I}(\mathcal{U}) \subset \mathcal{U}$, which is due to L. Barchini [Ba], is also proved in Section 7.2B. Thus one sees that $\mathcal{U}$ agrees with its Iwasawa envelope,

$$\mathcal{U} = \mathcal{E}_{\mathcal{I}}(\mathcal{U}).$$

It should be mentioned that, in the meantime, T. Matsuki has given a purely combinatorial proof of this fact [M4], i.e., a proof which avoids the use of analytic tools. This and other applications of his basic methods [M2, M3] could perhaps lead to complex analysis-free proofs of our results which are valid in more general settings. We prefer here to attempt to blend the geometric and analytic methods into the representation-theoretic setting.

The envelope $\mathcal{E}_{\mathcal{I}}(\mathcal{U})$ is a very natural object of consideration in harmonic analysis. In that context it is usually denoted by Ξ. It is known to be the maximal domain of analytic continuation of spherical functions, and for that reason its characterization as the universal domain is important. For this reason, using other methods which only handle the case of the classical groups, Krotz and Stanton have also shown that it agrees with $\mathcal{U}$ [KS].

As indicated above, our interest in the Iwasawa envelope originated from our effort to relate the Schubert domain $\mathcal{S}_D$ to $\mathcal{U}$. Due to the equality $\mathcal{U} = \mathcal{E}_{\mathcal{I}}(\mathcal{U})$ and the formal inclusion $\mathcal{E}_I(\mathcal{U}) \subset \mathcal{S}_D$, it then follows that $\mathcal{U} \subset \mathcal{S}_D$. In Chapter 9 we show that, in fact, $\mathcal{M}_D = \mathcal{S}_D$. This is done by constructing an incidence hypersurface H at every boundary point of $\mathcal{M}_D$. It should be remarked that using the results in Chapter 10 a simpler, but non constructive, proof of this can be given.

One of the essential concepts for our approach is that of a *Schubert slice*. In the case of an open G_0-orbit D is defined as follows.

Let C_0 be the base cycle in D, define $q := \dim C_0$, and let S be a q-codimensional Schubert variety defined by an Iwasawa–Borel group B such that $S \cap C_0 \neq \emptyset$. It is shown that this intersection is contained in the open B-orbit in S, and that it is finite and transversal (see Chapter 9). Also, for every $z \in S \cap C_0$ the orbit $\Sigma := A_0 N_0(z)$ is open in S and closed in D.

We refer to Σ as the *Schubert slice* determined by the base point z. In fact $S \cap D$ is the union of these Schubert slices. The closure $c\ell(\Sigma)$ of any Schubert slide Σ meets every G_0-orbit in D, and Σ meets every $C \in \mathcal{M}_D$ at exactly one point. See Chapter 9.

These results and the analogous statements for G_0-orbits of any dimension require a geometric understanding of the orbit duality theorem which was originally proved by combinatorial methods (in the case of open orbits in [W3] and in general by Matsuki [M2], [M3]).

The symplectic geometric approach to orbit duality is explained in Chapter 8. This is based on a fundamental idea of Uzawa, and in our context was first carried out in detail for the case of G/B in [MUV]. That was extended to the general case of $Z = G/Q$ by Bremigan and Lorch [BL], and in Chapter 8 we present their proof of the orbit duality theorem.

Having shown that $\mathcal{U} \subset \mathcal{M}_D$, in Chapter 10 we give a precise description of $\mathcal{M}_D$. In this matter the case of groups G_0 of hermitian type must be treated with care. The reason is that K is not a maximal (up to components) subgroup of G; it is the reductive part of two maximal parabolic subgroups $P_\pm = KS_\pm$, where G/P_+ and G/P_- are the compact duals of the bounded symmetric domains $\mathcal{B}$ and $\overline{\mathcal{B}}$ that correspond to G_0. If $\pi_\pm : G/K \to G/P_\pm$ are the natural projections and we regard $\mathcal{M}_D$ as a domain in $\Omega = G/K$ as above, then it is quite possible that $\mathcal{M}_D$ agrees with $\pi_-^{-1}(\mathcal{B})$ or $\pi_+^{-1}(\overline{\mathcal{B}})$. For example, if $D = \mathcal{B}$ or $D = \overline{\mathcal{B}}$, then this is obviously the case.

There are other interesting cases where this happens. However, in all such cases the base cycle is P_+- or P_--invariant, and it immediately follows that the true cycle space, i.e., the space $\mathcal{M}_D$ regarded in the cycle space $C_q(Z)$ and not lifted to G/K, is either $\mathcal{B}$ or $\overline{\mathcal{B}}$.

Otherwise it follows that there is some incidence hypersurface H which is neither a lift from a B-invariant hypersurface in G/P_+ nor from G/P_-. Note that for G_0 not of hermitian type, there is no possibility for such a lift. Thus, after this small discussion of the hermitian case, it is enough to consider $\mathcal{M}_D$ again as being lifted in G/K and being contained in an envelope $\mathcal{E}_\mathcal{H}(\mathcal{U})$, where H is not a lift.

In the language of algebraic geometry, this implies that the line bundle defined by the reduced divisor H in the cycle space $\mathcal{C}_q(Z)$ induces a finite G-equivariant map from the orbit $G.C_0$ onto $G/\check{K}$, where $\check{K}$ is at most a finite extension of K. Hence, up to a finite map, which in the end plays no role at all, one may regard $\mathcal{M}_D$ as being contained in G/K in the situation at hand.

In Chapter 10 it is shown that domains of the type $\mathcal{E}_\mathcal{H}$, where H is not a lift, are Kobayashi-hyperbolic. In particular there is no nonconstant map $f : \mathbb{C} \to \mathcal{M}_D$ in the situation where there is a hypersurface H in its complement which is not a lift. Then there is an exposition of the main result of [FH], which says that $\mathcal{U}$ is the maximal, G_0-invariant, Kobayashi-hyperbolic, Stein domain in G/K that contains the base point $x_0 = 1K$ corresponding to the base cycle C_0.

Now the envelope $\mathcal{E}_\mathcal{H}$ is clearly G_0-invariant and is shown to be Stein by classical methods. Since it contains the universal domain and the $\mathcal{U}$ is maximal with respect to the three basic properties, it follows that $\mathcal{E}_\mathcal{H} = \mathcal{U}$.

However, in this situation the cycle space has been trapped,

$$\mathcal{U} \subset \mathcal{M}_D = \mathcal{S}_D \subset \mathcal{E}_\mathcal{H}(\mathcal{U}) = \mathcal{U},$$

and therefore $\mathcal{M}_D = \mathcal{U}$ with the exception of the hermitian cases explained above.

The description of the cycles spaces associated to open G_0-orbits in Z is completed in Chapter 10, and in Chapter 11 we turn to the cycle space $\mathcal{C}(\gamma)$ of an arbitrary G_0-orbit $\gamma \in \mathrm{Orb}_Z(G_0)$. In order to define $\mathcal{C}(\gamma)$ it is important to recall that Matsuki duality states that for every $\gamma \in \mathrm{Orb}_Z(G_0)$ there exists $\kappa \in \mathrm{Orb}_Z(K)$ so that $\gamma \cap \kappa$ is compact, and conversely for every κ there exists such a γ. In this situation we refer to (γ, κ) as a *dual pair*.

In view of the duality it is natural to define the cycle space of γ to be

$$\mathcal{C}(\gamma) : \text{component of } 1K \text{ in } \{gK \in G/K \mid g(\kappa) \cap \gamma \text{ is nonempty and compact}\}.$$

This is well defined because κ is K-invariant. The point to having $\mathcal{C}(\gamma) \subset G/K$ is that we are going to compare it to $\mathcal{U}$ using the results of Chapter 10.

This definition of $\mathcal{C}(\gamma)$ (see [GM]) turns out to be the right one, but a priori there could be difficulties with it. The problem is that it is not clear whether the condition that $g(\kappa) \cap \gamma$ be compact, or even the condition that $g(c\ell(\kappa)) \cap \gamma$ be compact, is an open condition. Our classification work in Chapter 11 handles this difficulty and extends the open orbit result to this general case: *If G_0 is of hermitian type and $c\ell(\kappa)$ is P_- (respectively, P_+)-invariant, then $\mathcal{C}(\gamma)$ is $\mathcal{B}$ (respectively, $\overline{\mathcal{B}}$). Otherwise, $\mathcal{C}(\gamma) = \mathcal{U}$.*

The final chapter in Part II, Chapter 13, is devoted to three types of examples. First, we discuss very roughly the first results for $SL(n; \mathbb{R})$ which were proved by the Schubert slice method [HS]. Second, using Grassmann geometry, a differential geometric characterization of $\mathcal{U}$, and the characterization of $\mathcal{M}_D$ which was discussed above, it is shown that in many cases one can regard $\mathcal{M}_D$ as the differential-geometric product $\mathcal{B} \times \mathcal{B}$ of the riemannian symmetric space $\mathcal{B} = G_0/K_0$ with itself. The action on this realization is in very simple rational form. It should be emphasized that this holds, for example, for $\mathrm{Sp}(p, q)$, and not just for the hermitian case. (In the hermitian case $\mathcal{M}_D$ is biholomorphic to $\mathcal{B}$ or to $\mathcal{B} \times \overline{\mathcal{B}}$.) Third, we look at the simplest hermitian examples and compare the slice methods presented here with those coming from the classical theory (see [W12]).

Part III, "Analytic and Geometric Consequences," is where we apply the results of Parts I and II to the mechanism of the double fibration transform, as well as to certain other matters. We start in Chapter 14 with a general discussion of double fibration transforms. Under the proper assumptions the double fibration transform is a natural equivariant map $\mathcal{P} : H^q(D; \mathcal{O}(\mathbb{E})) \to H^0(\mathcal{M}_D; \mathcal{O}(\mathbb{E}'))$, where $\mathbb{E}$ is a homogeneous vector bundle on D and $\mathbb{E}'$ is a certain derived bundle on the cycle space.

It had long been conjectured that if $\mathbb{E}$ is sufficiently negative, then $\mathcal{P}$ should be injective, and would therefore realize the cohomological representation space in the simpler situation in a space of sections of a bundle. It had also been known that in order to prove this, it would be sufficient to prove that the fiber F of the map $\mu : \mathfrak{X} \to D$ from the universal family of cycles to D is contractible.

The main new development in Chapter 14 yields a proof of this fact. To do this one uses the characterization $\mathcal{M}_D = \mathcal{U}$ to show that $\mathcal{M}_D$ is a cell. Then one notes that every Schubert slice Σ defines an A_0N_0-equivariant holomorphic map $\mathcal{M}_D \to \Sigma$ which is a differentiable bundle and whose fiber is also F. Since Σ is a cell this bundle is trivial, and it then follows that F is also contractible.

With injectivity proved for the double fibration transform, one has another tool for studying possibly singular unitary representations of real reductive Lie groups. We end Chapter 14 with a sketch of the background and a number of references to the literature for this potential application.

One cannot say that so far there have been important applications of the theory of moduli of compact complex manifolds to the cycle space theory presented here, or conversely. However, period domains for moduli problems are quite often open G_0-orbits D in some flag manifold G/Q. It is quite possible that there will be stronger interaction between these subjects in the near future, and therefore we indicate the connections along with background material in Part III.

In Chapter 15 we present a very brief exposition of Griffiths' period map. The period domains in this case are closely related to hermitian symmetric spaces of noncompact type, i.e., to bounded symmetric domains. In the case of the moduli of marked K3 surfaces, the period domain is an open $SO(3, 19)$-orbit in a 20-dimensional quadric. The cycles which are in the G_0-orbit of the base cycle C_0, i.e., the set $\Omega_0 = G_0(1K)$ of real points in $\mathcal{M}_D$, can be interpreted in the context of quaternionic geometry. Further, this set of real points is a one-to-one correspondence with the set of Ricci-flat Calabi–Yau metrics on the underlying differentiable manifold of the K3 surface.

Part IV is devoted to considerations of the full cycle space $\mathcal{C}$ of an open G_0-orbit D, i.e., the irreducible component containing C_0 in $\mathcal{C}_q(D)$. This contains $\mathcal{M}_D$ as a closed submanifold [HoH]. We have already seen that it is possible that $\mathcal{C}$ is larger than $\mathcal{M}_D$. For example, if D is the unique open $SL(3;\mathbb{R})$-orbit on the full flag manifold of $SL(3;\mathbb{C})$, this is indeed the case; see Section 13.1C. On the other hand, in some cases it is quite clear that $\mathcal{M}_D = \mathcal{C}$, for example in the case of the open $SL(n+1;\mathbb{R})$-orbit in $\mathbb{P}_n$.

The results proved in Part IV give a local description of $\mathcal{C}$ at the base cycle C_0 in an arbitrary open G_0-orbit D in an arbitrary flag manifold $Z = G/Q$. The cases where $\mathcal{M}_D = \mathcal{C}$ are precisely described.

There are numerous series where $\mathcal{C}$ is larger, i.e., where $\mathcal{M}_D$ is a proper submanifold of $\mathcal{C}$, and where the generic cycle is not reachable from the base cycle by a transformation in G. In those cases it is shown that $\mathcal{C}$ is smooth at C_0, and its tangent space $T_{C_0}\mathcal{C}$ is precisely computed as a K-representation space. Although the calculations here are local, it is expected that together with certain fibration methods they will yield a global description of $\mathcal{C}$. These "new" cycle domains are the most natural ones from the complex geometric point of view, and we expect that they will be useful in representation theory.

The local description of $\mathcal{C}$ at C_0 actually has nothing to do with the domain D. This simply means that the K-module structure of the tangent space $T_{C_0}(\mathcal{C})$ is explicitly computed. Conceptually speaking, it is clear what needs to be done, but substantial technical work is required to reach the goal of precisely describing these spaces. Let us give a very rough summary of the necessary steps.

In Chapter 17 we compute the normal bundle $\mathbb{N}_Z(C_0)$ of a closed K-orbit C in Z in abstract terms, as a holomorphic K-homogeneous vector bundle. The goal is to prove the vanishing of $H^1(C_0; \mathcal{O}(\mathbb{N}_Z(C_0)))$, i.e., the smoothness of $\mathcal{C}_q(Z)$ at C_0, and to compute the tangent space $H^0(C_0; \mathcal{O}(\mathbb{N}_Z(C_0)))$ as a K-module.

In these introductory remarks we put aside the work required to prove the vanishing theorem. Let us only give an idea of computation of the tangent space in the (notationally simplest) measurable, non-hermitian case.

In the notation of Part IV, the tangent space in that case has a natural decomposition

$$H^0(C; \mathcal{O}(\mathbb{N}_Z(C))) = \mathfrak{s} \oplus H^1(C; \mathcal{O}(\mathbb{E}((\mathfrak{q} + \theta(\mathfrak{q}))_{\mathfrak{q}}))),$$

where $\mathfrak{g} = \mathfrak{k} + \mathfrak{s}$ is the decomposition of the Lie algebra $\mathfrak{g}$ of G under the Cartan involution θ of the real group G_0, and where $(\mathfrak{q} + \theta(\mathfrak{q}))_{\mathfrak{q}}$ is the $\mathfrak{s}$-component of $\mathfrak{q} + \theta(\mathfrak{q})$, and where $\mathbb{E}((\mathfrak{q} + \theta(\mathfrak{q}))_{\mathfrak{q}})$ is the K-homogeneous holomorphic vector bundle $K \times_{K \cap Q} (\mathfrak{q} + \theta(\mathfrak{q}))_{\mathfrak{q}})$ defined by the (isotropy) representation of $K \cap Q$ on $(\mathfrak{q} + \theta(\mathfrak{q}))_{\mathfrak{q}}$.

One is naturally led to apply the Bott–Borel–Weil Theorem to compute $H^1(C; \mathcal{O}(\mathbb{E}((\mathfrak{q}+\theta(\mathfrak{q}))_{\mathfrak{q}})$ as a K-module. Since the isotropy representation of $K \cap Q$ on $\mathfrak{q}_{\mathfrak{s}}$ is usually reducible, one must first compute the cohomology groups of the quotient bundles which arise from a natural filtration of $\mathbb{E}((\mathfrak{q}+\theta(\mathfrak{q}))_{\mathfrak{q}})$. The filtration is constructed so that the Bott–Borel–Weil Theorem can indeed be applied to these bundles. In the language of Section 18.1, one must compute the weights λ with $\lambda + \rho$ regular and of index 1 and 2.

A starting point for this computation is the description of the highest weights of the representation of $\mathfrak{k}$ on $\mathfrak{s}$. These highest weights are known, and for the convenience of the reader we recall the result in Section 17.4. In Chapter 18 we indicate the connection with weighted affine Dynkin diagrams.

Knowledge of these weights for a fixed Cartan subgroup (maximal complex torus) in the isotropy subgroup of K leads to knowledge of all weights of that torus on $\mathfrak{s}$. Then in turn one can calculate the weights λ on $(\mathfrak{q}_{\mathfrak{s}} + \theta(\mathfrak{q}_{\mathfrak{s}}))_{\mathfrak{s}}$ with $\lambda + \rho$ regular and of index 1 and 2. This work is carried out in Chapters 18 and 19. It gives an explicit method for computing the cohomology groups of the quotient bundles. The essential result is the String Lemma (see 18.4.4). The Cohomology Lemma (see 18.5.6) then gives the method of computing the original cohomology groups $H^*(C; \mathcal{O}(\mathbb{E}((\mathfrak{q}_{\mathfrak{s}} + \theta(\mathfrak{q}_{\mathfrak{s}}))_{\mathfrak{s}})))$ from those of the quotient bundles.

The String and Cohomology lemmas are quite explicit, and it is clear that within finite time one should be able compute the complementary space

$$H^1(C; \mathcal{O}(\mathbb{E}((\mathfrak{q}_{\mathfrak{s}} + \theta(\mathfrak{q}_{\mathfrak{s}}))_{\mathfrak{s}})))$$

to the tangent space $T_C(G.C)$ of the G-orbit $G.C$ in $T_C(\mathcal{C}_q(Z))$. Chapter 20 is devoted to these calculations. It is necessary to consider five infinite series and three exceptional cases. In most cases, depending on the weights which are determined by the base cycle at hand, both vanishing and nonvanishing occurs.

The work in Part IV is partially contained in the Tübinger Habilitationsschrift [Fe2] of the first author. There he carried out the computations for the case of the full flag manifold G/B. His work on the general case $Z = G/Q$ appears for the first time in this book.

We thank our many colleagues and collaborators over the years, too many to name here, who introduced us to the topics that come together to comprise this monograph. We thank Ann Kostant of Birkhäuser Boston for this opportunity to give a systematic presentation of the complex-geometric approach to cycle space theory.

Part I

Introduction to Flag Domain Theory

Overview

In this as well as the introductions to the other parts of the book we sketch our main goals and outline the structure of the part at hand. More detailed summaries are given at the beginning of every chapter.

Here we first present the basic background material on semisimple Lie algebras and Lie groups. Much of this is necessarily combinatorial in nature. Nevertheless, certain aspects are given from a complex geometric viewpoint which reflects the main spirit of this monograph.

Much of this book is devoted to the study of the action of a noncompact real form G_0 of a complex semisimple group G on a flag manifold $Z = G/Q$. The action of the complexification K of a maximal compact subgroup K_0 of G_0 is also of basic importance.

One special case is of particular interest in complex analysis and representation theory. This is the case where Z is a hermitian symmetric space of compact type and one of the open G_0-orbits on Z is the dual bounded symmetric domain of G_0. We go into substantial detail on this hermitian case in Part I because it provides an important class of explicit examples that serve to illustrate the general setting.

The closed K-orbits in Z play a fundamental role in our work. As a first example of duality, each is contained in a unique open G_0-orbit D, and there it is the unique complex K_0-orbit C. One is therefore led to study spaces of cycles in Z or D which contain C. The foundational material for the study of the open G_0-orbits and associated cycle spaces is provided in this part. First, complex analytic results are proved, e.g., on the holomorphic convexity of certain cycle spaces in the case where the open G_0-orbit possesses an invariant pseudo-Kähler metric.

Part I is organized as follows. Most of Chapter 1 is devoted to the Lie theoretic background which is used throughout the monograph. The last two sections, on homogeneous bundles and the Bott–Borel–Weil Theorem, are of particular importance both in our consideration of the double fibration transform (see Chapter 14 in Part III) and the cycle deformation theory of Part IV.

The brief Chapter 2 introduces the Bruhat decomposition. An important application, Corollary 2.1.3, proves the existence of certain Cartan algebras defined over the reals. That result is the key to understanding the isotropy subgroups of G_0 on Z.

Chapter 3 is devoted to the case where G_0 is of hermitian type. Substantial information on the action of G_0 on the associated compact hermitian symmetric space is given in this chapter. Later on in Sections 5.4 and 5.5 detailed results on the cycle spaces for the action of a hermitian real form on any of the associated flag manifolds Z are obtained.

The basic setup for cycle spaces associated to open G_0-orbits is presented in Chapter 4. In addition to providing foundational material, we discuss the holomorphic reduction and show that its base is a bounded symmetric domain (see Theorem 4.4.3). In this chapter we also introduce the notion of a *measurable* open G_0-orbit. There are a number of equivalent conditions for this (Theorem 14.5.1), one being the existence of an invariant pseudo-Kähler metric. This metric arises from invariant metrics on the anticanonical bundles of the open orbit and the ambient space Z, the latter being the

Kähler–Einstein metric on Z which is invariant under a compact real from of G. We then present a brief exposition of Levi geometry, oriented toward the notions of Stein manifold and various degrees of holomorphic convexity, notions which later translate into cohomology vanishing theorems for holomorphic vector bundles over open G_0-orbits. Chapter 4 ends with the construction of a certain exhaustion function for measurable open orbits, which makes those cohomology vanishing theorems explicit.

In Chapter 5 it is shown that the exhaustion of a measurable open G_0-orbit, which is constructed in Chapter 4 and is closely related to the above mentioned Kähler–Einstein metric, induces a K_0-invariant plurisubharmonic exhaustion of the cycle space $\mathcal{M}_D$. As a consequence it follows that in the measurable case $\mathcal{M}_D$ is a Stein manifold. The proof of this is given in Section 5.3. In fact, the Stein property, and much more, holds in general; see Theorems 11.3.1 and 11.3.7.

At the beginning of Chapter 5 we explain the *holomorphic hermitian* setting, the setting where the base cycle C is stabilized by a much larger group P than its defining group K. In fact P is a parabolic subgroup of G with reductive component K, the flag manifold G/P is the associated symmetric space of compact type, and the cycle space $\mathcal{M}_D$ is just the associated bounded domain. See Section 5.3 and Proposition 5.4.8.

1

Structure of Complex Flag Manifolds

In this chapter we recall some basic notions of semisimple Lie group structure, finite-dimensional representation theory, and Bott–Borel–Weil theory. The two objectives here are to establish a system of notation and to recall certain standard or near-standard results that we need later on. Section 1 consists of some basic definitions concerning Lie algebras, Lie groups, Cartan subalgebras, and Dynkin diagrams. Then in Section 2 we recall the Cartan highest weight theory for finite-dimensional representations of semisimple Lie algebras, and we record the Dynkin diagrams of the adjoint representations. We introduce parabolic subalgebras, parabolic subgroups, and complex flag manifolds, is Sections 3 and 4. Section 5 discusses homogeneous holomorphic vector bundles, and Section 6 shows how the Bott–Borel–Weil Theorem describes their Dolbeault (or sheaf) cohomologies.

1.1 Structure theory and the root decomposition

We assume that the reader has some familiarity with Lie groups and Lie algebras, but we recall some basic facts in order to establish notation and terminology. This material can, of course, be found in standard texts such as those of Hochschild [Hoch], Humphreys [Hu1], Knapp [Kna] and Varadarajan [V].

Let $\mathfrak{g}$ be a finite-dimensional Lie algebra over a field $\mathbb{F}$. In other words it is a finite-dimensional vector space over $\mathbb{F}$ with an algebra structure $\mathfrak{g} \times \mathfrak{g} \to \mathfrak{g}$, denoted $(\xi, \eta) \mapsto [\xi, \eta]$, that is alternating ($[\xi, \xi] = 0$) and satisfies the Jacobi identity

$$[\xi, [\eta, \zeta]] + [\eta, [\zeta, \xi]] + [\zeta, [\xi, \eta]] = 0.$$

The Lie algebra $\mathfrak{g}$ is

- **commutative** or **abelian** if the composition is identically zero, $[\mathfrak{g}, \mathfrak{g}] = 0$;
- **nilpotent** if the **descending central series** $\mathfrak{g}^0 = \mathfrak{g},\ \mathfrak{g}^{k+1} = [\mathfrak{g}, \mathfrak{g}^k]$ is eventually zero;
- **solvable** if the **derived series** $\mathfrak{g}^{(0)} = \mathfrak{g},\ \mathfrak{g}^{(k+1)} = [\mathfrak{g}^{(k)}, \mathfrak{g}^{(k)}]$ is eventually zero;
- **simple** if it has no proper ideal and is not commutative;

- **semisimple** if it has no nonzero solvable ideal;
- **reductive** if it is the direct sum of a semisimple ideal and a commutative ideal.

The following are equivalent: (i) $\mathfrak{g}$ is semisimple, (ii) $\mathfrak{g}$ is a direct sum of simple ideals, and (iii) the Killing form $\langle \xi, \eta \rangle$ of $\mathfrak{g}$ is nondegenerate as a bilinear form on the underlying vector space. Here recall that $\langle \xi, \eta \rangle = \text{trace}(\text{ad}(\xi)\, \text{ad}(\eta))$ where $\text{ad}(\xi)$ is the linear transformation $\zeta \mapsto [\xi, \zeta]$ of $\mathfrak{g}$.

From this point on, $\mathbb{F}$ is $\mathbb{R}$ or $\mathbb{C}$. Let G be a finite-dimensional Lie group over $\mathbb{F}$. We always denote the Lie algebra by the corresponding lower case German letter, in this case $\mathfrak{g}$. We say that G is **commutative** (respectively, **nilpotent**, respectively, **solvable**) if it has that property as a group. If G is connected it is equivalent to the corresponding property for $\mathfrak{g}$. In any case, we say that G is **simple** (respectively, **semisimple**, respectively, **reductive**) if $\mathfrak{g}$ is simple (respectively, semisimple, respectively, reductive), whether G is connected or not. In the setting of linear algebraic groups (and in much of this book) it is natural to add the condition that the component group G/G^0 be finite, but the greater generality of the definition above is more suitable in the representation theory of real reductive Lie groups.

There are two caveats here. First, a simple Lie group need not be simple as an abstract group; it may have (necessarily discrete) nontrivial center. Second, the meaning of *reductive* in the category of linear algebraic groups is different from the one we have just given, which applies to the category of finite-dimensional Lie groups. This distinction has no effect on the structure of complex flag manifolds or on the real group orbit structure of those flag manifolds, but it does have consequences for the representation theory of real reductive Lie groups; see [W2] and [W3].

The set of all linear transformations of a vector space V forms a Lie algebra with composition $[\xi, \eta] = \xi\eta - \eta\xi$. It is denoted $\mathfrak{gl}(V)$, and the ideal consisting of trace zero transformations is denoted $\mathfrak{sl}(V)$. If $V = \mathbb{F}^n$ one usually writes $\mathfrak{gl}(n; \mathbb{F})$ for $\mathfrak{gl}(V)$ and $\mathfrak{sl}(n; \mathbb{F})$ for $\mathfrak{sl}(V)$. The **adjoint representation** of $\mathfrak{g}$ is the Lie algebra homomorphism $\text{ad} : \mathfrak{g} \to \mathfrak{gl}(\mathfrak{g})$ given by $\text{ad}(\xi)(\eta) = [\xi, \eta]$.

If G is a Lie group with Lie algebra $\mathfrak{g}$, we denote the exponential map by $\exp : \mathfrak{g} \to G$. Here we view $\mathfrak{g}$ as the algebra of all right-invariant vector fields on the differentiable manifold G, and $t \mapsto \exp(t\xi)$ is the integral curve for ξ parameterized so that $\exp(0) = 1$. Now we have the **adjoint representation** G, the homomorphism of G into the group of invertible linear transformations of $\mathfrak{g}$ given by $\text{Ad}(g)(\xi) = \frac{d}{dt}|_{t=0}\, g \exp(t\xi) g^{-1}$. Evidently $\text{Ad}(G)$ is a subgroup of the group $\text{Aut}(\mathfrak{g})$ of automorphisms of $\mathfrak{g}$. If G is connected, then $\text{Ad}(G)$ is denoted $\text{Int}(\mathfrak{g})$ and called the group of **inner automorphisms** of $\mathfrak{g}$. Given $\mathfrak{g}$, we can express any connected G in the form $\widetilde{G}/Z$ where $\widetilde{G}$ is the connected simply connected Lie group with Lie algebra $\mathfrak{g}$ and Z is a discrete central subgroup; thus G and $\widetilde{G}$ lead to the same group $\text{Int}(\mathfrak{g})$, and $\text{Int}(\mathfrak{g})$ is well defined.

Since we will be dealing with real and complex Lie groups and Lie algebras at the same time, we need a way to distinguish them. We denote complex Lie groups by upper case Latin letters, and denote real Lie groups by upper case Latin letters with subscript 0, say G and G_0. We denote complex Lie algebras by lower case German letters, and denote real Lie algebras by lower case German letters with subscript 0, say $\mathfrak{g}$ and $\mathfrak{g}_0$. When a Lie group is denoted by some upper case Latin letter, its Lie algebra is denoted by the corresponding lower case German letter, so for example $\mathfrak{h}$ would be the Lie algebra of H and $\mathfrak{g}_0$ would be the Lie algebra of G_0. Given a Lie subalgebra $\mathfrak{h} \subset \mathfrak{g}$ where $\mathfrak{g}$ is the Lie algebra of G, we write H without comment for the analytic subgroup of G corresponding to H. Given Lie groups $G_0 \subset G$, we say that G_0 is a **real form** of G if $\mathfrak{g}$ is the complexification of $\mathfrak{g}_0$. And finally, given a real form G_0 of G and a Lie subgroup H of G stable under complex conjugation of G over G_0, we write H_0 for the real form $H \cap G_0$ of H.

There is one exception: we denote a compact real form of G (if it exists) by G_u and its Lie algebra by $\mathfrak{g}_u$. If G has only finitely many topological components, then it has a compact real form G_u if and only if it is reductive. In other words, G is reductive if and only if it is the complexification of a compact Lie group.

Let $\mathfrak{g}$ be a complex Lie algebra. A subalgebra $\mathfrak{h} \subset \mathfrak{g}$ is called a **Cartan subalgebra** if it is nilpotent and is equal to its own normalizer. Then $\mathfrak{g} = \mathfrak{h} + \sum_{\alpha \in \Sigma} \mathfrak{g}_\alpha$, where Σ is the set of all nonzero homomorphisms $\alpha : \mathfrak{h} \to \mathbb{C}$ such that the generalized joint eigenspace

$$\mathfrak{g}_\alpha = \{\eta \in \mathfrak{g} \mid (\mathrm{Ad}(\xi) - \alpha(\xi)\,\mathrm{Id})^n(\eta) = 0 \text{ for } n \gg 0 \text{ whenever } \xi \in \mathfrak{h}\}$$

is nonzero. Then the elements of $\Sigma = \Sigma(\mathfrak{g}, \mathfrak{h})$ are the **roots** of $\mathfrak{g}$ relative to $\mathfrak{h}$, the summands $\mathfrak{g}_\alpha$ are the **root spaces**, and $\mathfrak{g} = \mathfrak{h} + \sum_{\alpha \in \Sigma} \mathfrak{g}_\alpha$ is the **root space decomposition**.

Let $\mathfrak{g}_0$ be a real Lie algebra. Again, a subalgebra $\mathfrak{h}_0 \subset \mathfrak{g}_0$ is called a **Cartan subalgebra** if it is nilpotent and is equal to its own normalizer. Suppose that $\mathfrak{h}_0$ is a Cartan subalgebra. If the generalized joint eigenvalues for $\mathrm{Ad}(\mathfrak{h}_0)$ are real valued, then one has root space decomposition $\mathfrak{g}_0 = \mathfrak{h}_0 + \sum_{\alpha \in \Sigma} (\mathfrak{g}_0)_\alpha$ as above. In general, however, one must look at the way that $\mathfrak{g}_0$ meets the summands in the root space decomposition of $\mathfrak{g}$ for the Cartan subalgebra $\mathfrak{h}$.

Now let $\mathfrak{g}$ be a complex semisimple Lie algebra. Then the root space decomposition is much cleaner. First, the Cartan subalgebras are commutative, and they are all conjugate by the group $\mathrm{Int}(\mathfrak{g})$ of inner automorphisms of $\mathfrak{g}$. Fix one such Cartan subalgebra $\mathfrak{h}$. The root spaces $\mathfrak{g}_\alpha$ all are 1-dimensional. The Killing form is nondegenerate on $\mathfrak{h}$. In fact $\mathfrak{h}$ has a distinguished real form

$$\mathfrak{h}_{\mathbb{R}} = \{\xi \in \mathfrak{h} \mid \alpha(\xi) \in \mathbb{R} \text{ for all } \alpha \in \Sigma\},$$

and the Killing form is positive definite on $\mathfrak{h}_{\mathbb{R}}$.

The Killing form also satisfies several important conditions: $\langle \mathfrak{h}, \mathfrak{g}_\alpha \rangle = 0$ for all roots $\alpha \in \Sigma(\mathfrak{g}, \mathfrak{h})$; $\langle \mathfrak{g}_\alpha, \mathfrak{g}_\beta \rangle = 0$ if $\alpha + \beta \neq 0$; and $\langle \cdot, \cdot \rangle$ defines a nondegenerate pairing of $\mathfrak{g}_\alpha$ with $\mathfrak{g}_{-\alpha}$. This last condition can be formulated: the restriction of $\langle \cdot, \cdot \rangle$ to $\mathfrak{g}_\alpha + \mathfrak{g}_{-\alpha}$ is a nondegenerate bilinear form there. These conditions lead to an explicit classification of the complex reductive Lie algebras.

We just saw, implicitly, that $\alpha \in \Sigma = \Sigma(\mathfrak{g}, \mathfrak{h})$ implies $-\alpha \in \Sigma$. A **positive root system** is a subset $\Sigma^+ = \Sigma^+(\mathfrak{g}, \mathfrak{h})$ such that (i) Σ is the disjoint union of Σ^+ and $-\Sigma^+$, and (ii) if $\alpha, \beta \in \Sigma^+$ and $\alpha + \beta \in \Sigma$ then $\alpha + \beta \in \Sigma^+$. The **Weyl chambers** for $(\mathfrak{g}, \mathfrak{h})$ are the topological components of $\mathfrak{h}_{\mathbb{R}} \setminus (\cup_{\alpha \in \Sigma(\mathfrak{g}, \mathfrak{h})} \alpha^\perp)$ where

$\alpha^\perp$ denotes the root hyperplane $\{\xi \in \mathfrak{h}_\mathbb{R} \mid \alpha(\xi) = 0\}$. The Weyl chambers are convex open cones in $\mathfrak{h}_\mathbb{R}$, and they are in one-to-one correspondence $C \leftrightarrow \Sigma^+$ with the set of all positive root systems. If C and Σ^+ correspond then $C = \{\xi \in \mathfrak{h}_\mathbb{R} \mid \alpha(\xi) > 0 \text{ for all } \alpha \in \Sigma^+\}$ and $\Sigma^+ = \{\alpha \in \Sigma \mid \alpha(\xi) > 0 \text{ for all } \xi \in C\}$. Given Σ^+ we refer to the corresponding Weyl chamber as the **positive Weyl chamber**.

The **simple root system** or **root basis** corresponding to a positive root system $\Sigma^+ = \Sigma^+(\mathfrak{g}, \mathfrak{h})$ is the set of minimal positive roots, given by

$$\Psi = \{\psi \in \Sigma^+(\mathfrak{g}, \mathfrak{h}) \mid \psi \text{ is not a sum of two positive roots}\}.$$

A root $\alpha \in \Sigma^+$ is simple if and only if the corresponding root hyperplane $\alpha^\perp$ bounds the positive Weyl chamber at a nonzero point of its boundary. When Ψ is the simple root system for Σ^+, every root has form $\alpha = \sum_{\psi \in \Psi} n_\psi(\alpha)\psi$, where the $n_\psi(\alpha)$ are integers, either all $\geqq 0$ for $\alpha \in \Sigma^+$ or all $\leqq 0$ for $-\alpha \in \Sigma^+$. Evidently this gives a one-to-one correspondence $\Sigma^+ \leftrightarrow \Psi$ between the set of positive root systems and the set of simple root systems.

Let $\Psi = \{\psi_1, \dots, \psi_\ell\}$ denote a simple root system of $\mathfrak{g}$. The **Dynkin diagram** $\Delta = \Delta_\mathfrak{g}$ of $\mathfrak{g}$ is the graph whose set of vertices is Ψ, and in which vertices ψ_i and ψ_j are attached by $\frac{2\langle\psi_i,\psi_j\rangle}{\langle\psi_j,\psi_j\rangle}\frac{2\langle\psi_j,\psi_i\rangle}{\langle\psi_i,\psi_i\rangle}$ lines, with an arrowhead pointing toward ψ_i in case $\langle\psi_i, \psi_i\rangle < \langle\psi_j, \psi_j\rangle$. Simplicity of $\mathfrak{g}$ corresponds to connectedness of $\Delta_\mathfrak{g}$. Using the Bourbaki order on Ψ, the possibilities for connected Dynkin diagrams are

(1.1.1a)

(type A_ℓ, $\ell \geqq 1$)

(1.1.1b)

(type B_ℓ, $\ell \geqq 2$)

(1.1.1c)

(type C_ℓ, $\ell \geqq 3$)

(1.1.1d)
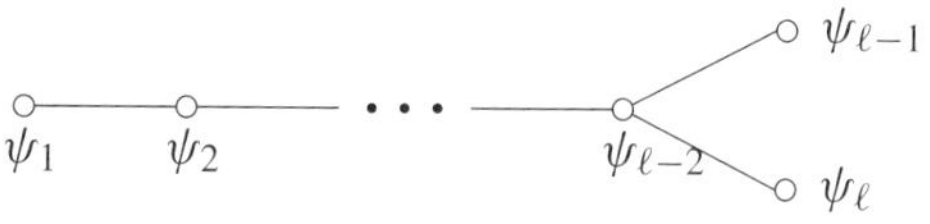

(type D_ℓ, $\ell \geqq 4$)

(1.1.1e)

(type G_2)

(1.1.1f)

(type F_4)

(1.1.1g)
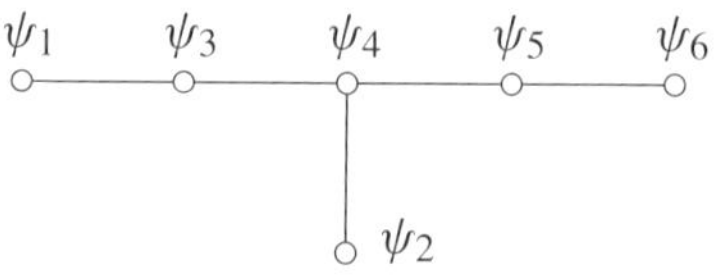

(type E_6)

(1.1.1h)

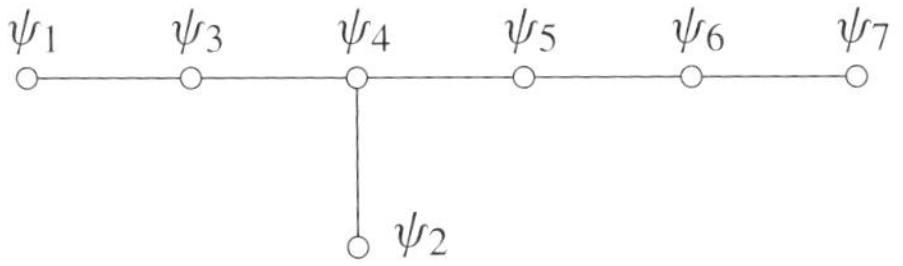

(type E_7)

(1.1.1i)

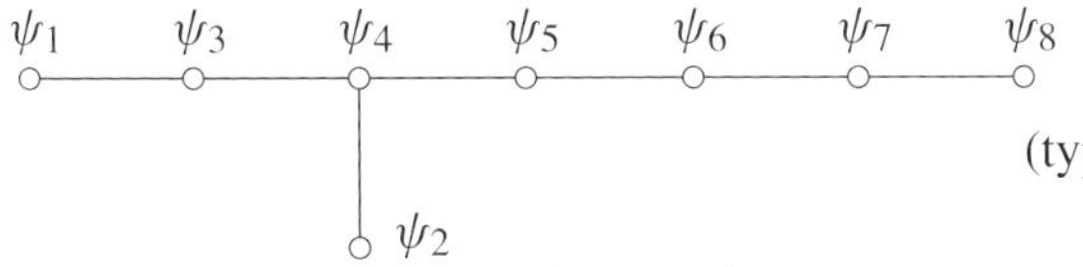

(type E_8)

If $\mathfrak{g}$ is a complex reductive Lie algebra, we decompose $\mathfrak{g} = \mathfrak{z} \oplus \mathfrak{g}'$ where $\mathfrak{z}$ is the center and $\mathfrak{g}' = [\mathfrak{g}, \mathfrak{g}]$ is semisimple, and we apply the above considerations to $\mathfrak{g}'$. Then the Cartan subalgebra $\mathfrak{h} = \mathfrak{z} \oplus \mathfrak{h}'$ where $\mathfrak{h}' = \mathfrak{h} \cap \mathfrak{g}'$ is a Cartan subalgebra of $\mathfrak{g}'$, the roots all vanish on $\mathfrak{z}$, $\Sigma(\mathfrak{g}, \mathfrak{h}) = \Sigma(\mathfrak{g}', \mathfrak{h}')$, and $\Sigma^+(\mathfrak{g}, \mathfrak{h}) = \Sigma^+(\mathfrak{g}', \mathfrak{h}')$. In this case $\mathfrak{h}_{\mathbb{R}}$ means $\mathfrak{h}'_{\mathbb{R}}$, and the definitions are exactly as above.

Let G be a complex reductive Lie group, and let $\mathfrak{h}$ be a Cartan subalgebra of $\mathfrak{g}$. The corresponding **Cartan subgroup** of G is the centralizer of $\mathfrak{h}$ in G, given by

$$H = Z_G(\mathfrak{h}) := \{g \in G \mid \mathrm{Ad}(g)\xi = \xi \text{ for all } \xi \in \mathfrak{h}\}. \tag{1.1.2}$$

If G is semisimple it is a linear algebraic group. If G is a connected algebraic group, then its Cartan subgroups are **algebraic tori**, groups isomorphic to $\mathbb{C}^* \times \cdots \times \mathbb{C}^*$. We reserve the term **tori** for (compact) products of circle groups.

One also has the normalizer of $\mathfrak{h}$ in G, given by

$$N_G(\mathfrak{h}) := \{g \in G \mid \mathrm{Ad}(g)\mathfrak{h} \subset \mathfrak{h}\}, \tag{1.1.3}$$

and the **Weyl group** of G is defined to be $W(G, H) = N_G(\mathfrak{h})/H$. The adjoint representation of G gives a faithful realization of $W(G, H)$ as a finite group of linear transformations of $\mathfrak{h}_{\mathbb{R}}$. If G is connected, then H is connected and $W(G, H)$ is realized as the group of linear transformations of $\mathfrak{h}_{\mathbb{R}}$ that is generated by the reflections[1] in the root hyperplanes $\alpha^\perp = \{\xi \in \mathfrak{h}_{\mathbb{R}} \mid \alpha(\xi) = 0\}$. That finite reflection group is defined, whether or not G is connected, to be the **Weyl group** of $\mathfrak{g}$ and denoted $W(\mathfrak{g}, \mathfrak{h})$.

Let G be a complex reductive Lie group, and let $\mathfrak{h}$ be a Cartan subalgebra of $\mathfrak{g}$. The Weyl group $W(\mathfrak{g}, \mathfrak{h})$ is simply transitive on the set of all positive root systems $\Sigma^+(\mathfrak{g}, \mathfrak{h})$. Now we have one to one correspondences between (i) the elements of the Weyl group $W(\mathfrak{g}, \mathfrak{h})$, (ii) the set of positive root systems $\Sigma^+(\mathfrak{g}, \mathfrak{h})$, (iii) the set of simple root systems Ψ, and (iv) the set of all Weyl chambers in $\mathfrak{h}_{\mathbb{R}}$.

1.2 Cartan highest weight theory

Let $\mathfrak{g}$ be a complex semisimple Lie algebra and $\mathfrak{h}$ a Cartan subalgebra. Let G be the (unique) connected simply connected Lie group with Lie algebra $\mathfrak{g}$. Let V be a

[1] Remember that the Killing form is positive definite on $\mathfrak{h}_{\mathbb{R}}$.

finite-dimensional complex vector space. We write $\mathfrak{gl}(V)$ for the complex Lie algebra of all linear transformations of V, with composition $[a, b] = ab - ba$. It is the Lie algebra of the complex reductive Lie group $GL(V)$, which consists of all invertible linear transformations of V with complex manifold structure given by the matrices for any choice of basis of V. By a **representation** of $\mathfrak{g}$ on V we mean a complex Lie algebra homomorphism $\nu : \mathfrak{g} \to \mathfrak{gl}(V)$. By a **representation** of G on V we mean a holomorphic homomorphism $\pi : G \to GL(V)$. In each case V is the **representation space**.

Since G is connected and simply connected, there is a one-to-one correspondence between representations of $\mathfrak{g}$ on V and representations of G on V. In effect, $\pi : G \to GL(V)$ defines the associated Lie algebra map $d\pi : \mathfrak{g} \to \mathfrak{gl}(V)$, and $\nu : \mathfrak{g} \to \mathfrak{gl}(V)$ integrates to a representation $\pi : G \to GL(V)$ such that $\nu = d\pi$. If G' is any connected Lie group with Lie algebra $\mathfrak{g}$, then it is a quotient G/Z of its universal covering group G by a finite central subgroup Z, $\pi' : G' \to GL(V)$ always defines a representation $d\pi'$ of $\mathfrak{g}$, but $\nu : \mathfrak{g} \to \mathfrak{gl}(V)$ defines a representation of G' if and only if the associated representation π of G annihilates Z.

A subspace of V is $\pi(G)$-invariant if and only if it is $d\pi(\mathfrak{g})$-invariant. Every (finite-dimensional complex) representation π of G is **semisimple** or **completely reducible** in the sense that every proper invariant subspace V_1 of V has an invariant complement V_2. So if π_i is the representation defined on the proper invariant subspace V_i then $V = V_1 \oplus V_2$, and we say that π is the **direct sum** $\pi_1 \oplus \pi_2$. That is equivalent to $d\pi = d\pi_1 \oplus d\pi_2$. If there is no proper invariant subspace we say that π and $d\pi$ are **irreducible**. The key fact here is the following.

Lemma 1.2.1. *Every finite-dimensional complex representation of a complex semisimple Lie algebra (or of a connected complex Lie group) is completely reducible, and thus is a direct sum of irreducible representations.*

This reduces the study of finite-dimensional complex representations of $\mathfrak{g}$ and G to the case of irreducible representations.

Two representations π_1 and π_2 of G are **equivalent** if there is an isomorphism $T : V_1 \cong V_2$ of the representation spaces that intertwines them in the sense that $T(\pi_1(g)(v)) = \pi_2(g)(T(v))$ for all $g \in G$ and all $v \in V_1$. Since G is connected, that is equivalent to the condition that T intertwines the Lie algebra representations $d\pi_1$ and $d\pi_2$. Therefore π_1 and π_2 are equivalent if and only if $d\pi_1$ and $d\pi_2$ are equivalent.

Fix a Cartan subalgebra $\mathfrak{h}$ of $\mathfrak{g}$ and an irreducible (finite-dimensional complex) representation ν of $\mathfrak{g}$ on V. Then $\nu(\mathfrak{h})$ is a commuting family of diagonalizable linear transformations, so $\nu(\mathfrak{h})$ is simultaneously diagonalizable. Thus $V = \sum_{\gamma \in M(\mathfrak{g},\nu)} V_\gamma$ where $M(\mathfrak{g}, \nu)$ is a set of linear functionals on $\mathfrak{h}$ and

$$V_\gamma = \{v \in V \mid \nu(\xi)(v) = \gamma(\xi)v \text{ for all } \xi \in \mathfrak{h}\} \neq 0.$$

$M(\mathfrak{g}, \nu)$ is the **weight system** of ν, its elements are the **weights** of ν, and if $\gamma \in M(\mathfrak{g}, \nu)$ then V_γ is the corresponding **weight space**. The elements of the various weight spaces are called **weight vectors**. Note that $M(\mathfrak{g}, \mathrm{ad}) = \Sigma(\mathfrak{g}, \mathfrak{h}) \cup \{0\}$.

In order to emphasize the weight system, rather than the representation, we sometimes write $M(\mathfrak{h}, V)$ for $M(\mathfrak{g}, \nu)$, where V is the representation space of ν.

Every weight takes real values on $\mathfrak{h}_{\mathbb{R}}$. Thus we may view $M(\mathfrak{g}, \nu) \subset \mathfrak{h}_{\mathbb{R}}^*$ and view $\mathfrak{h}_{\mathbb{R}}^*$ as the real span of the roots. More precisely, the set of all possible weights (ν variable) forms a lattice, the **weight lattice** Λ_{wt}, in $\mathfrak{h}_{\mathbb{R}}^*$. Similarly, the integer span of the roots is a lattice Λ_{rt} in $\mathfrak{h}_{\mathbb{R}}^*$ called the **root lattice**. If Ψ is the simple root system corresponding to any positive root system, then Ψ is an $\mathbb{R}$-basis of $\mathfrak{h}_{\mathbb{R}}^*$ and a $\mathbb{Z}$-basis of Λ_{rt}. Note $\Lambda_{\text{rt}} \subset \Lambda_{wt}$ by construction. The quotient $\Lambda_{wt}/\Lambda_{\text{rt}}$ is naturally isomorphic to the center of the connected simply connected complex Lie group G with Lie algebra $\mathfrak{g}$.

Now fix a positive root system $\Sigma^+ = \Sigma^+(\mathfrak{g}, \mathfrak{h})$. Choose a weight γ. Since $M(\mathfrak{g}, \nu)$ is finite we have a least integer $k \geqq 0$ such that $\nu(\sum_{\alpha\in\Sigma^+} \mathfrak{g}_\alpha)^{k+1}(V_\gamma) = 0$. Let v be a weight vector in $\nu(\sum_{\alpha\in\Sigma^+} \mathfrak{g}_\alpha)^k(V_\gamma)$, say $v = v_\lambda \in V_\lambda$. Then $\sum_{\ell\geqq 0} \nu(\mathfrak{h} + \sum_{\alpha\in\Sigma^+} \mathfrak{g}_{-\alpha})^\ell(v_\lambda)$ is an invariant subspace of V (this takes some computation), so it is equal to V by irreducibility. We say that a weight γ is **less than** a weight δ, and that δ is greater than γ, written $\gamma < \delta$, if we can go from γ to δ through a string of weights obtained by adding one simple root at a time, in other words if there is a string $\gamma = \gamma_0, \gamma_1, \dots, \gamma_r = \delta$ such that $\gamma_i = \gamma_{i-1} + \psi_i$ for $1 \leqq i \leqq r$ with every ψ_i simple. The weight λ is greater than any other weight of V, and we call it the **highest weight** of V. Analogously there is also a **lowest weight**. All this depends, of course, on the choice of $\Sigma^+(\mathfrak{g}, \mathfrak{h})$. The ingredients just described fit together as follows. This material can be found in standard texts such as [Hoch], [Hu1], [Kna] and [V], mentioned in connection with structure theory.

Theorem 1.2.2 (Cartan highest weight theory). *Let $\mathfrak{g}$ be a finite-dimensional complex semisimple Lie algebra, $\mathfrak{h}$ a Cartan subalgebra, and $\Sigma^+ = \Sigma^+(\mathfrak{g}, \mathfrak{h})$ a positive root system.*

- *Every irreducible (finite-dimensional complex) representation ν of $\mathfrak{g}$ has a unique highest weight.*
- *If two irreducible (finite-dimensional complex) representations have the same highest weight, then they are equivalent.*
- *By duality carry the Killing form inner product from $\mathfrak{h}_{\mathbb{R}}$ to an inner product $\langle \cdot, \cdot \rangle$ on $\mathfrak{h}_R^*$. Then a linear functional $\lambda \in \mathfrak{h}^*$ is the highest weight of some irreducible (finite-dimensional complex) representation of $\mathfrak{g}$ if and only if (i) $\lambda \in \mathfrak{h}_{\mathbb{R}}^*$, and (ii) $\frac{2\langle \lambda, \psi \rangle}{\langle \psi, \psi \rangle}$ is an integer $\geqq 0$ for every simple root ψ.*

We enumerate the simple root system, $\Psi = \{\psi_1, \dots, \psi_r\}$ where $r = \dim \mathfrak{h} = \operatorname{rank} \mathfrak{g}$. Define **fundamental highest weights** ξ_i by $\frac{2\langle \xi_i, \psi_j \rangle}{\langle \psi_j, \psi_j \rangle} = \delta_{i,j}$. Then the highest weights are just the linear combinations $\lambda = \sum_1^r n_i \xi_i$ where the n_i are nonnegative integers. It is standard to indicate the irreducible representation ν_λ of highest weight $\lambda = \sum n_i \xi_i$ by use of the Dynkin diagram $\Delta_{\mathfrak{g}}$ as follows. The vertices of $\Delta_{\mathfrak{g}}$ are the elements of Ψ, and we write n_i by the vertex ψ_i whenever $n_i \neq 0$. Thus the adjoint representations of the complex simple Lie algebras are indicated by the first three columns of Table 1.2.3.

Table 1.2.3. Adjoint Representations

Type of $\mathfrak{g}$	Adjoint Representation of $\mathfrak{g}$	Highest Weight	Other Dominant Weights $\neq 0$
$A_\ell, \ell \geqq 1$	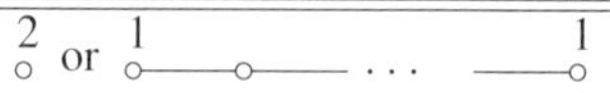	$\xi_1 + \xi_\ell$	
$B_\ell, \ell \geqq 2$	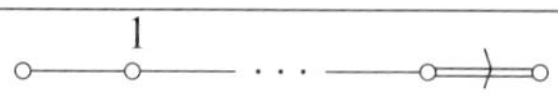	ξ_2	ξ_1
$C_\ell, \ell \geqq 3$	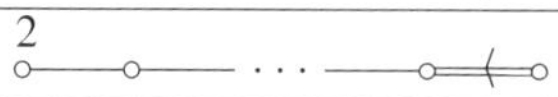	$2\xi_1$	ξ_2
$D_\ell, \ell \geqq 4$		ξ_2	
G_2	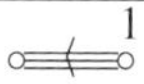	ξ_2	ξ_1
F_4	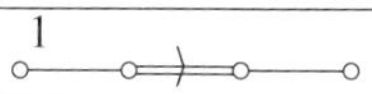	ξ_1	ξ_4
E_6	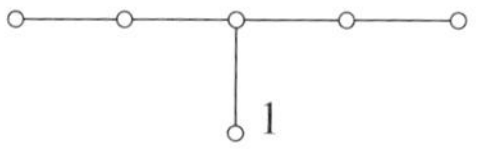	ξ_2	
E_7	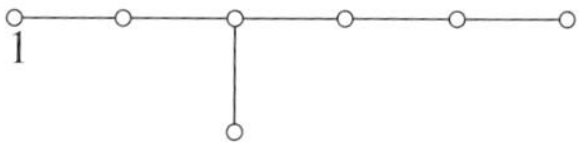	ξ_1	
E_8	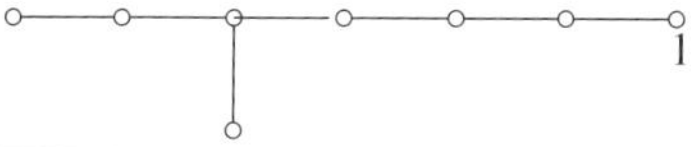	ξ_8	

Corollary 1.2.4. *Let ν_λ be the irreducible representation of $\mathfrak{g}$ with highest weight λ, and let V be its representation space. Then the dual (contragredient) ν_λ^*, representation of $\mathfrak{g}$ on the linear dual space V^*, is the irreducible representation of $\mathfrak{g}$ with lowest weight $-\lambda$.*

Proof. Compare the weight space decompositions $V = \sum_{\gamma \in M(\mathfrak{g},\nu_\lambda)} V_\gamma$ and $V^* = \sum_{\gamma \in M(\mathfrak{g},\nu_\lambda^*)} V_\gamma$. Since $(\nu(\xi)v, v^*) + (v, \nu^*(\xi)v^*) = 0$, the duality between V and V^* is $\mathfrak{h}$-equivariant, so it pairs V_γ with $V_{-\gamma}^*$. In particular the weight systems satisfy $M(\mathfrak{g}, \nu_\lambda^*) = -M(\mathfrak{g}, \nu_\lambda)$, and so $-\lambda$ is the lowest weight of ν_λ^*. □

Now let us note the extension of Theorem 1.2.2 from semisimple $\mathfrak{g}$ to reductive $\mathfrak{g}$. Decompose the complex reductive Lie algebra $\mathfrak{g} = \mathfrak{z} \oplus \mathfrak{g}'$ as before, where $\mathfrak{z}$ is its center and the semisimple $\mathfrak{g}'$ is derived algebra $[\mathfrak{g}, \mathfrak{g}]$. If $\mathfrak{z} \neq 0$ every linear functional ζ on $\mathfrak{z}$ defines a nonsemisimple representations of $\mathfrak{g}$ by functional ζ on $\mathfrak{z}$ defines a representation by $z + \xi' \mapsto \begin{pmatrix} 0 & \zeta(z) \\ 0 & 0 \end{pmatrix}$ of $\mathfrak{g}$ for $z \in \mathfrak{z}$ and $\xi' \in \mathfrak{g}'$. Thus we must explicitly require semisimplicity in order to reduce considerations to irreducible representations.

Now, essentially as in Theorem 1.2.2, let $\mathfrak{g}$ be a finite-dimensional complex reductive Lie algebra, $\mathfrak{h}$ a Cartan subalgebra, and Σ^+ a positive root system. Decompose $\mathfrak{g} = \mathfrak{z} \oplus \mathfrak{g}'$, and thus also $\mathfrak{h} = \mathfrak{z} \oplus \mathfrak{h}'$. In an irreducible representation ν of $\mathfrak{g}$, the center $\mathfrak{z}$ acts by scalars, so $\nu = \nu_{\mathfrak{z}} \otimes \nu'$, where

- $\nu_{\mathfrak{z}}$ is a linear functional on the commutative algebra $\mathfrak{z}$,
- ν' is an irreducible representation of the semisimple algebra $\mathfrak{g}'$, and
- we define $(\nu_{\mathfrak{z}} \otimes \nu')(z + \xi') = \nu_{\mathfrak{z}}(z) + \nu'(\xi')$ for $z \in \mathfrak{z}$ and $\xi' \in \mathfrak{g}'$.

Thus ν has a unique highest weight $\lambda = \nu_{\mathfrak{z}} + \lambda'$, where λ' is the highest weight of ν', and λ determines ν up to equivalence. Moreover, a linear functional $\lambda \in \mathfrak{h}^*$ is the highest weight of some irreducible (finite-dimensional complex) representation of $\mathfrak{g}$ if and only if (i) $\lambda \in \mathfrak{z}^* \oplus \mathfrak{h}^*_{\mathbb{R}}$, and (ii) $\frac{2\langle\lambda,\psi\rangle}{\langle\psi,\psi\rangle}$ is an integer $\geqq 0$ for every simple root ψ. Here note that the Killing form ignores the $\mathfrak{z}^*$-component of λ, so that $\mathfrak{z}^*$-component can be any linear functional on $\mathfrak{z}$.

Now we carry the Cartan highest weight theory over to the group level. There is not too much to that, except that we have to be careful that the group representations are well defined. Let G be a complex connected reductive Lie group and π an irreducible complex finite-dimensional representation of G. Then $d\pi$ is an irreducible representation of $\mathfrak{g}$, and consequently it has some highest weight λ as described just above. Since λ determines $d\pi$ up to equivalence, it also determines π up to equivalence. Thus we say that λ is the **highest weight** of π. If λ is the highest weight of an irreducible representation ν of $\mathfrak{g}$, we integrate ν to a representation π of the connected simply connected group $\widetilde{G}$ with Lie algebra $\mathfrak{g}$, obtaining a representation π of highest weight λ such that $d\pi = \nu$. Thus the Cartan highest weight theory is the same for complex reductive Lie algebras and connected complex reductive Lie groups. In general, when G is connected but not necessarily simply connected, we express $G = \widetilde{G}/Z$ where $\widetilde{G}$ is the connected simply connected group with Lie algebra $\mathfrak{g}$, and Z is a discrete central subgroup of $\widetilde{G}$. Then a linear functional $\lambda \in \mathfrak{h}^*$ is the highest weight of some irreducible (finite-dimensional complex) representation of G if and only if the associated representation $\widetilde{\pi}$ of $\widetilde{G}$ annihilates Z; that is an integrality condition on λ.

Let $\mathfrak{g}_0$ be a (finite-dimensional) real reductive Lie algebra. Let $\mathfrak{h}_0$ be a Cartan subalgebra of $\mathfrak{g}_0$ and choose a positive root system $\Sigma^+(\mathfrak{g}, \mathfrak{h})$. The decomposition $\mathfrak{g} = \mathfrak{z} \oplus \mathfrak{g}'$ of its complexification gives corresponding decompositions $\mathfrak{g}_0 = \mathfrak{z}_0 \oplus \mathfrak{g}_0'$ and $\mathfrak{h}_0 = \mathfrak{z}_0 \oplus \mathfrak{h}_0'$. If ν is an irreducible representation of $\mathfrak{g}$ with representation space $V = V_\nu$, then $\nu_0 := \nu|_{\mathfrak{g}_0}$ is an irreducible representation[2] of $\mathfrak{g}_0$. If ν_0 is an irreducible representation of $\mathfrak{g}_0$ on a complex vector space V, then the complex linear extension ν of $\nu_0 : \mathfrak{g}_0 \to \mathfrak{gl}(V)$ is an irreducible representation of $\mathfrak{g}$ on V. We have in effect proved the following:

> Let $\mathfrak{g}_0$ be a real form of a complex reductive Lie algebra $\mathfrak{g}$. Restriction $\nu \mapsto \nu|_{\mathfrak{g}_0}$ and complexification $\nu_0 \mapsto \nu_0 \otimes \mathbb{C}$ give a one-to-one correspondence between finite-dimensional representations of $\mathfrak{g}$ and finite-dimensional representations of $\mathfrak{g}_0$. This correspondence preserves equivalence, semisimplicity, and irreducibility.

[2] Although ν_0 is a real Lie algebra homomorphism, *irreducibility* means that there are no proper $\nu_0(\mathfrak{g}_0)$-invariant complex subspaces of V.

If ν is irreducible with highest weight λ, then we also refer to λ as the highest weight of $\nu|_{\mathfrak{g}_0}$. Thus the Cartan highest weight theory (Theorem 1.2.2) is available for real semisimple Lie algebras, and then for real reductive Lie algebras as well.

The situation is the same on the level of connected real reductive Lie groups. Let G_0 be a connected real reductive Lie group, $\widetilde{G}_0$ its universal covering group, and $\widetilde{G}$ the connected simply connected Lie group with Lie algebra $\mathfrak{g}$. Then $\widetilde{G}$ has a smallest discrete central subgroup Z, and quotient $G = \widetilde{G}/Z$, such that the inclusion $\mathfrak{g}_0 \hookrightarrow \mathfrak{g}$ induces a well-defined real Lie group homomorphism $G_0 \to G$. Write Z_0 for the kernel of this homomorphism, so $\mathfrak{g}_0 \hookrightarrow \mathfrak{g}$ defines an inclusion $G_0/Z_0 \hookrightarrow G$. The point is that $\widetilde{G}$ is universal for finite-dimensional representations of G_0:

> If π_0 is a representation of G_0 on a (finite-dimensional) complex vector space V, then π_0 annihilates Z_0 and gives a representations of G_0/Z_0, $\mathfrak{g}_0$ and $\mathfrak{g}$ on V. If ν is a representation of $\mathfrak{g}$ on V and $\widetilde{\pi}$ is the corresponding representation of $\widetilde{G}$ on V, then ν corresponds to a representation π_0 of G_0 if and only if $\widetilde{\pi}$ annihilates Z.

For example, let $\widetilde{SL(2;\mathbb{R})}$ denote the universal covering group of the group $SL(2;\mathbb{R})$ of all 2×2 real matrices of trace 0. Let $G_0 = \widetilde{SL(2;\mathbb{R})}$. Then Z_0 is an infinite cyclic group, and the irreducible finite-dimensional representations of G_0 are the ones that factor through $SL(2;\mathbb{R})$.

The real reductive group situation simplifies in the compact case. Let G_u be a compact connected Lie group, $\widetilde{G}_u$ its universal covering group, and $\widetilde{G}$ the connected simply connected Lie group with Lie algebra $\mathfrak{g}$. Then $\widetilde{G}_u$ is a real analytic subgroup of $\widetilde{G}$. If we express $G_u = \widetilde{G}_u/Z$ now G_u is a real analytic subgroup of the connected complex reductive Lie group $G := \widetilde{G}/Z$. Thus G_u has the same finite-dimensional representation theory and highest weight theory as the connected complex reductive Lie group G.

Also, since G_u is compact, one can prove that every continuous irreducible representation π_u of G_u on a Fréchet space V is finite-dimensional and is equivalent to a **unitary** representation. In other words, the representation space V has a positive definite hermitian form h such that $h(\pi_u(g)u, \pi_u(g)v) = h(u, v)$ for all $g \in G_u$ and $u, v \in V$. Recall that one can construct h by starting with any positive definite hermitian form h' on V and defining $h(u, v) = \int_{G_u} h(\pi_u(g)u, \pi_u(g)v)dg$.

The picture becomes complicated, and the highest weight theory is only part of the story when G (or G_0) is not connected. After all, the case $\dim G = 0$ is the finite-dimensional representation theory of discrete groups.

1.3 Borel subgroups and subalgebras

In this section and the next we recall the theory of parabolic subgroups and their Lie algebras, starting with Borel subalgebras and Borel subgroups. Much of this material can be found in standard texts, for example those of Knapp [Kna, Chapter VII] and Humphreys [Hu2, Part VIII], but the treatment in [W10, Part 1] is better oriented toward our needs here.

Let $\mathfrak{g}$ be a complex semisimple Lie algebra and $\mathfrak{h}$ a Cartan subalgebra. Let $\Sigma = \Sigma(\mathfrak{g}, \mathfrak{h})$ denote the corresponding root system, and fix a positive subsystem $\Sigma^+ = \Sigma^+(\mathfrak{g}, \mathfrak{h})$. The corresponding **Borel subalgebra** is

$$\mathfrak{b} = \mathfrak{h} + \sum_{\alpha \in \Sigma^+} \mathfrak{g}_{-\alpha} \subset \mathfrak{g}. \tag{1.3.1}$$

The Borel subalgebra $\mathfrak{b}$ is solvable. It is the semidirect sum of the nilpotent ideal $\sum_{\alpha \in \Sigma^+} \mathfrak{g}_{-\alpha}$ with a complementary reductive algebra $\mathfrak{h}$. In other words, it has its nilradical[3] $\mathfrak{b}^{-n} = \sum \mathfrak{g}_{-\alpha}$ and with Levi complement $\mathfrak{h}$.

For example, if $\mathfrak{g} = \mathfrak{sl}(n; \mathbb{C})$, Lie algebra of trace-free linear transformations of $\mathbb{C}^n$, we can choose $\mathfrak{h}$ to consist of the diagonal matrices of trace 0. We make the standard choice, $\Sigma^+(\mathfrak{g}, \mathfrak{h}) = \{\varepsilon_i - \varepsilon_j \mid 1 \leqq i < j \leqq n\}$, where ε_k on a diagonal matrix picks out the kth diagonal entry. Then $\mathfrak{b}$ consists of the lower triangular $n \times n$ complex matrices of trace 0.

In general a subalgebra $\mathfrak{b} \subset \mathfrak{g}$ is called a **Borel subalgebra** if it is $\mathrm{Int}(\mathfrak{g})$-conjugate to a subalgebra of the form (1.3.1), in other words if there exist choices of $\mathfrak{h}$ and $\Sigma^+(\mathfrak{g}, \mathfrak{h})$ such that $\mathfrak{b}$ is given by (1.3.1).

Let G denote the connected simply connected Lie group with Lie algebra $\mathfrak{g}$. The Cartan subgroup of G corresponding to $\mathfrak{h}$ is the centralizer of $\mathfrak{h}$ in G, given by (1.1.2). It has Lie algebra $\mathfrak{h}$, and it is connected because G is connected, complex and semisimple. The **Borel subgroup** $B \subset G$ corresponding to a Borel subalgebra $\mathfrak{b} \subset \mathfrak{g}$ is defined to be the normalizer of G in $\mathfrak{b}$, given by

$$B = N_G(\mathfrak{b}) := \{g \in G \mid \mathrm{Ad}(g)\mathfrak{b} = \mathfrak{b}\}. \tag{1.3.2}$$

As indicated by their names here, these notions are due to A. Borel (see [Bo1]; also see [Bo2]).

For example, let $G = SL(n; \mathbb{C})$ corresponding to $\mathfrak{g} = \mathfrak{sl}(n; \mathbb{C})$. Use the Cartan subalgebra $\mathfrak{h} \subset \mathfrak{g}$ and the positive root system $\Sigma^+(\mathfrak{g}, \mathfrak{h})$ mentioned just after (1.3.1). The corresponding Borel subgroup B consists of the lower triangular matrices in G, so it is just the G-stabilizer of the flag $\mathcal{F} = (F_1 \subsetneqq F_2 \subsetneqq \cdots \subsetneqq F_{n-1})$ of subspaces of $\mathbb{C}^n$ given by $F_j = \mathrm{Span}(e_{n-j+1}, \ldots, e_n)$ in the standard basis $\{e_i\}$. It follows that, in general, the Borel subalgebras of $G = SL(n; \mathbb{C})$ are just the G-stabilizers of flags $\mathcal{F} = (F_1 \subsetneqq F_2 \subsetneqq \cdots \subsetneqq F_{n-1})$ where F_j is a j-dimensional subspace of $\mathbb{C}^n$.

Here are the basic facts on Borel subgroups.

Lemma 1.3.3. *In the notation of* (1.3.1) *and* (1.3.2), *B has Lie algebra $\mathfrak{b}$, B is its own normalizer in G, B is a closed connected subgroup of G, and G/B is simply connected.*

Proof. The Borel subgroup B is closed in G by definition (1.3.2). It follows that the normalizer $E = N_G(B)$ is closed in G and consequently E is a Lie subgroup. Let $\mathfrak{e}$

[3] Here we describe the nilradical as a sum of negative root spaces, rather than positive, so that, in applications, positive functionals on $\mathfrak{h}$ will correspond to positive bundles (instead of negative bundles), and holomorphic discrete series representations will be highest weight (instead of lowest weight) representations.

denote the Lie algebra of E. Then $\mathfrak{b} \subset \mathfrak{e}$ and $[\mathfrak{e}, \mathfrak{b}] \subset \mathfrak{b}$. Any subalgebra of $\mathfrak{g}$ that properly contains $\mathfrak{b}$ must be of the form $\mathfrak{b} + \sum_{\sigma \in S} \mathfrak{g}_\alpha$ with $S \subset \Sigma^+$, because $\mathfrak{h} \subset \mathfrak{b}$. Thus it would contain a 3-dimensional simple subalgebra and could not normalize $\mathfrak{b}$. Now $\mathfrak{e} = \mathfrak{b}$, in particular E normalizes $\mathfrak{b}$, so $E = B$. This shows both that B is its own normalizer and that B has Lie algebra $\mathfrak{b}$.

Here is a complex-analytic proof that B is connected and G/B is simply connected. Let B^{op} denote the opposite Borel subgroup, corresponding to the Borel subalgebra $\mathfrak{b}^{\text{op}} = \mathfrak{h} + \mathfrak{b}^{+n}$ where $\mathfrak{b}^{+n} = \sum_{\alpha \in \Sigma^+} \mathfrak{g}_\alpha$. Then $B^{\text{op}} \cap B = H$, so the orbit $B^{\text{op}}(1B)$ in G/B is a submanifold of full dimension, thus open, and the unipotent radical $B^{+n} = \exp(\mathfrak{b}^{+n})$ is transitive on that orbit. An open orbit of a complex Lie group in a connected complex manifold is the complement of a closed analytic subset E. Thus the open B^{+n}-orbit, which is biholomorphic to $\mathbb{C}^{\dim G/B}$, is the complement of such a set E in G/B. Since E has real codimension $\geqq 2$ in G/B, it follows that G/B is simply connected, and thus that B is connected.

Here is the standard Lie structure theory proof that B is connected and G/B is simply connected. Let $b \in B$. All Levi complements to the nilradical B^{-n} (in other words all closed subgroups of B that project isomorphically onto B/B^{-n}) are conjugate by the identity component B^0 of B, so we have $b' \in bB^0$ that normalizes $\mathfrak{h}$. Let $w = \mathrm{Ad}(b')|_{\mathfrak{h}} \in W(\mathfrak{g}, \mathfrak{h})$. Then w preserves the positive root system Σ^+. Since the Weyl group $W(\mathfrak{g}, \mathfrak{h})$ is **simply** transitive on the set of all positive subsystems of $\Sigma(\mathfrak{g}, \mathfrak{h})$, now $w = 1$, in other words b' belongs to the Cartan subgroup H defined by $\mathfrak{h}$. Since G is connected, H is connected, so $b' \in H \subset B^0$. We have proved that B is connected, and it follows that G/B is simply connected. □

Some of the other basic facts are not quite as obvious.

Lemma 1.3.4 (See J. Tits [T1]). *Let $G_u \subset G$ be a compact real form. Then G_u is transitive on $X = G/B$, and X has a G_u-invariant Kähler metric. In particular X has the structure of a compact Kähler manifold.*

Remark. As the argument will show, the G_u-invariant Kähler metrics on X form a convex open cone in $\mathbb{R}^\ell$ where $\ell = \dim \mathfrak{h}$. This cone is sometimes called the *Kähler cone*. ◇

Indication of proof. It suffices to consider a G_u constructed by means of a "Weyl basis" of $\mathfrak{g}$ using $\mathfrak{h}$ and Σ^+. See, for example, [V, p. 280]. This yields a real form $\mathfrak{g}_u \subset \mathfrak{g}$ on which the Killing form is negative definite. Then the G-normalizer of $\mathfrak{g}_u$ coincides with the real analytic subgroup of G for $\mathfrak{g}_u$, that is, G_u. By construction $\mathfrak{h}_u = \mathfrak{g}_u \cap \mathfrak{h}$ is the real form of $\mathfrak{h}$ on which the roots take pure imaginary values, and $\mathfrak{g} \cap \mathfrak{b} = \mathfrak{h}_u$. Now a dimension count shows that the G_u-orbit of the identity coset $x_0 = 1B \in G/B = X$ is open in X. It is also closed in X because G_u is compact. This proves the transitivity and thus proves that X is compact.

Let $\lambda \in \mathfrak{h}^*$ such that $\langle \lambda, \alpha \rangle > 0$ for every $\alpha \in \Sigma^+$. Extend λ to a linear functional on $\mathfrak{g}$ by $\lambda(\mathfrak{g}_\gamma) = 0$ for every $\gamma \in \Sigma$. Define $d\lambda : \mathfrak{g} \times \mathfrak{g} \to \mathbb{C}$ by $d\lambda(\xi, \eta) = \lambda([\xi, \eta])$. The argument of [T1] shows that $d\lambda$ pushes down to a G_u-invariant closed 2-form ω of maximal rank on X, and ω combines with the complex structure to define a

G_u-invariant Kähler metric on X. Thus, for every λ in the positive Weyl chamber of $(\mathfrak{g}, \mathfrak{h}, \Sigma^+)$, we have a G_u-invariant Kähler metric on X. □

The above argument really is a Lie algebra cohomology argument. There, λ is a 1-cochain for Lie algebra cohomology of $(\mathfrak{g}, \mathfrak{h})$ and $d\lambda$ is a 2-cocycle.

Lemma 1.3.5. *There is a finite-dimensional irreducible representation π of G with the following property: Let $[v]$ be the image of a lowest weight vector in the projective space $\mathbb{P}(V_\pi)$ corresponding to the representation space of π. Then the action of G on V_π induces a holomorphic action of G on $\mathbb{P}(V_\pi)$, and B is the G-stabilizer of $[v]$. In particular $X = G/B$ is a complex projective variety.*

Proof. For example, let $\rho = \frac{1}{2}\sum_{\alpha\in\Sigma^+}\alpha$ as usual and let π be the irreducible representation of highest weight ρ. The lowest weight is $-\rho$ and the assertions are immediate. □

Lemma 1.3.6. *The group B is a maximal solvable subgroup of G.*

Proof. The argument of Lemma 1.3.3 shows that $\mathfrak{b}$ is a maximal solvable subalgebra of $\mathfrak{g}$ and that any Lie subgroup of G with Lie algebra $\mathfrak{b}$ is B itself. If $E \subset G$ is a solvable subgroup and $B \subset E$, then the closure of E in G has those same properties, we may assume E is closed in G. But then E is a solvable Lie subgroup of G. Thus its Lie algebra $\mathfrak{e}$ is solvable, so $\mathfrak{e} = \mathfrak{b}$ and $E = B$. We conclude that B is maximal solvable. □

A theorem of Borel (see [Bo1], or see [Bo2]) says that any connected Lie group R acting linearly on a complex projective variety X has a fixed point. Applying this to $X = G/B$, it follows that such a group R is contained in some conjugate of B. This shows that the Borel subgroups are exactly the maximal connected solvable subgroups of G. That's how Borel originally defined them. In view of Lemma 1.3.3, our definitions (1.3.1) and (1.3.2) are equivalent to those of Borel.

Given the Cartan subalgebra $\mathfrak{h}$, the Borel subalgebras and subgroups given by (1.3.1) and (1.3.2) are usually called the **standard** Borels.

Here is a proof of Borel's fixed point theorem mentioned above. Let X be a subvariety of $\mathbb{P}(V)$ where V is a complex vector space with a linear action of a Lie group R, where R stabilizes X. Then by Lie's Theorem R stabilizes a full flag $F_0 \subset F_1 \subset \cdots \subset F_n = \mathbb{P}(V)$ of projective linear subspaces. Note that $F_j \setminus F_{j-1} \cong \mathbb{C}^j$. Recall that every compact subvariety of $\mathbb{C}^j$ is finite. If X is a finite set, then the assertion follows from the fact that R is connected. Otherwise, F_{n-1} has nonempty intersection with X. So either $X \subset F_{n-1}$, and we apply induction on the dimension of the ambient projective space, or the dimension of some component of $X \cap F_{n-1}$ is smaller than that of X and we apply induction on the dimension of X to that component.

1.4 Parabolic subgroups and subalgebras

A subalgebra $\mathfrak{q} \subset \mathfrak{g}$ is called **parabolic** if it contains a Borel subalgebra. The concept, as specified by the root space decomposition below, is due to J. Tits [T0]. The relation with Borel subalgebras and Borel subgroups was discovered later. Recall the expository references [Hu2], [Kna], and [W10] at the beginning of Section 1.3.

Let Ψ be the simple root system corresponding to Σ^+ and let Φ be an arbitrary subset of Ψ. Every $\alpha \in \Sigma$ has a unique expression $\alpha = \sum_{\psi \in \Psi} n_\psi(\alpha)\psi$, where the $n_\psi(\alpha)$ are integers, all $\geqq 0$ if $\alpha \in \Sigma^+$ and all $\leqq 0$ if $\alpha \in \Sigma^- = -\Sigma^+$. Set

(1.4.1) $$\Phi^r = \{\alpha \in \Sigma \mid n_\psi(\alpha) = 0 \text{ whenever } \psi \notin \Phi\}$$

and

(1.4.2) $$\Phi^n = \{\alpha \in \Sigma^+ \mid \alpha \notin \Phi^r\} = \{\alpha \in \Sigma \mid n_\psi(\alpha) > 0 \text{ for some } \psi \notin \Phi\}.$$

Now define

(1.4.3) $$\mathfrak{q}_\Phi = \mathfrak{q}_\Phi^r + \mathfrak{q}_\Phi^{-n} \text{ with } \mathfrak{q}_\Phi^r = \mathfrak{h} + \sum\nolimits_{\alpha \in \Phi^r} \mathfrak{g}_\alpha \text{ and } \mathfrak{q}_\Phi^{-n} = \sum\nolimits_{\alpha \in \Phi^n} \mathfrak{g}_{-\alpha}.$$

Then $\mathfrak{q}_\Phi$ is a subalgebra of $\mathfrak{g}$ that contains the Borel subalgebra (1.3.1), so it is a parabolic subalgebra of $\mathfrak{g}$.

Proposition 1.4.4. *Let $\mathfrak{q} \subset \mathfrak{g}$ be a subalgebra that contains the Borel subalgebra given by $\mathfrak{b} = \mathfrak{h} + \sum_{\alpha \in \Sigma^+} \mathfrak{g}_{-\alpha}$ of $\mathfrak{g}$. Then there is a set Φ of simple roots such that $\mathfrak{q} = \mathfrak{q}_\Phi$.*

Proof. Define $\Phi = \{\psi \in \Psi \mid \mathfrak{g}_\psi \subset \mathfrak{q}\}$. Then $\mathfrak{q}_\Phi \subset \mathfrak{q}$, and we must prove $\mathfrak{q} \subset \mathfrak{q}_\Phi$. Both contain $\mathfrak{b}$, so this comes down to showing that $\alpha \in \Sigma^+$, $\mathfrak{g}_\alpha \subset \mathfrak{q}$ implies $n_\psi(\alpha) = 0$ whenever $\psi \in \Psi \setminus \Phi$. We will prove this by induction on the **level** $\ell(\alpha) = \sum n_\psi(\alpha)$.

If $\ell(\alpha) = 1$ then α is simple, and so $\mathfrak{g}_\alpha \subset \mathfrak{q}$ implies $\alpha \in \Phi$. Then $\psi \notin \Phi$ implies $\psi \neq \alpha$. Thus $n_\psi(\alpha) = 0$.

Now let $\ell(\alpha) = \ell_0 > 1$ and suppose that $n_{\psi'}(\gamma) = 0$ for all $\psi' \in \Psi \setminus \Phi$, whenever $\gamma \in \Sigma^+$ and $\mathfrak{g}_\gamma \subset \mathfrak{q}$ with $\ell(\gamma) < \ell_0$. Suppose first that we can (and do) choose $\psi \in \Phi$ such that $\gamma = \alpha - \psi$ is a root. Then

$$\mathfrak{g}_\gamma = [\mathfrak{g}_\alpha, \mathfrak{g}_{-\psi}] \subset [\mathfrak{q}, \mathfrak{b}^{-n}] \subset [\mathfrak{q}, \mathfrak{q}] \subset \mathfrak{q}.$$

If $\psi' \in \Psi \setminus \Phi$, then $n_{\psi'}(\alpha) = n_{\psi'}(\gamma)$, which is zero by the induction hypothesis. Secondly, suppose that we cannot (and do not) choose ψ from among the elements of Φ. Then

$$\mathfrak{g}_\psi = [\mathfrak{g}_\alpha, \mathfrak{g}_{-\gamma}] \subset [\mathfrak{q}, \mathfrak{b}^{-n}] \subset [\mathfrak{q}, \mathfrak{q}] \subset \mathfrak{q}.$$

Therefore $\psi \in \Phi$ which is a contradiction. We have proved $n_{\psi'}(\gamma) = 0$ for all $\psi' \in \Psi \setminus \Phi$. Proposition 1.4.4 is proved. □

The **parabolic subgroup** $Q \subset G$ corresponding to a given parabolic subalgebra $\mathfrak{q} \subset \mathfrak{g}$ is defined to be the G-normalizer of $\mathfrak{q}$, that is,

(1.4.5) $$Q = \{g \in G \mid \mathrm{Ad}(g)\mathfrak{q} = \mathfrak{q}\}.$$

When $\mathfrak{q} = \mathfrak{q}_\Phi$ we also write $Q = Q_\Phi$. The basic facts on parabolic subgroups are most easily derived from the corresponding results for Borel subgroups.

For example, when $G = SL(n; \mathbb{C})$ we can interpret Proposition 1.4.4 as follows. The Borel subgroup B is the stabilizer of a particular flag

$$\mathcal{F} = (F_1 \subsetneqq F_2 \subsetneqq \cdots \subsetneqq F_{n-1})$$

of subspaces of $\mathbb{C}^n$. Then the parabolic subgroups Q of G that contain B are in one-to-one correspondence with the dimension sequences $0 < d_1 < \cdots < d_q < n$. There $0 \leqq q < n$. The parabolic Q corresponds to $0 < d_1 < \cdots < d_q < n$ if and only if it is the G-stabilizer of the partial flag $(F_{d_1} \subsetneqq \cdots \subsetneqq F_{d_q})$.

Lemma 1.4.6. *The parabolic subgroup $Q \subset G$ defined by* (1.4.5) *has Lie algebra $\mathfrak{q}$. That group Q is a closed connected subgroup of G, and is equal to its own normalizer in G. In particular, a Lie subgroup of G is parabolic if and only if it contains a Borel subgroup.*

Proof. The argument of Lemma 1.3.3 shows that Q has Lie algebra $\mathfrak{q}$, is closed and connected, and is equal to its own G-normalizer. Let $S \subset G$ be a Lie subgroup that contains a Borel subgroup B. Then its Lie algebra $\mathfrak{s}$ contains $\mathfrak{b}$ and is hence parabolic. Because S is pinched between the analytic subgroup of G for $\mathfrak{s}$ and the G-normalizer of $\mathfrak{s}$, which coincide because parabolic subgroups are closed and connected, S is the parabolic subgroup of G for $\mathfrak{s}$. □

Lemma 1.4.7. *The parabolic subgroup $Q = Q_\Phi$ is the semidirect product $Q_\Phi^{-n} \cdot Q_\Phi^r$ of its unipotent radical Q_Φ^{-n} with a reductive (Levi) complement Q_Φ^r. Here Q_Φ^{-n} is the analytic group* $\exp(\mathfrak{q}_\Phi^{-n})$ *and Q_Φ^r is the connected reductive group with Lie algebra $\mathfrak{q}_\Phi^r$.*

Proof. Let Q_Φ^{-n} denote the analytic subgroup of G with Lie algebra $\mathfrak{q}_\Phi^{-n}$ and let Q_Φ^r denote the analytic subgroup for $\mathfrak{q}_\Phi^r$. They have trivial intersection and therefore we have the semidirect product group $Q_\Phi^{-n} \cdot Q_\Phi^r$. That semidirect product group is an open subgroup of the connected group $Q = Q_\Phi$ and consequently they are equal. □

Let $B \subset Q \subset G$ consist of a Borel subgroup contained in a parabolic subgroup. Then we have complex homogeneous quotient spaces $X = G/B$ and $Z = G/Q$ and a G-equivariant holomorphic projection $X \to Z$ given by $gB \mapsto gQ$. In particular, transitivity of G_u on X gives transitivity of G_u on Z in

Lemma 1.4.8 (J. Tits [T1]). *Let $G_u \subset G$ be a compact real form. Then G_u is transitive on $Z = G/Q$, and Z has a G_u-invariant Kähler metric. In particular Z has the structure of a compact Kähler manifold.*

The argument of Lemma 1.3.4 is easily modified to prove the Kähler statement in Lemma 1.4.8. Just take λ in the dual space of the center of $\mathfrak{q}^r$ such that $\langle\lambda, \alpha\rangle > 0$ for all $\alpha \in \Phi^n$. Note that, as before, the G_u-invariant Kähler metrics on Z form a nonempty convex open cone.

Lemma 1.4.9. *Fix a standard parabolic subgroup $Q = Q_\Phi$ in G. Then there is a finite-dimensional irreducible representation π of G with the following property: Let $[v]$ be the image of a lowest weight vector in the projective space $\mathbb{P}(V_\pi)$ corresponding to the representation space of π. Then the action of G on V_π induces a holomorphic action of G on $\mathbb{P}(V_\pi)$, and Q is the G-stabilizer of $[v]$. In particular $Z = G/Q$ is a complex projective variety.*

Proof. We use the argument of Lemma 1.3.5, with a different choice of highest weight. Recall $\rho = \frac{1}{2}\sum_{\alpha\in\Sigma^+}\alpha$ and set $\rho_\Phi = \frac{1}{2}\sum_{\alpha\in(\Phi^r\cap\Sigma^+)}\alpha$. Let $\psi \in \Psi$. Then $\frac{2\langle\rho_\Phi,\psi\rangle}{\langle\psi,\psi\rangle} = 1$ for $\psi \in \Phi$, and $\frac{2\langle\rho_\Phi,\psi\rangle}{\langle\psi,\psi\rangle} = 0$ for $\psi \notin \Phi$. Now let π be the irreducible representation of G with lowest weight $-(\rho - \rho_\Phi)$, in other words highest weight $w(\rho - \rho_\Phi)$, where w is the element of the Weyl group that sends Σ^+ to its negative. Then the assertions are immediate. □

At this point we summarize as follows.

Proposition 1.4.10 ([T0], [T1]). *Let Q be a complex algebraic group that is a Lie subgroup of G. Then the following conditions are equivalent:*

(1) *G/Q is a compact complex manifold.*
(2) *G/Q is a complex projective variety.*
(3) *If G_u denotes a compact real form of G, then G/Q is a G_u-homogeneous compact Kähler manifold.*
(4) *G/Q is the projective space orbit of an extremal weight vector in an irreducible finite-dimensional representation of G.*
(5) *G/Q is a G-equivariant quotient manifold of G/B, for some Borel subgroup $B \subset G$.*
(6) *Q is a parabolic subgroup of G.*

If those conditions hold, then Q is connected and G/Q is simply connected.

We will simply refer to these spaces $Z = G/Q$ as **complex flag manifolds**.

1.5 Homogeneous holomorphic vector bundles

In this section we look at the structure of homogeneous holomorphic vector bundles over complex flag manifolds $Z = G/Q = G_u/L_u$. Let $p : \mathbb{E} \to Z$ be a fiber bundle with typical fiber E and structure group J. We say that the fiber bundle $p : \mathbb{E} \to Z$ is a **vector bundle** if E is a (real or complex) vector space and J is a subgroup of $GL(V)$. It is **holomorphic** if $\mathbb{E}$ is a complex manifold, the fibers are closed complex submanifolds, and the action $J \times E \to E$ is holomorphic. Combining these concepts, $p : \mathbb{E} \to Z$ is a **holomorphic vector bundle** if it is a complex vector bundle and is a holomorphic fiber bundle.

There is a slightly delicate point here. In order for $J \times E \to E$ to be holomorphic one wants the vector space E to be complex and the structure group J to be a complex Lie group. In fact we will often meet situations where the structure group initially is a real Lie group and the holomorphic vector bundle structure is defined by an extension of that real structure group to a complex Lie group. This is based on [TW] and explained in Lemma 1.6.1 below.

We say that $p : \mathbb{E} \to Z$ is G-**homogeneous** (respectively, G_u-**homogeneous**) if the action of G (respectively, G_u) on Z lifts to an action on $\mathbb{E}$, and the stabilizer of the typical fiber E is contained in the structure group J. Finally, $p : \mathbb{E} \to Z$ is a G-**homogeneous** (or just **homogeneous**) **holomorphic vector bundle** if it is a holomorphic vector bundle that is G-homogeneous in such a way that the action $G \times \mathbb{E} \to \mathbb{E}$ is holomorphic.

The G-homogeneous holomorphic vector bundles over Z are constructed as follows. Fix a holomorphic representation χ of Q on a finite-dimensional complex vector space $E = E_\chi$. Define a complex manifold $\mathbb{E} = \mathbb{E}_\chi := G \times_Q E_\chi$, where $G \times_Q E_\chi$ is the quotient of $G \times E_\chi$ by the equivalence relation

$$(gq, e) \sim (g, \chi(q)e) \quad \text{for } g \in G, q \in Q, \text{and } e \in E_\chi.$$

Write $[g, e]$ for the equivalence class of (g, e). The projection $p : \mathbb{E}_\chi \to Z$ is given by $p[g, e] = gQ$. The fiber $p^{-1}(gQ) = \{[g, e] \mid e \in E_\chi\} \cong E_\chi$ is a complex submanifold of $\mathbb{E}_\chi$. The natural action of G on $\mathbb{E}_\chi$ is given by $g[g', e] = [gg', e]$. For the holomorphic local trivializations, consider the holomorphic principal bundle $G \to G/Q = Z$ with structure group Q and note that $p : \mathbb{E}_\chi \to Z$ is bundle associated by the action χ of Q on E_χ.

Proposition 1.5.1. *Let Z be a complex flag manifold G/Q. Every G-homogeneous holomorphic vector bundle $p : \mathbb{E} \to Z$ is holomorphically equivalent to a vector bundle $\mathbb{E}_\chi \to Z$ constructed as just above, where χ is the action of Q on the fiber $E = E_\chi := p^{-1}(1Q)$. Two such bundles, $\mathbb{E}_{\chi'} \to Z$ and $\mathbb{E}_{\chi''}$, are equivalent by a G-equivariant holomorphic bundle map if and only if the representations χ' and χ'' of Q are equivalent.*

Proof. Let $p : \mathbb{E} \to Z$ be a G-homogeneous holomorphic vector bundle and let χ denote the action of Q on the fiber $E = E_\chi := p^{-1}(1Q)$. Then χ is a holomorphic representation of Q, and we have the bundle $\mathbb{E}_\chi \to Z$ constructed just above. Let U be a small open neighborhood of 0 in a complex vector space complement to $\mathfrak{q}$ in $\mathfrak{g}$. Then $(\xi, e) \mapsto [\exp(\xi), e]$ gives a holomorphic trivialization of $\mathbb{E}$ over the open set $V = \exp(U)Q \subset Z$. This is the same holomorphic local trivialization as the one given by the associated bundle construction. Now both $\mathbb{E} \to Z$ and $\mathbb{E}_\chi \to Z$ have the same holomorphic local trivialization over V, so they are holomorphically equivalent by G-homogeneity. □

Let $E_{\chi'}$ be a $\chi(Q)$-invariant subspace of E_χ, where χ' denotes the corresponding subrepresentation of Q. The constructions used in Proposition 1.5.1 exhibit $\mathbb{E}_{\chi'}$ as a G-homogeneous holomorphic vector subbundle of $\mathbb{E}_\chi$. Conversely a G-homogeneous holomorphic vector subbundle $\mathbb{E}$ of $\mathbb{E}_\chi$ gives a subrepresentation χ' of χ such that $\mathbb{E} = \mathbb{E}_{\chi'}$. Thus, composition series (or direct sum decompositions) for χ correspond to those for $\mathbb{E}_\chi$. In particular, when χ is completely reducible (respectively, irreducible) we will refer to $\mathbb{E}_\chi$ as **completely reducible** (respectively, **irreducible**). Here note

Lemma 1.5.2. *Let χ be an irreducible complex finite-dimensional representation of Q. Then χ annihilates the nilradical unipotent radical Q^{-n} of Q, so it factors through an irreducible representation χ^r of the reductive component $Q^r \cong Q/Q^{-n}$.*

Corollary 1.5.3. *The holomorphic bundle equivalence classes of irreducible G-homogeneous holomorphic vector bundles $\mathbb{E} \to Z$ are in natural one-to-one correspondence with the equivalence classes of finite-dimensional irreducible representations of Q^r, and in particular are parameterized by highest weights of those representations.*

Now consider holomorphic vector bundles $\mathbb{E} \to Z$ that are G_u-homogeneous. It is automatic here that $\mathbb{E} \to Z$ is G-homogeneous.

Proposition 1.5.4. *Let $p : \mathbb{E} \to Z$ be a holomorphic vector bundle that is G_u-homogeneous. Then it carries a unique structure of G-homogeneous holomorphic vector bundle such that the original action of G_u on $\mathbb{E}$ is the restriction of the action of G on $\mathbb{E}$.*

Proof. We can assume that the real form G_u is the one defined by a complex conjugation σ that stabilizes the Cartan subalgebra used to define $\mathfrak{q}$. Then the isotropy subgroup $G_u \cap Q$ of G_u at $1Q$ is a compact real form Q_u^r of Q^r. It is connected by Lemma 1.4.7. As in the proof of Proposition 1.5.1, let χ_u denote the action of Q_u^r on the fiber $E = p^{-1}(1Q)$. Then we obtain $p : \mathbb{E} \to Z$ as $\mathbb{E}_{\chi_u} \to Z$ using the construction of Proposition 1.5.1. However, χ_u extends uniquely to a holomorphic representation of Q^r on E, and that representation lifts uniquely to a holomorphic representation χ of Q on E. That defines the G-homogeneous holomorphic vector bundle $\mathbb{E}_\chi$ with the required relation to $\mathbb{E}$. □

In view of Proposition 1.5.4, we can now speak of G_u-homogeneous holomorphic vector bundles over Z.

Corollary 1.5.5. *The holomorphic bundle equivalence classes of irreducible G_u-homogeneous holomorphic vector bundles $\mathbb{E} \to Z$ are in natural one-to-one correspondence with the equivalence classes of irreducible representations of Q_u^r, and in particular are parameterized by highest weights of those representations.*

1.6 The Bott–Borel–Weil Theorem

Let $p : \mathbb{E}_\chi \to Z$ be an irreducible G-homogeneous holomorphic vector bundle. Write $\mathcal{O}(\mathbb{E})$ for the sheaf of (germs of) holomorphic sections. Then the natural action of G on $\mathbb{E}_\chi$ and Z induces a representation of G on each of the Dolbeault cohomology spaces $H^k(Z; \mathcal{O}(\mathbb{E}))$. The Bott–Borel–Weil Theorem gives a precise description of these cohomologies as G-modules in terms of the highest weight of χ.

It does not matter whether one describes the $H^k(Z; \mathcal{O}(\mathbb{E}))$ as G-modules or as G_u-modules, because the continuous representation theory of a compact group G_u is the same as the completely reducible holomorphic representation theory of its complexification G. In other words, the Cartan highest weight theory is exactly the same for G and G_u.

A **section** of $p : \mathbb{E}_\chi \to Z$ over an open set $U \subset Z$ is a function $s : U \to \mathbb{E}_\chi$ such that $s(z) \in p^{-1}(z)$ for all $z \in U$. Expressing $\mathbb{E}_\chi = G \times_Q E_\chi$, the section s takes the form $s(gQ) = [g, f_s(g)]$, where $f_s : \{g \in G \mid gQ \in U\} \to E_\chi$ satisfies $f_s(gq) = \chi(q)^{-1} f_s(g)$ for $g \in G$, $gQ \in U$, $q \in Q$. In particular, the space of all global sections is identified with the space of all functions $f : G \to E_\chi$ such that $f(gq) = \chi(q)^{-1} f(g)$ for $g \in G$ and $q \in Q$. The point is that now the natural action (for the moment denote it by π) of G on the space of global sections is given by $(\pi(g)f)(g') = f(g^{-1}g')$.

A section s of $p : \mathbb{E}_\chi \to Z$ over an open set $U \subset Z$ is smooth of class C^k if and only if the corresponding function $f_s : \{g \in G \mid gQ \in U\} \to E_\chi$ is C^k. Similarly, s is **holomorphic** if and only if f_s is holomorphic.

Lemma 1.6.1 (Compare [TW]). *View Z as G_u/L_u, where $L_u = Q^r_u$, and view $p : \mathbb{E}_\chi \to Z$ as a G_u-homogeneous holomorphic vector bundle. Then a function $f : G_u \to E_\chi$ is a holomorphic section of $p : \mathbb{E}_\chi \to Z$ if and only if*

$$f(g; \xi) + d\chi(\xi)(f(g)) = 0 \text{ for } g \in G_u \text{ and } \xi \in \mathfrak{q}, \tag{1.6.2}$$

where the right derivative $f(g;\xi)$ is given by

$$f(g; \xi_1 + i\xi_2) = \left.\frac{d}{dt}\right|_{t=0} f_s(g \exp(t\xi_1)) + i \left.\frac{d}{dt}\right|_{t=0} f_s(g \exp(t\xi_2)) \tag{1.6.3}$$

for $g \in G_u$ and $\xi_1, \xi_2 \in \mathfrak{g}_u$.

Proof. We only need to consider the case where f is smooth. Since L_u is connected, the section condition $f(g\ell) = \chi(\ell)^{-1} f(g)$ is equivalent to

$$f(g; \xi) + d\chi_\lambda(\xi) f(g) = 0 \quad \text{for all } \xi \in \mathfrak{l} = \mathfrak{q}^r.$$

As ξ ranges over the antiholomorphic tangent space $\mathfrak{q}^{-n}$ of Z, the equation $f(g;\xi) + d\chi_\lambda(\xi) f(g) = 0$ for all $\xi \in \mathfrak{q}^{-n}$ is the Cauchy–Riemann system for the section s with $f = f_s$. Thus (1.6.2) characterizes holomorphic sections. □

Recall the Peter–Weyl Theorem. Let J be a compact topological group. Let $J \times J$ act on $L^2(J)$ by left and right translations: $(t(j_1, j_2)f)(j) = f(j_1^{-1} j j_2)$ Write $\widehat{J}$ for the set of all equivalence classes of irreducible unitary representations of J. In the connected Lie group case it is given by the highest weights of irreducible representations. If $\gamma \in \widehat{J}$, and if V_γ denotes its representation space, then the space of matrix coefficients is spanned by the functions $f_{u,v}(j) := v(\gamma(j)u)$ for $j \in J$, $v \in V_\gamma$ and $u \in V_\gamma^*$. This identifies that space of matrix coefficients with $V_\gamma \otimes V_\gamma^*$. The Peter–Weyl Theorem says that $L^2(J)$ is the Hilbert space direct sum of those subspaces,

$$L^2(J) = \sum_{\gamma \in \widehat{J}} V_\gamma \otimes V_\gamma^*,$$

as $J \times J$ module, where on the left translation action $t(j_1, 1)$ preserves each summand $V_\gamma \otimes V_\gamma^*$ and acts there on the factor V_γ, and the right translation action $t(1, j_2)$ preserves each summand $V_\gamma \otimes V_\gamma^*$ and acts there on the factor V_γ^*.

We apply the Peter–Weyl Theorem to the irreducible G_u-homogeneous holomorphic vector bundle $\mathbb{E}_{\chi_\lambda} \to Z$, where χ_δ denotes the irreducible complex representation of L_u of highest weight δ. The space of E_{χ_λ}-valued square integrable functions on G_u is

$$\text{(1.6.4)} \qquad \begin{aligned} L^2(G_u) \otimes E_{\chi_\lambda} &= \Big(\sum_{\pi_\gamma \in \widehat{G_u}} V_{\pi_\gamma} \otimes V_{\pi_\gamma}^*\Big) \otimes E_{\chi_\lambda} \\ &= \sum_{\pi_\gamma \in \widehat{G_u}} V_{\pi_\gamma} \otimes (V_{\pi_\gamma}^* \otimes E_{\chi_\lambda}), \end{aligned}$$

where π_γ denotes the irreducible representation of G_u (or of G) of highest weight γ. Here the left action of G_u is on the first factor V_{π_γ} and the right action of its isotropy subgroup L_u is on the factor $V_{\pi_\gamma}^* \otimes E_{\chi_\lambda}$. A function $f \in L^2(G_u, E_{\chi_\lambda})$ is a section of $\mathbb{E}_{\chi_\lambda} \to Z$. Thus it satisfies $f(g\ell) = \chi(\ell)^{-1} f(g)$. That says that f is fixed by the right action of L_u. We have seen that the space of square integrable sections of $\mathbb{E}_{\chi_\lambda} \to Z$ is the G_u-module

$$\text{(1.6.5)} \qquad L^2(Z; \mathbb{E}_{\chi_\lambda}) = \sum_{\pi_\gamma \in \widehat{G_u}} V_{\pi_\gamma} \otimes (V_{\pi_\gamma}^* \otimes E_{\chi_\lambda})^{L_u}.$$

In general, write $m(\delta, \gamma) = \text{mult}(\chi_\delta, \pi_\gamma)$ for the **multiplicity** of $\chi_\delta \in \widehat{L_u}$ as a summand of $\pi_\gamma|_{L_u}$. Using Lemma 1.2.4 we now have the highest weight formulation of the Frobenius Reciprocity Theorem:

$$\text{(1.6.6)} \qquad L^2(Z; \mathbb{E}_{\chi_\lambda}) = \sum_{\pi_\gamma \in \widehat{G_u}} V_{\pi_\gamma} \otimes \mathbb{C}^{m(\lambda,\gamma)}.$$

Thus π_γ occurs exactly $m(\lambda, \gamma)$ times as a subrepresentation of the left regular representation of G_u on the space $L^2(Z; \mathbb{E}_{\chi_\lambda})$ of L^2 sections of $\mathbb{E}_{\chi_\lambda} \to Z$.

Continuous sections, in particular holomorphic sections, are automatically L^2. Let $H^0(Z; \mathcal{O}(\mathbb{E}_{\chi_\lambda}))$ denote the space of holomorphic sections of $\mathbb{E}_{\chi_\lambda} \to Z$. Now $H^0(Z; \mathcal{O}(\mathbb{E}_{\chi_\lambda}))$ is given by

Theorem 1.6.7 (Borel–Weil Theorem [Ser]). *If λ is dominant for G_u, that is, if $\langle \lambda, \psi \rangle \geqq 0$ for every simple root ψ of $\Sigma^+(\mathfrak{g}, \mathfrak{h})$, then G and G_u act on $H^0(Z; \mathcal{O}(\mathbb{E}_{\chi_\lambda}))$ by the irreducible representation of highest weight λ. If λ is not dominant for G_u, then $H^0(Z; \mathcal{O}(\mathbb{E}_{\chi_\lambda}) = 0$.*

Proof. Combine Lemma 1.6.1 with (1.6.5) to see that the space of (automatically L^2) holomorphic sections is

$$H^0(Z; \mathcal{O}(\mathbb{E}_{\chi_\lambda})) = \sum_{\pi_\gamma \in \widehat{G_u}} V_{\pi_\gamma} \otimes \Big((V_{\pi_\gamma}^* \otimes E_{\chi_\lambda})^{L_u} \cap (V_{\pi_\gamma}^* \otimes E_{\chi_\lambda})^{\mathfrak{q}^{-n}}\Big).$$

Here $\mathfrak{q}^{-n}$ acts trivially on E_{χ_λ}. Thus $(V^*_{\pi_\gamma} \otimes E_{\chi_\lambda})^{\mathfrak{q}^{-n}} = (V^*_{\pi_\gamma})^{\mathfrak{q}^{-n}} \otimes E_{\chi_\lambda}$, and $d\pi^*_\gamma(\mathfrak{q}^{-n})$ permutes the irreducible summands of $d\pi^*_\gamma|_{\mathfrak{q}^r}$, lowering their lowest weights and annihilating only the summand whose lowest weight is lowest of all of them. But Lemma 1.2.4 says that the lowest one is $-\gamma$. Thus $(V^*_{\pi_\gamma} \otimes E_{\chi_\lambda})^{\mathfrak{q}^{-n}} = E^*_{\chi_\gamma} \otimes E_{\chi_\lambda}$. Now

$$H^0(Z; \mathcal{O}(\mathbb{E}_{\chi_\lambda})) = \sum\nolimits_{\pi_\gamma \in \widehat{G_u}} V_{\pi_\gamma} \otimes (E^*_{\chi_\gamma} \otimes E_{\chi_\lambda})^{L_u}$$

as G-module. Since Lemma 1.2.4 says that $(E^*_{\chi_\gamma} \otimes E_{\chi_\lambda})^{L_u}$ is 0 for $\gamma \neq \lambda$ and is $\mathbb{C}$ for $\gamma = \lambda$. That proves the Borel–Weil Theorem. □

The Bott–Borel–Weil Theorem interprets the space of holomorphic sections as 0-cohomology, as we have done by using the notation $H^0(Z; \mathcal{O}(\mathbb{E}_{\chi_\lambda}))$, and describes the action of G and G_u on the higher cohomologies as well. There are two approaches, Lie algebra cohomology and the Leray spectral sequence [Bot2], [Kos2]. The argument given above for Theorem 1.6.7 is really a Lie algebra cohomology argument. A proof of the Bott–Borel–Weil Theorem would take us too far away from our goal of understanding cycle spaces, so we only state the result. See Warner's book [War, Section 3.1.2] for a complete concise exposition.

Theorem 1.6.8 (Bott–Borel–Weil Theorem [Bot2]). *Let $\mathbb{E}_{\chi_\lambda} \to Z$ be a G_u-homogeneous holomorphic vector bundle. Let $\rho = \frac{1}{2}\sum_{\alpha \in \Sigma^+(\mathfrak{g},\mathfrak{h})} \alpha$, half the sum of the positive roots.*

1. *If $\lambda + \rho$ is singular ($\langle \alpha + \rho, \alpha \rangle = 0$ for some $\alpha \in \Sigma(\mathfrak{g}, \mathfrak{h})$), then every $H^k(Z; \mathcal{O}(\mathbb{E}_{\chi_\lambda})) = 0$.*
2. *Suppose that $\lambda + \rho$ is regular, i.e., $\langle \alpha + \rho, \alpha \rangle \neq 0$ for all $\alpha \in \Sigma(\mathfrak{g}, \mathfrak{h})$. Let w be the unique element of the Weyl group such that $\langle w(\lambda + \rho), \psi \rangle > 0$ for every simple root ψ. Define[4] $q = q(\lambda + \rho) = \#\{\alpha \in \Sigma^+(\mathfrak{g}, \mathfrak{h}) \mid \langle \lambda + \rho, \alpha \rangle < 0\}$. Then $H^k(Z; \mathcal{O}(\mathbb{E}_{\chi_\lambda})) = 0$ for $k \neq q$, and G and G_u act irreducibly on $H^q((Z; \mathcal{O}(\mathbb{E}_{\chi_\lambda}))$ by the irreducible representation of highest weight $w(\lambda+\rho)-\rho$.*

Later we will need to know the expression of the Bott–Borel–Weil theorem in the framework of Lie algebra cohomology. Compare [War, Section 2.5].

We first recall the basic definitions for relative Lie algebra cohomology. Let π be a representation of $\mathfrak{g}$ on a complex vector space V_π. The space of V_π-valued **cochains** is $C^q(\mathfrak{g}; V_\pi) := \mathrm{Hom}(\bigwedge^q \mathfrak{g}, V_\pi)$ for $q > 0$, and $C^0(\mathfrak{g}; V_\pi) := V_\pi$. The **coboundary operator** $\delta : C^q(\mathfrak{g}; V_\pi) \to C^{q+1}(\mathfrak{g}; V_\pi)$ is defined by

$$\delta(\omega)(\xi_0, \ldots, \xi_q) = \sum_{j=0}^{q} (-1)^j \pi(\xi_j)\big(\omega(\xi_0, \ldots, \widehat{\xi_j}, \ldots, \xi_q)\big)$$
$$+ \sum_{0 \leqq r < s \leqq q} (-1)^{r+s} \omega([\xi_r, \xi_s], \xi_0, \ldots, \widehat{\xi_r}, \ldots, \widehat{\xi_s}, \ldots, \xi_q).$$

[4] One can see that q is the minimal length of a word expressing w as a product of simple root reflections.

(Here the notation $\widehat{w}$ means that w is omitted.) It has the property that $\delta^2 = 0$. The **cocycles** are $Z^q(\mathfrak{g}; V_\pi)$ kernel of $\delta : C^q(\mathfrak{g}; V_\pi) \to C^{q+1}(\mathfrak{g}; V_\pi)$ and the **coboundaries** are $B^q(\mathfrak{g}; V_\pi)$: image of $\delta : C^{q-1}(\mathfrak{g}; V_\pi) \to C^q(\mathfrak{g}; V_\pi)$ Then the **Lie algebra cohomology** is

(1.6.9) $$H^q(\mathfrak{g}; V_\pi) := Z^q(\mathfrak{g}; V_\pi)/B^q(\mathfrak{g}; V_\pi).$$

Now let $\mathfrak{h}$ be a subalgebra of $\mathfrak{g}$. Then we have a subcomplex $\{C^q(\mathfrak{g}, \mathfrak{h}; V_\pi), \delta\}$ of $\{C^q(\mathfrak{g}; V_\pi), \delta\}$ as follows. The space of V_π-valued **relative cochains** is

$$\begin{aligned} C^q(\mathfrak{g}, \mathfrak{h}; V_\pi) &:= \mathrm{Hom}_{\mathfrak{h}}(\textstyle\bigwedge^q(\mathfrak{g}/\mathfrak{h}), V_\pi) \\ &= \{\omega \in C^q(\mathfrak{g}; V_\pi) \mid L_\xi\omega = 0 = \iota_\xi\omega \text{ for all } \xi \in \mathfrak{h}\} \quad \text{for } q > 0, \\ C^0(\mathfrak{g}; \mathfrak{h}; V_\pi) &:= V_\pi^{\mathfrak{h}} = \{v \in V_\pi \mid \pi(\xi)v = 0 \text{ for all } \xi \in \mathfrak{h}\}, \end{aligned}$$

where L_ξ is Lie derivative and ι_ξ is interior product. Essentially as before, the space of **relative cocycles** is

$$Z^q(\mathfrak{g}, \mathfrak{h}; V_\pi): \quad \text{kernel of } \delta : C^{q-1}(\mathfrak{g}, \mathfrak{h}; V_\pi) \to C^q(\mathfrak{g}, \mathfrak{h}; V_\pi)$$

and the space of **relative coboundaries** is

$$B^q(\mathfrak{g}, \mathfrak{h}; V_\pi): \quad \text{image of } \delta : C^{q-1}(\mathfrak{g}, \mathfrak{h}; V_\pi) \to C^q(\mathfrak{g}, \mathfrak{h}; V_\pi).$$

The **relative Lie algebra cohomology** is

(1.6.10) $$H^q(\mathfrak{g}, \mathfrak{h}; V_\pi) := Z^q(\mathfrak{g}, \mathfrak{h}; V_\pi)/B^q(\mathfrak{g}, \mathfrak{h}; V_\pi).$$

Note in particular that $H^0(\mathfrak{g}, \mathfrak{h}; V_\pi) = V_\pi$. The Lie algebra cohomology formulation of the Bott–Borel–Weil Theorem is the following.

Theorem 1.6.11 ([Bot2, Section 1.6]). *Let χ_ν be an irreducible representation of Q with representation space E_{χ_ν} and highest weight ν. Let π_λ be an irreducible representation of G with representation space V_{π_λ} and highest weight λ. Then* $\mathrm{Hom}_G(V_\lambda, H^\bullet(G/Q; \mathcal{O}(\mathbb{E}_{\chi_\nu})) = H^\bullet(\mathfrak{q}, \mathfrak{q}^r; \mathrm{Hom}_{\mathfrak{q}}(V_{\pi_\lambda}, E_{\chi_\nu}))$. *Here the cohomology on the right is relative Lie algebra cohomology.*

2

Real Group Orbits

In this chapter we record the basic facts used to study a real group orbit $G_0(z)$ in a complex flag manifold $Z = G/Q$. We start with the Bruhat decomposition as completed by Harish-Chandra, and use it to analyze the root structure of the real isotropy algebra $\mathfrak{g}_0 \cap \mathfrak{q}_z$. This leads directly to a codimension formula for $G_0(z)$ in Z, to a (finite) bound on the number of G_0-orbits, and to the existence of open orbits.

2.1 Bruhat Lemma and an application

The Bruhat Lemma for the complex flag manifold $X = G/B$ is as follows. We may assume that B is given by (1.3.1) and (1.3.2). Consider the Weyl group $W = W(\mathfrak{g}, \mathfrak{h}) := N_G(H)/H$. Given $w \in W$ choose a representative $s_w \in N_G(H)$. Let $x_0 = 1B \in G/B = X$. The weakest form of the Bruhat decomposition [HC0] is sufficient for our needs. It is the following.

Lemma 2.1.1 (Bruhat [Bru], Harish-Chandra [HC0]). *The manifold X is the disjoint union of the B-orbits $B(s_w x_0)$, $w \in W$. The B-orbit*

$$\mathcal{O} = B(s_w x_0) \cong B/(B \cap \mathrm{Ad}(s_w)B)$$

is holomorphically (in fact, $\mathbb{C}$-algebraically) equivalent to $\mathbb{C}^{\dim \mathcal{O}}$, and $\mathcal{O}$ has dimension $\#\{\alpha \in \Sigma^+ \mid w(\alpha) \in \Sigma^+\}$.

This decomposes X as a union of cells, as follows. The isotropy subgroup of B at $s_w x_0$ is $B \cap B_w$, where B_w is the analytic subgroup of G with Lie algebra $\mathfrak{b}_w = \mathfrak{h} + \sum_{\beta \in w(\Sigma^+)} \mathfrak{g}_{-\beta}$. This decomposes B as $N_w(B \cap B_w)$, where N_w is the unipotent analytic subgroup of G with Lie algebra $\mathfrak{n}_w = \sum_{\alpha \in \Sigma^+ \cap w(\Sigma^-)} \mathfrak{g}_{-\alpha}$. Now $\mathfrak{n}_w^+ := \sum_{\alpha \in \Sigma^+ \cap w(\Sigma^+)} \mathfrak{g}_\alpha$ represents the holomorphic tangent space of $B(s_w x_0)$, and $\xi \mapsto \exp(\xi)(s_w x_0)$ is a biholomorphic map of $\mathfrak{n}_w^+$ onto $B(s_w x_0)$.

Lemma 2.1.2. *If Q_1 and Q_2 are parabolic subgroups of G, then $Q_1 \cap Q_2$ contains a Cartan subgroup of G.*

Proof. Let $\mathfrak{b}$ and $\mathfrak{b}'$ be Borel subalgebras of $\mathfrak{g}$. We will show that $\mathfrak{b} \cap \mathfrak{b}'$ contains a Cartan subalgebra of $\mathfrak{g}$. For this, we may assume that $\mathfrak{b}$ is our standard Borel $\mathfrak{h} + \sum_{\alpha\in\Sigma^+} \mathfrak{g}_{-\alpha}$. Let B and B' be the corresponding Borel subgroups of G. Then B' is the G-stabilizer of a point $x' \in X = G/B$. Following the Bruhat Lemma 2.1.1, we may take $x' = bs_w x_0$ for some $b \in B$ and $w \in W$. Without loss of generality we conjugate by b^{-1}. Now we may assume $x' = s_w x_0$. Then $B' = \mathrm{Ad}(s_w)B$ and therefore $\mathfrak{b}' = \mathrm{ad}(s_w)\mathfrak{b}$, which contains $\mathfrak{h}$.

Every $h \in H$ normalizes both $\mathfrak{b}$ and $\mathfrak{b}'$, so $h \in B \cap B'$. Thus the intersection of two Borel subgroups contains a Cartan subgroup. The lemma follows. □

Let G_0 be a real form of G. In other words, G_0 is a Lie subgroup of G whose Lie algebra $\mathfrak{g}_0$ is a real form of $\mathfrak{g}$. Although G is connected, G_0 does not have to be connected. We always write τ both for the complex conjugation of $\mathfrak{g}$ over $\mathfrak{g}_0$ and for the corresponding conjugation of G over G_0. So τ denotes both the antiholomorphic involution of $\mathfrak{g}$ with fixed point set $\mathfrak{g}_0$ and the antiholomorphic involution of G such that G_0 is an open subgroup of the fixed point set of τ.

We recall some key notation: θ is the Cartan involution of $\mathfrak{g}_0$ and G_0 that commutes with τ. We extend θ to holomorphic involutions of $\mathfrak{g}$ and G. Now $\sigma := \tau\theta$ is the complex conjugation of G over its θ-stable compact real form G_u and is also complex conjugation of $\mathfrak{g}$ over $\mathfrak{g}_u$.

Corollary 2.1.3. *If $\mathfrak{q}$ is a parabolic subalgebra of $\mathfrak{g}$, then $\mathfrak{q} \cap \tau\mathfrak{q}$ contains a τ-stable Cartan subalgebra of $\mathfrak{g}$.*

Proof. Set $\mathfrak{r} = \mathfrak{q} \cap \tau\mathfrak{q}$. It is a τ-stable complex subalgebra of $\mathfrak{g}$, and therefore $\mathfrak{r}_0 = \mathfrak{g}_0 \cap \mathfrak{r}$ is a real form of $\mathfrak{r}$ and τ induces the complex conjugation of $\mathfrak{r}$ over $\mathfrak{r}_0$. Choose a Cartan subalgebra $\mathfrak{j}_0$ of $\mathfrak{r}_0$. Its complexification $\mathfrak{j}$ is a Cartan subalgebra of $\mathfrak{r}$. Lemma 2.1.2 says that $\mathfrak{r}$ contains Cartan subalgebras of $\mathfrak{g}$. Thus $\mathfrak{j}$ is a τ-stable Cartan subalgebra of $\mathfrak{g}$. □

We have the parabolic subgroup $Q \subset G$ and the complex flag manifold $Z = G/Q$. Since Q is its own normalizer in G, we may view Z as the space of all G-conjugates of $\mathfrak{q}$, by the correspondence $gQ \leftrightarrow \mathrm{Ad}(g)\mathfrak{q}$. We will write $\mathfrak{q}_z$ for the parabolic subalgebra of $\mathfrak{g}$ corresponding to $z \in Z$, and will write Q_z for the corresponding parabolic subgroup of G.

2.2 Real isotropy

Here is the principal strategy for dealing with G_0-orbits on Z. We will use it constantly. Consider the orbit $G_0(z)$. The isotropy subgroup of G_0 at z is $G_0 \cap Q_z$. That isotropy subgroup has Lie algebra $\mathfrak{g}_0 \cap \mathfrak{q}_z$, which is a real form of $\mathfrak{q}_z \cap \tau\mathfrak{q}_z$. Corollary 2.1.3 says that $\mathfrak{q}_z \cap \tau\mathfrak{q}_z$ contains a τ-stable Cartan subalgebra $\mathfrak{h}$ of $\mathfrak{g}$. Now $\mathfrak{q}_z$ contains a Borel subalgebra of $\mathfrak{g}$ that contains $\mathfrak{h}$. Express that Borel subalgebra as $\mathfrak{b} = \mathfrak{h} + \sum_{\alpha\in\Sigma^+} \mathfrak{g}_\alpha$ for an appropriate choice of positive root system $\Sigma^+ = \Sigma^+(\mathfrak{g}, \mathfrak{h})$. We have proved

Theorem 2.2.1. *Let G_0 be a real form of the complex semisimple Lie group G, let τ denote complex conjugation of $\mathfrak{g}$ over $\mathfrak{g}_0$, and consider an orbit $G_0(z)$ on a complex flag manifold $Z = G/Q$. Then there exist a τ-stable Cartan subalgebra $\mathfrak{h} \subset \mathfrak{q}_z$ of $\mathfrak{g}$, a positive root system $\Sigma^+ = \Sigma^+(\mathfrak{g}, \mathfrak{h})$, and a set Φ of simple roots, such that $\mathfrak{q}_z = \mathfrak{q}_\Phi$ and $Q_z = Q_\Phi$.*

Corollary 2.2.2. *In the notation of Theorem 2.2.1, $\mathfrak{q}_z \cap \tau\mathfrak{q}_z$ is the semidirect sum of its nilpotent radical*

$$(\mathfrak{q}_\Phi^{-n} \cap \tau\mathfrak{q}_\Phi^{-n}) + (\mathfrak{q}_\Phi^{r} \cap \tau\mathfrak{q}_\Phi^{-n}) + (\mathfrak{q}_\Phi^{-n} \cap \tau\mathfrak{q}_\Phi^{r})$$

with the Levi complement

$$\mathfrak{q}_\Phi^{r} \cap \tau\mathfrak{q}_\Phi^{r} = \mathfrak{h} + \sum_{\Phi^r \cap \tau\Phi^r} \mathfrak{g}_\alpha.$$

In particular, $\dim_{\mathbb{R}} \mathfrak{g}_0 \cap \mathfrak{q}_z = \dim_{\mathbb{C}} \mathfrak{q}_\Phi^r + |\Phi^n \cap \tau\Phi^n|$.

Proof. The subspace $(\mathfrak{q}_\Phi^{-n} \cap \tau\mathfrak{q}_\Phi^{-n}) + (\mathfrak{q}_\Phi^{r} \cap \tau\mathfrak{q}_\Phi^{-n}) + (\mathfrak{q}_\Phi^{-n} \cap \tau\mathfrak{q}_\Phi^{r})$ of $\mathfrak{q}_\Phi \cap \tau\mathfrak{q}_\Phi$ is the sum of all root spaces $\mathfrak{g}_\alpha \subset \mathfrak{q}_\Phi \cap \tau\mathfrak{q}_\Phi$ such that $\mathfrak{g}_{-\alpha} \not\subset \mathfrak{q}_\Phi \cap \tau\mathfrak{q}_\Phi$. So it is the nilradical of $\mathfrak{q}_\Phi \cap \tau\mathfrak{q}_\Phi$. The subspace $\mathfrak{q}_\Phi^r \cap \tau\mathfrak{q}_\Phi^r = \mathfrak{h} + \sum_{\Phi^r \cap \tau\Phi^r} \mathfrak{g}_\alpha$ is a reductive subalgebra that is a vector space complement. Hence it is a Levi complement. Now compute

$$\begin{aligned}
&\dim_{\mathbb{R}} \mathfrak{g}_0 \cap \mathfrak{q}_z \\
&\quad = \dim_{\mathbb{C}} \mathfrak{q}_\Phi \cap \tau\mathfrak{q}_\Phi = \dim_{\mathbb{C}} \mathfrak{h} + |(\Phi^r \cup \Phi^n) \cap \tau(\Phi^r \cup \Phi^n)| \\
&\quad = \big(\dim_{\mathbb{C}} \mathfrak{h} + |\Phi^r \cap \tau\Phi^r| + |\Phi^n \cap \tau\Phi^r| + |\Phi^r \cap \tau\Phi^n|\big) + |\Phi^n \cap \tau\Phi^n| \\
&\quad = \dim_{\mathbb{C}} \mathfrak{q}_\Phi^r + |\Phi^n \cap \tau\Phi^n|,
\end{aligned}$$

as asserted. □

Corollary 2.2.3. *In the notation of Theorem 2.2.1,*

$$\operatorname{codim}_{\mathbb{R}}(G_0(z) \subset Z) = |\Phi^n \cap \tau\Phi^n|.$$

In particular, $G_0(z)$ is open in Z if and only if $\Phi^n \cap \tau\Phi^n$ is empty.

Proof. In view of Corollary 2.2.2, the codimension in question is given by

$$\begin{aligned}
&\operatorname{codim}_{\mathbb{R}}(G_0(z) \subset Z) \\
&\quad = \dim_{\mathbb{R}} Z - \dim_{\mathbb{R}} G_0(z) \\
&\quad = 2|\Phi^n| - [\dim_{\mathbb{R}} G_0 - \dim_{\mathbb{R}}(G_0 \cap Q_z)] \\
&\quad = 2|\Phi^n| - [(\dim_{\mathbb{R}} \mathfrak{h} + |\Phi^r| + 2|\Phi^n|) - (\dim_{\mathbb{R}} \mathfrak{h} + |\Phi^r| + |\Phi^n \cap \tau\Phi^n|)] \\
&\quad - |\Phi^n \cap \tau\Phi^n|,
\end{aligned}$$

as asserted. □

Corollary 2.2.4. *There are only finitely many G_0-orbits on Z. The maximal-dimensional orbits are open and the minimal-dimensional orbits are closed.*

Proof. There are only finitely many G_0-conjugacy classes of Cartan subalgebras $\mathfrak{h}_0 \subset \mathfrak{g}_0$, so the number of G_0-conjugacy classes of τ-stable Cartan subalgebras $\mathfrak{h} \subset \mathfrak{g}$ is finite. Given such an $\mathfrak{h}$, the number of positive root systems Σ^+ is finite. Given $(\mathfrak{h}, \Sigma^+)$, the number of sets Φ of simple roots is finite. Thus the number of possibilities for Q_Φ is finite up to G_0-conjugacy. This proves that the number of G_0-orbits on Z is finite. It also gives a (very) rough upper bound on the number. The other statements follow because the closure of an orbit is a union of orbits. □

We refer to the open G_0-orbits on Z as **flag domains**. As G_0-invariant open subsets of Z, the flag domains $D \subset Z$ are G_0-homogeneous complex manifolds. At the other extreme, we will see that the G_0-closed orbit is unique, and that it sometimes has an interpretation as the Shilov boundary of a particular flag domain.

3

Orbit Structure for Hermitian Symmetric Spaces

In this chapter we look at the case where Z is an irreducible hermitian symmetric space G_u/K_0 of compact type, viewed as a complex flag manifold G/Q, and G_0 is the real form of G that is hermitian in the sense that $D_0 = G_0/K_0$ is a bounded symmetric domain. In Section 1 we construct the maximal set of strongly orthogonal noncompact positive roots, the partial Cayley transforms, and the conjugacy classes of Cartan subalgebras of $\mathfrak{g}_0$. These constructions are the foundation for the special features of the hermitian symmetric spaces. Then in Section 2 we indicate the G_0-orbit structure of Z in the hermitian symmetric space case.

Since Z is an irreducible hermitian symmetric space, G is a connected complex simple Lie group, and for convenience we suppose that G is simply connected.

3.1 Strongly orthogonal roots

First, recall our standard notation. Fix a Cartan involution θ of G_0. Thus θ is an involutive automorphism of G_0 whose fixed point set $K_0 = G_0^\theta$ is a maximal compact subgroup. Since G is simply connected we can extend θ holomorphically to G. The corresponding decompositions into (± 1)-eigenspaces of $d\theta$ are $\mathfrak{g}_0 = \mathfrak{k}_0 + \mathfrak{s}_0$ and $\mathfrak{g} = \mathfrak{k} + \mathfrak{s}$ where $\mathfrak{k}_0$ is the Lie algebra of $K_0 = G_0^\theta$. Then $G_u \subset G$ is the compact real form of G that is the analytic subgroup for the compact real form $\mathfrak{g}_u = \mathfrak{k}_0 + \mathfrak{s}_u$ of $\mathfrak{g}$ where $\mathfrak{s}_u = \sqrt{-1}\mathfrak{s}_0$ of $\mathfrak{g}$. Recall that τ is complex conjugation of G over G_0 and of $\mathfrak{g}$ over $\mathfrak{g}_0$. Now $\sigma := \theta\tau$ is complex conjugation of G over G_u and of $\mathfrak{g}$ over $\mathfrak{g}_u$.

There is a compact Cartan subalgebra $\mathfrak{t}_0 \subset \mathfrak{k}_0$ of $\mathfrak{g}_0$. If $\alpha \in \Sigma(\mathfrak{g}, \mathfrak{t})$, then either $\mathfrak{g}_\alpha \subset \mathfrak{k}$ and we say that the root α is **compact**, or $\mathfrak{g}_\alpha \subset \mathfrak{s}$ and we say that α is **noncompact**. There is a simple root system $\Psi = \{\psi_0, \dots, \psi_m\}$ such that ψ_0 is noncompact and the other ψ_i are compact. Furthermore, ψ_0 is a long root, and every noncompact positive root is of the form $\psi_0 + \sum_{1 \leqq i \leqq m} n_i \psi_i$ with each integer $n_i \geqq 0$. Thus $\mathfrak{g} = \mathfrak{k} + \mathfrak{s}_+ + \mathfrak{s}_-$ where

(3.1.1) $$\mathfrak{k} = \mathfrak{t} + \sum_{n_0=0} \mathfrak{g}_\alpha, \quad \mathfrak{s}_+ = \sum_{n_0=1} \mathfrak{g}_\alpha, \quad \text{and } \mathfrak{s}_- = \sum_{n_0=-1} \mathfrak{g}_\alpha.$$

Here $\mathfrak{q} = \mathfrak{q}_{\{\psi_1,\dots,\psi_m\}}$. In other words

$$\mathfrak{q}^r = \mathfrak{k},\ \mathfrak{q}^n = \mathfrak{s}_+,\ \mathfrak{q}^{-n} = \mathfrak{s}_-, \text{ and therefore } \mathfrak{q} = \mathfrak{k} + \mathfrak{s}_-. \tag{3.1.2}$$

The Cartan subalgebras of $\mathfrak{g}_0$ all are $\mathrm{Ad}(G_0)$-conjugate to one of the $\mathfrak{h}_{\Gamma,0}$ given as follows. Let $\Gamma = \{\gamma_1, \dots, \gamma_r\}$ be a set of noncompact positive roots that is **strongly orthogonal** in the sense that

$$\text{if } 1 \leqq i < j \leqq r \text{ then none of } \pm\gamma_i \pm \gamma_j \text{ is a root.} \tag{3.1.3}$$

Then each $\mathfrak{g}[\gamma_i] = [\mathfrak{g}_{\gamma_i}, \mathfrak{g}_{-\gamma_i}] + \mathfrak{g}_{\gamma_i} + \mathfrak{g}_{-\gamma_i} \cong \mathfrak{sl}(2, \mathbb{C})$, say with

$$h_{\gamma_i} \leftrightarrow \begin{pmatrix} 1 & 0 \\ 0 & -1 \end{pmatrix}, \quad e_{\gamma_i} \leftrightarrow \begin{pmatrix} 0 & 1 \\ 0 & 0 \end{pmatrix}, \quad f_{\gamma_i} \leftrightarrow \begin{pmatrix} 0 & 0 \\ 1 & 0 \end{pmatrix},$$

where $h_{\gamma_i} \in [\mathfrak{g}_{\gamma_i}, \mathfrak{g}_{-\gamma_i}]$, $e_{\gamma_i} \in \mathfrak{g}_{\gamma_i}$ and $f_{\gamma_i} \in \mathfrak{g}_{-\gamma_i}$ as usual, and such that $\mathfrak{g}_0[\gamma_i] = \mathfrak{g}_0 \cap \mathfrak{g}_{\gamma_i} \cong \mathfrak{su}(1,1)$ is spanned by

$$\sqrt{-1}h_{\gamma_i}, \quad e_{\gamma_i} + f_{\gamma_i}, \quad \text{and} \quad \sqrt{-1}(e_{\gamma_i} - f_{\gamma_i}).$$

Thus $\sqrt{-1}h_{\gamma_i}$ spans the compact Cartan subalgebra $\mathfrak{t}_{\gamma_i} = \mathfrak{g}_0[\gamma_i] \cap \mathfrak{t}$ of $\mathfrak{g}_0[\gamma_i]$ and $e_{\gamma_i} + f_{\gamma_i}$ spans the noncompact Cartan subalgebra $\mathfrak{a}_{\gamma_i} = \mathfrak{g}_0[\gamma_i] \cap \mathfrak{s}$ of $\mathfrak{g}_0[\gamma_i]$. Strong orthogonality (3.1.3) says $[\mathfrak{g}[\gamma_i], \mathfrak{g}[\gamma_j]] = 0$ for $1 \leqq i < j \leqq r$. Define

$$\mathfrak{t}_\Gamma = \sum_{1 \leqq i \leqq r} \mathfrak{t}_{\gamma_i} \text{ and } \mathfrak{a}_\Gamma = \sum_{1 \leqq i \leqq r} \mathfrak{a}_{\gamma_i}. \tag{3.1.4}$$

Then $\mathfrak{g}$ has Cartan subalgebras

$$\mathfrak{t} = \mathfrak{t}_\Gamma + (\mathfrak{t} \cap \mathfrak{t}_\Gamma^\perp) \quad \text{and} \quad \mathfrak{h}_\Gamma = \mathfrak{a}_\Gamma + (\mathfrak{t} \cap \mathfrak{t}_\Gamma^\perp). \tag{3.1.5}$$

They are $\mathrm{Int}(\mathfrak{g})$-conjugate, for the **partial Cayley transform**

$$c_\Gamma = \prod_{1 \leqq i \leqq r} \exp\left(\tfrac{\pi}{4}\sqrt{-1}(e_{\gamma_i} - f_{\gamma_i})\right) \text{ satisfies } \mathrm{Ad}(c_\Gamma)\mathfrak{t}_\Gamma = \mathfrak{a}_\Gamma. \tag{3.1.6}$$

However, their real forms

$$\mathfrak{t}_0 = \mathfrak{g}_0 \cap \mathfrak{t} \text{ and } \mathfrak{h}_{\Gamma,0} = \mathfrak{g}_0 \cap \mathfrak{h}_\Gamma \tag{3.1.7}$$

are not $\mathrm{Ad}(G_0)$-conjugate except in the trivial case where Γ is empty, for the Killing form has rank $m = \dim \mathfrak{t}_0$ and signature $2|\Gamma| - m$ on $\mathfrak{h}_{\Gamma,0}$. More precisely, we have the following.

Proposition 3.1.8. *Every Cartan subalgebra of* $\mathfrak{g}_0$ *is* $\mathrm{Ad}(G_0)$*-conjugate to one of the* $\mathfrak{h}_{\Gamma,0}$*, and Cartan subalgebras* $\mathfrak{h}_{\Gamma,0}$ *and* $\mathfrak{h}_{\Gamma',0}$ *are* $\mathrm{Ad}(G_0)$*-conjugate if and only if the cardinalities* $|\Gamma| = |\Gamma'|$.

We recall Kostant's **cascade construction** of a maximal set of strongly orthogonal noncompact positive roots in $\Sigma(\mathfrak{g}, \mathfrak{t})$. This set has cardinality $\ell = \operatorname{rank}_{\mathbb{R}} \mathfrak{g}_0$ and is given by

$$\begin{aligned} \Xi = &\{\xi_1, \dots, \xi_\ell\}, \text{ where} \\ &\xi_1 \text{ is the maximal (necessarily noncompact positive) root and} \\ &\xi_{m+1} \text{ is a maximal noncompact positive root } \perp \{\xi_1, \dots, \xi_m\}. \end{aligned} \tag{3.1.9}$$

The roots ξ_i are long, and any set of strongly orthogonal noncompact positive long roots in $\Sigma(\mathfrak{g}, \mathfrak{t})$ is $W(G_0, T_0)$-conjugate to a subset of Ξ. Further, the Weyl group $W(G_0, T_0)$ induces every permutation of Ξ.

The notion of strong orthogonality of roots was developed in various forms: by Harish-Chandra [HC1] for the study of the holomorphic discrete series, by B. Kostant [Kos1] and later by M. Sugiura [Su] for the construction and classification of real Cartan subalgebras, and by C. C. Moore [Mo] for an analysis of compactifications of the bounded symmetric domains.

3.2 Orbits and Cayley transforms

Let $z_0 = 1Q \in G/Q = Z$ denote the base point of our flag manifold Z when Z is viewed as a homogeneous space. The Cartan subalgebra $\mathfrak{h}_{\Gamma,0} \subset \mathfrak{g}_0$ leads to the orbits $G_0(c_\Gamma c_\Delta^2 z_0) \subset Z$, where $\Gamma \cup \Delta$ is a set of strongly orthogonal noncompact positive roots in $\Sigma(\mathfrak{g}, \mathfrak{t})$ with Γ and Δ disjoint. In view of the Weyl group equivalence just discussed, we may take $\Gamma = \{\xi_1, \dots, \xi_r\}$ and $\Delta = \{\xi_{r+1}, \dots, \xi_{r+s}\}$, both inside Ξ. Using $G_0 = K_0 \exp(\mathfrak{a}_{\Xi,0}) K_0$ one arrives at

Theorem 3.2.1. *The G_0-orbits on Z are just the orbits $D_{\Gamma,\Delta} = G_0(c_\Gamma c_\Delta^2 z_0)$ where Γ and Δ are disjoint subsets of Ξ. Two such orbits $D_{\Gamma,\Delta}$ and $D_{\Gamma',\Delta'}$ are equal if and only if cardinalities $|\Gamma| = |\Gamma'|$ and $|\Delta| = |\Delta'|$. An orbit $D_{\Gamma,\Delta}$ is open if and only if Γ is empty, closed if and only if $(\Gamma, \Delta) = (\Xi, \emptyset)$. An orbit $D_{\Gamma',\Delta'}$ is in the closure of $D_{\Gamma,\Delta}$ if and only if $|\Delta'| \leqq |\Delta|$ and $|\Gamma \cup \Delta| \leqq |\Gamma' \cup \Delta'|$.*

Example 3.2.2. Before indicating the proof of Theorem 3.2.1, we illustrate it with the case $G_0 = SU(m, n)$. Thus $G = SL(m + n; \mathbb{C})$, Z is the complex Grassmann manifold of n-dimensional linear subspaces of $\mathbb{C}^{m+n}$, and G acts on Z by $g([v_1 \wedge \dots \wedge v_n]) = [gv_1 \wedge \dots \wedge gv_n]$. (As usual, $[w_1 \wedge \dots \wedge w_\ell]$ denotes the linear span of $\{w_1, \dots, w_\ell\}$.)

Fix an ordered basis $\mathbf{e} = \{e_1, \dots, e_{m+n}\}$ of $\mathbb{C}^{m+n}$. Then

$$z_0 = [e_{m+1} \wedge \dots \wedge e_{m+n}] \in Z$$

will be our base point. We write linear transformations of $\mathbb{C}^{m+n}$ in matrix form relative to $\mathbf{e}$, often in block matrix form $\left(\begin{smallmatrix} A & B \\ C & D \end{smallmatrix}\right)$, where A is $m \times m$, B is $m \times n$, C is $n \times m$ and D is $n \times n$. In particular,

$$Q = \{g \in G \mid gz_0 = z_0\} = \left\{\left(\begin{smallmatrix} A & B \\ C & D \end{smallmatrix}\right) \mid B = 0 \text{ and } (\det A)(\det D) = 1\right\}.$$

We fix the hermitian form $h(u, v) = -\sum_{j=1}^{m} u^j \overline{v^j} + \sum_{k=1}^{n} u^{m+k} \overline{v^{m+k}}$ on $\mathbb{C}^{m+n}$, where $u = \sum u^i e_i$ and $v = \sum v^i e_i$. Then our real form is

$$G_0 = SU(m, n) = \{g \in G \mid h(gu, gv) = h(u, v) \text{ for all } u, v \in \mathbb{C}^{m+n}\}.$$

The Cartan involution of G_0 is $\theta : g \mapsto \begin{pmatrix} I_m & 0 \\ 0 & -I_n \end{pmatrix} g \begin{pmatrix} I_m & 0 \\ 0 & -I_n \end{pmatrix}$. It defines the maximal compact subgroup

$$K_0 = G_0^\theta = \left\{ \left(\begin{smallmatrix} A & 0 \\ 0 & D \end{smallmatrix} \right) \mid A \in U(m), D \in U(n), \text{ and } (\det A)(\det D) = 1 \right\}.$$

The corresponding Lie algebra decompositions $\mathfrak{g}_0 = \mathfrak{k}_0 + \mathfrak{s}_0$ and $\mathfrak{g} = \mathfrak{k} + \mathfrak{s}$ are given by

$$\begin{aligned} \mathfrak{k}_0 &= \left\{ \left(\begin{smallmatrix} A & 0 \\ 0 & D \end{smallmatrix} \right) \mid A + A^* = 0, D + D^* = 0, \operatorname{trace}(A) + \operatorname{trace}(D) = 0 \right\}, \\ \mathfrak{k} &= \left\{ \left(\begin{smallmatrix} A & 0 \\ 0 & D \end{smallmatrix} \right) \mid \operatorname{trace}(A) + \operatorname{trace}(D) = 0 \right\}, \\ \mathfrak{s}_0 &= \left\{ \left(\begin{smallmatrix} 0 & B \\ B^* & 0 \end{smallmatrix} \right) \right\} \text{ and } \mathfrak{s} = \mathfrak{s}_+ + \mathfrak{s}_- \text{ where } \mathfrak{s}_+ = \left\{ \left(\begin{smallmatrix} 0 & B \\ 0 & 0 \end{smallmatrix} \right) \right\} \text{ and } \mathfrak{s}_- = \left\{ \left(\begin{smallmatrix} 0 & 0 \\ C & 0 \end{smallmatrix} \right) \right\}. \end{aligned}$$

In particular the compact real form of G with Lie algebra $\mathfrak{g}_u = \mathfrak{k}_0 + \sqrt{-1}\mathfrak{s}_0$ is

$$G_u = SU(m+n) = \{g \in G \mid h_u(gu, gv) = h_u(u, v) \text{ for all } u, v \in \mathbb{C}^{m+n}\},$$

where $h_u(u, v) = \sum_{j=1}^{m+n} u^j \overline{v^j}$. That exhibits Z as the irreducible compact hermitian symmetric space $G_u/K_0 = SU(m+n)/S(U(m) \times U(n))$.

The symmetric space rank of Z is $r = \min\{m, n\}$. The maximal set $\Xi = \{\xi_i, \ldots, \xi_r\}$ of strongly orthogonal noncompact positive roots, relative to the diagonal Cartan subalgebra, is given by

$$\xi = \varepsilon_i - \varepsilon_{m+n-i}, \quad \text{where } \varepsilon_j(\operatorname{diag}\{a_1, \ldots, a_{m+n}\}) = a_j.$$

This is the cascade construction (3.1.9). Let $G[\xi_i]$ be the $SL(2; \mathbb{C})$ of the plane $[e_i \wedge e_{m+n-i}]$, extended to act trivially on the span of the other e_j, and $G_0[\xi_i] = G[\xi_i] \cap G_0 \cong SL(2; \mathbb{R})$. We have **partial Cayley transforms**

$$c_i(e_i) = \frac{1}{\sqrt{-2}}(\sqrt{-1}e_i + e_{m+n-i}), c_i(e_{m+n-i}) = \frac{1}{\sqrt{-2}}(e_i + \sqrt{-1}e_{m+n-i}),$$

and $c_i(e_j) = e_j$ for $i \neq j \neq m+n-i$. We now check that the points

$$z_{s,t} = c_1 c_2 \ldots c_s c_{s+1}^2 c_{s+2}^2 \ldots c_{s+t}^2 z_0 \in Z, \quad 0 \leqq s \leqq t \leqq r,$$

play the role of (and in fact are) the $c_\Gamma c_\Delta^2 z_0$ of Theorem 3.2.1.

To $z \in Z$ we associate the triple (a, b, c) where $h|_z$ has rank $a + b$ with a positive and b negative signs, and nullity $c = n - (a + b)$. The only restrictions are $a + b + c = n$, $0 \leqq a + c \leqq n$, $0 \leqq b + c \leqq n$ and $0 \leqq c \leqq r$. For example, we associate $(t, n - s - t, s)$ to $z_{s,t}$. Thus each admissible triple is associated to exactly one of the $z_{s,t}$. If the same triple is associated to z and z', then the hermitian version of Witt's theorem shows that $z' \in G_0(z)$. Thus

(3.2.3) the G_0-orbits on Z are just the $G_0(z_{s,t})$, $0 \leqq s \leqq t \leqq r$.

In particular,

(3.2.4)
there are precisely $\frac{1}{2}(r+1)(r+2)$ G_0-orbits on Z,
the open orbits are the $r+1$ orbits $G_0(z_{0,t})$, $0 \leqq t \leqq r$, and
$G_0(z_{r,r})$ is the unique closed orbit.

The orbit $G_0(z_0) = G_0(z_{0,0})$ consists of the negative definite n-dimensional subspaces of $(\mathbb{C}^{m+n}, h)$. Any such subspace z has a basis $\mathbf{b}(z)$ of the form $\{v_1, \ldots, v_n\}$, where $v_k = (\sum_{j=1}^{m} x_{i,j} e_j) + e_{m+k}$. Hence z is represented by the $m \times n$ matrix $X = (x_{j,k})$. The condition that z be negative definite is $I - XX^* \gg 0$, where $\gg 0$ means positive definite. Thus $G_0(z_0)$ is the bounded symmetric domain of $m \times n$ matrices X such that $I - XX^* \gg 0$. Its boundary consists of the $m \times n$ matrices X such that $I - XX^*$ is positive semidefinite but not positive definite, in other words the union of the $G_0(x_s, 0)$, $1 \leqq s \leqq r$. The closed orbit

$$G_0(z_{r,r}) = \{m \times n \text{ matrices } X \mid XX^* = I\}$$

is the Bergman–Shilov boundary of $G_0(z_0)$ (see [KW]). In the tube domain case, $m = n = r$, the Bergman–Shilov boundary is frequently viewed as the unitary group $U(r)$. ◇

3.2.5. Indication of proof of Theorem 3.2.1. The idea of the proof of Theorem 3.2.1 is the use of the polysphere $G[\Xi](z_0)$ in Z. Here, if Φ is a subset of the maximal set Ξ of strongly orthogonal noncompact roots, then $G[\Phi]$ denotes $\prod_{\xi \in \Phi} G[\xi]$. Thus $G[\Xi](z_0)$ is the product $\prod_{\xi \in \Xi} G[\xi](z_0)$ of Riemann spheres. The orbits of $G_0[\xi] = G[\xi] \cap G_0$ on the Riemann sphere $S[\xi] = G[\xi](z_0)$ are the lower hemisphere $G_0[\xi](z_0)$, the equator $G_0[\xi](c_\xi(z_0)$, and the upper hemisphere $G_0[\xi](c_\xi^2(z_0)$. Now $K_0 G[\Xi](z_0) = Z$ tells us that every G_0-orbit on Z is of the form $D_{\Gamma,\Delta} = G_0(c_\Gamma c_\Delta^2 z_0)$, where Γ and Δ are disjoint subsets of Ξ, as asserted. Since $\{w \in W(G_0.T_0) \mid w\Xi = \Xi\}$ induces every permutation of Ξ, the orbit $D_{\Gamma,\Delta}$ depends only on the cardinalities of Γ and Δ in the sense that $|\Gamma| = |\Gamma'|$ and $|\Delta| = |\Delta'|$ imply $D_{\Gamma,\Delta} = D_{\Gamma',\Delta'}$.

The delicate point now is the converse. It states that if $D_{\Gamma,\Delta} = D_{\Gamma',\Delta'}$, then $|\Gamma| = |\Gamma'|$ and $|\Delta| = |\Delta'|$. Normalize the G_u-invariant Kähler metric on Z so that the spheres $S[\xi]$ have circumference 2π. Denote

$$z_\infty = c_\Xi^2(z_0) \quad \text{and} \quad \Phi = \Xi \setminus (\Gamma \cup \Delta).$$

Then one expects that $\inf\{\mathrm{dist}(z, z_0)^2 \mid z \in D_{\Gamma,\Delta}\} = |\Gamma \cup \Delta|(\frac{\pi}{2})^2$ and $\inf\{\mathrm{dist}(z, z_\infty)^2 \mid z \in D_{\Gamma,\Delta}\} = |\Gamma \cup \Phi|(\frac{\pi}{2})^2$. From those numbers one easily computes $|\Gamma|$ and $|\Delta|$. While the expectation is correct, it has to be done carefully.

Theorem 3.2.1 follows by looking at the closure and boundary properties of the $G_0[\Xi]$-orbits in the polysphere $\prod_{\xi \in \Xi} S[\xi]$. □

Strongly orthogonal roots and partial Cayley transforms were used by Korányi and Wolf [KW], [WK] to describe the G_0-orbit structure of the closure of the bounded symmetric domain $D_{\emptyset,\emptyset}$ in its compact dual, and then by Wolf [W2] for all the G_0-orbits. See [W4] for a detailed exposition. The special root orders here are based on results of Borel and de Siebenthal [BoS]. Strongly orthogonal roots and the Borel-de Siebenthal root order have since become a standard tool in the structure and representation theories of semisimple Lie algebras.

4

Open Orbits

In this chapter we describe the basic structure of flag domains, i.e., of open real group orbits $D = G_0(z)$ in complex flag manifolds $Z = G/Q$. In Section 1 we prove that every elliptic automorphism of $\mathfrak{g}_0$ fixes a regular semisimple element, using a variation of de Sibenthal's technique for the case of $\mathfrak{g}_u$. Then in Section 2 we establish the relation between fundamental Cartan subalgebras $\mathfrak{h}_0 \subset \mathfrak{g}_0$ and open G_0-orbits in Z. In Section 3 we prove the existence and uniqueness (relative to a choice of Cartan involution) of the base cycle in D. That is the starting point for the theory of our cycle space $\mathcal{M}_D$. Section 4 is a digression in which we describe the ring of holomorphic functions on a flag domain. In Section 5 we introduce the concept of measurable open orbit. Section 6 is an exposition of Levi geometry, oriented toward the notions of Stein manifold and various degrees of holomorphic convexity. Those concepts later translate into cohomology vanishing theorems for holomorphic vector bundles over flag domains. In Section 7 we see that measurable open orbits carry certain canonical exhaustion functions, and that those exhaustion functions give a measure of the holomorphic convexity/concavity of the open orbit, thus making the cohomology vanishing theorems explicit for the case of measurable flag domains.

4.1 Automorphisms and regular elements

Fix a Cartan involution θ of $\mathfrak{g}_0$ and G_0. It is an automorphism of square 1, and $K_0 = G_0^\theta$ is a maximal compact subgroup of G_0. Decompose $\mathfrak{g}_0 = \mathfrak{k}_0 + \mathfrak{s}_0$, where $\mathfrak{k}_0$ is the Lie algebra of K_0 and is the $(+1)$-eigenspace of θ on $\mathfrak{g}_0$, and $\mathfrak{s}_0$ is the (-1)-eigenspace. The Killing form of $\mathfrak{g}_0$ is negative definite on $\mathfrak{k}_0$ and positive definite on $\mathfrak{s}_0$, and $\mathfrak{k}_0 \perp \mathfrak{s}_0$ under the Killing form.

An automorphism of $\mathfrak{g}_0$ is called **semisimple** if its complex linear extension to $\mathfrak{g}$ is diagonalizable. A semisimple automorphism of $\mathfrak{g}_0$ is called **elliptic** if all its eigenvalues have absolute value 1. For example, automorphisms of finite order are elliptic. An element $\xi \in \mathfrak{g}_0$ is called **semisimple** if the linear transformation $\mathrm{ad}(\xi)$ of $\mathfrak{g}$ is diagonalizable, in other words if the inner automorphisms $\mathrm{Ad}(\exp(t\xi))$ are semisimple. A semisimple element $\xi \in \mathfrak{g}_0$ is called **regular** if its centralizer

$\mathfrak{z}_{\mathfrak{g}_0}(\xi) := \{\eta \in \mathfrak{g}_0 \mid [\xi, \eta] = 0\}$ is a Cartan subalgebra of $\mathfrak{g}_0$. This section is devoted to the following.

Theorem 4.1.1 (Essentially [dS, Proposition 2, p. 56]). *Let $\mathfrak{g}_0$ be a real reductive Lie algebra and let α be an elliptic (e.g., finite order) automorphism of $\mathfrak{g}_0$. Then $\mathfrak{g}_0$ has a regular semisimple element fixed by α.*

Proof. It suffices to consider the case where the adjoint representation of G_0 is faithful, so G_0 is identified with a finite index subgroup of $\mathrm{Aut}(\mathfrak{g}_0)$, and there it suffices to consider the case where G_0 is identified with all of $\mathrm{Aut}(\mathfrak{g}_0)$. Thus, in the proof we write G_0 for $\mathrm{Aut}(\mathfrak{g}_0)$. Now α belongs to some maximal compact subgroup of G_0. All maximal compact subgroups of G_0 are conjugate. Thus we may replace α within its conjugacy class in $\mathrm{Aut}(\mathfrak{g}_0)$ and suppose that it preserves our given $K_0 = G_0^\theta$. Thus α commutes with the Cartan involution θ and preserves the Cartan decomposition $\mathfrak{g}_0 = \mathfrak{k}_0 + \mathfrak{s}_0$. We extend α by complex linearity to an automorphism of $\mathfrak{g}$, and it also preserves the compact real form $\mathfrak{g}_u = \mathfrak{k}_0 + i\mathfrak{s}_0$. The result [dS, Proposition 2, p. 56] of J. de Siebenthal says that the fixed point set $\mathfrak{g}_u^\alpha$ contains a regular element of $\mathfrak{g}_u$.

Since θ commutes with α it preserves $\mathfrak{g}_u^\alpha$. Thus θ preserves some Cartan subalgebra $\mathfrak{t}_u$ of $\mathfrak{g}_u^\alpha$. We may assume that α-fixed $\mathfrak{g}_u$-regular element to be contained in $\mathfrak{t}_u$. Write it as $\xi + i\eta$, where $\xi \in \mathfrak{t}_u \cap \mathfrak{k}_0$ and $i\eta \in \mathfrak{t}_u \cap i\mathfrak{s}_0$. Now $\alpha(\xi) = \xi$ and $\alpha(\eta) = \eta$, and α fixes the element $\xi + \eta$ of $\mathfrak{g}_0$. Thus we need only prove that $\xi + \eta$ is $\mathfrak{g}_0$-regular.

Let $\mathfrak{h}_u$ denote the centralizer of $\mathfrak{t}_u$ in $\mathfrak{g}_u$. It is a Cartan subalgebra because $\mathfrak{g}_u$ contains a regular element. Its complexification $\mathfrak{h}$ is thus a Cartan subalgebra of $\mathfrak{g}$, and $\mathfrak{h}_0 := \mathfrak{h} \cap \mathfrak{g}_0$ is a Cartan subalgebra of $\mathfrak{g}_0$. Since $\xi + i\eta$ is $\mathfrak{g}_u$-regular, $\mathfrak{h}_u$ is its centralizer in $\mathfrak{g}_u$, and thus $\mathfrak{h}$ is its centralizer of $\mathfrak{g}$. As ξ and $i\eta$ both belong to the commutative algebra $\mathfrak{t}_u$, they commute and if follows that

$$\mathfrak{h} \subset \mathfrak{z}_{\mathfrak{g}}(\xi) \cap \mathfrak{z}_{\mathfrak{g}}(\eta) \subset \mathfrak{z}_{\mathfrak{g}}(\xi + i\eta) = \mathfrak{h}.$$

Also, since $\mathrm{ad}(\xi)$ has all eigenvalues pure imaginary and $\mathrm{ad}(\eta)$ has all eigenvalues real, there is no cancellation of eigenvalues between them in expressing $\mathrm{ad}(\xi + \eta) = \mathrm{ad}(\xi) + \mathrm{ad}(\eta)$. Thus $\mathfrak{h} \subset \mathfrak{z}_{\mathfrak{g}}(\xi + \eta) \subset \mathfrak{z}_{\mathfrak{g}}(\xi + i\eta) = \mathfrak{h}$. Now $\mathfrak{z}_{\mathfrak{g}}(\xi + \eta) = \mathfrak{h}$. Therefore $\mathfrak{z}_{\mathfrak{g}_0}(\xi + \eta) = \mathfrak{h}_0$, and $\xi + \eta$ is an α-fixed regular element of $\mathfrak{g}_0$. □

4.2 Fundamental Cartan subalgebras and open G_0-orbits

Every Cartan subalgebra of $\mathfrak{g}_0$ is $\mathrm{Ad}(G_0)$-conjugate to a θ-stable Cartan subalgebra of $\mathfrak{g}_0$. A θ-stable Cartan subalgebra $\mathfrak{h}_0 \subset \mathfrak{g}_0$ is called **fundamental** or **maximally compact** if it maximizes $\dim(\mathfrak{h}_0 \cap \mathfrak{k}_0)$, in other words if $\mathfrak{h}_0 \cap \mathfrak{k}_0$ is a Cartan subalgebra of $\mathfrak{k}_0$. It is called **compact** if it is contained in $\mathfrak{k}_0$, which is a more stringent condition. More generally, a Cartan subalgebra of $\mathfrak{g}_0$ is called **fundamental** or **maximally compact** if it is conjugate to a θ-stable fundamental Cartan subalgebra.

Lemma 4.2.1. *The following conditions are equivalent for a θ-stable Cartan subalgebra $\mathfrak{h}_0 \subset \mathfrak{g}_0$.*

(i) $\mathfrak{h}_0$ *is a fundamental Cartan subalgebra of* $\mathfrak{g}_0$,
(ii) $\mathfrak{h}_0 \cap \mathfrak{k}_0$ *contains a regular element of* $\mathfrak{g}_0$, *and*
(iii) *there is a positive root system* $\Sigma^+ = \Sigma^+(\mathfrak{g}, \mathfrak{h}), \mathfrak{h} = \mathfrak{h}_0 \otimes \mathbb{C}$, *with* $\tau\Sigma^+ = \Sigma^-$.

A θ*-stable Cartan subalgebra* $\mathfrak{h}_0 \subset \mathfrak{g}_0$ *is compact if and only if* $\tau\Sigma^+ = \Sigma^-$ *for every positive root system* $\Sigma^+(\mathfrak{g}, \mathfrak{h})$. *Any two fundamental Cartan subalgebras of* $\mathfrak{g}_0$ *are conjugate.*

Proof. Split $\mathfrak{h}_0 = \mathfrak{t}_0 + \mathfrak{a}_0$ where $\mathfrak{t}_0 = \mathfrak{h}_0 \cap \mathfrak{k}_0$ and $\mathfrak{a}_0 = \mathfrak{h}_0 \cap \mathfrak{s}_0$. Then (i) implies (ii) because $\mathfrak{k}_0$ contains a regular element ζ of $\mathfrak{g}_0$, by Theorem 4.1.1. Conjugating, we may suppose that $\zeta \in \mathfrak{t}_0$. Then $\{\alpha \in \Sigma(\mathfrak{g}, \mathfrak{h}) \mid \alpha(\sqrt{-1}\zeta) > 0\}$ is a positive root system sent to its negative by τ. Thus (ii) implies (iii). If (i) fails, then we have a root α such that $\alpha(\mathfrak{t}_0) = 0$. Therefore $\tau(\alpha) = \alpha$ and (iii) fails, and thus (iii) implies (i). That proves the equivalence statement.

If $\mathfrak{h}_0 \subset \mathfrak{k}_0$, then τ sends every root to its negative. So $\tau\Sigma^+ = \Sigma^-$ for every positive root system $\Sigma^+ = \Sigma^+(\mathfrak{g}, \mathfrak{h})$. If $\tau\Sigma^+ = \Sigma^-$ for every positive root system $\Sigma^+(\mathfrak{g}, \mathfrak{h})$, then τ sends every root to its negative. Hence $\mathfrak{a}_0 = 0$ and $\mathfrak{h}_0$ is compact.

Let $\mathfrak{h}_0$ and $'\mathfrak{h}_0$ be fundamental Cartan subalgebras of $\mathfrak{g}_0$. In order to prove that they are are $\mathfrak{g}_0$-conjugate we may assume that they are both θ-stable. Any two Cartan subalgebras of $\mathfrak{k}_0$ are conjugate, so we may assume $\mathfrak{t}_0 = {}'\mathfrak{t}_0$. Then they have a regular element ζ in common, and consequently each is the $\mathfrak{g}_0$-centralizer of ζ, forcing $\mathfrak{h}_0 = {}'\mathfrak{h}_0$. □

Theorem 4.2.2. *Let* $Z = G/Q$ *be a complex flag manifold, G semisimple and simply connected, and let* G_0 *be a real form of G. The orbit* $G_0(z)$ *is open in Z if and only if* $\mathfrak{q}_z = \mathfrak{q}_\Phi$, *where*

(i) $\mathfrak{q}_z \cap \mathfrak{g}_0$ *contains a fundamental Cartan subalgebra* $\mathfrak{h}_0 \subset \mathfrak{g}_0$ *and*
(ii) Φ *is a set of simple roots for a system* $\Sigma^+ = \Sigma^+(\mathfrak{g}, \mathfrak{h}), \mathfrak{h} = \mathfrak{h}_0 \otimes \mathbb{C}$, *such that* $\tau\Sigma^+ = \Sigma^-$.

Let $Q_1 \subset Q_2$ *be parabolic subgroups of G,* $Z_i = G/Q_i$ *the resulting complex flag manifolds,* $\pi : Z_1 \to Z_2$ *the natural projection* $gQ_1 \mapsto gQ_2$, *and* $z_i = 1Q_i \in Z_i$. *Then*

(1) *if* $G_0(x_1)$ *is open in* Z_1, *then* $\pi(G_0(x_1)) = G_0(x_2)$ *is open in* Z_2, *and*
(2) *if* $G_0(x_2)$ *is open in* Z_2, *then* $\pi^{-1}(G_0(x_2))$ *contains an open* G_0*-orbit.*

Proof. We start with the assertions on $\pi : Z_1 \to Z_2$. If $G_0(z_2)$ is open in Z_2, then $\pi^{-1}(G_0(z_2))$ is open in Z_1. Hence it contains a (necessarily open) orbit of full dimension. If $G_0(z_1)$ is open in Z_1, then $\pi(G_0(z_1) = G_0(z_2)$ is open in Z_2 because π is an open map.

Suppose for the moment that Q is a Borel subgroup of G. Then there is just one simple root system Ψ such that Q has form Q_Φ. There Φ is empty and $\Phi^n = \Sigma^+$. Thus the open orbit criterion $\Phi^n \cap \tau\Phi^n = \emptyset$ of Corollary 2.2.3 says $\tau\Sigma^+ = \Sigma^-$. Using Lemma 4.2.1, it follows that $G_0(z)$ is open in Z if and only if (i) and (ii) hold.

In the general case, we apply (1) and (2) with Q_1 Borel and $Q_2 = Q$. Thus (i) and (ii) imply that $G_0(z)$ is open in Z, and if $G_0(z)$ is open in Z, then we use the positive root system for the Borel $(Q_1)_{z_1'}$, where $G_0(z_1') \subset \pi^{-1}(G_0(z))$ is open, in order to verify (i) and (ii). □

Computing first on G/B, and then pushing down to G/Q, we obtain

Corollary 4.2.3. *Fix* $\mathfrak{h}_0 = \theta\mathfrak{h}_0$, $\Sigma^+(\mathfrak{g}, \mathfrak{h})$ *and* Φ *as in* Theorem 4.2.2. *Let* $W(\mathfrak{g}, \mathfrak{h})^{\mathfrak{h}_0}$ *and* $W(\mathfrak{q}^r_\Phi, \mathfrak{h})^{\mathfrak{h}_0}$ *denote the respective subgroups of Weyl groups that stabilize* $\mathfrak{h}_0$. *Then the open* G_0*-orbits on* Z *are parameterized by the double coset space* $W(\mathfrak{k}, \mathfrak{h} \cap \mathfrak{k})\backslash W(\mathfrak{g}, \mathfrak{h})^{\mathfrak{h}_0}/W(\mathfrak{q}^r_z, \mathfrak{h})^{\mathfrak{h}_0}$.

Corollary 4.2.4. *Suppose that* G_0 *has a compact Cartan subgroup, i.e., that* $\mathfrak{k}_0$ *contains a Cartan subalgebra of* $\mathfrak{g}_0$. *Then an orbit* $G_0(z)$ *is open in* Z *if and only if* $\mathfrak{g}_0 \cap \mathfrak{q}_z$ *contains a compact Cartan subalgebra* $\mathfrak{h}_0$ *of* $\mathfrak{g}_0$, *and then, in the notation of Theorem* 4.2.2, *the open* G_0*-orbits on* Z *are parameterized by the double coset space* $W(\mathfrak{k}, \mathfrak{h})\backslash W(\mathfrak{g}, \mathfrak{h})/W(\mathfrak{q}^r_z, \mathfrak{h})$.

4.3 Compact subvarieties of open orbits

Here we examine the way $\mathfrak{k}_0$ sits in both $\mathfrak{k}$ and $\mathfrak{g}_0$, and prove prove the following

Theorem 4.3.1. *Let* $Z = G/Q$ *be a complex flag manifold,* G *semisimple and simply connected, and let* G_0 *be a real form of* G. *Let* $z \in Z$ *such that* $D = G_0(z)$ *is open in* Z, *and suppose that* $\mathfrak{h}_0 \subset \mathfrak{g}_0 \cap \mathfrak{q}_z$ *is a* θ*-stable fundamental Cartan subalgebra of* $\mathfrak{g}_0$. *Then* $K_0(z)$ *is a compact complex submanifold of* $G_0(z)$. *Let* K *be the complexification of* K_0. *It is an analytic subgroup of* G *with Lie algebra* $\mathfrak{k} = \mathfrak{k}_0 \otimes \mathbb{C}$. *Then* $K_0(z) = K(z) \cong K/(K \cap Q_z)$ *is a complex flag manifold of* K.

Let $C \subset D$ *be a compact complex submanifold. Then the following are equivalent:* (i) C *is a* K_0*-orbit,* (ii) C *is a* K*-orbit,* (iii) $C = K_0(z)$.

Remark. Theorem 4.3.1 is the first instance of the duality between the sets of G_0-orbits and K-orbits on Z. This is made precise in Proposition 4.3.3. See Theorem 8.3.2 below for the geometric formulation of that duality. ◇

Proof. Suppose that we know the theorem for $X = G/B$ where $B \subset Q$ is a Borel subgroup of G. Let $\pi : X \to Z$ be the natural projection. Choose $x \in \pi^{-1}(z)$ such that $G_0(x)$ is open and $\mathfrak{h}_0 \subset \mathfrak{b}_x$. Since $K_0(x)$ is a compact complex submanifold of $G_0(x)$, also $K_0(z) = \pi(K_0(x))$ is a compact complex submanifold of D, and $K_0(x) = K(x)$ implies $K_0(z) = \pi(K_0(x)) = \pi(K(x)) = K(z)$. Of course $K(z) \cong K/(K \cap Q_z)$. Since $K(z)$ is closed in Z it is a projective variety. Therefore $K \cap Q_z$ is a parabolic subgroup of K and $K(z)$ is a complex flag manifold. That reduces the first assertion to the case where Q is a Borel subgroup of G.

Now we prove the first assertion in the case where Q is a Borel subgroup of G. In that case $\tau\Sigma^+ = \Sigma^-$ and therefore $\mathfrak{q}_z \cap \tau\mathfrak{q}_z = \mathfrak{h}$. Choose $\zeta \in (\mathfrak{h}_0 \cap \mathfrak{k}_0)$ such that $\Sigma^+ = \{\alpha \in \Sigma \mid \alpha(\sqrt{-1}\zeta) > 0\}$. The holomorphic tangent space to Z (and thus to

D) at z is the sum of the $\mathfrak{g}_\alpha$ for $\alpha \in \Sigma^+$, in other words it is the sum of the positive eigenspaces of $\mathrm{ad}(\sqrt{-1}\zeta)$, and the antiholomorphic tangent space is the sum of the negative eigenspaces of $\mathrm{ad}(\sqrt{-1}\zeta)$. By construction $\mathrm{ad}(\zeta)$ preserves $\mathfrak{k}_0$. So now the complexification of the real tangent space of $K_0(z)$ at z is the sum of its intersections with the holomorphic and the antiholomorphic tangent spaces of D. Thus $K_0(z)$ is a complex submanifold of D. It is compact because K_0 is compact, so it is a projective algebraic subvariety of Z, and thus is K-stable with K-isotropy parabolic in K. That proves the first assertion when Q is a Borel subgroup of G, and thus proves it in general.

The argument of the first assertion shows that (i) and (ii) are equivalent in the second assertion, and that (iii) implies (i) and (ii). Now suppose that we are given (i) and (ii), say with $C = K_0(w) = K(w) \subset D$. Then $\mathfrak{k}_0 \cap \mathfrak{q}_w$ contains a Cartan subalgebra $'\mathfrak{h}_0^k$ of $\mathfrak{k}_0$, and moving w within the K_0-orbit we may assume that $'\mathfrak{h}_0^k = \mathfrak{h}_0 \cap \mathfrak{k}_0$. As $'\mathfrak{h}_0^k$ contains a regular element of $\mathfrak{g}_0$, it follows that $\mathfrak{h}_0 \subset \mathfrak{q}_w$. Thus $w = g(z)$ for some element $g \in K_0$ that normalizes $\mathfrak{h}_0$. Now g represents an element of the Weyl group $W(\mathfrak{k}_0, \mathfrak{h}_0 \cap \mathfrak{k}_0)$. Hence $g \in K_0$, and $K_0(w) = K_0(z) = C$. □

Lemma 4.3.2. *Let $\mathfrak{b}_\mathfrak{k}$ be a Borel subalgebra of $\mathfrak{k}$, and let $\mathfrak{b}$ be a Borel subalgebra of $\mathfrak{g}$ that contains $\mathfrak{b}_\mathfrak{k}$. Then $\theta(\mathfrak{b}) = \mathfrak{b}$.*

Proof. Choose a τ-stable Cartan subalgebra $\mathfrak{t}$ in $\mathfrak{b}_\mathfrak{k}$. There is just one Cartan subalgebra $\mathfrak{h}$ of $\mathfrak{g}$ that contains $\mathfrak{t}$, because $\mathfrak{t}$ contains a regular element of $\mathfrak{g}$ and $\mathfrak{h}$ must be its centralizer. We can assume that this regular element is contained in $i\mathfrak{t}_0$, and hence it is automatic that $\mathfrak{h}$ is τ-stable and θ-stable. By construction, $\mathfrak{h}_0 := \mathfrak{h} \cap \mathfrak{g}_0$ is a fundamental Cartan subalgebra of $\mathfrak{g}_0$. Since $\mathfrak{t}$ extends to a Cartan subalgebra of $\mathfrak{g}$ that is contained in $\mathfrak{b}$, we have $\mathfrak{h} \subset \mathfrak{b}$.

Choose $\xi \in \mathfrak{h}$ such that $\mathfrak{b} = \mathfrak{h} + \sum_{\alpha(\xi)>0} \mathfrak{g}_{-\alpha}$. Since $\mathfrak{t}$ contains regular elements of $\mathfrak{g}$ we have $\mathfrak{b}_\mathfrak{k} = \mathfrak{b} \cap \mathfrak{k} = \sum_{\nu(\xi)>0} \mathfrak{k}_{-\nu}$. Now suppose α is a root of $(\mathfrak{g}, \mathfrak{h})$ such that $\mathfrak{g}_{-\alpha} \subset \mathfrak{b}$ but $\mathfrak{g}_{-\theta(\alpha)} \not\subset \mathfrak{b}$. In particular, then, $\alpha \neq \theta(\alpha)$. Denote $\mathfrak{v} = \mathfrak{g}_{-\alpha} + \mathfrak{g}_{-\theta(\alpha)}$. Then $\dim \mathfrak{v} = 2$, $\dim(\mathfrak{v} \cap \mathfrak{k}) = 1$, and $\dim(\mathfrak{v} \cap \mathfrak{s}) = 1$. Also, $\alpha \neq \theta(\alpha)$ says that neither of $\mathfrak{g}_{-\alpha}$ and $\mathfrak{g}_{-\theta(\alpha)}$ is contained in $\mathfrak{k}$ nor in $\mathfrak{s}$. Now any two of $\mathfrak{g}_{-\alpha}$, $\mathfrak{g}_{-\theta(\alpha)}$, $\mathfrak{v} \cap \mathfrak{k}$ and $\mathfrak{v} \cap \mathfrak{s}$ spans $\mathfrak{v}$. Set $\nu = \alpha|_\mathfrak{t}$. Then $\mathfrak{v} \cap \mathfrak{k} = \mathfrak{k}_{-\nu}$. If $\mathfrak{k}_{-\nu} \subset \mathfrak{b}_\mathfrak{k}$, then $\mathfrak{v} = \mathfrak{g}_{-\alpha} + (\mathfrak{v} \cap \mathfrak{k}) \subset \mathfrak{b}$. Hence $\mathfrak{g}_{-\theta(\alpha)} \subset \mathfrak{b}$ and therefore we need only prove that $\mathfrak{k}_{-\nu} \subset \mathfrak{b}_\mathfrak{k}$.

Suppose $\mathfrak{k}_{-\nu} \not\subset \mathfrak{b}_\mathfrak{k}$. Denote $\mathfrak{v}' = \mathfrak{g}_\alpha + \mathfrak{g}_{\theta(\alpha)}$. Then $\mathfrak{v}' \cap \mathfrak{k} = \mathfrak{k}_\nu \subset \mathfrak{b}_\mathfrak{k} \subset \mathfrak{b}$. Also, $\mathfrak{g}_{-\theta(\alpha)} \not\subset \mathfrak{b}$ implies $\mathfrak{g}_{\theta(\alpha)} \subset \mathfrak{b}$. The two spaces $\mathfrak{v}' \cap \mathfrak{k}$ and $\mathfrak{g}_{\theta(\alpha)}$ span $\mathfrak{v}'$, in particular $\mathfrak{v}' \subset \mathfrak{b}$. That forces $\mathfrak{g}_\alpha \subset \mathfrak{b}$, in other words $\mathfrak{g}_{-\alpha} \not\subset \mathfrak{b}$, which contradicts our initial assumption on α. Thus $\mathfrak{k}_{-\nu} \subset \mathfrak{b}_\mathfrak{k}$. That completes the proof. □

Proposition 4.3.3. *Every open G_0-orbit $D \subset Z$ contains a unique closed K-orbit. Every closed K-orbit on Z is contained in a (necessarily unique) open G_0-orbit. This sets up a duality between the set of open G_0-orbits on Z and the set of closed K-orbits on Z.*

Proof. The first assertion is contained in Theorem 4.3.1. Now let $F = K(z)$ be a closed orbit in $Z = G/Q$. Apply Proposition 1.4.10 to $F = K/(K \cap Q_z)$ to see that $K \cap Q_z$ is a parabolic subgroup of K, and that the compact real form K_0 of K acts transitively on F. Let $T_0 \subset K_0$ be a Cartan subgroup and extend it to a θ-stable fundamental Cartan subgroup H_0 of G_0.

Let $\mathfrak{b}$ be a Borel subalgebra of $\mathfrak{g}$ such that $\mathfrak{b}_{\mathfrak{k}} \subset \mathfrak{b} \subset \mathfrak{q}_z$. Then there is a regular element $\xi \in \mathfrak{h}$ such that $\mathfrak{b} = \mathfrak{h} + \sum_{\alpha(\xi)>0} \mathfrak{g}_{-\alpha}$. Lemma 4.3.2 says that $\theta(\mathfrak{b}) = \mathfrak{b}$. Thus it also follows that $\mathfrak{b} = \mathfrak{h} + \sum_{\alpha(\theta(\xi))>0} \mathfrak{g}_{-\alpha}$. Now $\mathfrak{b} = \mathfrak{h} + \sum_{\alpha(\eta)>0} \mathfrak{g}_{-\alpha}$, where $\eta = \xi + \theta(\xi)$. Thus the positive root system $\Sigma^+(\mathfrak{g}, \mathfrak{h})$ that defines $\mathfrak{b}$ is given by the regular element $\eta \in i\mathfrak{t}_0$ of $\mathfrak{g}$, and in particular $\tau\Sigma^+(\mathfrak{g}, \mathfrak{h}) = -\Sigma^+(\mathfrak{g}, \mathfrak{h})$. Theorem 4.2.2 now says that $G_0(z)$ is open in Z, and $K(z) = K_0(z) \subset G_0(z)$. □

As we saw before, or by Corollary 4.2.4, the compact real form $G_u \subset G$ is transitive on Z. This gives us a realization $Z = G_u/L_u$, where $L_u \subset G_u$ is the centralizer of a torus subgroup S_u. Spaces of that form have some beautiful topological properties. For example, Z is compact and simply connected, and its integral cohomology is torsion free and is nonzero only in even degrees. In view of Theorem 4.3.1, this also holds for the compact subvariety $K_0(z) \subset G_0(z)$:

Lemma 4.3.4. *$K_0(z)$ is compact and simply connected, and its integral cohomology is torsion free and is nonzero only in even degrees.*

Theorem 8.3.1 below shows that $K_0(z)$ is a critical set of a certain flow on $G_0(z)$ in such a way that $K_0(z)$ is a deformation retraction of $G_0(z)$. See the third paragraph of Chapter 8. Thus $G_0(z)$ has the same homotopy and cohomological properties (except, of course, compactness) as $K_0(z)$. The integral cohomology statement is known from the work of Bott and Samelson in the 1950s, and here it will follow from discussion of Section 7.1A. in particular we will prove the somewhat stronger Proposition 7.1.7. With Theorem 8.3.1 one then arrives at

Proposition 4.3.5. *Let $Z = G/Q$ be a complex flag manifold, G semisimple and simply connected, and let G_0 be a real form of G. Let $z \in Z$ such that $D = G_0(z)$ is open in Z. Then D is simply connected and the isotropy subgroup $(Q_z \cap \tau Q_z)_0$ of G_0 at z is connected. The integral cohomologies $H^q(D; \mathbb{Z})$ are torsion free, and $H^q(D; \mathbb{Z}) = 0$ whenever q is odd.*

The material of this section is taken from [W2], except for Lemma 4.3.2 and the resulting extension of Theorem 4.3.1 in Proposition 4.3.3, which are taken from [Fe2]. Theorem 4.3.1 was the first instance of a phenomenon now called Matsuki duality. Further developments of that phenomenon are due to Matsuki [M3], to Mirkovič, Uzawa and Vilonen [MUV], and to Bremigan and Lorch [BL], in that chronological order. See Theorem 8.3.2 below.

4.4 Holomorphic functions

The compact subvariety $C = K_0(z)$ also has a strong influence on the function theory for an open orbit $D = G_0(z) \subset Z$. The idea is that a holomorphic function on D

must be constant on gC whenever $g \in G$ and $gC \subset D$. Hence, if there are too many translates of C inside D, then that holomorphic function must be constant on D. But this has to be formulated carefully.

Definition 4.4.1. Let $Z = G/Q$ be a complex flag manifold, G semisimple and simply connected, and let G_0 be a real form of G. Let $z \in Z$ such that $G_0(z)$ is open in Z. Then there are decompositions $G = G_1 \times \cdots \times G_m$ and $Q = Q_1 \times \cdots \times Q_m$ with $Q_i = Q \cap G_i$ and every G_i simple. Consider the corresponding decompositions $Z = Z_1 \times \cdots \times Z_m$ with $Z_i = G_i/Q_i$ and $z = (z_1, \ldots, z_m)$, $G_0 = G_{1,0} \times \cdots \times G_{m,0}$, $G_0(z) = G_{1,0}(z_1) \times \cdots \times G_{m,0}(z_m)$ and $K_0(z) = K_{1,0}(z_1) \times \cdots \times K_{m,0}(z_m)$. If

(i) $G_{i,0} \cap (Q_i)_{z_i} = ((Q_i)_{z_i} \cap \tau(Q_i)_{z_i})_0$ is compact, thus contained in $K_{i,0}$,
(ii) $G_{i,0}/K_{i,0}$ is a hermitian symmetric coset space, and
(iii) $G_{i,0}(z_i) \to G_{i,0}/K_{i,0}$ is holomorphic for one of the two invariant complex structures on $G_{i,0}/K_{i,0}$,

then we set $L_i = K_i$ so that $L_{i,0} = K_{i,0}$. Otherwise we set $L_i = G_i$ and therefore $L_{i,0} = G_{i,0}$. Note that each $G_{i,0}/L_{i,0}$ is a bounded symmetric domain, irreducible or reduced to a point. Set $L = L_0 \times \cdots \times L_m$ so $L_0 = L_{1,0} \times \cdots \times L_{m,0}$. Then we say that

$$D(G_0, z) := G_0/L_0 = (G_{1,0}/L_{1,0}) \times \cdots \times (G_{m,0}/L_{m,0}) \tag{4.4.2}$$

is the **bounded symmetric domain subordinate to** $G_0(z)$.

Now we can state a precise result for holomorphic functions on $G_0(z)$.

Theorem 4.4.3. *Let $Z = G/Q$ be a complex flag manifold, G semisimple and simply connected, and let G_0 be a real form of G. Let $z \in Z$ with $G_0(z)$ be open in Z. Let $D(G_0, z)$ be the bounded symmetric domain subordinate to $G_0(z)$. Then $\pi : g(z) \mapsto gL_0$ is a holomorphic map of $G_0(z)$ onto $D(G_0, z)$, and the holomorphic functions on $G_0(z)$ are just the $\widetilde{f} = f \cdot \pi$, where $f : D(G_0, z) \to \mathbb{C}$ is holomorphic.*

Proof. It suffices to consider the case where G_0 is simple. Let $\mathcal{A}$ denote the algebra of all holomorphic functions on $G_0(z)$, and consider the quotient $W = G_0(z)/\mathcal{A}$ of $G_0(z)$ by the relation $z_1 \sim z_2 \Leftrightarrow f(z_1) = f(z_2)$ for every $f \in \mathcal{A}$. Then W is of the form G_0/M_0 for some subgroup $M_0 \subset G_0$ that contains $G_0 \cap Q_z$. Since holomorphic functions separate points on the maximal hermitian symmetric quotient $D(G_0, z) = G_0/L_0$ of (4.4.2), we have $M_0 \subset L_0$. Every holomorphic function is constant on the compact complex manifold $C = K_0(z)$ and on its G_0-translates. Thus $K_0 \subset M_0 \subset L_0$. As K_0 is a maximal subgroup of G_0, the only possibilities are (i) $K_0 = M_0 = L_0$, (ii) $K_0 = M_0$ and $L_0 = G_0$, and (iii) $M_0 = L_0 = G_0$.

The content of Theorem 4.4.3 is that $M_0 = L_0$. If $M_0 \neq L_0$, then $K_0 = M_0$ and $L_0 = G_0$. But $G_0 \cap Q_z \subset M_0$, and G_0/M_0 has an invariant complex structure. Therefore $K_0 = M_0$ implies that $G_0(z)$ fibers over the hermitian symmetric space $G_0/K_0 = W$, and this forces $L_0 = K_0$. That excludes possibility (ii), so either (i) or (iii) holds, and Theorem 4.4.3 is proved. □

One can prove a result for holomorphic sections of line bundles $\mathbb{L} \to G_0(z)$ that is similar to Theorem 4.4.3. It is somewhat more complicated and depends on the double fibration transform described in Chapter 14 below, in particular on the notion of cycle space described in Chapter 5. Roughly speaking, it says that holomorphic sections of $\mathbb{L} \to G_0(z)$ are determined by smooth families of holomorphic sections of the restrictions $\mathbb{L}|_C \to C$ where C ranges over the family of G-translates of $K_0(z)$ that remain inside $G_0(z)$.

4.5 Measurability of open orbits

A flag domain $D = G_0(z) \subset Z$ is called **measurable** if it carries a G_0-invariant volume element. This is the type of flag domain currently of most interest in representation theory. The basic structure is given by the following.

Theorem 4.5.1 ([W2]). *Let $G_0(z)$ be an open orbit in the complex flag manifold $Z = G/Q$. Then the following conditions are equivalent:*

(4.5.2a) *The orbit $G_0(z)$ is measurable.*
(4.5.2b) *$G_0 \cap Q_z$ is the G_0-centralizer of a (compact) torus subgroup of G_0.*
(4.5.2c) *D has a G_0-invariant, possibly indefinite, Kähler metric, thus a G_0-invariant measure obtained from the volume form of that metric.*
(4.5.2d) *$\tau\Phi^r = \Phi^r$, and $\tau\Phi^n = -\Phi^n$, where $\mathfrak{q}_z = \mathfrak{q}_\Phi$.*
(4.5.2e) *$\mathfrak{q}_z \cap \tau\mathfrak{q}_z$ is reductive, i.e., $\mathfrak{q}_z \cap \tau\mathfrak{q}_z = \mathfrak{q}_z^r \cap \tau\mathfrak{p}_z^r$.*
(4.5.2f) *$\mathfrak{q}_z \cap \tau\mathfrak{q}_z = \mathfrak{q}_z^r$.*
(4.5.2g) *$\tau\mathfrak{q}$ is G-conjugate to the parabolic subalgebra $\mathfrak{q}^r + \mathfrak{q}^n$ opposite to $\mathfrak{q}$.*

In particular, since (4.5.2g) *is independent of choice of z, it follows that if one open G_0-orbit on Z is measurable, then all open G_0-orbits are measurable.*

Condition (4.5.2d) is also automatic if Q is a Borel subgroup of G, and more generally Condition (4.5.2g) provides a quick test for measurability.

Condition (4.5.2d) holds whenever the Cartan subalgebra $\mathfrak{h}_0 = \mathfrak{h} \cap \mathfrak{g}_0$ of $\mathfrak{g}_0$ corresponds to a compact Cartan subgroup $H_0 \subset G_0$, where $\mathfrak{h} = \tau\mathfrak{h}$ is the Cartan subalgebra relative to which $\mathfrak{q}_z = \mathfrak{q}_\Phi$. For in that case $\tau\alpha = -\alpha$ for every α in $\Delta(\mathfrak{g}, \mathfrak{h})$. (In particular, if G_0 has discrete series representations, so that by a result of Harish-Chandra it has a compact Cartan subgroup, then every open G_0-orbit on Z is measurable.) Thus we have the following.

Corollary 4.5.3. *If G_0 has a compact Cartan subgroup, for example, if G_0 is hermitian, then every open G_0-orbit on Z is measurable.*

Example 4.5.4. Let Z be the complex projective space $\mathbb{P}^n(\mathbb{C})$. Then $Z = G/Q$, where $G = SL(n; \mathbb{C})$ and Q is the G-stabilizer of a line in $\mathbb{C}^{n+1}$. Now let $G_0 = SL(n+1; \mathbb{R})$. This means that we have a basis $\{e_1, \ldots, e_{n+1}\}$ in which G_0 consists of the real matrices of determinant 1. There are two G_0-orbits: the closed orbit

$G_0([v]) = G_0([e_{n+1}])$ whenever $[v] = [\overline{v}] \in \mathbb{P}^n(\mathbb{C})$ and the open orbit $G_0([v]) = G_0([e_1 + \sqrt{-1}e_{n+1}])$ whenever $[v] \neq [\overline{v}] \in \mathbb{P}^n(\mathbb{C})$. If $n > 1$, then the isotropy subgroup of G_0 at $[e_1 + \sqrt{-1}e_{n+1}]$ is not reductive. Hence the open orbit is not measurable. ◇

Example 4.5.5. Example 4.5.4 is a special case of the following. Let $n > 1$ and consider a dimension sequence $\delta : 0 < d_1 < \cdots < d_k < n$. Then we have the classical complex flag manifold Z_δ consisting of flags

$$z = (\{0\} \subset \Lambda_{d_1} \subset \cdots \subset \Lambda_{d_k} \subset \mathbb{C}^n),$$

where Λ_d is a d-dimensional linear subspace of $\mathbb{C}^n$. According to [HS], the orbit $G_0(z)$ is open if and only if, for each $\{i, j\}$, $\dim(\Lambda_{d_i} \cap \tau(\Lambda_{d_j}))$ is minimal, i.e., is equal to $\max\{d_i + d_j - n, 0\}$. By [HW2, Proposition 2.4], the open G_0-orbits on Z_δ are measurable if and only if δ is symmetric in the sense that $(d_1, \ldots, d_k) = (n - d_k, \ldots, n - d_1)$. ◇

Example 4.5.6. Let G_0 be the underlying real Lie group structure of a complex Lie group J. In other words G_0 is a complex Lie group J viewed as a real Lie group. Then $\mathfrak{g} = \mathfrak{j} \oplus \bar{\mathfrak{j}}$ with $\mathfrak{g}_0$ embedded as the diagonal. Here $\bar{\mathfrak{j}}$ is the Lie algebra $\mathfrak{j}$ with the conjugate complex structure. Thus $G \cong G_0 \times G_0$ with G_0 embedded diagonally, and $\tau(g_1, g_2) = (g_2, g_1)$. Consider a complex flag manifold $Z = G/Q$. Then Q is of the form $Q_1 \times Q_2$ with $\tau(Q) = Q_2 \times Q_1$, where Q_1 is parabolic in the first factor G_0 and Q_2 is parabolic in the second factor. Thus $\tau(Q)$ is conjugate to the parabolic subgroup of G opposite to Q if and only if Q_2 is conjugate to the parabolic subgroup of G_0 opposite to Q_1. Now the open G_0-orbits on Z are measurable if and only if Q_1 and Q_2 are (up to conjugacy) opposite parabolic subgroups of G_0. ◇

4.5.7. Proof of Theorem 4.5.1. Since $G_0(z)$ is open, Corollary 2.2.2 says that $\mathfrak{q}_z \cap \tau\mathfrak{q}_z$ has nilpotent radical $(\mathfrak{q}_z^r \cap \tau\mathfrak{q}_z^{-n}) + (\tau\mathfrak{q}_z^r \cap \mathfrak{q}_z^{-n})$ and Levi complement $\mathfrak{q}_z^r \cap \tau\mathfrak{q}_z^r$. Equivalence of (4.5.2d), (4.5.2e) and (4.5.2f) follows directly, and of course (4.5.2d) implies (4.5.2g). Two parabolic subalgebras containing the same Borel are conjugate if and only if they are equal. So in fact (4.5.2g) implies (4.5.2d) as well. Consequently (4.5.2d), (4.5.2e), (4.5.2f) and (4.5.2g) are equivalent.

The orbit $G_0(z)$ is measurable if and only if the action of the isotropy subgroup $G_0 \cap Q_z$ on the real tangent space $\mathfrak{g}_0/(\mathfrak{g}_0 \cap \mathfrak{q}_z)$ is by linear transformations of determinant ± 1. That condition is that $\mathrm{ad}(\mathfrak{q}_z \cap \tau\mathfrak{q}_z)$ acts on $\mathfrak{g}/(\mathfrak{q}_z \cap \tau\mathfrak{q}_z)$ with trace 0. In other words $\mathrm{ad}(\mathfrak{q}_z \cap \tau\mathfrak{q}_z)$ acts on $\mathfrak{q}_z \cap \tau\mathfrak{q}_z$ itself with trace 0. Using Corollary 2.2.2, now $\mathfrak{q}_z \cap \tau\mathfrak{q}_z$ is reductive. Thus (4.5.2a) and (4.5.2e) are equivalent.

Suppose that $G_0(z)$ is measurable. Now we know that $G_0 \cap Q_z$ is the complexification of the Levi component $Q_z^r = \tau Q_z^r$ of Q_z. If we express $\mathfrak{q}_z = \mathfrak{q}_\Phi$ as usual, then $\mathfrak{q}_z^r$ is the centralizer of $\mathfrak{v} = \{\xi \in \mathfrak{h} \mid \psi(\xi) = 0 \text{ for all } \psi \in \Phi\}$. Now $\mathfrak{q}_z^r$ is the $\mathfrak{g}$-centralizer of $\mathfrak{t} = \mathfrak{v} \cap \mathfrak{k}$, and $G_0 \cap Q_z$ is the G_0-centralizer of the (compact) torus subgroup $T_0 = \exp(\mathfrak{t} \cap \mathfrak{g}_0)$ of G_0 and (4.5.2a) implies (4.5.2b).

Suppose that $G_0 \cap Q_z$ is the centralizer of a torus $T_0 \subset G_0$. Then it is the centralizer of some element $\zeta \in \mathfrak{t}_0$, and now we have the 1-form η on $G_0(z)$ given at

z by the Killing form inner product formula $\eta_z(h) = \langle \zeta, h \rangle$. By construction, $\omega = d\eta$ is a closed nondegenerate G_0-invariant 2-form on $G_0(z)$. Hence it is the imaginary part of the (possibly indefinite) G_0-invariant Kähler metric on $G_0(z)$ whose real part is given by the Killing form of $\mathfrak{g}_0$. Thus (4.5.2b) implies (4.5.2c), which of course implies (4.5.2a). This completes the proof. □

4.6 Background on Levi geometry

In Section 4.7 below, we see that measurable open G_0-orbits possess exhaustion functions have properties which are important from the viewpoint of complex analysis. It is therefore appropriate at this stage to sketch some background information on the geometry of such exhaustions and the implications, e.g., for analytic continuation and the vanishing of certain cohomology groups. See, for example, [Gu], [GF], [GR1], and [GR2] for systematic treatments.

4.6A Hartogs Theorems

The first hint that a sort of curvature is relevant for analytic continuation theorems in several complex variables can be found in theorems of *Hartogs type*. These certainly reflect the fact that complex analysis in dimensions greater than one is really quite different from the one variable version. The following is such a theorem. Let (z, w) be the standard coordinates in $\mathbb{C}^2$ and think of the z-axis as a parameter space for geometric figures in the w-plane. For example, let $D_z := \{(z, w) \mid |w| < 1\}$ be a disk and $A_z := \{(z, w) : 1 - \varepsilon < |w| < 1\}$ be an annulus. An example of a **Hartogs figure** H in $\mathbb{C}^2$ is the union of the family of disks D_z for $|z| < 1-\delta$ with the family A_z of annuli for $1-\delta \leqq |z| < 1$. One should visualize the moving disks which suddenly change to moving annuli. One speaks of filling in the Hartogs figure to obtain the polydisk $\widehat{H} := \{(z, w) \mid |z| < 1 \text{ and } |w| < 1\}$. Hartogs' continuation theorem states that a function which is holomorphic on H extends holomorphically to $\widehat{H}$.

4.6B Holomorphic convexity

One of the major developments in complex analysis in several variables was the realization that certain convexity concepts lie behind the strong continuation properties. At the analytic level one such is defined as follows by the full algebra of holomorphic functions $\mathcal{O}(D)$ on a domain D. If K is a compact subset of D, then its **holomorphic convex hull** $\widehat{K}$ is defined as the intersection of the sets $P(f) := \{p \in D \mid |f(p)| \leqq |f|_K\}$ as f runs through $\mathcal{O}(D)$. One says that D is **holomorphically convex** if $\widehat{K}$ is compact for every compact subset K of D.

A theorem of H. Cartan and P. Thullen relates this concept to analytic continuation phenomena as follows. A domain D is said to be a **domain of holomorphy** if, given a divergent sequence $\{z_n\} \subset D$, there exists $f \in \mathcal{O}(D)$ which is unbounded along it. In other words, one cannot extend all holomorphic functions on D to a truly larger domain $\widehat{D}$.

Theorem 4.6.1 (Cartan–Thullen). *A domain D in $\mathbb{C}^n$ is a domain of holomorphy if and only if it is holomorphically convex.*

4.6C Levi convexity

There is a very close relationship between holomorphic convexity and a certain complex geometric convexity of the boundary $\mathrm{bd}(D)$. For this, consider a smooth (local) real hypersurface Σ containing $0 \in \mathbb{C}^n$ with $n > 1$. It is the zero set $\{\rho = 0\}$ in some neighborhood U of 0 of a smooth function with $d\rho \neq 0$ on U. This is viewed as a piece of a boundary of a domain D, where $U \cap D = \{\rho < 0\}$. The real tangent space $T_0\Sigma = \mathrm{Ker}(d\rho(0))$ contains a unique maximal (1-codimensional) complex subspace $T_0^{\mathbb{C}}\Sigma = \mathrm{Ker}(\partial\rho(0)) = H$. The signature of the restriction of the complex Hessian (or **Levi form**) $i\partial\bar{\partial}\rho$ to H is a biholomorphic invariant of Σ. In this notation the Hessian is a real alternating 2-form which is compatible with the complex structure, and its *signature* is defined to be the signature of the associated symmetric form.

If the restriction of this Levi form to the complex tangent space has a negative eigenvalue, i.e., if the boundary $\mathrm{bd}(D)$ has a certain degree of concavity, then there is a map map $F : \Delta \to U$ of the unit disk Δ which is biholomophic onto its image with $F(0) = 0$, and otherwise $F(c\ell(\Delta)) \subset D$. The reader can imagine pushing the image of this map into the domain to obtain a family of disks that are in the domain, and pushing it in the outward pointing direction to obtain annuli that are also in the domain. Making this precise, one builds a (higher-dimensional) Hartogs figure H at the base point 0 so that $\widehat{H}$ is an open neighborhood of 0. In particular this proves the Theorem of E. E. Levi: *Every function holomorphic on $U \cap D$ extends to a neighborhood of* 0. This can be globally formulated as follows.

Theorem 4.6.2. *If D is a domain of holomorphy with smooth boundary in $\mathbb{C}^n$, then* $\mathrm{bd}(D)$ *is Levi pseudoconvex.*

Here the terminology **Levi pseudoconvex** is used to denote the condition that the restriction of the Levi form to the complex tangent space of every boundary point is positive semidefinite.

One of the guiding problems of complex analysis in higher dimensions is the Levi problem. This is the question of just when the converse of Levi's Theorem is valid:

Levi Problem. *Is a domain D with smooth Levi pseudoconvex boundary in a complex manifold necessarily a domain of holomorphy?*

Stated in this form it is not true, but for domains in $\mathbb{C}^n$ or in Stein manifolds, which are discussed below, it is true.

4.6D Stein manifolds

The founding fathers of the first phase of "modern complex analysis" (Cartan, Oka, and Thullen) realized that domains of holomorphy form the basic class of spaces where it would be possible to solve the important problems of the subject concerning the existence of holomorphic or meromorphic functions with reasonably prescribed properties. In fact, Oka formulated a principle which more or less states that if a complex analytic problem which is well formulated on a domain of holomorphy has a continuous solution, then it should have a holomorphic solution. Given the flexibilty of continous functions and the rigidity of holomorphic functions, this would seem doubtful, but in fact is true.

Beginning in the late 1930s, Stein worked on problems related to this "Oka Principle," in particular, on those related to what we would now call the algebraic topological aspects of the subject, and he was led to formulate conditions on a general complex manifold X which should hold if problems of the above type are to be solved. First, his axiom of *holomorphic convexity* was simply that, given a divergent sequence $\{x_n\}$ in X, there should be a function $f \in \mathcal{O}(X)$ such that $\{f(x_n)\}$ is unbounded. Secondly, holomorphic functions should *separate points* in the sense that, given distinct points $x_1, x_2 \in X$, there exists $f \in \mathcal{O}(X)$ with $f(x_1) \neq f(x_2)$. Finally, *globally defined holomorphic functions should give local coordinates*. Assuming that X is n-dimensional, this means that, given a point $x \in X$, there exist $f_1, \ldots, f_n \in \mathcal{O}(X)$ such that $df_1(x) \wedge \cdots \wedge df_n(x) \neq 0$.

This definition is so important that we restate it formally.

Definition 4.6.3. A complex analytic space X is a **Stein** space if it satisfies the following:

- *Separation property*: Given a point $x_0 \in X$ there exists a holomorphic map $F : X \to \mathbb{C}^N$ such that x_0 is isolated in its fiber.
- *Holomorphic convexity*: If $\{p_n\} \subset X$ is a divergent sequence, then there exists $f \in \mathcal{O}(X)$ such that $\limsup_{n\to\infty} |f(p_n)| = \infty$.

Assuming Stein's axioms, Cartan and Serre then produced a powerful theory in the context of sheaf cohomology which proved certain vanishing theorems that led to the desired existence theorems. It should be mentioned that Grauert's version of the Cartan–Serre theory requires only very weak versions of Stein's axioms: (1) the connected component containing K of the holomorphic convex hull $\widehat{K}$ of every compact set should be compact, and (2) given $x \in X$, there are functions $f_1, \ldots, f_m \in \mathcal{O}(X)$ so that x is an isolated point in the fiber of the map $F := (f_1, \ldots, f_m) : X \to \mathbb{C}^m$. Of course the results also hold for complex spaces.

The following are examples of the strong properties of Stein manifolds X for holomorphic vector bundles $\mathbb{E} \to X$. The Dolbeault cohomology $H^q(X; \mathbb{E})$ of $\bar{\partial}$-closed $\mathbb{E}$-valued $(0, q)$-forms modulo $\bar{\partial}$-exact forms of the same type vanishes for all $q \geqq 1$ (a special case of Theorem B). Global sections of $\mathbb{E}$ separate points and give coordinates on X (an indication of Theorem A). The bundle $\mathbb{E}$ is holomorphically trivial if and only if it is topologically trivial (an indication of Grauert's Oka Principle).

Holomorphically convex domains in $\mathbb{C}^n$ are Stein manifolds, and since closed complex submanifolds of Stein manifolds are Stein, it follows that any closed complex submanifold of $\mathbb{C}^n$ is Stein. In particular, affine varieties are Stein spaces.

Remmert's theorem states the converse: An n-dimensional Stein manifold can be embedded as a closed complex submanifold of $\mathbb{C}^{2n+1}$. A nontrivial result of Behnke and Stein implies that every noncompact Riemann surface is also Stein.

4.6E Strongly pseudoconvex domains

A relatively compact domain D with smooth boundary in a complex manifold is said to be **strongly pseudoconvex** if the restriction of the Levi form of a local defining function to the complex tangent space of $\mathrm{bd}(D)$ at each boundary point is positive definite. This condition turns out to be equivalent to $\mathrm{bd}(D)$ being defined as the 0-set of a smooth strictly plurisubharmonic function $\rho : U \to \mathbb{R}$, where U is some neighborhood of $\mathrm{bd}(D)$. The condition that ρ be **strict plurisubharmonic** is that the *full* complex Hessian $i\partial\bar{\partial}\rho$ is positive definite. Grauert showed that such a domain is almost Stein:

Theorem 4.6.4. *D has a compact subvariety E, and a holomorphic map $R : D \to D'$ onto a Stein space D', such that E is mapped to a finite subset E' and $R|D \setminus E \to D' \setminus E'$ is biholomorphic.*

This solution of the Levi Problem leads to another important characterization of Stein manifolds:

Theorem 4.6.5. *A complex manifold is Stein if and only if it possesses a strictly plurisubharmonic exhaustion function.*

4.6F The conditions of q-convexity and q-completeness

If Z is a compact complex manifold embedded in a projective space $\mathbb{P}(V)$, and H is the intersection of Z and the 0-set of a homogeneous polynomial which is nowhere locally identically zero on Z, then H is 1-codimensional in Z and the complement $Z \setminus H$ is Stein. Thus we think of $Z \setminus H$ as having maximal positive Levi curvature.

For considerations of complements of higher codimensional varieties, and in particular for our study of the Levi geometry of open G_0-orbits in flag manifolds, it is important to understand the situation where the positive definite condition is weakened. The works of Andreotti and Grauert (see [AnG]) are aimed at understanding such manifolds.

A complex manifold X is said to be q-**convex** if it possesses an exhaustion $\rho : X \to \mathbb{R}^{\geqq 0}$ such that every point outside of a compact subset of X the complex Hessian of ρ has at least $n - q$ positive eigenvalues. It is difficult to keep the numbers in mind, but the following might help: 0-convex is the same as "almost Stein," i.e., outside of a compact set the exhaustion is strongly plurisubharmonic.

Caveat: In the older literature (e.g., starting with [AnG] and as late as [W8]) one finds "$(q + 1)$-convex" used for the condition of $n - q$ positive eigenvalues outside a compact set, where the current convention (as here) is to write "q-convex."

One says that X is q-**complete** if the condition on the compact set can be dropped, i.e., there is an exhaustion ρ whose Hessian at every point point of X has at least $n - q$ positive eigenvalues. Thus 0-complete means that X has a strictly plurisubharmonic exhaustion, in other words that it is Stein. Caveat: in the older literature one finds "$(q + 1)$-complete" used to express the condition of $n - q$ positive eigenvalues everywhere, while now we write "q-complete."

As we will see in the next section, measurable open G_0-orbits in Z are q-complete, where the exhaustion which exhibits this property is of a natural group-theoretic nature. For such manifolds Andreotti and Grauert have proved the following fundamental vanishing theorem.

Theorem 4.6.6. *If X is q-complete and $\mathcal{F}$ is a coherent analytic sheaf on X, then $H^k(X; \mathcal{F}) = 0$ for all $k > q$.*

Here $H^k(X, \mathcal{F})$ is the Cech-cohomology with coefficients in $\mathcal{F}$. If $\mathcal{F}$ is the sheaf of germs of local sections in a holomorphic vector bundle $\mathbb{E}$, then this cohomology is just the above mentioned Dolbeault coholomology. Note that if $q = 0$, then this is exactly the vanishing theorem of Stein theory. In many situations one also has the q-nonvanishing type of theorem that is reflected by Theorem A in the Stein case. Roughly stated, this means that one can expect that $H^q(X, \mathcal{F})$ is a very rich space, for example one that is appropriate for geometric realization of group representations.

In closing we should note that Andreotti and Grauert also prove vanishing theorems and finiteness theorems under assumptions of concavity as well as just under the assumption of a Levi condition outside a compact subset. For example, if X is only known to be q-convex, then nevertheless $H^k(X; \mathcal{F})$ is finite dimensional for every coherent analytic sheaf $\mathcal{F}$, for all $k > q$.

4.7 The exhaustion function for measurable open orbits

Measurable open orbits $D = G_0(z) \subset Z$ carry an especially useful real analytic exhaustion function $\varphi : D \to \mathbb{R}$ whose Levi form $\mathcal{L}(\varphi)$ has at least $n - q$ positive eigenvalues at every point of D, where $n = \dim_{\mathbb{C}} D$, $C_0 = K_0(z)$ is the compact subvariety of Theorem 4.3.1, and $q = \dim_{\mathbb{C}} C_0$. In this section we construct that exhaustion function and look at some of its consequences.

Let $\mathbb{K}_Z \to Z$ and $\mathbb{K}_D = \mathbb{K}_Z|_D \to D$ denote the canonical line bundles. Those are the top exterior powers $\bigwedge^n(\mathbb{T}_Z^{1,0})$ and $\bigwedge^n(\mathbb{T}_D^{1,0})$, $n = \dim_{\mathbb{C}} Z = \dim_{\mathbb{C}} D$, of the holomorphic tangent bundles. Their dual bundles

$$\mathbb{L}_Z = \mathbb{K}_Z^* \to Z \quad \text{and} \quad \mathbb{L}_D = \mathbb{K}_D^* \to D \tag{4.7.1}$$

are the homogeneous holomorphic line bundles over Z associated to the character

$$e^{\lambda} : Q_z \to \mathbb{C} \text{ defined by } e^{\lambda}(p) = \det \mathrm{Ad}(p)|_{\mathfrak{q}_z^n}. \tag{4.7.2}$$

Write $D = G_0/V_0$, where V_0 is the real form $G_0 \cap Q_z$ of Q_z^r. Write V for the complexification Q_z^r of V_0, $\rho_{G/V}$ for half the sum of the roots that occur in $\mathfrak{q}_z^n$, and $\lambda = 2\rho_{G/V}$. If $\alpha \in \Sigma(\mathfrak{g}, \mathfrak{h})$, then (i) $\langle \alpha, \lambda \rangle = 0$ and $\alpha \in \Phi^r$, or (ii) $\langle \alpha, \lambda \rangle > 0$ and $\alpha \in \Phi^n$, or (iii) $\langle \alpha, \lambda \rangle < 0$ and $\alpha \in \Phi^{-n}$. Now $\tau\lambda = -\lambda$. Decompose $\mathfrak{g}_0 = \mathfrak{k}_0 + \mathfrak{s}_0$ under the Cartan involution with fixed point set $\mathfrak{k}_0$, thus decomposing the Cartan subalgebra $\mathfrak{h}_0 \subset \mathfrak{g}_0 \cap \mathfrak{q}_z$ as $\mathfrak{h}_0 = \mathfrak{t}_0 + \mathfrak{a}_0$ with $\mathfrak{t}_0 = \mathfrak{h}_0 \cap \mathfrak{k}_0$ and $\mathfrak{a}_0 = \mathfrak{h}_0 \cap \mathfrak{s}_0$. Then $\lambda(\mathfrak{a}_0) = 0$.

View $D = G_0/V_0$ and $Z = G_u/V_u$ where G_u is the analytic subgroup of G for the compact real form $\mathfrak{g}_u = \mathfrak{k}_0 + \sqrt{-1}\mathfrak{s}_0$ and V_u is the compact real form $G_u \cap Q_z$ of Q_z^r. Then $e^\lambda : V \to \mathbb{C}$ is unitary on both V_u and V_0. The Killing form and complex structure of $\mathfrak{g}$ define indefinite-hermitian metrics on the holomorphic tangent spaces. Thus

$$\begin{aligned} &\mathbb{L}_Z \to Z = G_u/V_u \text{ has a } G_u\text{-invariant hermitian metric } h_u, \\ &\mathbb{L}_D \to D = G_0/V_0 \text{ has a } G_0\text{-invariant hermitian metric } h_0. \end{aligned} \tag{4.7.3}$$

Lemma 4.7.4. *The hermitian form $\sqrt{-1}\partial\bar\partial \log h_u$ on the holomorphic tangent bundle of Z is negative definite. The hermitian form $\sqrt{-1}\partial\bar\partial \log h_0$ on the holomorphic tangent bundle of D has signature $(n-q, q)$ where $n = \dim_{\mathbb{C}} D$ and $q = \dim_{\mathbb{C}} C$.*

Proof. Let $\mathfrak{q}_z = \mathfrak{q}_\Phi$ as usual. If $\alpha \in \Sigma(\mathfrak{g}, \mathfrak{h})$ let $0 \neq e_\alpha \in \mathfrak{g}_\alpha$.

The hermitian metric h_0 on $\mathbb{L}_D \to D$ has connection form λ and curvature form $\omega_0 = 2\pi\sqrt{-1}\, d\lambda = -\partial\bar\partial \log h_0$. Let $\alpha, \beta \in \Phi^n$. Then

- if $\tau\beta \neq -\alpha$: $\lambda([e_\alpha, \tau e_\beta]) = 0$ and $\langle e_\alpha, \tau e_\beta \rangle = 0$,
- if $\tau\beta = -\alpha$: there exists $c \neq 0$ such that $[e_\alpha, \tau e_\beta] = c h_\alpha$ and $\lambda([e_\alpha, \tau e_\beta]) = (\lambda, \alpha)c = (\lambda, \alpha)\langle e_\alpha, \tau e_\beta \rangle$.

Let $x = \sum x_\alpha$, $y = \sum y_\beta \in \mathfrak{q}^n$ where $x_\gamma, y_\gamma \in \mathfrak{g}_\gamma$. Then

$$\begin{aligned} (\sqrt{-1}\partial\bar\partial \log h_0)(x, y) &= 2\pi d\lambda(x, \tau y) = \pi\lambda([x, \tau y]) \\ &= \pi \sum\nolimits_{\alpha,\beta \in \Phi^n} (\lambda, \alpha)\langle e_\alpha, \tau e_\beta \rangle. \end{aligned}$$

Each $(\lambda, \alpha) > 0$, and $\langle x, \tau y \rangle$ is positive definite on $\mathfrak{s} \cap \mathfrak{q}_z^n$, negative definite on $\mathfrak{k} \cap \mathfrak{q}_z^n$. This completes the proof of the h_0-statement of Lemma 4.7.4.

The calculation for h_u is essentially the same, except that we use the complex conjugation $\sigma = \theta\tau$ of $\mathfrak{g}$ over $\mathfrak{g}_u$ rather than the complex conjugation τ of $\mathfrak{g}$ over $\mathfrak{g}_0$:

$$\begin{aligned} (\sqrt{-1}\partial\bar\partial \log h_u)(x, y) &= 2\pi d\lambda(x, \sigma y) = \pi\lambda([x, \sigma y]) \\ &= \pi \sum\nolimits_{\alpha,\beta \in \Phi^n} (\lambda, \alpha)\langle e_\alpha, \sigma e_\beta \rangle. \end{aligned}$$

The difference here is that $\langle x, \sigma y \rangle$ is negative definite on all of $\mathfrak{q}^n$. This proves the h_u-statement of Lemma 4.7.4. □

Lemma 4.7.5. *Define $\varphi : D \to \mathbb{R}$ by $\varphi = \log(h_0/h_u)$. Then the Levi form $\mathcal{L}(\varphi)$ has at least $n - s$ positive eigenvalues at every point of D.*

Proof. Compute $\mathcal{L}(\varphi) = \sqrt{-1}\partial\bar\partial\varphi = \sqrt{-1}\partial\bar\partial \log h_0 - \sqrt{-1}\partial\bar\partial \log h_u$ and apply Lemma 4.7.4. □

Lemma 4.7.6. *$\varphi : D \to \mathbb{R}$ is an exhaustion function.*

Proof. Since φ is a strictly positive function, we need only check that the set $\{z \in D \mid \varphi(z) \leqq c\}$ is compact for every $c \in \mathbb{R}$. For that, it suffices to show that $e^{-\varphi}$ has a continuous extension from D to the compact manifold Z that vanishes on the topological boundary $\mathrm{bd}(D)$ of D in Z.

Choose a G_0-invariant metric h_0^* on $\mathbb{L}_D^* = \mathbb{K}_D \to D$ and a G_u-invariant metric h_u^* on $\mathbb{L}_Z^* = \mathbb{K}_Z \to Z$, normalized so that $h_0 h_0^* = 1$ on D and $h_u h_u^* = 1$ on Z. Then $e^{-\varphi} = h_0^*/h_u^*$. So it suffices to show that h_0^*/h_u^* has a continuous extension from D to Z that vanishes on $\mathrm{bd}(D)$.

The holomorphic cotangent bundle $\mathbb{T}_Z^* \to Z$ has fiber given by $\mathrm{Ad}(g)(\mathfrak{q}_z^n)^* = \mathrm{Ad}(g)(\mathfrak{q}_z^{-n})$ at $g(z)$. Thus its G_u-invariant hermitian metric is given on the fiber $\mathrm{Ad}(g)(\mathfrak{q}_z^{-n})$ at $g(z)$ by $F_u(\xi, \eta) = -\langle \xi, \tau\theta\eta \rangle$ where $\langle , \rangle$ is the Killing form. Similarly the G_0-invariant indefinite-hermitian metric on $\mathbb{T}_D^* \to D$ is given on the fiber $\mathrm{Ad}(g)(\mathfrak{q}_z^{-n})$ at $g(z)$ by $F(\xi, \eta) = -\langle \xi, \tau\eta \rangle$. But $\mathbb{K}_Z = \det \mathbb{T}_Z^*$ and $\mathbb{K}_D = \det \mathbb{T}_D^*$. Thus

$$h_0^*/h_u^* = c \cdot (\text{determinant of } F \text{ with respect to } F_u)$$

for some nonzero real constant c. This extends from D to a C^∞ function on Z given by

$$f(g(z)) = c \cdot (\det F|_{\mathrm{Ad}(g)(\mathfrak{q}_z^{-n})} \text{ relative to } \det F_u|_{\mathrm{Ad}(g)(\mathfrak{q}_z^{-n})}). \tag{4.7.7}$$

It remains only to show that f vanishes on $\mathrm{bd}(D)$. If $g(z) \in \mathrm{bd}(D)$, then $G_0(g(z))$ is not open in Z, and $\mathrm{Ad}(g)(\mathfrak{q}_z) + \tau\,\mathrm{Ad}(g)(\mathfrak{q}_z) \neq \mathfrak{g}$. Thus there exists an $\alpha \in \Sigma(\mathfrak{g}, \mathrm{Ad}(g)\mathfrak{h})$ such that $\mathfrak{g}_{-\alpha} \not\subset \mathrm{Ad}(g)(\mathfrak{q}_z) + \tau\,\mathrm{Ad}(g)(\mathfrak{q}_z)$, in other words $\mathfrak{g}_\alpha \subset \mathrm{Ad}(g)(\mathfrak{q}_z^{-n}) \cap \tau\,\mathrm{Ad}(g)(\mathfrak{q}_z^{-n})$. If $\beta \in \Sigma(\mathfrak{g}, \mathrm{Ad}(g)\mathfrak{h})$ with $\mathfrak{g}_\beta \subset \mathrm{Ad}(g)(\mathfrak{q}_z^{-n})$, then $F(\mathfrak{g}_\alpha, \mathfrak{g}_\beta) = 0$, and so $f(g(z)) = 0$. Thus φ is an exhaustion function for D in Z. □

Combining Lemmas 4.7.5 and 4.7.6, we see that φ is an exhaustion function on D whose Levi form has at least $n - s$ positive eigenvalues at every point of D. This proves the following.

Theorem 4.7.8. *Let $Z = G/Q$ be a complex flag manifold, G semisimple and simply connected, and let G_0 be a real form of G. Let $D = G_0(z) \subset Z = G/Q$ be a measurable open orbit. Let $C = K_0(z)$, maximal compact subvariety of D. Then D is q-complete where $q = \dim_{\mathbb{C}} C$. In particular, if $\mathcal{F} \to D$ is a coherent analytic sheaf, then $H^k(D; \mathcal{F}) = 0$ for $k > q$.*

Corollary 4.7.9. *If $\mathbb{E} \to D$ is a holomorphic vector bundle, then $H^r(D; \mathcal{O}(\mathbb{E})) = 0$ for $r > q$.*

On the other hand, if the bundle $\mathbb{E} \to D$ is (sufficiently) negative [GrS], then methods based on the Bott–Borel–Weil Theorem show that $H^r(D; \mathcal{O}(\mathbb{E})) = 0$ for $r < q$. As a consequence we have vanishing except for $r = q$.

Corollary 4.7.10. *If $\mathbb{E} \to D$ is (sufficiently) negative, then $H^r(D; \mathcal{O}(\mathbb{E})) = 0$ for $r \neq q$.*

Corollary 4.7.10 will be very important when we discuss double fibration transforms.

The material of this section is developed in [S1] for the setting where G_0 has a compact Cartan subgroup and $Z = G/B$, where B is a Borel subgroup of G, then in [WeW] for the more general setting in which the isotropy subgroup V_0 of G_0 at a point $z \in D$ is compact, and finally in [SW] for the situation considered here where V_0 has only to be reductive.

5

The Cycle Space of a Flag Domain

In this chapter we assemble the basic facts on cycle spaces of flag domains. For the most part they are based on the structure theory of semisimple Lie groups and Lie algebras. The more subtle results, based on new complex-geometric methods, will be studied in Part II.

Section 1 consists of the definition of the cycle space $\mathcal{M}_D$ and its complex structure. In Section 2 we pin down the degenerate case where G_0 is noncompact but acts transitively on Z, and we describe the three cases of interest. For reasons involving bounded symmetric domains, those three cases are the "hermitian holomorphic" case, the "hermitian nonholomorphic" case, and the "generic" (or "non-hermitian") case. Sections 3, 4 and 5 are concerned with special situations. In Section 3 we sketch the proof that if the open G_0-orbit $D \subset Z$ is measurable, then the exhaustion function of Section 4.7 lifts to a strictly plurisubharmonic exhaustion function on the cycle space $\mathcal{M}_D$, and consequently $\mathcal{M}_D$ is Stein. In Section 4 we look at special features of the hermitian case (where G_0/K_0 is a bounded symmetric domain $\mathcal{B}$). In the hermitian holomorphic case one easily sees that $\mathcal{M}_D = \mathcal{B}$. In the hermitian nonholomorphic case we see some natural inclusions $\mathcal{M}_D \subset \mathcal{B} \times \overline{\mathcal{B}} \subset G/K$. In Section 5 we describe $\mathcal{M}_D$ explicitly, in terms of Grassmann manifolds, in the hermitian cases with G_0 classical. There the distinction between the holomorphic cases (where $\mathcal{M}_D = \mathcal{B}$) and the nonholomorphic cases (where we verify $\mathcal{M}_D = \mathcal{B} \times \overline{\mathcal{B}}$) is clear.

For several years the understanding of $\mathcal{M}_D$ was limited to the examples just described. The complete analysis of $\mathcal{M}_D$ requires the complex-geometric methods of Part II.

5.1 Definitions and first properties

Fix $z \in Z$ such that $D = G_0(z)$ is open in Z. For convenience we suppose that z is the base point in $Z = G/Q$. Thus $Q = Q_z$ and $\mathfrak{q} = \mathfrak{q}_z$, and we fix a fundamental Cartan subalgebra $\mathfrak{h}_0 \in \mathfrak{g}_0 \cap \mathfrak{q}_z$. For notational consistency with many papers in this area, we write L for the Levi component Q^r of Q. So D is measurable if and only if $Q \cap G_0$ is a real form L_0 of L, and in this case $D \cong G_0/L_0$.

Fix a Cartan involution θ of G_0 that stabilizes the Cartan subgroup $H_0 \subset G_0$, and denote its fixed point sets on G_0 and G by $K_0 = G_0^\theta$ and $K = G^\theta$. Then K_0 is a maximal compact subgroup of G_0 and K is its complexification; $L \cap K_0$ is a compact real form of $L \cap K$ and $K_0(z) \cong K_0/(L \cap K_0)$. Now $H_0 \cap K_0$ is a Cartan subgroup of K_0.

Theorem 4.3.1 tells us that $C_0 = K_0(z) = K(z) \cong K/(K \cap Q_z)$ is a compact complex submanifold of D and is essentially unique. We define C_0 to be the **base cycle** in D.

Example 5.1.1. Let Z be the complex projective space $\mathbb{P}^n(\mathbb{C})$ and let $G_0 = SU(n, 1)$. Let $\{e_1, \dots, e_{n+1}\}$ denote the standard basis of $\mathbb{C}^{n+1}$ relative to which the hermitian form defining G_0 is $\langle u, v\rangle = \left(\sum_{1 \leqq a \leqq n} u_a \overline{v_a}\right) - u_{n+1}\overline{v_{n+1}}$ where $u = \sum u_a e_a$ and $v = \sum v_a e_a$. Then G_0 has three orbits on Z: (i) the (open) unit ball $\mathcal{B}$ in $\mathbb{C}^n$ inside Z, consisting of the negative definite lines, (ii) the $(2n-1)$-sphere S which is the boundary of $\mathcal{B}$, consisting of the null lines, and (iii) the complement D of $\mathcal{B} \cup S$, consisting of the positive definite lines. The flag domain D is the nonconvex open G_0-orbit on Z in the following sense. From the complex analytic viewpoint it is not pseudoconvex. From the euclidean viewpoint of $\mathbb{C}^n \subset \mathbb{P}^n(\mathbb{C})$, $D \cap \mathbb{C}^n$ is nonconvex in the usual sense. Here C_0 is the hyperplane at infinity, complement to $\mathbb{C}^n$ in Z. In homogeneous coordinates $[z_1, \dots, z_{n+1}]$, $\mathcal{B}$ is given by $\sum_{1 \leqq a \leqq n} |z_a|^2 < |z_{n+1}|^2$, S is given by $\sum_{1 \leqq a \leqq n} |z_a|^2 = |z_{n+1}|^2$, D is given by $\sum_{1 \leqq a \leqq n} |z_a|^2 > |z_{n+1}|^2$, and C_0 is given by $|z_{n+1}|^2 = 0$. ◇

Example 5.1.2. Let Z be the complex projective space $\mathbb{P}^n(\mathbb{C})$ and let $G_0 = SL(n+1; \mathbb{R})$. Let $\{e_1, \dots, e_{n+1}\}$ denote the standard basis of $\mathbb{C}^{n+1}$ relative to which G_0 is given by the action of the real matrices, as in Example 4.5.4. Recall that there are just two G_0-orbits, the closed orbit $G_0([e_{n+1}]) \cong \mathbb{P}^n(\mathbb{R})$ and the open orbit $G_0([e_1 + \sqrt{-1}e_{n+1}])$. The special orthogonal group $K_0 = SO(n+1)$ and its complexification $K = SO(n+1; \mathbb{C})$ are defined by the bilinear form $b(u, v) = \sum_{1 \leqq a \leqq n+1} u_a v_a$, where $u = \sum u_a e_a$ and $v = \sum v_a e_a$. Here

$$\{[u + iv] \in \mathbb{P}^n(\mathbb{C}) \mid u \in \mathbb{R}^{n+1}, v \in \mathbb{R}^{n+1}, 0 \neq b(u, u) = b(v, v), \text{ and } b(u, v) = 0\}$$

is the base cycle C_0 for the open orbit. ◇

Lemma 5.1.3. *Fix an open G_0-orbit $D = G_0(z) \subset Z$ and let C_0 be the base cycle. Let $J = \{g \in G \mid gC_0 = C_0\}$. Then J is a closed complex subgroup of G, the quotient manifold $\mathcal{M}_Z := \{gC_0 \mid g \in G\} \cong G/J$ has a natural structure of G-homogeneous complex manifold, and the subset $\{gC_0 \mid g \in G$ and $gC_0 \subset D\}$ is open in $\mathcal{M}_Z$.*

Proof. The group J is a closed complex subgroup of G, because it is the stabilizer of a subvariety in a holomorphic action. That defines the G-invariant complex structure on $\mathcal{M}_Z \cong G/J$. The subset $\{gC_0 \mid g \in G \text{ and } gC_0 \subset D\}$ is open in $\mathcal{M}_Z$, because D is open and C_0 is compact. □

In [WeW] the *group-theoretic* cycle space was introduced as follows.

Definition 5.1.4. Recall from Lemma 5.1.3 that $\{gC_0 \mid g \in G \text{ and } gC_0 \subset D\}$ is an open subset of the complex manifold $\mathcal{M}_Z$. The **cycle space** of D is

$$\mathcal{M}_D : \text{ topological component of } C_0 \text{ in } \{gC_0 \mid g \in G \text{ and } gC_0 \subset D\}. \tag{5.1.5}$$

with the natural complex manifold structure as open submanifold of $\mathcal{M}_Z$.

5.2 The three cases

In order to understand the structure of Z, D and $\mathcal{M}_D$ we may assume that G_0 is simple, because G_0 is a local direct product of simple groups, and Z, D and $\mathcal{M}_D$ break up as global direct products along the local direct product decomposition of G_0. From this point on G_0 is simple unless we say otherwise.

Since $\mathfrak{g}_0$ is simple and $\mathfrak{j}$ contains $\mathfrak{k}$, there are only a few possibilities, one trivial. The trivial one is the case $\mathfrak{j} = \mathfrak{g}$, in other words the case where G_0 acts transitively on Z, and $\mathcal{M}_D$ is reduced to a single point. The following describes the cases where this occurs.

Proposition 5.2.1 ([O1], [O2]). *The (connected) noncompact real forms $G_0 \subset G$ transitive on Z are precisely those given as follows.*

1. $Z = \mathbb{P}^{2n-1}(\mathbb{C})$, *complex projective* $(2n-1)$*-space;* $G = SL(2n;\mathbb{C})$ *and* G_0 *is the quaternion linear group* $SL(n;\mathbb{H})$*, which has maximal compact subgroup* $\mathrm{Sp}(n)$.
2. $Z = SO(2r+2)/U(r+1)$*, unitary structures on* $\mathbb{R}^{2r+2}$*;* $G = SO(2r+2;\mathbb{C})$ *and* G_0 *is the Lorenz group* $SO(1, 2r+1)$*, which has maximal compact subgroup* $SO(2r+1)$.

The Lie algebra $\mathfrak{j}$ contains $\mathfrak{k}$. Except when G_0/K_0 is a bounded symmetric domain, $\mathfrak{k}$ is a maximal subalgebra of $\mathfrak{g}$. Therefore either $\mathfrak{j} = \mathfrak{g}$ as in Proposition 5.2.1, or $\mathfrak{j} = \mathfrak{k}$. If G_0/K_0 is a bounded symmetric domain, there is another case, $\mathfrak{j} = \mathfrak{k} + \mathfrak{s}_-$ or $\mathfrak{j} = \mathfrak{k} + \mathfrak{s}_+$. From now on we ignore the trivial case of Proposition 5.2.1 and concentrate on the other cases:

1. HERMITIAN HOLOMORPHIC CASE. G_0/K_0 is a bounded symmetric domain $\mathcal{B}$, we have the usual decomposition $\mathfrak{g} = \mathfrak{s}_+ + \mathfrak{k} + \mathfrak{s}_-$, and $\mathfrak{j}$ is one of $\mathfrak{k} + \mathfrak{s}_\pm$.
2. HERMITIAN NONHOLOMORPHIC CASE. G_0/K_0 is a bounded symmetric domain $\mathcal{B}$ and $\mathfrak{j} = \mathfrak{k}$.
3. GENERIC (OR NON-HERMITIAN) CASE. G_0/K_0 does not have a G_0-invariant complex structure.

The two hermitian cases are distinguished as follows:

Lemma 5.2.2. *Suppose that G_0/K_0 is a bounded symmetric domain $\mathcal{B}$. Then D is measurable, say $D = G_0/L_0$, and there is a double fibration*

$$\begin{array}{ccc} & G_0/(L_0 \cap K_0) & \\ \swarrow & & \searrow \\ D & & \mathcal{B} \end{array}$$

We are in the hermitian holomorphic case if and only if the double fibration is holomorphic,[1] *in other words if the two projections are simultaneously holomorphic for some choice between* $\mathcal{B}$ *and the complex conjugate structure* $\overline{\mathcal{B}}$ *and some choice of invariant complex structure on* $G_0/(L_0 \cap K_0)$. (If the latter choice exists, it is unique because $G_0/(L_0 \cap K_0) = \{(z, C) \in D \times \mathcal{B} \mid z \in C \in D\}$.)

Proof. Since G_0 is of hermitian type, $\mathfrak{k}_0$ contains a fundamental Cartan subalgebra $\mathfrak{h}_0$ of $\mathfrak{g}_0$. Thus D is measurable and we may assume $\mathfrak{h}_0 \subset \mathfrak{l}_0 = (\mathfrak{g}_0 \cap \mathfrak{q}_z)$.

Suppose that the double fibration is holomorphic. Then there is a choice of positive root system $\Sigma^+(\mathfrak{g}, \mathfrak{h})$ such that the corresponding Borel subalgebra

$$\mathfrak{b} = \mathfrak{h} + \mathfrak{b}^{-n} \subset ((\mathfrak{k} + \mathfrak{s}_-) \cap \mathfrak{q}_z).$$

Now $\mathfrak{b}_\mathfrak{k} + \mathfrak{s}_- = \mathfrak{b} = \mathfrak{b}_\mathfrak{l} + \mathfrak{q}_z^{-n}$, where $\mathfrak{b}_\mathfrak{k}$ and $\mathfrak{b}_\mathfrak{l}$ are the Borel subalgebras of $\mathfrak{k}$ and $\mathfrak{l}$ given by intersection with $\mathfrak{b}$. In particular $\mathfrak{s}_- \subset \mathfrak{q}_z$. If $z' \in C_0$, say $z' = k(z)$ where $k \in K$, then $S_-(z') = S_-(k(z)) = k(S_-(z)) = k(z) = z'$. Hence S_- fixes every point of C_0, and in particular $S_- \subset J$. Thus we are in the hermitian holomorphic case.

Conversely, suppose that we are in the hermitian holomorphic case, say $J = KS_-$. Since S_- is solvable and preserves the projective variety C_0, it has a fixed point on C_0. The argument just above shows that S_- fixes every point of C_0, in particular that $S_- \subset Q_z$. Now the projections of the double fibration simultaneously holomorphic, in other words the double fibration is holomorphic. □

In Part III we will study double fibration transforms, the most famous case of which is the Penrose transform, using results of Part II on the structure of the cycle space. Then we will describe some connections between cycle spaces and the unitary representation theory of semisimple Lie groups. Finally we will indicate some of the applications of cycle spaces to variation of Hodge structure, specifically to period matrix domains and construction of automorphic cohomology classes by Poincaré ϑ-series.

5.3 Cycle spaces of measurable open orbits are Stein

In connection with construction of automorphic cohomology, Wells [We1] showed by direct computation that $\mathcal{M}_D$ is a Stein manifold in the particular case $D \cong SO(2h, k)/(U(h) \times SO(k))$. That result was extended in Wells–Wolf [WeW] to

[1] By **holomorphic fibration** we mean a holomorphically locally trivial fiber space, essentially a holomorphic fiber bundle, except perhaps lacking a complex structure group. This is an important distinction, because the presence of a complex structure group would have many implications that are not available otherwise.

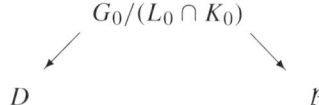

the more general situation of open G_0-orbits D of the form G_0/L_0 with L_0 compact, using a special case of the double fibration transform together with somewhat general methods of complex analysis (Andreotti–Grauert [AnG], Andreotti–Norguet [AN], Docquier–Grauert [DG]) associated to questions of holomorphic convexity and the Levi problem. Finally, when D is a measurable open orbit, Wolf [W8] combined his extension of boundary component theory of bounded symmetric domains (see [W2], or see [W4]) with the exhaustion function $\varphi : D \to \mathbb{R}$, to construct an essentially canonically defined strictly plurisubharmonic exhaustion function on $\mathcal{M}_D$. In view of Grauert's solution [Gra] to the Levi Problem—a complex manifold is Stein if and only if it has a C^∞ strictly plurisubharmonic exhaustion function—it follows that if D is measurable, then $\mathcal{M}_D$ is a Stein manifold. In Part II below we will use complex-geometric methods to prove in general that $\mathcal{M}_D$ is a Stein manifold, but in the general situation we do not construct a canonical strictly plurisubharmonic exhaustion.

CASE: D IS OF HERMITIAN HOLOMORPHIC TYPE. Here we may assume $D = G_0(z)$ and $J = KQ_-$, and therefore $\mathcal{M}_Z = G/J$ is the compact hermitian symmetric space dual to the bounded symmetric domain $\mathcal{B}$. Now $\mathcal{B} \subset \mathcal{M}_D \subset \mathcal{M}_Z$ because $\mathcal{M}_D$ is invariant by the action of G_0 on $\mathcal{M}_Z$. In Theorem 3.2.1 we give a precise description of the G_0-orbit structure of $\mathcal{M}_Z$, and the closure relations among the G_0-orbits, in terms of partial Cayley transforms. If $\mathcal{M}_D$ contains a G_0-orbit $\mathcal{O}$, then it contains every open G_0-orbit whose closure contains $\mathcal{O}$, because $\mathcal{M}_D$ is open in $\mathcal{M}_Z$.

We indicate the operator norm argument of [W8] which shows that $\mathcal{M}_D$ cannot contain an open orbit different from $\mathcal{B}$; a detailed argument based on Schubert cycles will be given below in Proposition 9.1.7. A convexity theorem of R. Hermann (see [W4, p. 286]) says that the bounded symmetric domain is given by

$$\mathcal{B} = G_0KS_-/KS_- = \{\exp(\zeta)(z_0) \mid \zeta \in \mathfrak{s}_+ \text{ with } ||\xi||_{\mathfrak{g}} < 1\},$$

where $||\xi||_{\mathfrak{g}}$ is the operator norm of $\mathrm{ad}(\xi)$. Let $||\xi||$ denote the norm on $\mathfrak{g}$ associated to the positive definite hermitian inner product $\langle \xi, \eta \rangle = -b(\xi, \tau\theta\eta)$, where b is the Killing form. Let V_0 denote the isotropy subgroup of G_0 at z. Using Harish-Chandra's result $G_0 \subset \exp(\mathfrak{s}_+)K\exp(\mathfrak{s}_-)$ and comparing the norms $||\xi||_{\mathfrak{g}}$ and $||\xi||$, Herman's result tells us that every $g \in G_0$ has expression $g = \exp(\zeta_1 + \zeta_2)k\exp(\eta)$, where $\eta \in \mathfrak{s}_-$, $k \in K$, $\zeta_2 \in \mathrm{Ad}(k)(\mathfrak{v} \cap \mathfrak{s}_+)$, and where $\zeta_1 \in \mathfrak{s}_+$ is orthogonal to $\mathrm{Ad}(k)(\mathfrak{v} \cap \mathfrak{s}_+)$. It follows that there is a number $a = a_G > 0$ such that, in this expression, $||\zeta_1||_{\mathfrak{g}} < a_G$. From this one proves that $\widetilde{f}(g) = ||\zeta_1||_{\mathfrak{g}}$ pushes down to a well-defined function $f : D \to \mathbb{R}$, and $0 \leqq f(gz_0) < a_G$. If $\mathcal{M}_D$ contains an open orbit $D_{\Sigma,\emptyset} \neq \mathcal{B}$, the roots in Σ define a product of $|\Sigma|$ Riemann spheres in Z, and the restriction of f to the diagonal in that product is unbounded. From that contradiction one concludes $\mathcal{M}_D = \mathcal{B}$. In particular one sees that $\mathcal{M}_D$ is Stein.

CASE: D IS NOT OF HERMITIAN HOLOMORPHIC TYPE. Then J has identity component K. Hence, J is reductive and $\mathcal{M}_Z = G/J$ is affine. Define

$$\beta : \mathcal{M}_D \to \mathbb{R}^+ \quad \text{by } \beta(gC_0) = \sup_{y \in C_0} \varphi(g(y)).$$

Since φ is an exhaustion function and the gC_0 are compact, one can prove that $\beta : \mathcal{M}_D \to \mathbb{R}^+$ blows up at every boundary point of $\mathcal{M}_D$. From the specific construction of φ, and a close look at the real analytic variety given by $d\varphi = 0$, one sees that β is continuous, piecewise C^ω and plurisubharmonic. Now a modification suggested by results of Docquier and Grauert [DG] gives a C^ω strictly plurisubharmonic exhaustion function $\psi = \varphi + \nu$ constructed as follows. Since $\mathcal{M}_Z$ is Stein there is a proper holomorphic embedding $f : \mathcal{M}_Z \to \mathbb{C}^{2n+1}$ with closed image, by Remmert's theorem. Define $\nu(C) := ||f(C)||^2$ for $C \in \mathcal{M}_D$. Since $\mathcal{M}_D$ carries a strictly plurisubharmonic exhaustion function, Grauert's solution [Gra] to the Levi Problem shows that $\mathcal{M}_D$ is Stein.

Using Grauert's solution of the q-Levi Problem and a natural integral transform, Andreotti and Norguet proved that $\mathcal{C}_q(D)$ is Stein under the conditions that D is q-complete and contained in the smooth points of a projective variety, and that the Levi form of the exhaustion function is nondegenerate at every point of D [AN]. The first two conditions are satisfied in our case, but it is not known whether the Levi form is nondegenerate at all points.

In [NS] the nongeneracy condition was eliminated, but it was still required that D be contained in the smooth points of a projective variety. Finally, Barlet proved the result with no superfluous conditions [B2]: *If D is a q-complete complex space, then $\mathcal{C}_q(D)$ is Stein.* Furthermore, if D is only q-convex, then it still follows that $\mathcal{C}_q(D)$ is holomorphically convex.

The methods of proof of these results are a bit technical, in part due to the fact that the cycle space can be singular. However, the basic approach which was intiated by Andreotti and Norguet (also see [NS]) is particularly interesting for representation theory, and therefore we outline it here. For the details on the full space $\mathcal{C}_q(D)$ of cycles see Section 7.4A.

If α is a $\bar{\partial}$-closed (q, q)-form and ξ is its cohomology class in $H^q(D; \Omega^q)$, then

$$AN(\xi)(C) := \int_C \alpha$$

defines a linear map $AN : H^q(D; \Omega^q) \to \mathcal{O}(\mathcal{C}_q(D))$. If D is a q-complete domain in a projective variety, then using the solution of the above mentioned Levi Problem one builds sufficiently many cohomology classes in $H^q(D; \Omega^q)$ to prove that $\mathcal{C}_q(D)$ is Stein with respect to the image of AN. In particular, given a divergent sequence $\{C_n\}$ of cycles there exists $\xi \in H^q(D; \Omega^q)$ so that $\lim_{n\to\infty} AN(\xi)(C_n) = \infty$, and, if C_1 and C_2 are different cycles in $\mathcal{C}_q(D)$, there exists $\xi \in H^q(D; \Omega^q)$ with $AN(\xi)(C_1) \neq AN(\xi)(C_2)$.

5.4 The cycle space in the hermitian case

In this section G_0 is of hermitian type, and we relate the cycle space $\mathcal{M}_D$ of an open orbit $G_0(z_0) = D \subset Z = G/Q$ to the bounded symmetric domain $\mathcal{B} \cong G_0/K_0$ associated to G_0. We just saw that $\mathcal{M}_D$ is $\mathcal{B}$ or $\overline{\mathcal{B}}$ when D is of holomorphic type, and Proposition 5.4.8 gives yet another proof of that. Here we show that if D is not of holomorphic type, then $\mathcal{M}_D \subset \mathcal{B} \times \overline{\mathcal{B}}$ where both are appropriately embedded in G/K. The opposite containment will be proved in Part II below.

In order to distinguish our basic flag manifold $Z = G/Q$ from the hermitian symmetric flag manifolds containing $\mathcal{B}$ and the complex conjugate structure $\overline{\mathcal{B}}$, we modify the notation of Chapter 3 as follows. The parabolic subgroups of G containing K_0 are $P_+ = KS_+$ and $P_- = KS_-$, the corresponding flag manifolds are $X_+ = G/P_+$ and $X_- = G/P_-$, the base points there are the $x_\pm = 1P_\pm \in X_\pm$, and the choices of complex structures are normalized by $\mathcal{B} = G_0(x_-) \subset X_-$ and $\overline{\mathcal{B}} = G_0(x_+) \subset X_+$.

5.4A $\mathcal{B} \times \overline{\mathcal{B}} \subset G/K$

Lemma 5.4.1. *We have $(G_0 \times G_0)(x_-, x_+) \subset \delta G(x_-, x_+) \subset X_- \times X_+$, where δG is the diagonal in $G \times G$, and both of these orbits are open in $X_- \times X_+$.*

Proof. Let $g_1, g_2 \in G_0$. Use $G_0 \subset S_+KS_-$ to write $g_2^{-1}g_1 = \exp(\eta_+)k\exp(\eta_-)$ with $k \in K$ and $\eta_\pm \in \mathfrak{s}_\pm$. Then

$$\begin{aligned}(g_1x_-, g_2x_+) &= \delta g_2(g_2^{-1}g_1x_-, x_+) = \delta g_2(\exp(\eta_+)x_-, x_+)\\ &= \delta g_2(\exp(\eta_+)x_-, \exp(\eta_+)x_+)\\ &= \delta g_2\, \delta\exp(\eta_+)(x_-, x_+) \in \delta G(x_-, x_+)\end{aligned}$$

shows that $(G_0 \times G_0)(x_-, x_+) \subset \delta G(x_-, x_+) \subset X_- \times X_+$. They are open because $G_0(x_-) = \mathcal{B}$ is open in X_- and $G_0(x_+) = \overline{\mathcal{B}}$ is open in X_+. Hence they all have full dimension. □

The isotropy subgroup of δG at (x_-, x_+) is

$$\{(g, g) \in G \times G \mid gx_- = x_- \text{ and } gx_+ = x_+\};$$

in other words,

$$\{(g, g) \in G \times G \mid g \in P_- \cap P_+ = K\}.$$

Thus

(5.4.2) δG has isotropy subgroup δK at (x_-, x_+), i.e., $\delta G(x_-, x_+) \cong G/K$.

We combine (5.4.2) with Lemma 5.4.1. This gives us the first part of the following.

Proposition 5.4.3. *There is a natural holomorphic embedding of $\mathcal{B} \times \overline{\mathcal{B}}$ into G/K. Let $\pi : G/K \to G/J = \mathcal{M}_Z$ be the natural projection. If the open G_0-orbit $D \subset Z$ is not of holomorphic type, then π is injective on $\mathcal{B} \times \overline{\mathcal{B}}$.*

Remark. In Part II we will discuss a certain universal domain $\mathcal{U} \subset G/K$, which in the hermitian nonholomorphic case is $\mathcal{B} \times \overline{\mathcal{B}}$, and prove (Corollary 11.3.6) that $\pi : G/K \to G/J$ is injective on $\mathcal{U}$ whenever D is not of holomorphic type. The proof just below is valid only in the hermitian nonholomorphic case, but it is considerably simpler than the proof in the general nonholomorphic case. ◇

Proof. Suppose that D is not of holomorphic type. Let $g_1, g_1', g_2, g_2' \in G_0$ and suppose $\pi(g_1 x_-, g_2 x_+) = \pi(g_1' x_-, g_2' x_+)$. As in the argument of Lemma 5.4.1, write

$$g_2^{-1} g_1 = \exp(\eta_+) k \exp(\eta_-) \text{ so } (g_1 x_-, g_2 x_+) = \delta g_2 \, \delta \exp(\eta_+)(x_-, x_+).$$

Similarly, this time reversing roles of the two factors,

$$g_1'^{-1} g_2' = \exp(\eta_-') k' \exp(\eta_+') \text{ so } (g_1' x_-, g_2' x_+) = \delta g_1' \, \delta \exp(\eta_-')(x_-, x_+).$$

The hypothesis $\pi(g_1 x_-, g_2 x_+) = \pi(g_1' x_-, g_2' x_+)$ now provides $j \in J$ such that $g_2 \exp(\eta_+) = g_1' \exp(\eta_-) j$. In other words, $(g_1')^{-1} g_2 \in S_- j S_+$.

Let $\{w_i\}$ be a set of representatives of the double coset space $W_K \backslash W_G / W_K$ for the Weyl groups of G and K. The Bruhat decomposition of G for X_+ is $G = \bigcup_i P_- w_i P_+$, the real group G_0 is contained in the cell $P_- P_+$ for $w_i = 1$, and G_0 does not meet any other cell $P_- w_i P_+$.

Since D is of nonholomorphic type $J \subset N_{G_u}(K_0) K$, we may write $j = nk$ with $n \in N_{G_u}(K_0)$ and $k \in K$. Express $n = w k_0$ with $w \in \{w_i\}$ and $k_0 \in K_0$. Then $j = k'' w k''' \in K w K$ with $k,'' k''' \in K$. Therefore $(g_1')^{-1} g_2 = \exp(\eta_-') k'' w k''' \exp(-\eta_+) \in P_- w P_+$. In particular G_0 meets $P_- w P_+$, and consequently $w = 1 \in W_K$ and $j \in K$. This shows $g_2 \exp(\eta_+) K = g_1' \exp(\eta_-') K$. Now

$$\begin{aligned}(g_1 x_-, g_2 x_+) &= \delta g_2 \, \delta \exp(\eta_+)(x_-, x_+) \\ &= \delta g_1' \, \delta \exp(\eta_-')(x_-, x_+) = (g_1' x_-, g_2' x_+)\end{aligned}$$

as asserted. That completes the proof. □

5.4B $\mathcal{M}_D \subset \mathcal{B} \times \overline{\mathcal{B}}$

If $\mathfrak{v}$ is a subspace of $\mathfrak{g}$ normalized by the Cartan subalgebra $\mathfrak{h}$, then

$$\mathfrak{v} = (\mathfrak{v} \cap \mathfrak{h}) + \sum_{\alpha \in \Sigma(\mathfrak{v}, \mathfrak{h})} \mathfrak{g}_\alpha,$$

and that defines a set $\Sigma(\mathfrak{v}, \mathfrak{h})$ of roots. For example $\Sigma(\mathfrak{g}, \mathfrak{h})$ is, as already defined, the whole root system; $\Sigma(\mathfrak{k}, \mathfrak{h})$ consists of the compact (relative to $\mathfrak{g}_0$) roots, $\Sigma(\mathfrak{s}_+, \mathfrak{h})$ consists of the noncompact positive roots, and $\Sigma(\mathfrak{s}_-, \mathfrak{h})$ consists of the noncompact negative roots.

Lemma 5.4.4. *One or both of* $\Delta(\mathfrak{q}_{z_0}^{+n} \cap \mathfrak{s}_\pm, \mathfrak{h})$ *contains a long root of* $\mathfrak{g}$.

Proof. If all the roots of $\mathfrak{g}$ are of the same length there is nothing to prove. If there are two root lengths, the only possibilities are (i) $G_0 = \mathrm{Sp}(n; \mathbb{R})$ up to a covering and (ii) $G_0 = SO(2, 2k+1)$ up to a covering.

Case (i). $D = G_0(z) \subset Z$ is open and $\mathfrak{q} = \mathfrak{q}_{z_0}$. The positive root system is adapted to $\mathfrak{q}$. Therefore $\mathfrak{q}_{z_0}^{-n}$ is spanned by negative root spaces. Let γ be the maximal root.

Then $\gamma \in \Sigma(\mathfrak{q}_{z_0}^{+n}, \mathfrak{h})$ and γ is long. Every compact root of $\mathfrak{g}_0 = \mathfrak{sp}(n; \mathbb{R})$ is short. So γ is noncompact and is hence contained in one of $\mathfrak{s}_\pm$. Lemma 5.4.4 is proved in case (i).

CASE (ii). In this case $\mathfrak{g}$ has a simple root system of the form $\{\alpha_1, \ldots, \alpha_{k+1}\}$ with α_1 noncompact and the other α_i compact. Here $\alpha_i = \varepsilon_i - \varepsilon_{i+1}$ for $1 \leqq i \leqq k$ and $\alpha_{k+1} = \varepsilon_{k+1}$ with the ε_i mutually orthogonal and of the same length. The noncompact positive roots are the $\alpha_1 + \cdots + \alpha_m$ with $1 \leqq m \leqq k+1$ and the $(\alpha_1 + \cdots + \alpha_m) + 2(\alpha_{m+1} + \cdots + \alpha_{k+1})$. All are long except $\alpha_1 + \cdots + \alpha_{k+1} = \varepsilon_1$, which is short. Now at least one of $\Sigma(\mathfrak{q}_{z_0}^{+n} \cap \mathfrak{s}_\pm, \mathfrak{h})$ contains a long root unless both $\mathfrak{q}_{z_0}^{+n} \cap \mathfrak{s}_+ = \mathfrak{g}_{\varepsilon_1}$ and $\mathfrak{q}_{z_0}^{+n} \cap \mathfrak{s}_- = \mathfrak{g}_{-\varepsilon_1}$. That is impossible because $\mathfrak{q}_{z_0}^{+n}$ is nilpotent. Lemma 5.4.4 is proved in case (ii). □

Interchange $\mathfrak{s}_+$ and $\mathfrak{s}_-$ if necessary so that $\Sigma(\mathfrak{q}_{z_0}^{+n} \cap \mathfrak{s}_+, \mathfrak{h})$ contains at least one long root. Recall the G_0-orbit structure of $X_\pm$ from Theorem 3.2.1 and the sketch of its proof. In particular we use the notation $\Xi = \{\xi_1, \ldots, \xi_\ell\}$ for the maximal set (3.1.9) of strongly orthogonal roots in $\Sigma(\mathfrak{s}_+, \mathfrak{h})$ and c_Γ for the partial Cayley transform defined by a subset $\Gamma \subset \Xi$.

Lemma 5.4.5. *Suppose* $\Gamma \subset \Xi \cap \Delta(\mathfrak{q}_{z_0}^{+n}, \mathfrak{h})$. *If* $\Gamma \cap \Delta(\mathfrak{q}_{z_0}^{+n}, \mathfrak{h})$ *is nonempty, then* $c_\Gamma(z_0)$ *is not contained in any open* G_0*-orbit on* Z.

Proof. The isotropy subgroup of G_0 at $c_\Gamma(z)$ has Lie algebra $\mathfrak{g}_0 \cap \mathfrak{q}'$, where $\mathfrak{q}' = \mathrm{Ad}\,(c_\Gamma)\mathfrak{q}$. If $\gamma \in \Gamma \cap \Delta(\mathfrak{q}_{z_0}^{+n}, \mathfrak{h})$ then, by [WZ0, (3.5)], $\mathrm{Ad}\,(c_\Gamma)(e_{-\gamma}) = \mathrm{Ad}\,(c_\Gamma)(\frac{1}{2}(x_\gamma + \sqrt{-1}y_\gamma)) = \frac{1}{2}(x_\gamma - \sqrt{-1}h_\gamma)$. But $x_\gamma, \sqrt{-1}h_\gamma \in \mathfrak{g}_0$, and so now $\mathrm{Ad}\,(c_\Gamma)(e_{-\gamma}) \in \mathfrak{g}_0 \cap \mathfrak{q}'$. Evidently $\mathrm{Ad}\,(c_\Gamma)(e_\gamma) \notin \mathfrak{g}_0 \cap \mathfrak{q}'$. Conclusion: $\mathfrak{g}_0 \cap \mathfrak{q}'$ is not reductive. As the G_0-orbits on Z are measurable, now $G_0(c_\Gamma(z))$ cannot be open in Z [W2, Theorem 6.3]. □

We also need the following topological lemma.

Lemma 5.4.6. *Let* X_1 *and* X_2 *be topological spaces, let* $B_i \subset X_i$ *be open subsets, and let* $M \subset (X_1 \times X_2)$ *be a connected open subset such that* (i) M *meets* $B_1 \times B_2$ *and* (ii) $M \cap (\mathrm{bd}(B_1) \times B_2) = \emptyset = M \cap (B_1 \times \mathrm{bd}(B_2))$. *Then* $M \subset (B_1 \times B_2)$.

Proof. $(X_1 \times X_2) \setminus M$ is closed in $(X_1 \times X_2)$ because M is open, contains $(\mathrm{bd}(B_1) \times B_2) \cup (B_1 \times \mathrm{bd}(B_2))$ by (ii), and thus contains the closure of the set $(\mathrm{bd}(B_1) \times B_2) \cup (B_1 \times \mathrm{bd}(B_2))$. That closure contains the boundary of the open set $B_1 \times B_2$. Thus

$$M = \Big(M \cap (B_1 \times B_2)\Big) \cup \Big(M \cap \big((X_1 \times X_2) \setminus \mathrm{closure}\,(B_1 \times B_2)\big)\Big).$$

As M is connected and meets $B_1 \times B_2$, now $M \subset (B_1 \times B_2)$. □

Now we come to the main result of this section.

Theorem 5.4.7. *Let* G_0 *be of hermitian type, let* $Z = G/Q$ *be a complex flag manifold, and let* $D = G_0(z) \subset Z = G/Q$ *be an open* G_0 *orbit that is not of holomorphic type. View* $\mathcal{B} \times \overline{\mathcal{B}} \subset \mathcal{M}_Z$ *as in Proposition* 5.4.3 *and* $\mathcal{M}_D \subset \mathcal{M}_Z$ *as usual. Then* $\mathcal{M}_D \subset \mathcal{B} \times \overline{\mathcal{B}}$.

Proof. Retain the notation of the proof of Lemma 5.4.1. Suppose that (g_1x_-, g_2x_+) belongs to the boundary of $\mathcal{B}\times\overline{\mathcal{B}}$ in $X_-\times X_+$. The closure of G_0KS_- in G is contained in S_+KS_-, and similarly the closure of G_0KS_+ in G is contained in S_-KS_+. That allows us to write $g_2^{-1}g_1 = \exp(\eta_+)k\exp(\eta_-)$ with $\eta_\pm \in \mathfrak{s}_\pm$ and $k \in K$, as before. We will prove that $g_2\exp(\eta_+)C_0 \not\subset D$, that is, $g_2\exp(\eta_+)C_0 \notin \mathcal{M}_D$. The theorem will follow. The proof breaks into three cases, according to the way (g_1x_-, g_2x_+) sits in the boundary of $\mathcal{B}\times\overline{\mathcal{B}}$.

Case i. Here $g_1x_- \in \mathrm{bd}(\mathcal{B})$ and $g_2x_+ \in \overline{\mathcal{B}}$ with $g_1, g_2 \in G$. We may suppose $g_2 \in G_0$. Then $g_2^{-1}g_1x_-$ also belongs to the boundary of $\mathcal{B}$ in X_-. So

$$g_2^{-1}g_1x_- \in k_0G_0[\Xi\setminus\Gamma](c_\Gamma(x_-))$$

for some $k_0 \in K_0$ and $\Gamma \subset \Xi$, because

$$G_0(c_\Gamma x_-) = K_0G_0[\Xi\setminus\Gamma](c_\Gamma x_-).$$

Thus $g_2^{-1}g_1(x_+) = k_0g_0c_\Gamma(x_-)$, $g_0 \in G_0[\Xi\setminus\Gamma]$. From its definition, $\mathrm{Ad}(c_\gamma)$ maps x_γ to itself, y_γ to $-h_\gamma$, and h_γ to y_γ; so $\mathrm{Ad}(c_\gamma^2)\mathfrak{h}_0 = \mathfrak{h}_0$. Now, using strong orthogonality of Ξ, we decompose

$$g_0 = \prod_{\Xi\setminus\Gamma}\Big(\exp(\eta_{+,\xi})k_\xi\exp(\eta_{-,\xi})\Big)$$

and

$$c_\Gamma = \prod_{\Gamma}\Big(\exp(\sqrt{-1}e_\gamma)\exp(\sqrt{2}h_\gamma)\exp(\sqrt{-1}e_{-\gamma})\Big)$$

with $\eta_{\pm,\xi} \in \mathfrak{g}_{\pm\xi}$ for $\xi \in \Xi\setminus\Gamma$. Set $\eta_{\pm,\gamma} = \sqrt{-1}e_{\pm\gamma}$ for $\gamma \in \Gamma$. Now

$$(g_1x_-, g_2x_+) = \delta g_2\,\delta\exp(\mathrm{Ad}(k_0)\eta'_+)(x_-, x_+), \quad \text{where } \eta'_+ = \textstyle\sum_{\xi\in\Xi}\eta_{+,\xi}.$$

By Lemma 5.4.4, $\Sigma(\mathfrak{q}_{z_0}^{+n}, \mathfrak{h})\cap\Sigma(\mathfrak{s}_+, \mathfrak{h})$ contains a long root. Thus, changing k_0 within K_0, in other words modifying the choice of z_0 within $C_0 = K(z_0)$, we may assume that Ξ meets $\Sigma(\mathfrak{q}_{z_0}^{+n}, \mathfrak{h})$. Any two subsets of Ξ with the same cardinality are $W(K_0, H_0)$-conjugate. In particular, changing k_0 within K_0, in other words, by modifying the choice of z_0 within $C_0 = K(z_0)$, we may assume that Γ meets $\Sigma(\mathfrak{q}_{z_0}^{+n}, \mathfrak{h})$. By Lemma 5.4.5 $c_{\Gamma\cap\Sigma(\mathfrak{q}_{z_0}^{+n}, \mathfrak{h})}(z_0)$ is not contained in any open G_0-orbit on Z.

Whenever $\xi \in \Xi$ and $g \in G[\xi]$ has the decomposition $g = \exp(\eta_+)k\exp(\eta_-)$, $\eta_\pm \in \mathfrak{g}_{\pm\xi}$ and $k \in K\cap G[\xi]$. Then $k \in H \subset Q_{z_0}$ and

(1) if $\xi \in \Sigma(\mathfrak{q}_{z_0}))$, then $\eta_+ \in \mathfrak{g}_\xi \subset \mathfrak{q}_{z_0}$ so $\exp(\eta_+)(z_0) = z_0$, and
(2) if $\xi \in \Sigma(\mathfrak{q}_{z_0}^{+n})$, then $\eta_- \in \mathfrak{g}_{-\xi} \subset \mathfrak{q}_{z_0}^{-n}$ and thus

$$\exp(\eta_+)(z_0) = \exp(\eta_+)k\exp(\eta_-)(z_0) = g(z_0).$$

Now

$$\begin{aligned}\exp(\eta_+)(k_0 z_0) &= \exp(\mathrm{Ad}(k_0)(\eta'_+))(k_0 z_0)\\ &= k_0 \exp(\eta'_+)(z_0) = k_0 g_0 c_\Gamma(z_0) \notin D,\end{aligned}$$

and we conclude that $g_2 \exp(\eta_+) C_0 \not\subset D$.

CASE 2. Here $g_1 x_- \in \mathcal{B}$ and $g_2 x_+ \in \mathrm{bd}(\overline{\mathcal{B}})$. The argument is exactly as in Case 1, but with the roles of $\mathcal{B}$ and $\overline{\mathcal{B}}$ reversed. Here note that this reversal of roles replaces Ξ by $-\Xi$ and c_Γ by $c_{-\Gamma}$.

CASE 3. Here $g_1 x_- \in \mathrm{bd}(\mathcal{B})$ and $g_2 x_+ \in \mathrm{bd}(\overline{\mathcal{B}})$. Then $\mathcal{M}_D$ is connected, $\mathcal{M}_D$ meets $\mathcal{B} \times \overline{\mathcal{B}}$ because $C_0 \in \mathcal{M}_D \cap (\mathcal{B} \times \overline{\mathcal{B}})$, and $\mathcal{M}_D \cap \big(\mathrm{bd}(\mathcal{B}) \times \overline{\mathcal{B}}\big) = \emptyset = \mathcal{M}_D \cap \big(\mathcal{B} \times \mathrm{bd}(\overline{\mathcal{B}})\big)$ by Cases 1 and 2. Case 3 now follows from Lemma 5.4.6. □

The same type of argument gives a short proof of the following result from [W8].

Proposition 5.4.8. *If D is of holomorphic type, then $\mathcal{M}_D$ is biholomorphic to either $\mathcal{B}$ or $\overline{\mathcal{B}}$.*

Proof. We may assume that $\mathcal{M}_Z = X_- = G/KS_-$ by switching $\mathfrak{s}_\pm$ if necessary. It is clear that $gC \subset D$ for $g \in G_0$, and as a result $\mathcal{B} \subset \mathcal{M}_D$. Now suppose that gx_- (for some $g \in G$) is in the boundary of $\mathcal{B} \subset X_-$. Then $gx_- = g_0 c_\Gamma(x_-)$ for some $g_0 \in G_0$ and some $\Gamma \neq \emptyset$. Conjugating by an element of K_0 we may assume $\Gamma \cap \Delta(\mathfrak{r}_+, \mathfrak{h}) \neq \emptyset$. Now, for $\Gamma' = \Gamma \cap \Delta(\mathfrak{r}_+, \mathfrak{h})$, gC contains $g_0 c_\Gamma(z) = g_0 c_{\Gamma'}(z)$. By Lemma 5.4.5 that is not in an open orbit. □

5.4C A reduction for $\mathcal{B} \times \overline{\mathcal{B}} \subset \mathcal{M}_D$

We prove a reduction result that is used in Chapter 5.5D below (see [WZ1]).

Proposition 5.4.9. *Suppose that $\mathcal{B} \times \overline{\mathcal{B}} \subset \mathcal{M}_D$ whenever D is an open G_0-orbit on G/Q that is not of holomorphic type, in the case where Q is a Borel subgroup of G. Then the same is true when Q is any parabolic subgroup of G.*

Proof. The base cycle in the open orbit $D = G(z) \subset Z$ is $C = K(z) = K_0(z)$. We may, and do, take Q to be the G-stabilizer of z; in other words we assume that $\mathfrak{q} = \mathfrak{q}_z$. Let $Q' \subset Q$ be any parabolic subgroup of G contained in Q such that $G_0 \cap Q'$ contains a compact Cartan subgroup $H_0 \subset K_0$ of G_0, let $Z' = G/Q'$ be the corresponding flag manifold, and let $\pi : Z' \to Z$ denote the associated G-equivariant projection $gQ' \mapsto gQ$. Write $z' \in Z'$ for the base point $1Q'$. Then $D' = G_0(z')$ is open in Z', because $\mathfrak{g}_0 \cap \mathfrak{q}'$ contains a compact Cartan subalgebra of $\mathfrak{g}_0$. We have set things up so that $Y' = K(z') = K_0(z')$ is a maximal compact subvariety of D'.

Since D is not of holomorphic type, both intersections $\mathfrak{r}_- \cap \mathfrak{s}_\pm$ are nonzero. But $\mathfrak{r}_-$ is contained in the nilradical $\mathfrak{r}'_-$ of $\mathfrak{q}'$. Now both intersections $\mathfrak{r}'_- \cap \mathfrak{s}_\pm$ are nonzero, and so D' is not of holomorphic type.

If $g \in G$ with $gY' \subset D'$ then $gK_0 \subset G_0 Q'$. Therefore $gK_0 \subset G_0 Q$ and thus $gY \subset D$. In other words, π maps $\mathcal{M}_{D'}$ to $\mathcal{M}_D$. It follows from Proposition 5.4.3 that this map is an injection. Thus, inside G/K we have $\mathcal{M}_{D'} \subset \mathcal{M}_D$. If $\mathcal{B} \times \mathcal{B} \subset \mathcal{M}_{D'}$ then it follows that $\mathcal{B} \times \overline{\mathcal{B}} \subset \mathcal{M}_D$. The assertion of the Proposition is the special case where Q' is a Borel subgroup. □

5.5 The classical hermitian case

As an example of the structure of cycle spaces, we look at the case where G_0 is a classical group of hermitian type. In this case the structure of $\mathcal{M}_D$ was worked out in [WZ1] by elementary means. The result is the following

Theorem 5.5.1. *Let G_0 be a classical simple Lie group of hermitian type. Let $D = G_0(z) \subset Z = G/Q$ be an open G_0-orbit. If D is of holomorphic type, then the cycle space $\mathcal{M}_D$ is biholomorphic either to $\mathcal{B}$ or to $\overline{\mathcal{B}}$. If D is not of holomorphic type, then $\mathcal{M}_D$ is biholomorphic to $\mathcal{B} \times \overline{\mathcal{B}}$.*

As we will see later, Theorem 5.5.1 holds without the requirement that G_0 be classical, and even that is a special case of the general result. But in the present case one has an elementary treatment.

We run through the classical cases. In each case, the standard basis of $\mathbb{C}^m$ will be denoted $\{e_1, \ldots, e_m\}$. Without further comment we will decompose vectors as $v = \sum v_j e_j$. We will have symmetric bilinear forms $(\cdot, \cdot)$ or antisymmetric bilinear forms $\omega(\cdot, \cdot)$ on $\mathbb{C}^m$ and the term **isotropic** will refer only to those bilinear forms. We will also have hermitian forms $\langle \cdot, \cdot \rangle$ on $\mathbb{C}^m$, and the term **signature** will refer only to those hermitian forms. In each case the flag manifold Z and the bounded symmetric domains $\mathcal{B}$ and $\overline{\mathcal{B}}$ are described in terms of flags

$$\mathcal{F} = (F_{d_1} \subsetneqq \cdots \subsetneqq F_{d_r})$$

of subspaces of $\mathbb{C}^n$, where $0 < d_1 < \cdots < d_r < n$ and $\dim F_j = j$, and of block form matrices. We write $\mathbb{C}^{p \times q}$ for the space of $p \times q$ complex matrices.

In the first three classical cases we work directly, but the orbit descriptions can be very complicated in the fourth classical case. So in that case we use Proposition 5.4.9 to reduce considerations to the case where Q is a Borel subgroup of G.

5.5A Type I: $\mathcal{B} = \{Z \in \mathbb{C}^{p \times q} \mid I - Z^*Z \gg 0\}$

Here $G = SL(n; \mathbb{C})$ and $G_0 = SU(p, q)$, indefinite unitary group defined by the hermitian form $\langle u, v \rangle = \sum_{j=1}^{p} v_j \overline{w_j} - \sum_{j=1}^{q} v_{p+j} \overline{w_{p+j}}$ with $p + q = n$.

The hermitian symmetric flag $X_- = G/KS_-$ is identified with the Grassmannian of q-planes in $\mathbb{C}^n$, the base point $x_- = [e_{p+1} \wedge \cdots \wedge e_{p+q}]$, and $\mathcal{B} = G_0(x_-)$ consists of the negative definite q-planes. Similarly, $X_+ = G/KS_+$ is identified with the Grassmannian of p-planes in $\mathbb{C}^n$, $x_+ = [e_1 \wedge \cdots \wedge e_p]$, and $\overline{\mathcal{B}} = G_0(x_+)$ consists of the positive definite p-planes. The embedding

$$\mathcal{B} \times \overline{\mathcal{B}} \subset G/K = G(x_-, x_+) \subset X_- \times X_+$$

of Proposition 5.4.3 is given by

$$\mathcal{B} \times \overline{\mathcal{B}} = \{(V, W) \subset (X_- \times X_+) \mid V \ll 0 \text{ and } W \gg 0\} \text{ and}$$
$$G/K = G(x_-, x_+) = \{(V, W) \in (X_- \times X_+) \mid V \cap W = 0\}.$$

The flag manifold $Z = Z_d$ consists of all flags $\mathcal{F} = (F_{d_1} \subsetneqq \cdots \subsetneqq F_{d_r})$ of subspaces of $\mathbb{C}^n$ for some fixed dimension sequence $d : 0 < d_1 < \cdots < d_r < n$. In view of Witt's theorem, the G_0-orbits on Z_d are the $D_{a,b}$ defined as follows. We have sequences $a : 0 \leqq a_1 \leqq \cdots \leqq a_r \leqq p$ and $b : 0 \leqq b_1 \leqq \cdots \leqq b_r \leqq q$ with $a_i + b_i = d_i$, and $D_{a,b}$ consists of all flags $\mathcal{F} \in Z_d$ such that F_{d_i} has signature (a_i, b_i) relative to $\langle \cdot, \cdot \rangle$ for $1 \leqq i \leqq r$. Then $D_{a,b} = G_0(z_{a,b})$, where

$$\begin{aligned} z_{a,b} = (z_{a,b,1} \subsetneqq \cdots \subsetneqq z_{a,b,r}) \text{ with} \\ z_{a,b,i} &= \mathrm{Span}(e_1, e_2, \ldots, e_{a_i}; e_n, e_{n-1}, \ldots, e_{n-b_i+1}) \text{ if } a_i, b_i > 0, \\ &= \mathrm{Span}(e_1, e_2, \ldots, e_{a_i}) \text{ if } a_i > 0 = b_i, \\ &= \mathrm{Span}(e_n, e_{n-1}, \ldots, e_{n-b_i+1}) \text{ if } a_i = 0 < b_i. \end{aligned}$$

Open orbits $D_{a,b} = D_{a',b'}$ if and only if $a = a'$ and $b = b'$; and $D_{a,b}$ is biholomorphic to $D_{a',b'}$ if and only if either (i) $a = a'$ and $b = b'$ or (ii) $p = q$, $a = b'$ and $a' = b$.

Now fix the open orbit $D_{a,b}$. Given $(V, W) \in G/K \subset (X_- \times X_+)$ we define

$$C_{V,W} = \{\mathcal{F} \in Z \mid \dim F_j \cap V = a_j \text{ and } \dim F_j \cap W = b_j \text{ for all } j\}.$$

Our base cycle $C_0 = K(z_{a,b}) = C_{x_-,x_+}$. If $g \in G$, then $gC_0 = C_{gx_-,gx_+}$. If $(V, W) \in \mathcal{B} \times \overline{\mathcal{B}}$, then $C_{V,W} \subset D_{a,b}$, in other words $C_{V,W} \in \mathcal{M}_{D_{a,b}}$. Thus

$$(V, W) \mapsto C_{V,W}$$

defines a map $\eta : \mathcal{B} \times \overline{\mathcal{B}} \to \mathcal{M}_{D_{a,b}}$. If $a_1 = \cdots = a_r = 0$, then $b_i = d_i$ for $1 \leqq i \leqq r$, and $\eta(V, W)$ depends only on V; then $\eta : \mathcal{B} \cong \mathcal{M}_{D_{a,b}}$. If $s_1 = \cdots = s_r = 0$, then $a_i = d_i$ for $1 \leqq i \leqq r$, and $\eta(V, W)$ depends only on W; then $\eta : \overline{\mathcal{B}} \cong \mathcal{M}_{D_{a,b}}$. Those are the cases where $D_{a,b}$ is of holomorphic type. In the nonholomorphic cases, $\eta : \mathcal{B} \times \overline{\mathcal{B}} \cong \mathcal{M}_{D_{r,s}}$. Theorem 5.5.1 is verified when $\mathcal{B}$ is of Type I.

5.5B Type II: $\mathcal{B} = \{Z \in \mathbb{C}^{n\times n} \mid Z = {}^tZ \text{ and } I - Z \cdot Z^* \gg 0\}$

Here $G = \mathrm{Sp}(n; \mathbb{C})$ and $G_0 = \mathrm{Sp}(n; \mathbb{R})$. These are the complex and real symplectic groups, defined by the antisymmetric bilinear form

$$\omega(v, w) = \sum_{j=1}^{n} (v_j w_{n+j} - v_{n+j} w_j)$$

on $\mathbb{C}^{2n}$ and $\mathbb{R}^{2n}$, respectively. For convenience we realize G_0 as $G \cap U(n, n)$, where $U(n, n)$ is the unitary group of the hermitian form

$$\langle v, w \rangle = \sum_{j=1}^{n} v_j \overline{w_j} - \sum_{j=1}^{n} v_{n+j} \overline{w_{n+j}}.$$

The hermitian symmetric flag $X_- = G/KS_-$ is identified with the Grassmannian of ω-isotropic n-planes in $\mathbb{C}^{2n}$, the base point $x_- = \mathrm{Span}(e_{n+1}, \ldots, e_{2n})$, and $\mathcal{B} =$

$G_0(x_-)$ consists of the negative definite ω-isotropic n-planes. Similarly, $X_+ = G/KS_+$ is identified with the Grassmannian of ω-isotropic n-planes in $\mathbb{C}^{2n}$, $x_+ = \mathrm{Span}(e_1, \ldots, e_n)$, and $\overline{\mathcal{B}} = G_0(x_+)$ consists of the positive definite ω-isotropic n-planes. The embedding

$$\mathcal{B} \times \overline{\mathcal{B}} \subset G/K = G(x_-, x_+) \subset X_- \times X_+$$

of Proposition 5.4.3 is given by

$$\begin{aligned}\mathcal{B} \times \overline{\mathcal{B}} &= \{(V, W) \subset (X_- \times X_+) \mid V \ll 0 \text{ and } W \gg 0\} \text{ and}\\ G/K = G(x_-, x_+) &= \{(V, W) \in (X_- \times X_+) \mid V \cap W = 0\}.\end{aligned}$$

The flag manifold $Z = Z_d$ consists of all flags $\mathcal{F} = (F_{d_1} \subsetneqq \cdots \subsetneqq F_{d_r})$ of subspaces of $\mathbb{C}^{2n}$ for some fixed dimension sequence $d : 0 < d_1 < \cdots < d_r \leqq n$. The variation on Witt's theorem that we need here is the following.

Lemma 5.5.2. *Let $U_1, U_2 \subset \mathbb{C}^{2n}$ be ω-isotropic subspaces of the same nondegenerate signature for $\langle \cdot, \cdot \rangle$. Then there exists $g \in G_0$ with $gU_1 = U_2$.*

The G_0-orbits on Z_d are again determined by the signature sequence of the flag. Therefore the open G_0-orbits are the $D_{a,b}$ which are defined as follows. We have sequences $a : 0 \leqq a_1 \leqq \cdots \leqq a_r$ and $b : 0 \leqq b_1 \leqq \cdots \leqq b_r$ with $a_i + b_i = d_i$, and $D_{a,b}$ consists of all flags $\mathcal{F} \in Z_d$ such that F_{d_i} has signature (a_i, b_i) relative to $\langle \cdot, \cdot \rangle$, for $1 \leqq i \leqq r$. Then $D_{a,b} = G_0(z_{a,b})$, where

$$\begin{aligned} z_{a,b} = (z_{a,b,1} &\subsetneqq \cdots \subsetneqq z_{a,b,r}) \text{ with}\\ z_{a,b,i} &= \mathrm{Span}(e_1, \ldots, e_{a_i}; e_{2n}, e_{2n-1}, \ldots, e_{2n-b_i+1}) \text{ if } a_i, b_i > 0,\\ &= \mathrm{Span}(e_1, e_2, \ldots, e_{a_i}) \text{ if } a_i > 0 = b_i,\\ &= \mathrm{Span}(e_{2n}, e_{2n-1}, \ldots, e_{2n-b_i+1}) \text{ if } a_i = 0 < b_i.\end{aligned}$$

Open orbits $D_{a,b} = D_{a',b'}$ if and only if $a = a'$ and $b = b'$; $D_{a,b}$ is biholomorphic to $D_{a',b'}$ if and only if either (i) $a = a'$ and $b = b'$ or (ii) $a = b'$ and $a' = b$.

Now fix the open orbit $D_{a,b}$. Given $(V, W) \in G/K \subset (X_- \times X_+)$ we again define

$$C_{V,W} = \{\mathcal{F} \in Z \mid \dim F_j \cap V = a_j \text{ and } \dim F_j \cap W = b_j \text{ for all } j\}.$$

Our base cycle $C_0 = K(z_{a,b}) = C_{x_-,x_+}$. If $g \in G$, then $gC_0 = C_{gx_-,gx_+}$. If $(V, W) \in \mathcal{B} \times \overline{\mathcal{B}}$, then $C_{V,W} \subset D_{a,b}$, so $C_{V,W} \in \mathcal{M}_{D_{a,b}}$. Thus $(V, W) \mapsto C_{V,W}$ defines a map $\eta : \mathcal{B} \times \overline{\mathcal{B}} \to \mathcal{M}_{D_{a,b}}$. If $a_1 = \cdots = a_r = 0$, then $b_i = d_i$ for $1 \leqq i \leqq r$ and $\eta(V, W)$ depends only on V; then $\eta : \mathcal{B} \cong \mathcal{M}_{D_{a,b}}$. If $s_1 = \cdots = s_r = 0$, then $a_i = d_i$ for $1 \leqq i \leqq r$ and $\eta(V, W)$ depends only on W; then $\eta : \overline{\mathcal{B}} \cong \mathcal{M}_{D_{a,b}}$. Those are the cases where $D_{a,b}$ is of holomorphic type. In the nonholomorphic cases, $\eta : \mathcal{B} \times \overline{\mathcal{B}} \cong \mathcal{M}_{D_{r,s}}$. Theorem 5.5.1 is verified when $\mathcal{B}$ is of type II.

5.5C Type III: $\mathcal{B} = \{Z \in \mathbb{C}^{n\times n} \mid Z = -{}^tZ \text{ and } I - Z\cdot Z^* \gg 0\}$

Here $G = SO(2n;\mathbb{C})$, special orthogonal group defined by the symmetric bilinear form $(v,w) = \sum_{j=1}^n (v_j w_{n+j} + v_{n+j} w_j)$ on $\mathbb{C}^{2n}$, and $G_0 = SO^*(2n)$, the real form with maximal compact subgroup $U(n)$. We realize G_0 as $G \cap U(n,n)$, where $U(n,n)$ is the unitary group of $\langle v, w\rangle = \sum_{j=1}^n v_j \overline{w_j} - \sum_{j=1}^n v_{n+j}\overline{w_{n+j}}$.

The hermitian symmetric flags $X_\pm = G/KS_\pm$ are identified with the two choices of connected component in the Grassmannian of isotropic (relative to $(\cdot,\cdot)$) n-planes in $\mathbb{C}^{2n}$. The components in question are distinguished by orientation. X_- has base point $x_- = \mathrm{Span}(e_{n+1},\dots,e_{2n})$, $X_- = G(x_-)$, and $\mathcal{B} = G_0(x_-)$ consists of the negative definite isotropic n-planes in X_-. Similarly, X_+ has base point $x_+ = \mathrm{Span}(e_1,\dots,e_n)$, $X_+ = G(x_+)$, and $\overline{\mathcal{B}} = G_0(x_+)$ consists of the positive definite isotropic n-planes in X_+. The embedding

$$\mathcal{B}\times\overline{\mathcal{B}} \subset G/K = G(x_-,x_+) \subset X_- \times X_+$$

of Proposition 5.4.3 is given by

$$\begin{aligned}\mathcal{B}\times\overline{\mathcal{B}} &= \{(V,W)\subset (X_-\times X_+) \mid V \ll 0 \text{ and } W \gg 0\} \text{ and}\\ G/K = G(x_-,x_+) &= \{(V,W)\in (X_-\times X_+) \mid V\cap W = 0\}.\end{aligned}$$

The flag manifold $Z = Z_d$ consists of all flags $\mathcal{F} = (F_{d_1} \subsetneqq \cdots \subsetneqq F_{d_r})$ of subspaces of $\mathbb{C}^{2n}$ for some fixed dimension sequence $d : 0 < d_1 < \cdots < d_r \leqq n$. The variation on Witt's theorem that we need here is the following.

Lemma 5.5.3. *Let $U_1, U_2 \subset \mathbb{C}^{2n}$ be $(\cdot,\cdot)$-isotropic subspaces of the same nondegenerate signature for $\langle\cdot,\cdot\rangle$. If $\dim U_i = n$, then assume also that the U_i are contained in the same $X_\pm$. Then there exists $g \in G_0$ with $gU_1 = U_2$.*

As in the Type II case, it follows that the open G_0-orbits in $Z = G/Q$ are determined by the signature sequences of the subspaces in the flag. Hence the open G_0-orbits are the $D_{a,b}$ which are defined as follows. We have sequences $a : 0 \leqq a_1 \leqq \cdots \leqq a_r \leqq n$ and $b : 0 \leqq b_1 \leqq \cdots \leqq b_r \leqq n$ subject to the conditions (i) $a_i + b_i = d_i$ for $1 \leqq i \leqq r$, (ii) if $r = n$, in which case $a_n + b_n = n$, then $\mathrm{Span}(e_1, e_2, \dots, e_{a_n}; e_{2n}, e_{2n-1}, \dots, e_{2n-b_n+1}) \in X_-$. Here (ii) is a parity condition on a_n. Then $D_{a,b}$ consists of all $\mathcal{F} \in Z$ such that (i) F_{d_i} has signature (a_i, b_i) for $1 \leqq i \leqq r$ and (ii) if $r = n$ then $F_n \in X_-$. In other words $D_{a,b} = G_0(z_{a,b})$, where

$$\begin{aligned} z_{a,b} = (z_{a,b,1} &\subsetneqq \cdots \subsetneqq z_{a,b,r}) \text{ with}\\ z_{a,b,i} &= \mathrm{Span}(e_1,\dots,e_{a_i}; e_{2n}, e_{2n-1},\dots,e_{2n-b_i+1}) \text{ if } a_i, b_i > 0,\\ &= \mathrm{Span}(e_1, e_2,\dots,e_{a_i}) \text{ if } a_i > 0 = b_i,\\ &= \mathrm{Span}(e_{2n}, e_{2n-1},\dots,e_{2n-b_i+1}) \text{ if } a_i = 0 < b_i.\end{aligned}$$

Again, $D_{a,b} = D_{a',b'}$ just when $a = a'$ and $b = b'$, but $D_{a,b}$ is biholomorphic to $D_{b,a}$.

Now fix the open orbit $D_{a,b}$. Given $(V,W) \in G/K \subset (X_- \times X_+)$ we yet again define

$$C_{V,W} = \{\mathcal{F} \in Z \mid \dim F_j \cap V = a_j \text{ and } \dim F_j \cap W = b_j \text{ for all } j\}.$$

Our base cycle $C_0 = K(z_{a,b}) = C_{x_-,x_+}$. If $g \in G$, then $gC_0 = C_{gx_-,gx_+}$. If $(V, W) \in \mathcal{B} \times \overline{\mathcal{B}}$, then $C_{V,W} \subset D_{a,b}$ so $C_{V,W} \in \mathcal{M}_{D_{a,b}}$. Thus $(V, W) \mapsto C_{V,W}$ defines a map $\eta : \mathcal{B} \times \overline{\mathcal{B}} \to \mathcal{M}_{D_{a,b}}$. If $a_1 = \cdots = a_r = 0$, then $b_i = d_i$ for $1 \leqq i \leqq r$ and $\eta(V, W)$ depends only on V; then $\eta : \mathcal{B} \cong \mathcal{M}_{D_{a,b}}$. If $s_1 = \cdots = s_r = 0$, then $a_i = d_i$ for $1 \leqq i \leqq r$, and $\eta(V, W)$ depends only on W; then $\eta : \overline{\mathcal{B}} \cong \mathcal{M}_{D_{a,b}}$. Those are the cases where $D_{a,b}$ is of holomorphic type. In the nonholomorphic cases $\eta : \mathcal{B} \times \overline{\mathcal{B}} \cong \mathcal{M}_{D_{r,s}}$. Theorem 5.5.1 is verified when $\mathcal{B}$ is of type III.

5.5D Type IV: $\mathcal{B} = \{Z \in \mathbb{C}^n \mid 1 + (|{}^tZ \cdot Z|^2 - 2Z^* \cdot Z) > 0, I - Z^* \cdot Z > 0\}$

Here $G = SO(2 + n; \mathbb{C})$, the special orthogonal group defined by the symmetric bilinear form $(v, w) = \sum_{j=1}^{2} v_j w_j - \sum_{j=3}^{2+n} v_j w_j$ on $\mathbb{C}^{2+n}$, and G_0 is the identity component of $SO(2, n)$. We view G_0 as the identity component of $G \cap U(2, n)$, where $U(2, n)$ is defined by the hermitian form $\langle v, w\rangle = (v, \overline{w})$.

The hermitian symmetric flags $X_\pm = G/KS_\pm$ are each identified with the space of $(\cdot,\cdot)$ isotropic lines in $\mathbb{C}^{2+n}$. Thus $X_\pm$ are nondegenerate quadrics in $\mathbb{P}_{n+1}(\mathbb{C})$. $X_\pm$ has base point $x_\pm = (e_1 \pm ie_2)\mathbb{C}$. The bounded domains $\mathcal{B} = G_0(x_-)$ and $\overline{\mathcal{B}} = G_0(x_+)$, and each consists of the $\langle\cdot,\cdot\rangle$ positive definite $(\cdot,\cdot)$ isotropic lines. The embedding

$$\mathcal{B} \times \overline{\mathcal{B}} \subset G/K = G(x_-, x_+) \subset X_- \times X_+$$

of Proposition 5.4.3 is given by

$$\begin{aligned} \mathcal{B} \times \overline{\mathcal{B}} &= \{(V, W) \in (X_- \times X_+) \mid V \gg 0 \text{ and } W \gg 0\} \text{ and} \\ G/K = G(x_-, x_+) &= \{(V, W) \in (X_- \times X_+) \mid V \not\perp W\}. \end{aligned}$$

Here $\gg$ (positive definite) refers to the hermitian form $\langle\cdot,\cdot\rangle$ and $\perp$ refers to the symmetric bilinear form $(\cdot,\cdot)$.

The flag manifold $Z = Z_d$ is a connected component of the space

$$\widetilde{Z} = \{\mathcal{F} = (F_{d_1} \subsetneqq \cdots \subsetneqq F_{d_r}) \mid \dim F_i = d_i \text{ and } (F_{d_i}, F_{d_i}) = 0\ \forall i\}$$

of isotropic flags in $\mathbb{C}^{2+n}$ for the dimension sequence $d : 0 < d_1 < \dots d_r \leqq m$, where $m = [\frac{n}{2}] + 1$. If n is even and $r = m$, then $\widetilde{Z}$ has two topological components; otherwise $Z = \widetilde{Z}$. In any case

$$Z_+ = G([(e_1 + ie_2) \wedge (e_3 + ie_4) \wedge \cdots \wedge (e_{2m-1} + ie_{2m})])$$

is a connected component in the variety of all maximal isotropic subspaces of $\mathbb{C}^{2+n}$, and

$$Z = \left\{ \mathcal{F} = (F_{d_1} \subsetneqq \cdots \subsetneqq F_{d_r}) \,\middle|\, \begin{array}{l} \dim F_{d_i} = d_i \text{ and } (F_{d_i}, F_{d_i}) = 0\ \forall i, \\ \text{and further if } r = m \text{ then } F_{d_r} \in Z_+ \end{array} \right\}.$$

The appropriate variation on Witt's Theorem for our considerations here is the following.

Lemma 5.5.4. *Let $U_1, U_2 \subset \mathbb{C}^{2+n}$ be $(\cdot,\cdot)$-isotropic subspaces of the same nondegenerate signature for $\langle\cdot,\cdot\rangle$. Then there exists $g \in O(2+n;\mathbb{C}) \cap U(2,n)$ with $gU_1 = U_2$.*

It will follow that the open G_0-orbits on the Z are essentially determined by the signature sequences of the subspaces in the flag. It is a bit complicated to make this precise. Thus we now assume that Q is a Borel subgroup of G, in other words, that the dimension sequence $d : 1 < 2 < \cdots < m-1 < m$. Write Z for Z_d.

More or less as in the earlier cases it follows that the open G_0-orbits on the full flag $Z = G/Q$ are essentially determined by the signature sequences of the subspaces in the flag. Let $1 \leqq k \leqq m$, and define points $z_k^{\pm} \in Z$ and G_0-orbits $D_k^{\pm} = G_0(z_k^{\pm}) \subset Z$ by

$$z_k^{\pm} = (z_{k,1}^{\pm} \subset \cdots \subset z_{k,m}^{\pm}), \text{ where}$$
$$z_{k,j}^{\pm} = \mathrm{Span}(e_3 + ie_4, \ldots, e_{2j+1} + ie_{2j+2}) \text{ for } j < k,$$
$$z_{k,j}^{\pm} = \mathrm{Span}(e_1 \pm ie_2, \ldots, e_3 + ie_4, \ldots, e_{2j-1} + ie_{2j}) \text{ for } j \geqq k.$$

Then $D_k^{\pm} = G_0(z_k^{\pm})$ consists of all $\mathcal{F} \in Z$ such that F_j has signature $(0, j)$ for $j < k$, signature $(1, j-1)$ for $j \geqq k$, and F_j meets $G_0(x_{\pm})$ for $j \geqq k$. The open G_0-orbits on Z are the $D_k^{\pm}$, and they are distinct. The open $(O(2+n;\mathbb{C}) \cap U(2,n))$-orbits on Z are the $D_k^+ \cup D_k^-$.

Fix k and ε with $1 \leqq k \leqq m$ and $\varepsilon = \pm$. Let

$$(V, W) \in G/K = (X_- \times X_+).$$

So $V = v\mathbb{C}$ and $W = w\mathbb{C}$, where $v, w \in \mathbb{C}^{2+n}$ are isotropic vectors with $(v, w) \neq 0$. Define $C_{V,W}$ to be

$$\left\{ \mathcal{F} \in Z \;\middle|\; \begin{array}{l} \dim F_j \cap \mathrm{Span}(v,w) = 0 \text{ and } \dim F_j \cap \mathrm{Span}(v,w)^{\perp} = j,\ j < k, \\ \dim F_j \cap \mathrm{Span}(v,w) = 1 \text{ and } \dim F_j \cap \mathrm{Span}(v,w)^{\perp} = j-1,\ j \geqq k, \\ v \in F_j \text{ if } \varepsilon = + \text{ and } j \geqq k;\quad w \in F_j \text{ if } \varepsilon = - \text{ and } j \geqq k \end{array} \right\}.$$

As above, $\perp$ refers to the symmetric bilinear form. Note that the only isotropic vectors in $\mathrm{Span}(v, w)$ are the multiples of v and the multiples of w.

Denote $D = D_k^{\pm}$ so that $C = K(z_k^{\pm}) = C_{x_-,x_+}$. If $g \in G$, then $gC = C_{gx_-,gx_+}$. Arguing as in [WZ1, Lemma 5.1], we check

$$\text{if } (V, W) \in \mathcal{B} \times \overline{\mathcal{B}}, \quad \text{then } C_{V,W} \subset D_k^{\pm}, \quad \text{so } C_{V,W} \in \mathcal{M}_{D_k^{\pm}}.$$

Now $(V, W) \mapsto C_{V,W}$ defines a map $\eta : \mathcal{B} \times \overline{\mathcal{B}} \to \mathcal{M}_{D_k^{\pm}}$. If $k = 1$ and if $\varepsilon = +$, then $\eta(V, W)$ depends only on V; if $k = 1$ and $\varepsilon = -$, then $\eta(V, W)$ depends only on W; those are the cases where $D_k^{\pm}$ is of holomorphic type. In the nonholomorphic cases, η injects $\mathcal{B} \times \overline{\mathcal{B}}$ into $\mathcal{M}_{D_k^{\pm}}$ and we have $\mathcal{B} \times \overline{\mathcal{B}} \subset \mathcal{M}_{D_k^{\pm}}$.

Now Theorem 5.5.1 is verified when $\mathcal{B}$ is of type IV and Q is a Borel subgroup of G. In view of Proposition 5.4.9, that verifies it whenever $\mathcal{B}$ is of type IV. This completes the proof of Theorem 5.5.1.

Part II

Cycle Spaces as Universal Domains

Overview

In this part we study various G_0-invariant domains that are contained in the complex affine symmetric space $\Omega = G/K$ and that contain the riemannian symmetric space G_0/K_0. Except in the holomorphic hermitian case, which is discussed in sufficient detail in Part I, one such domain is the cycle space of *any* G_0-orbit in any G-flag manifold Z. Although up to this point we have restricted our attention to open G_0-orbits, the case of lower-dimensional orbits is also treated here in Chapter 12.

It turns out that the cycle domains have many different faces in the sense that they agree with various other domains which are defined in completely different ways. For example, these arise in the study of cut point loci of the dual compact symmetric space G_u/K_0, which is also embedded in Ω. From another differential or symplectic geometric viewpoint, these are domains of existence of the *adapted complex structure* on the tangent bundle of the riemannian symmetric space. As a basic definition for this *universal domain* $\mathcal{U}$ we use the one given by Akhiezer and Gindikin which guarantees a proper G_0-action. Other definitions involve, for example, maximal domains of existence for holomorphic extension of functions of representation-theoretic importance on G_0/K_0.

The new tool which is implemented here is the use of special Schubert varieties which cut the base cycle C in a perfect way. Associated to such varieties are incidence hypersurfaces (or divisors) which are contained in the complement of the given cycle space in Ω. Meromorphic functions with poles along these hypersurfaces, and the envelopes which are constructed by removing all appropriate translates of them, are of particular importance.

The main result of this part states that all of these domains are the same. In particular, they agree with $\mathcal{U}$ and as a result only depend on the real form G_0, and not on the flag manifold Z or the orbit under consideration.

The proofs utilize both complex analytic and group-theoretic techniques. The key final result which yields the classification states that $\mathcal{U}$ can be characterized as the maximal G_0-invariant, Kobayashi hyperbolic Stein domain in Ω that contains the riemannian symmetric space G_0/K_0. Its proof involves, for example, a rather detailed analysis of the boundary of $\mathcal{U}$.

This part is structured as follows. In Chapter 6 we introduce in detail three of the domains mentioned above. These are the domains which arise through metric considerations and the universal domain $\mathcal{U}$. The basic relation between G_0-invariant plurisubharmonic functions $\mathcal{U}$ and convex functions on a certain Weyl chamber are proved.

Chapter 7 is devoted to certain aspects of the complex geometry which comes from complex hypersurfaces H in Ω that are invariant under Borel groups B which contain a factor A_0N_0 of an Iwasawa decomposition $G_0 = K_0A_0N_0$. Removing these divisors, we define the resulting the connected component which contains G_0/K_0 to be the Iwasawa envelope $\mathcal{E}_{\mathcal{I}}$. Theorem 7.2.7 shows that $\mathcal{U} = \mathcal{E}_{\mathcal{I}}$.

In Chapter 7 we also introduce the reader to Barlet's construction of the cycle space (see Section 7.4). This, along with the principle of the trace transform, is used to show that B-invariant incidence varieties defined by certain Schubert varieties are in fact particular examples of the hypersurfaces that go into the construction of $\mathcal{E}_{\mathcal{I}}$. These then define another Stein G_0-invariant envelope which by definition contains both $\mathcal{U}$ and the cycle space $\mathcal{M}_D$ (see Corollary 7.4.13).

In order to derive precise information on the Schubert varieties that cut K-orbits transversally we need information arising in the symplectic geometric proof of the duality between the G_0-orbits and K-orbits in Z. This Matsuki duality is proved in Chapter 8, and the precise information on the Schubert intersection theory is given in Chapter 9 (see Theorem 9.1.1). As a result we construct complex supporting varieties at every boundary point of every G_0-orbit (Proposition 9.1.4). These yield incidence varieties in Ω on the boundary of every given cycle space. With a bit of care, given a boundary point of the cycle space $\mathcal{M}_D$ of an open G_0-orbit, we construct in an algorithmic way a supporting B-invariant hypersurface at that point (see Section 9.2). This already pins down the location of $\mathcal{M}_D$ as itself being an envelope which contains $\mathcal{U}$. This result is also a consequence of the considerations of hyperbolicity in Chapter 11.

In Chapter 10 we analyze the boundary of $\mathcal{U}$. This is done from the point of view of the G_0-invariant theory. In particular, for a point p in a generic nonclosed orbit, we construct a three-dimensional semisimple subgroup S of G_0 with a two-dimensional orbit $S(p)$ which closes up in a controlled way to a point of the closed orbit in the closure of $G_0(p)$ (see Section 10.6B). We refer to the orbits $S(p)$ as $SL(2)$ models. These are up to a 2:1 cover two-dimensional affine quadrics, i.e., the affine symmetric space associated to $SL_2(\mathbb{C})$. They cut $\mathcal{U}$ in the universal domain of $SL(2;\mathbb{R})$ (Theorem 10.6.9). This tool is used in a fundamental way for the characterization of $\mathcal{U}$ as the maximal G_0-invariant, hyperbolic Stein domain containing the riemannian symmetric space G_0/K_0. This result is stated and proved in Chapter 11 (see Theorems 11.3.1 and 11.3.7).

Chapter 12 is devoted to implementing the methods developed in the previous sections as well as a number of new techniques to prove that, with the usual exception of the hermitian holomorphic case, even for lower-dimensional orbits the cycle space agrees with the universal domain $\mathcal{U}$ (Theorem 12.1.3).

In the last chapter of this part we display our general methods in the case of $SL(n;\mathbb{R})$. We also give concrete topological realizations of $\mathcal{U}$ for a number of series of classical groups. Finally, we compare the Schubert slice method with another slice method which works quite well in the special case where Z is a compact hermitian symmetric space.

6

Universal Domains

One of the purposes of this monograph is to show that an open G_0-orbit D in Z either is of hermitian holomorphic type in which case the cycle space $\mathcal{M}_D$ is the associated bounded symmetric domain $\mathcal{B}$ or $\overline{\mathcal{B}}$, or is not of hermitian holomorphic type in which case $\mathcal{M}_D$ is naturally isomorphic to a certain universal domain $\mathcal{U}$. Here we introduce $\mathcal{U}$ from several viewpoints and derive certain of its basic properties. The main aspects of this chapter can be summarized as follows.

In Section 1 we discuss certain differential geometric properties of the compact symmetric space G_u/K_0, which is closely related to $\mathcal{U}$. This leads to a naturally defined G_0-invariant tubular neighborhood Ω_C of the 0-section of the tangent bundle of the riemannian symmetric space G_0/K_0. It is then shown that the polar map defines a diffeomorphism $\Pi : \Omega_C \to \Omega_{AG} =: \mathcal{U}$ onto a neighborhood Ω_{AG} in the affine symmetric space G/K (Proposition 6.1.1). The domain Ω_{AG} was defined in [AkG] as a thickening of G_0/K_0 which is appropriate for the study of proper actions of G_0. It is concretely defined by a certain restricted root polytope P in the $\mathfrak{a}_0$-part of an Iwasawa decomposition $\mathfrak{g}_0 = \mathfrak{k}_0 + \mathfrak{a}_0 + \mathfrak{n}_0$: $\mathcal{U} = G_0 \exp(iP)(x_0)$, where x_0 is the base point in G/K. Section 1 closes with a proof that $\mathcal{U} = \mathcal{B} \times \overline{\mathcal{B}}$ in the case where G_0 is of hermitian type (Proposition 6.1.9).

Section 2 begins with a brief discussion of the adapted complex structure on neighborhoods of the 0-section of the tangent bundle of a C^ω (real analytic) riemannian manifold (M, g). In the case of the symmetric space $M = G_0/K_0$ it is shown that the maximal domain of definition Ω_{adpt} of this structure is biholomorphically equivalent to $\mathcal{U}$, also by the polar map (Proposition 6.2.3).

The main point of Section 3 is to prove a characterization of G_0-invariant strictly plurisubharmonic functions on $\mathcal{U}$ as the G_0-invariant functions that pullback to strictly convex functions on P, by the map $\xi \mapsto \exp(i\xi)(x_0)$ (Theorem 6.3.1). In order to prove this basic result of [BHH], we derive detailed information on the Cauchy–Riemann structure of the G_0-orbits in $\mathcal{U}$ and on the Levi form of G_0-invariant hypersurfaces in $\mathcal{U}$.

6.1 Definitions and first properties

Throughout this chapter we assume that the complex semisimple Lie group G is simply connected. Note that under this assumption the set of fixed points, G^{ν}, is connected for every continuous involutive group automorphism $\nu : G \to G$. As before τ denotes the complex conjugation of G over G_0 and θ is a Cartan involution of G_0, which we extend holomorphically to G. That extension is also denoted by θ and it commutes with τ. Note that $\sigma = \tau\theta = \theta\sigma$ is the Cartan involution of the complex group G, where $G^{\sigma} = G_u$ is a θ-stable compact real form of G. The fixed point set G^{θ} is the connected complex subgroup K of G which is isomorphic to the universal complexification of the maximal compact subgroup $K_0 = G_0^{\theta}$.

Unless otherwise stated we assume (without loss of generality) that the real Lie group G_0 is simple.

Let Ω denote the complex homogeneous space G/K. Select the base point $x_0 = 1K$. The stabilizers $G_{x_0} = K$ and $(G_0)_{x_0} = G_0 \cap K = K_0$. Equipped with its invariant metric, the G_0-orbit of x_0 is the riemannian symmetric space $\Omega_0 = G_0/K_0$, embedded as a closed totally real subspace with $\dim_{\mathbb{R}} \Omega_0 = \dim_{\mathbb{C}} \Omega$.

If $\mathfrak{g}_0 = \mathfrak{k}_0 + \mathfrak{s}_0$ is the Cartan decomposition defined by θ, then the tangent space $T_{x_0}\Omega_0$ is the K_0-module $\mathfrak{g}_0/\mathfrak{k}_0$, which we identify with $\mathfrak{s}_0$, and the tangent bundle of Ω_0 is the G_0-homogeneous vector bundle

$$T\Omega_0 = G_0 \times_{K_0} \mathfrak{s}_0 \to G_0/K_0 = \Omega_0.$$

Consider the **polar coordinate** mapping

$$\Pi : G_0 \times_{K_0} \mathfrak{s}_0 \to G/K, \quad [(g, \xi)] \mapsto g \cdot \exp(i\xi)(x_0).$$

In the following sections we will discuss basic properties of Π on canonically defined neighborhoods of the 0-section of $T\Omega_0$.

6.1A Differential geometric viewpoint

In $\Omega = G/K$ we consider the orbit $\Omega_u := G_u(x_0) = G_u/K_0$. Equipped with its invariant metric, it is a compact riemannian symmetric space and is embedded as a totally real submanifold of Ω. The complex structure of the complex symmetric space yields a canonical identification $i : \mathfrak{s}_0 = T_{x_0}\Omega_0 \to T_{x_0}\Omega_u = i\mathfrak{s}_0$, and we have $T_{x_0}\Omega = T_{x_0}\Omega_0 \oplus T_{x_0}\Omega_u$. The riemannian exponential map $\exp : T_{x_0}\Omega_u \to \Omega_u$ is surjective and can be written as the composition of the Lie group exponential map $\exp : i\mathfrak{s}_0 \to G_u$ and the projection $G_u \to G_u/K_0$. Hence, $\exp : T_{x_0}\Omega_u \to \Omega_u$ coincides with the restriction of the polar mapping: $\Pi : T_{x_0}\Omega_0 \to \Omega_u \subset \Omega$. In [C] Crittenden describes a basic differential geometric property of this map.

Before we state the relevant results, we recall some basic differential geometric constructions. Given a complete riemannian manifold M let $\exp : T_{x_0}M \to M$ denote the exponential map at some point $x_0 \in M$ and

$$d\exp_w : T_w(T_{x_0}M) \to T_{\exp(w)}M$$

its differential at $w \in T_{x_0}M$. Let $W = W(x_0) \subset T_{x_0}M$ be the connected component containing $0 \in T_{x_0}$ of the set of tangent vectors w such that $(d\exp)_w$ is invertible. The set $\mathrm{Conj}(x_0) := \exp(\mathrm{bd}(W))$ is referred to as the **conjugate locus**. Let $V = V(x_0)$ be the set of all $v \in T_{x_0}M$ such that $t \mapsto \exp tv\, t \in [0,1]$, is the unique length minimizing geodesic segment connecting x_0 and $\exp v$. The set $\mathrm{Cut}(x_0) := \exp(\mathrm{bd}(V))$ is the **cut locus** defined by x_0.

It should be noted that $\exp : V(x_0) \to \exp(V(x_0))$ is a diffeomorphism and $M = \exp V(x_0) \dot{\cup} \mathrm{Cut}(x_0)$. The example of a flat torus shows that, in general, $V(x_0) \neq W(x_0)$, and $\exp : W(x_0) \to \exp(W(x_0))$ need not be injective. However, in our situation we have the following.

Proposition 6.1.1 ([C]). *Let $M = \Omega_u$ be a compact simply connected riemannian symmetric space. Then the sets $V(x_0)$ and $W(x_0)$ are equal. In particular $\Pi|_W$ is a diffeomorphism onto its image.*

Now let $V = W \subset T_{x_0}\Omega_u \cong T_{x_0}\Omega_0$ as above. The open set

$$\tfrac{1}{2}\Omega_u := \exp(\tfrac{1}{2}V) = \Pi(\tfrac{1}{2}V) \tag{6.1.2}$$

is a sort of hemisphere in Ω_u, half way to the cut locus,

$$\tfrac{1}{2}\Omega_u = \{\exp(t\xi) \mid \xi \in \mathrm{bd}(V(x_0)) \text{ and } |t| < \tfrac{1}{2}\}. \tag{6.1.3}$$

It can be computed explicitly in terms of roots.

To do this, regard V as a K_0-invariant domain in $\mathfrak{s}_0$, where $K_0 \times \mathfrak{s}_0 \to \mathfrak{s}_0$ is the restriction of the adjoint representation of G_0 on $\mathfrak{g}_0$. Let $\mathfrak{a}_0$ be a maximal abelian subalgebra of $\mathfrak{s}_0$. Recall that any two such algebras $\mathfrak{a}_0, \mathfrak{a}_0'$ are conjugate by an element of K_0. Fix such $\mathfrak{a}_0$ and let $\Sigma(\mathfrak{g}_0, \mathfrak{a}_0)$ denote the corresponding restricted root system, i.e., the set of all nonzero weights of the $\mathbb{R}$-diagonalizable representation $\mathrm{ad}_{\mathfrak{g}_0}|_{\mathfrak{a}_0}$ of $\mathfrak{a}_0$ on $\mathfrak{g}_0$. Let ξ be a regular element in $\mathfrak{a}_0$, i.e., $\alpha(\xi) \neq 0$ for all $\alpha \in \Sigma(\mathfrak{g}_0, \mathfrak{a}_0)$. If $k_0 \in K_0$ and $k_0(\xi) \in \mathfrak{a}_0$, then $k_0(\mathfrak{a}_0) = \mathfrak{a}_0$ and $k_0 \in N_{K_0}(\mathfrak{a}_0)$. Thus

$$\mathfrak{s}_0^{\mathrm{reg}} = K_0 \times_{N_{K_0}} \mathfrak{a}_0^{\mathrm{reg}}.$$

As a result $V = K_0 \cdot \omega_0$, where ω_0 is a certain domain in $\mathfrak{a}_0$ which is invariant under the Weyl group $W := N_{K_0}(\mathfrak{a}_0)/Z_{K_0}(\mathfrak{a}_0)$. Crittenden [C] showed that ω_0 is the connected component containing 0 of

$$\mathfrak{a}_0 \setminus \bigcup_{\alpha \in \Sigma(\mathfrak{g}_0, \mathfrak{a}_0)} H_\alpha,$$

where $H_\alpha := \{\xi \in \mathfrak{a}_0 : \alpha(\xi) = \frac{\pi}{2}\}$. Thus we now define ω_0 to be this set, i.e., $\omega_0 = \{\xi \in \mathfrak{a}_0 : |\alpha(\xi)| < \frac{\pi}{2} \text{ for all } \alpha \in \Sigma(\mathfrak{g}_0, \mathfrak{a}_0)\}$ and state the following result for further reference in our context.

Proposition 6.1.4 ([C]). *The restriction*

$$\Pi|_{K_0 \cdot \omega_0} : K_0 \cdot \omega_0 \to \tfrac{1}{2}\Omega_u = K_0 \cdot \exp(i\omega_0)(x_0)$$

is a diffeomorphism.

6.1B Proper actions

The set $\Omega_{AG} := G_0 \cdot \exp(i\omega_0)(x_0) = G_0 \cdot \frac{1}{2}\Omega_u \subset \Omega$ was introduced from the point of view of complex geometry and representation theory in [AkG]. For many things, for example the existence of G_0-invariant metrics, it is important to consider G_0-invariant domains in Ω on which the action of G_0 is proper. One of the main goals of [AkG] is to define such a neighborhood of Ω_0 in Ω.

Recall that a continuous action of a topological group L on a Hausdorff space X is **proper** if the induced map $L \times X \to X \times X$, $(\ell, x) \mapsto (\ell(x), x)$ is a proper mapping. In terms of sequences this means that if $\ell_n(x_n) \to y$ and $x_n \to x$, then, after going to a subsequence, $\ell_n \to \ell \in L$. All isotropy groups L_x are compact for a proper action of L, and all orbits $L(x)$ are closed.

Proposition 6.1.5 ([AkG]). *If $\xi \in \omega_0$ and $x = \exp(i\xi)(x_0)$, then the isotropy group $(G_0)_x$ is the (compact) centralizer $Z_{K_0}(\xi)$. If $x \in \exp(i\mathfrak{a}_0)(x_0)$ is in the boundary of $\exp(i\omega_0)(x_0)$, then $(G_0)_x$ is noncompact.*

This implies that for every $\xi \in \omega_0 \subset T_{x_0}\Omega_0$ the restriction

$$\Pi|_{G_0 \cdot \xi} : G_0 \cdot \xi \to G_0 \cdot \exp(i\xi)(x_0)$$

is an (equivariant) diffeomorphism. To simplify notation, let $\Omega_C := G_0 \cdot \omega_0$ denote an open invariant neighborhood of the 0-section Ω_0 in the tangent bundle $T\Omega_0 = G_0 \times_{K_0} \mathfrak{s}_0$. Using Proposition 6.1.4 and the fact that $T_x(\frac{1}{2}\Omega_u) + T_x(G_0(x)) = T_x\Omega_{AG}$ for all $x \in \frac{1}{2}\Omega_u$ (see [AkG]), one can summarize the above results as follows.

Proposition 6.1.6. *The restriction*

$$\Pi|_{\Omega_C} : \Omega_C \to \Omega_{AG}$$

is a diffeomorphism, the action of G_0 on Ω_{AG} is proper, the differential $d\Pi_x$ at $x \in \mathrm{bd}(\Omega_C)$ is not invertible.

Proof. If $V_{\mathfrak{s}} := K_0 \cdot \omega_0$, then the fact that Π is injective along $V_{\mathfrak{s}}$ and all G_0-orbits, along with its equivariance, implies that it is injective on $\Omega_C = G_0 \times_{K_0} V_{\mathfrak{s}}$. By definition $\Omega_{AG} = G_0 \cdot \Pi(V_{\mathfrak{s}})$ and therefore it is surjective. The above transversality statement, along with the fact that $d\Pi_x$ has maximal rank both along $V_{\mathfrak{s}}$ and all G_0-orbits shows that it is a local diffeomorphism at all points of $V_{\mathfrak{s}}$. Since $\Omega_C = G_0 \cdot V_{\mathfrak{s}}$, Π is equivariant, and the injectivity has already been proved, it follows that it is a global diffeomorphism.

Since the G_0-action on $\Omega_0 = G_0/K_0$ is proper, the induced action on $T\Omega_0$ is likewise proper. Since $\Pi|_{\Omega_C} : \Omega_C \to \Omega_{AG}$ is an equivariant diffeomorphism it follows that the action of G_0 on Ω_{AG} is proper.

Finally, since the isotropy subgroups of G_0 at points of the boundary of $\exp(i\omega_0).x_0$ are noncompact and the boundary of ω_0 in $T_{x_0}\Omega_0$ is mapped to this boundary, it follows that Π is not a local diffeomorphism at any $x \in \mathrm{bd}(\omega_0)$. Therefore the same statement is true for every point of $\mathrm{bd}(\Omega_C) = G_0.\,\mathrm{bd}(\omega_0)$. □

Remark. It should be underlined that $\Pi(\mathrm{bd}\,\Omega_C) \subsetneq \mathrm{bd}\,\Omega_{AG}$. ◇

In its form $G_0 \cdot \exp(i\omega_0)(x_0)$ the domain discussed above was first brought to our attention by the work in [AkG]. As a consequence we originally denoted it by Ω_{AG}. It turns out that it is naturally equivalent to a number of other domains, including Ω_C and the cycle spaces, which are defined from a variety of viewpoints. So now, unless we have a particular construction in mind, we denote

$$\mathcal{U} = \mathcal{U}(G_0) = G_0 \cdot \exp(i\omega_0)(x_0) \tag{6.1.7}$$

to underline its universal character.

Before proceeding to other characterizations of $\mathcal{U}$, we consider a basic concrete example, and indicate the identification of $\mathcal{U}$ with $\mathcal{B} \times \overline{\mathcal{B}}$ in the hermitian case.

6.1C The universal domain in the hermitian case

In order to describe $\mathcal{U}$ in the general hermitian case, it is essentially enough to understand the simplest example.

Example 6.1.8. Let $\langle \cdot, \cdot \rangle$ denote the standard hermitian product of signature (1,1), $\|\cdot\|$ the corresponding norm, $G := SL(2;\mathbb{C})$ and G_0 the real form $SU(1,1) = \mathrm{Isom}(\mathbb{C}^2, \langle \cdot, \cdot \rangle) \cap G$ of G. Identify $X_- \times X_+$ with $\mathbb{P}_1(\mathbb{C}) \times \mathbb{P}_1(\mathbb{C})$ equipped with the diagonal action of G_0. In this way we may assume that $\mathcal{B}$ is the set of negative lines in the first factor and $\overline{\mathcal{B}}$ is the set of positive lines in the second factor. The natural base point x_0 is $([0:1],[1:0]) \in \mathcal{B} \times \overline{\mathcal{B}}$. Here $G_u = SU(2)$ is also acting diagonally and, if $h : \mathbb{C}^2 \times \mathbb{C}^2 \to \mathbb{C}$ denotes the corresponding standard positive definite hermitian product, then $G_u(x_0) = \{([v],[w]) \mid h(v,w) = 0\}$.

The isotropy subgroups of G_0 at points of $\Sigma := G_u(x_0) \cap (\mathcal{B} \times \overline{\mathcal{B}})$ are compact, and therefore $\Sigma \subset \mathcal{U}$. Further, $\Omega := G(x_0)$ is the complement of the diagonal in $X_- \times X_+$ and the Shilov boundary of the polydisk $\mathcal{B} \times \overline{\mathcal{B}}$ intersects Ω in a cylinder, i.e., the complement of the diagonal circle in the 2-torus. This is, in fact, a G_0-orbit with noncompact isotropy. Since $\mathrm{bd}(\Sigma)$ is contained in this orbit, it follows that $\mathcal{U} = G_0.\Sigma$.

In this special case we now show that $\mathcal{U} = \mathcal{B} \times \overline{\mathcal{B}}$, i.e., $G_0.\Sigma = \mathcal{B} \times \overline{\mathcal{B}}$. For this it is convenient to introduce the G_0-invariant function (which looks rather like the Fubini–Study metric)

$$\alpha : \mathcal{B} \times \overline{\mathcal{B}} \to \mathbb{R}^{\geq 0} \text{ defined by } ([v],[w]) \mapsto -\tfrac{|\langle v,w\rangle|^2}{\|v\|^2\|w\|^2}.$$

At the vector space level, if $\|v_1\|^2 = \|v_2\|^2$, $\|w_1\|^2 = \|w_2\|^2$ and $\langle v_1, w_1\rangle = \langle v_2, w_2\rangle$, then there is an isometry $g \in G_0$ for which $g(v_1) = v_2$ and $g(w_1) = w_2$. At the level of projective spaces, for $(([v_1],[w_1]),([v_2],[w_2])) \in \mathcal{B} \times \overline{\mathcal{B}}$ we may choose representatives so that $\|v_1\|^2 = \|v_2\|^2 = 1$, $\|w_1\|^2 = \|w_2\|^2 = -1$ and $\langle v_i, w_i\rangle \in \mathbb{R}^{\geq 0}$ in both cases. Thus $\alpha([v_1],[w_1]) = \alpha([v_2],[w_2])$ if and only if there exists $g \in G_0$ with $g([v_1],[w_1]) = ([v_2],[w_2])$.

Note that $([a:b],[-\overline{b}:\overline{a}])$ with $|a|^2-|b|^2>0$ is a general point of Σ and that $\alpha|_\Sigma:\Sigma\to\mathbb{R}^{\geq 0}$ is surjective. Consequently, α takes on all of its values on Σ and so $G_0.\Sigma=\mathcal{B}\times\overline{\mathcal{B}}$. $\diamond$

The general hermitian case reduces to the above example by applying Harish-Chandra's polydisk slice theorem, which we now recall (also see Chapter 3). Let $\mathfrak{g}_0$ be of hermitian type with Cartan decomposition $\mathfrak{g}_0=\mathfrak{k}_0+\mathfrak{s}_0$. Then every maximal abelian subalgebra $\mathfrak{a}_0\subset\mathfrak{s}_0$ has a basis $\{\xi_1,\ldots,\xi_r\}$ with the following properties.

1. There exist three-dimensional θ-stable commuting subalgebras $\mathfrak{l}_{i,0}\subset\mathfrak{g}_0$ with $\xi_i\in\mathfrak{l}_{i,0}$ for all i.
2. Each $\mathfrak{l}_{i,0}\cong\mathfrak{su}(1,1)$ has induced Cartan decomposition $\mathfrak{l}_{i,0}=\mathfrak{k}_{i,0}+\mathfrak{s}_{i,0}$ with $\mathfrak{a}_i=\langle\xi_i\rangle\subset\mathfrak{s}_{i,0}$.
3. L_0 denotes th analytic subgroup of G_0 with Lie algebra $\mathfrak{l}_0=\mathfrak{l}_{1,0}\oplus\cdots\oplus\mathfrak{l}_{r,0}$. Then $L_0(z_0)$ is a polydisk $D=D_1\times\cdots\times D_r$ embedded as a closed complex submanifold of $\mathcal{B}=G_0/K_0$. Here the disks D_i are orbits $L_{i,0}(z_0)$ of the groups associated to the individual factors.
4. Let L and L_i denote the respective complexifications of L_0 and $L_{i,0}$. Regard $\mathcal{B}$ in its compact dual $X=G/P_-$, where P_- is the isotropy subgroup of G at z_0. Then the orbit $L_i(z_0)=:X_i$ is complex analytically isomorphic to $\mathbb{P}_1(\mathbb{C})$ and $(X_1\times\ldots X_r)\cap\mathcal{B}=D_1\times\ldots D_r$.
5. $K_0.D=\mathcal{B}$.

The following is proved from the point of view of the adapted complex structure in [BHH]. The proof here is essentially that given in [Ha].

Proposition 6.1.9. *If G_0 is of hermitian type, then $\mathcal{U}=\mathcal{B}\times\overline{\mathcal{B}}$.*

Proof. Let X_- and X_+ be the associated compact hermitian symmetric spaces with $\mathcal{B}=G_0(x_-)$ and $\overline{\mathcal{B}}=G_0(x_+)$. If $x_0:=(x_-,x_+)$, then $\Omega=G(x_0)=G/K$ and the product $\mathcal{B}\times\overline{\mathcal{B}}$ is embedded in Ω by Proposition 5.4.3.

Let ω_0 be the defining polyhedron for $\mathcal{U}=\mathcal{U}(G_0)$ and ω_i that for each $\mathcal{U}_i=D_i\times\overline{D}_i$. Since the isotropy subgroup of L_0 at every point of the boundary of $\Sigma=\Sigma_1\times\cdots\times\Sigma_r$ is noncompact and L_0 is closed in G_0, it follows that $\exp(i\omega_0)(x_0)$ is contained in $\exp(i\omega_1)\cdot\cdots\cdot\exp(i\omega_r)(x_0)$.

Conversely, since the action of G_0 on $\mathcal{B}\times\overline{\mathcal{B}}$ is proper, its isotropy groups along the latter set are compact and consequently

$$\exp(i\omega_0)(x_0)=\exp(i\omega_1)\cdots\exp(i\omega_r)(x_0).$$

Therefore, if $D=D_1\times\cdots\times D_r$ and $\mathcal{U}(L_0)=D\times\overline{D}$ is embedded in $\mathcal{B}\times\overline{\mathcal{B}}$ in the obvious way, it is enough to show that $G_0.\mathcal{U}(L_0)=\mathcal{B}\times\overline{\mathcal{B}}$. For this, just notice that if $x\in\mathcal{B}\times\overline{\mathcal{B}}$ is arbitrary, then, after moving it appropriately with an element of G_0, we may assume that its projection in $\mathcal{B}$ is the base point x_-.

Now the isotropy subgroup K_0 of G_0 at x_- acts on the fiber $\{x_-\}\times\overline{\mathcal{B}}$ as the G_0-isotropy at x_- on $\overline{\mathcal{B}}$. Thus, by the Polydisk Theorem, x may be further moved to a point of $\{x_-\}\times\overline{D}$ by an element of K_0. $\square$

Recall $\frac{1}{2}\Omega_u$ from (6.1.2), and its geometric characterization (6.1.3) in the compact symmetric space $\Omega_u = G_u(x_0) \cong G_u/K_0$. That compact symmetric space sits in its complexification $\Omega = G(x_0) \cong G/K$. With a glance at the polydisk one sees that, in Proposition 6.1.9, $\overline{\mathcal{B}} = \frac{1}{2}\Omega_u$ as a G_0–orbit in $\Omega_u = G_u(x_0) = X_+$ and $\mathcal{U} = G_0.\frac{1}{2}\Omega_u \cong \mathcal{B} \times \overline{\mathcal{B}}$ by $g_0 g_u(x_0) = (g_0(x_0), g_u(x_0)) = (g_0(x_-), g_u(x_+))$. The projection $\mathcal{B} \times \overline{\mathcal{B}} \to \overline{\mathcal{B}}$ is given by $g_0 g_u(x_0) \mapsto g_u(x_0)$.

6.2 Adapted structure

The adapted complex structure and its maximal domains of existence in the tangent bundle of a riemannian manifold N are, in general, difficult to compute. Here we describe these in the case $N = \Omega_0 = G_0/K_0$ via a natural identification with $\mathcal{U}$.

6.2A Background

If (N, g) is a riemannian manifold, which for simplicity we first assume to be complete, then a parameterized geodesic $\gamma : \mathbb{R} \to \Omega_u$ induces a map $\gamma' : T\mathbb{R} \to TN$. We refer to its image as the riemannian leaf through $\gamma'(0)$ in TN.

Let us identify $T\mathbb{R}$ with the complex numbers so that the base, or 0-section, corresponds to the $\mathbb{R}$-axis and the fiber over $0 \in \mathbb{R}$, i.e., $T_0\mathbb{R}$, to the $i\mathbb{R}$-axis. In this way a point $z = t + is \in \mathbb{C}$ corresponds to the tangent vector $s\frac{d}{dt}\big|_t$.

An integrable complex structure on a domain D in $T\Omega_u$ is said to be **adapted** if, after identifying $T\mathbb{R}$ with $\mathbb{C}$, wherever it makes sense, every lifted geodesic γ' is holomorphic. This definition is clearly local and therefore γ need not be globally defined.

Assuming that N and g are real analytic, it is known that on sufficiently small neighborhoods of points in the 0-section the adapted complex structures exist and are unique. These can also be characterized as the unique complex structures with the property

$$d^c \|\cdot\|_g^2 = \lambda_g \text{ where, as usual, } d^c f := J(df) = df \circ J^{-1}.$$

Here $\|\cdot\|_g^2$ is the norm squared function on TN which is defined by g, and λ_g is the pullback of the standard Liouville form on T^*N via the identification $TN \cong T^*N$ defined by g. It follows that $dd^c\|\cdot\|_g^2$ is the pullback ω_g of the standard symplectic form and $\|\cdot\|_g^2$ is strictly plurisubharmonic, i.e., ω_g is kählerian.

We refer to a domain in TN as **starlike** if it is invariant under the contractions defined by scalar multiplication. The following was noted in [Sz] in a slightly restricted setting, but the proof is valid in the general starlike case (see [Ha]).

Proposition 6.2.1. *Let (N, g) be real analytic and D be a starlike domain equipped with an adapted complex structure J. Then J is real analytic with respect to the standard real analytic structure of TN.*

Corollary 6.2.2. *An adapted structure on a starlike domain D is unique and there exists a unique maximal starlike domain of existence of the adapted structure.*

Proof. This follows from the identity principle and the fact that adapted structure is unique near the 0-section. The second statement follows from the first and the fact that intersections and unions of starlike domains are starlike. □

6.2B $\Pi : \Omega_{\text{adpt}} \to \mathcal{U}$ is biholomorphic

Let us restrict to the context of $N = \Omega_0 = G_0/K_0$ equipped with the invariant metric defined by the Killing form. Define Ω_{adpt} to be the maximal starlike domain of existence for the adapted complex structure J.

Since geodesics γ are simply orbits of one-parameter groups, it is clear that the restriction $\Pi|_{L_\gamma}$ to the riemannian leaf L_γ equipped with its adapted complex structure is holomorphic. Thus, by the uniqueness of the adapted structure on Ω_{adpt} it follows that

$$\Pi|_{\Omega_{\text{adpt}}} : \Omega_{\text{adpt}} \to \Omega$$

is holomorphic.

Using the fact that $\Pi|_{\Omega_C}$ is a diffeomorphism onto its image, Ω_C possesses the pullback complex structure coming from Ω. Since the riemannian leaves correspond to the orbits of the complex one-parameter groups in Ω, this structure is clearly adapted. Thus $\Omega_C \subset \Omega_{\text{adpt}}$.

The following was proved in [BHH] via an explicit calculation of J in a natural basis of Jacobi fields. Here we follow another proof in [Ha].

Proposition 6.2.3. *The domains Ω_{adpt} and Ω_C are the same and*

$$\Pi : \Omega_{\text{adpt}} \to \mathcal{U}$$

is biholomorphic.

Proof. By Proposition 6.1.6 the differential $\Pi_*(x)$ has nonmaximal rank for all $x \in \text{bd}(\Omega_C)$. If $\Omega_{\text{adpt}} \supsetneq \Omega_C$, this would be contrary to Π being holomorphic at such points, because the set of degeneracy of Π would be a complex variety of real codimension two, whereas $\text{bd}(\Omega_C)$ has codimension 1. □

6.3 Invariant CR structure and pseudoconvexity

A main goal of this section is to prove a basic result of [BHH] on the existence of G_0-invariant strictly plurisubharmonic functions on $\mathcal{U}$. For this it we must introduce some notation.

Let ω_0 be as above with $\mathcal{U} := G_0.\exp(i\omega_0).x_0$. The Weyl-group $W := N_{K_0}(\mathfrak{a}_0)/Z_{K_0}(\mathfrak{a}_0)$ acts on ω_0 in the usual way.

Let $C(\mathcal{U})^{G_0}$ denote the space of G_0-invariant real-valued continuous functions on $\mathcal{U}$, and let $C(\omega_0)^W$ be the algebra of continuous W-invariant functions. Restriction canonically defines an injective map

$$\mathcal{R} : C(\mathcal{U}) \to C(\omega_0)^W.$$

Its inverse is the *extension* map

$$\mathcal{E} : C(\omega_0)^W \to C(\mathcal{U})^{G_0}$$

which is defined by $\mathcal{E}(f)(g(x)) = f(x)$ for $x \in \omega_0$ and $g \in G$. Thus these maps are easily seen to be (topological) isomorphisms.

Here we are interested in restriction of strictly plurisubharmonic functions and extension of strictly convex functions. In principle such functions need only be continuous (or just semicontinuous), but we wish to compute derivatives, e.g., Hessians, and therefore we must consider the restrictions of $\mathcal{R}$ and $\mathcal{E}$ to C^∞ functions. In fact they are also isomorphisms (see, e.g., [HH]). In this case something must be proved, because it is not completely clear that the extension of a smooth function is smooth.

Let $SPSH$ denote the cone of smooth G_0-invariant strictly plurisubharmonic functions on $\mathcal{U}$, and SC the smooth W-invariant strictly convex functions on ω. Recall a smooth function f is strictly plurisubharmonic is the complex Hessian $dd^c f$ is positive definite, and that strong convexity on ω_0 is defined by the positive definiteness of the real Hessian in the vector space coordinates given by the open embedding of ω_0 in $\mathfrak{a}_0$.

If A is as usual the complexification of A_0, then, wherever it is defined, the restriction of a strictly plurisubharmonic function on $\mathcal{U}$ is strictly plurisubharmonic on the orbit $A(x_0)$. A comparison of the real and complex Hessians on $A(x_0)$ shows that an A_0-invariant function is strictly plurisubharmonic on $A(x_0)$ if and only if it is strongly convex when regarded as a function on $i\mathfrak{a}_0$. This is a local statement, and consequently restriction defines a mapping

$$\mathcal{R} : SPSH \to SC.$$

The important question for us goes in the converse direction. In this regard, the goal of this section is to show that the image of $\mathcal{E} : SC \to C^\infty(\mathcal{U})^{G_0}$ is exactly the space $SPSH$. This yields the following result.

Theorem 6.3.1 ([BHH]). *The mapping $\mathcal{E} : SC \to SPSH$ is an isomorphism.*

We prove this by analyzing the Levi geometry of G_0-invariant hypersurfaces in $\mathcal{U}$, and hope that this viewpoint is of interest in its own right. Before getting on with the proof let us note an important consequence.

For this observe that since the G_0-action on $\mathcal{U}$ is proper, it makes sense to consider G_0-invariant functions on $\mathcal{U}$ as functions on the Hausdorff orbit space $\mathcal{U}/G_0$. If ρ is G_0-invariant on $\mathcal{U}$ and the induced function on $\mathcal{U}/G_0$ is an exhaustion, then we refer to ρ as an **exhaustion modulo** G_0.

The restriction of a G_0-invariant strictly plurisubharmonic function ρ on $\mathcal{U}$ to the part of the orbit $A(x)$ which is contained in $\mathcal{U}$ is strictly plurisubharmonic and strongly convex along $\exp(i\mathfrak{a}_0)$-orbits. Furthermore, since such a restriction is K_0-invariant, it must be symmetric along the lines orthogonal to the reflection hyperplanes, and therefore has an absolute minimum along these lines at the base point x_0. If $\xi \in \mathfrak{a}_0$ it follows that the restriction $\rho|_{\exp(i\mathbb{R}\xi)(x_0)\cap\mathcal{U}}$ is an exhaustion with absolute minimum at the base point.

Such exhaustions along lines are in general not exhaustions of the polyhedral domain $\exp(\omega_0)$. However, applying the above theorem to a function $u \in SC$ which *is* an exhaustion, we have the following result.

Corollary 6.3.2. *There exist strictly plurisubharmonic functions on $\mathcal{U}$ that are exhaustions modulo G_0.*

Corollary 6.3.3 ([BHH]). *The domain $\mathcal{U}$ is Stein.*

Proof. Let Γ be a cocompact discrete subgroup of G_0 acting freely on G_0/K_0. Then Γ acts properly and freely on $\mathcal{U}$. A strictly plurisubharmonic function ρ on $\mathcal{U}$ which is an exhaustion modulo G_0 induces a strictly plurisubharmonic exhaustion of $\Gamma\backslash\mathcal{U}$. Thus this manifold is Stein. Since covering spaces of Stein manifolds are Stein, it follows that $\mathcal{U}$ is likewise Stein. □

6.3A The G_0-orbits as CR submanifolds

We now turn to a brief study of the invariant CR structures on the G_0-orbits in $\mathcal{U}$ and related results which then yield a proof of Theorem 6.3.1. The proof here emphasizes aspects which are slightly different from those in [BHH].

Observe that if ξ is a regular element in $\mathfrak{a}_0$ and $x = \exp(i\xi).x_0$, then the centralizer $\mathfrak{m}_0$ of $\mathfrak{a}_0$ in $\mathfrak{k}_0$ agrees with the isotropy algebra $(\mathfrak{g}_0)_x$. Therefore it is convenient to write the Iwasawa decomposition of $\mathfrak{g}_0$ as

$$\mathfrak{g}_0 = \theta\mathfrak{n}_0 + (\mathfrak{m}_0 + \mathfrak{a}_0) + \mathfrak{n}_0.$$

This is a vector space direct sum, where $\mathfrak{n}_0$ denotes the direct sum $\sum_{\alpha>0} \mathfrak{g}_\alpha$ of positive root spaces. Here the positive $\mathfrak{a}_0$-root system corresponds to a Weyl chamber that contains all of the $\xi \in \mathfrak{a}_0$ under discussion.

For $x = \exp(i\xi)(x_0)$ we consider the orbit $M := G_0(x)$ as a CR submanifold of $\mathcal{U}$ and compute its tangent and CR tangent spaces at x. The CR tangent space of a real submanifold M in a complex manifold Z is defined to be the intersection $T_x^{CR}(M) := T_x(M) \cap JT_x(M)$, where J is the complex structure of Z. If the dimension of this space is constant along M, then one refers to M as a Cauchy–Riemann or **CR manifold**. If M is a G_0-orbit as in the case of present interest, then, since the complexification G of G_0 acts transitively on Z, it follows that $T_xZ = T_xM + JT_xM$. Such CR submanifolds of Z are called **generic**. In the present monograph we use the notation and various points of view of the subject of CR manifolds, but in fact use none of its basic results, e.g., on analytic continuation (see [BER] for a systematic treatment and

[BF, KZ] for exemplary analytic results in a group action context). These manifolds have long been useful in the representation theory of real semisimple Lie groups [W3].

Throughout, if $\eta \in \mathfrak{g}_0$, then $\widehat{\eta}$ denotes the associated field on $\mathcal{U}$ and $\widehat{\eta}(x)$ its evaluation at x.

Lemma 6.3.4. *If $\eta \in \mathfrak{g}_\alpha \subset \mathfrak{n}_0$, then*

$$\widehat{\eta}(x) = e^{-i\alpha(\xi)} \exp(i\xi)_*(\widehat{\eta}(x_0)) \quad \text{and} \quad \widehat{\theta\eta}(x) = e^{i\alpha(\xi)}(\exp(i\xi)_*(\widehat{\theta\eta}(x_0)).$$

If $\alpha(\xi) \neq 0$, then $\widehat{\eta}(x)$ and $\widehat{\theta\eta}(x)$ span a complex line in $T_x\Omega$ which is contained in T_xM. If $\alpha(\xi) = 0$, then $\widehat{\eta}(x) = -\widehat{\theta\eta}(x)$.

Proof. The formulas are immediate consequence of $\mathrm{Ad} \circ \exp = e^{\mathrm{ad}}$. The conclusion concerning the complex line then follows from the fact that $\eta + \theta\eta \in \mathfrak{k}_0$ and therefore the corresponding field vanishes at x_0. □

Now define

$$\mathfrak{n}^0_{\xi,0} = \sum\nolimits_{\alpha>0,\alpha(\xi)=0} \mathfrak{g}_\alpha \text{ and } \mathfrak{n}^1_{\xi,0} = \sum\nolimits_{\alpha>0,\alpha(\xi)\neq 0} \mathfrak{g}_\alpha.$$

Both are subalgebras of $\mathfrak{n}_0$, $\mathfrak{n}^1_{\xi,0}$ is an ideal in $\mathfrak{n}_0$, $\mathfrak{n}_0 = \mathfrak{n}^1_{\xi,0} + \mathfrak{n}^0_{\xi,0}$ semidirect sum, and $\mathfrak{a}_0 + \mathfrak{n}^0_{\xi,0} + \theta(\mathfrak{n}^0_{\xi,0})$ is the centralizer of ξ in $\mathfrak{g}_0$.

It follows that

$$(\mathfrak{g}_0)_x = \mathfrak{z}_{\mathfrak{k}_0}(\xi) = \mathfrak{m}_0 + \mathfrak{d}_\xi,$$

where $\mathfrak{d}_\xi$ is the span of the $\eta + \theta\eta$ for $\eta \in \mathfrak{n}^0_{\xi,0}$. In particular

$$\mathrm{codim}_\Omega G_0(x) = \dim\ \mathfrak{a}_0 + \dim \mathfrak{n}^0_{\xi,0}.$$

Now let N^1_ξ be the complex Lie group associated to $\mathfrak{n}^1_\xi := \mathfrak{n}^1_{\xi,0} + i\mathfrak{n}^1_{\xi,0}$. Observe that $T_x(N^1_\xi(x))$ is also generated by the complex lines in the above lemma, and therefore

$$T_x(N^1_\xi(x)) \subset T^{CR}_x(M_x) := T_x(M_x) \cap J(T_x(M_x)).$$

We identify elements of $\mathfrak{g}_0$ with tangent vectors at x by $\eta \mapsto \widehat{\eta}(x)$.

Proposition 6.3.5. *If $\xi \in \omega_0$ and $x = \exp(i\xi)(x_0)$, then the Cauchy–Riemann tangent space $T^{CR}_x(M_x)$ agrees with $T_x(N^1_\xi(x))$ which in turn is identified with $\eta^1_{\xi,0} \oplus \theta\eta^1_{\xi,0}$. Furthermore,*

$$T_x(M_x) \cong \mathfrak{a}_0 + \mathfrak{n}^0_{\xi,0} + (\mathfrak{n}^1_{\xi,0} + \theta\mathfrak{n}^1_{\xi,0}).$$

Proof. The expression for T_xM_x follows from the above formula for the isotropy algebra $(\mathfrak{g}_0)_x$, and therefore we must only check that $T^{CR}_xM_x = T_xN^1_\xi(x)$. If $\eta \in \mathfrak{a}_0 + \mathfrak{n}^0_{\xi,0}$, then $i\widehat{\eta}(x)$ is tangent to the fiber $K_0 \exp(i\omega).x_0$. Since all G_0-orbits in $\mathcal{U}$ are transversal to this fiber, it follows that under the identification with a subspace of T_xM_x, the space $\mathfrak{a}_0 + \mathfrak{n}^0_{\xi,0}$ has empty intersection with $T^{CR}_xM_x$. But it is complementary to a space which is contained in $T^{CR}_xM_x$, and therefore the desired result follows. □

Remark 6.3.6. N^1_ξ is the unipotent radical of the parabolic group $P_\xi = Z_G(\xi).N^1_\xi$. ◇

If M is a CR manifold, then its intrinsic vector-valued Levi form

$$\mathcal{L}_M(m) : T^{CR}_m M \times T^{CR}_m M \to T_m M / T^{CR}_m M$$

at a point $m \in M$ is defined by

$$\mathcal{L}_M(m)(v, w) = [v, Jw](m) \ (\mathrm{mod}\ T^{CR}_m M).$$

Here $[v, Jw]$ is computed as follows. Extend v and w to vector fields $v^\dagger$ and $w^\dagger$ on M, let $[v^\dagger, Jw^\dagger]$ be the usual Lie bracket of vector fields, and then $[v, Jw](m) := [v^\dagger, Jw^\dagger](m)$.

Proposition 6.3.7. *For every $x \in \mathcal{U}$ the Levi form $\mathcal{L}_M(x)$ of the G_0-orbit $M := M_x$ is nondegenerate.*

Proof. The orbit $M = G_0(x)$ is contained in the level set of the norm function $\|\cdot\|^2$ at x. Since this function is strictly plurisubharmonic, it follows that its Levi form is positive definite. Thus $L_M(x)(v, v) \neq 0$ for all $v \in T^{CR}_x M$, and in particular $\mathcal{L}_M(x)$ is nondegenerate. □

Our eventual goal is to describe certain properties the Levi form of a G_0-invariant real hypersurface in $\mathcal{U}$. For this it is important to observe that the CR tangent space $T^{CR}_x M = T_x(N^1_\xi(x))$ has a natural complement in $T_x\Omega$, namely the tangent space of the orbit $Z_G(\xi)(x)$ of the reductive part of the parabolic group $P = Z_G(\xi).N^1_\xi$. In fact the real points $Z_{G_0}(\xi)(x)$ of this orbit define a natural complement to $T^{CR}_x(M_x)$ in $T_x(M_x)$.

To see this, first observe that $Z_G(\xi) = Z_G(i\xi)$ is τ-invariant; in particular, it has $Z_{G_0}(\xi)$ as a real form. Furthermore, since $x = \exp(i\xi)(x_0)$ and $G_{x_0} = K$, it is immediate that $Z_G(\xi)_x = Z_K(\xi)$. Since $Z_{G_0}(\xi)_x = Z_{K_0}(\xi)$, which we saw in the proof that $(\mathfrak{g}_0)_x = \mathfrak{z}_{\mathfrak{k}_0}(\xi_{reg}) + \mathfrak{d}_{\xi,0}$, we are in the same situation as our initial one with $Z_{G_0}(\xi)(x) = Z_{G_0}(\xi)/Z_{K_0}(\xi)$ being the real points of $Z_G(\xi)(x) = Z_G(\xi)/Z_K(\xi)$.

Proposition 6.3.8. *The tangent spaces split as follows:*

$$T_x\Omega = T_x(Z_G(\xi)(x)) \oplus T^{CR}_x M_x = T_x(Z_G(\xi)(x)) \oplus T_x(N^1_\xi(x))$$
$$\text{and } T_x(M_x) = T_x(Z_{G_0}(\xi)) \oplus T^{CR}_x M_x.$$

Proof. $T_x\Omega = T_x(A(x)) \oplus T_x(N(x)) = T_x(A(x)) \oplus T_x(N^0_\xi(x)) \oplus T_x(N^1_\xi(x)) = T_x(Z_G(\xi)(x)) \oplus T_x(N^1_\xi(x))$. Since $T^{CR}_x(M_x) = T_x(N^1_\xi(x))$, the first equality follows. The second then follows from the identifications

$$T_x(Z_{G_0}(\xi)(x)) = \widehat{\mathfrak{a}}_0(x) + \widehat{\mathfrak{n}}^0_0$$

and

$$T^{CR}_x(M_x) = T_x(N^1_\xi(x)) = \widehat{\mathfrak{n}}^1_{\xi,0}(x) + \widehat{\theta\mathfrak{n}}^1_{\xi,0}(x)$$

together with Remark 6.3.6. □

6.3B Invariant CR hypersurfaces

Whenever $\rho : \mathcal{U} \to \mathbb{R}$ is a G_0-invariant smooth function, $x = \exp(i\xi)(x_0)$, and $d\rho(x) \neq 0$, we refer to the level set $H = \{\rho = \rho(x)\}$ as an **invariant hypersurface**. We now compute the part of $T_x^{CR} H$ determined by the action of G_0.

If $\widetilde{A}$ is the center of the centralizer $Z_G(\xi)$, then $Z_G(\xi) = Z'_G(\xi).\widetilde{A}$ with $\widetilde{A} \cap Z'_G(\xi)$ finite. Of course $\widetilde{A}$ contains A and $\mathfrak{z}'_{\mathfrak{g}_0}(\xi)$ is generated as a Lie algebra by $\mathfrak{n}^0_{\xi,0} + \theta\mathfrak{n}^0_{\xi,0}$.

For $\eta \in \mathfrak{n}^0_{\xi,0}$, let $\mathfrak{l}_0(\eta)$ be the three-dimensional algebra generated by η and $\theta\eta$, and let $L_0(\eta)$ be the associated subgroup of G_0. Since in this case $\eta + \theta\eta \in (\mathfrak{g}_0)_x$, its orbit $L_0(\eta)(x)$ is a totally real copy of the unit complex disk contained in $G_0(x)$. The orbit $L(\eta)(x)$ of the associated complex group is a two-dimensional complex affine quadric which we denote by $Q_2(\eta)$.

Proposition 6.3.9. *If H is a G_0-invariant CR hypersurface in $\mathcal{U}$, then*

$$T_x Q_2(\eta) \subset T_x^{CR} H.$$

Proof. Since $L_0(\eta)(x) \subset H$, the real space $T_x(L_0(\eta))$ is contained in $T_x H$. Note that for any Iwasawa decomposition $\mathfrak{l}_0(\eta) = \mathfrak{k}_0(\eta) + \mathfrak{a}_0(\eta) + \mathfrak{n}_0(\eta)$ the space $i\widehat{\mathfrak{a}}_0(\eta)$ is invariant with respect to the Cartan involution of $\mathfrak{l}_0$, which is conjugation by an element of L_0. Thus the defining function ρ is symmetric along the orbit $\exp(i\mathfrak{a}_0(\eta))(x)$ and consequently $d\rho$ also vanishes on $i\widehat{\mathfrak{a}}_0(\eta)$. Since linear combinations of spaces of this type fill out the remaining directions in $T_x(L(\eta)(x))$, the desired result follows. □

Corollary 6.3.10. *The tangent space $T_x(Z'_G(\xi).N^1_\xi(x))$ is contained in the CR tangent space of every invariant hypersurface H.*

Proof. The algebra $\mathfrak{z}'_{\mathfrak{g}}(\xi)$ is spanned by the three-dimensional algebras $\mathfrak{l}(\eta)$. □

Let $C_x := T_x^{CR} H \cap T_x(Z_G(\xi)(x))$ and note that by the splitting results above it follows that $T_x^{CR} H = C_x \oplus T_x^{CR} M_x$.

Proposition 6.3.11. *The splitting $T_x^{CR} H = C_x \oplus T_x^{CR} M_x$ is orthogonal with respect to the intrinsic Levi form $\mathcal{L}_H$.*

Proof. Extend $v \in C_x$ to $v^\dagger = \sum(a_i v_i^\dagger + b_i J v_i^\dagger)$, where the $v_i^\dagger$ are invariant fields coming from $\mathfrak{z}_{\mathfrak{g}_0}(\xi)$. (For $w \in T_x(N^1_\xi(x))$ we recall that it is a linear combination of vectors $\widehat{\eta}(x)$ and $\widehat{\theta\eta}(x)$, where $\eta \in \mathfrak{n}^1_0$. Thus we extend it as a $\mathfrak{g}_0$-field $w^\dagger$ which is the same linear combination.)

To show that $\mathcal{L}_H(v, w) = 0$ we observe that $[v^\dagger, Jw^\dagger](x)$ is contained in $T_x(N^1_\xi(x))$. This follows immediately from the way the vectors have been extended as fields and the fact that $\mathfrak{z}_{\mathfrak{g}_0}(\xi)$ normalizes the vector space $\mathfrak{n}^1_{\xi,0} + \theta\mathfrak{n}^1_{\xi,0}$, which when evaluated at x becomes the (J-invariant) CR tangent space $T_x(N^1_\xi(x))$. □

6.3C Geometry of the full Levi form $dd^c\rho$

As above, we consider a smooth G_0-invariant function ρ on $\mathcal{U}$ and $x = \exp(i\xi).x_0$ with $\xi \in \omega_0$ and $d\rho(x) \neq 0$. Note that in the case of a CR hypersurface, the vector-valued Levi form takes values in a one-dimensional vector space and therefore is a bilinear form in the usual sense; in that case it is of the form $dd^c\rho$.

Assume further that $\exp(i\xi t)$ is transversal to the level set $H := \{\rho = \rho(x)\}$ at x and define P to be the complex curve $\exp(i\xi z)(x_0)$, $z \in \mathbb{C}$.

Proposition 6.3.12. *The decomposition $T_x\Omega = T_xP \oplus T_x^{CR}H$ is orthogonal with respect to the full Levi form $dd^c\rho(x)$.*

Proof. In order to show that $dd^c\rho(x)(v, w) = 0$ for $v \in T_xP$ and $w \in T_x^{CR}H$ we must extend these vectors to vectors $v^\dagger$ and $w^\dagger$ and prove that

$$v^\dagger Jw^\dagger(\rho) + w^\dagger Jv^\dagger(\rho) - J[v^\dagger, Jw^\dagger](\rho) \tag{6.3.13}$$

vanishes at x. Take $v^\dagger$ to be a G-field coming from the center $\widetilde{\mathfrak{a}}$ of $\mathfrak{z}_\mathfrak{g}(\xi)$. Decompose $w = y + z$ according to the splitting in Proposition 6.3.11.

For $\widetilde{x} \in A(x_0)$ we extend y to a $Z_G(\xi)$-field $y^\dagger$ in the complementary space $C_{\widetilde{x}}$. In particular $y^\dagger(\rho) = Jy^\dagger(\rho) = 0$ along $A(x_0)$. Similarly we extend z to a N_ξ^1-field which is CR tangent to every ρ-level set along $A(x)$. Since the one-parameter group which defines $v^\dagger$ commutes with $Z_G(\xi)$ and normalizes N_ξ^1, all terms of (6.3.13) vanish. □

Recall the extension isomorphism $\mathcal{E} : C^\infty(\omega_0)^W \to C^\infty(\mathcal{U})^{G_0}$.

Theorem 6.3.14. *If u is strictly convex and $\rho = \mathcal{E}(u)$, then $dd^c\rho(x) > 0$.*

Proof. Observe that $f : \mathbb{C} \to \mathbb{R}$, $z \mapsto \rho(\exp(z\xi))$, is $\mathbb{R}$-invariant and strictly convex along $i\mathbb{R}$. Thus $dd^c\rho(x) > 0$ on the perpendicular space P_x.

Secondly, we note that ρ is strictly plurisubharmonic at $x = \exp(i\xi)$ if ξ is sufficiently small. To see this, it is convenient to consider $\mathcal{U}$ as a neighborhood of $\Omega_0 = G_0/K_0$ in the tangent bundle $T\Omega_0$ equipped with the adapted structure. For $\xi = 0$, i.e., along the 0-section Ω_0, we use the identification of $J(T_{x_0}\Omega_0)$ with the tangent space of the fiber of $T\Omega_0$, i.e., with $T_{x_0}\Omega_0$, and use the invariance of ρ to identify the complex Hessian $dd^c\rho(x_0)$ with the real Hessian of the restriction of ρ to that fiber.

Now this restriction is strongly convex on ω_0, is K_0-invariant and $K_0.\omega_0 = T_{x_0}\Omega_0$. Thus ρ is strictly convex along every line in the fiber $T_{x_0}\Omega_0$ and therefore the Hessian of the restriction of ρ is positive definite. Hence, as we indicated before, ρ is indeed strictly plurisubharmonic in some neighborhood of Ω_0. Hence, it is enough to show that the intrinsic Levi form $\mathcal{L}_H$ is nondegenerate for all such x.

For this we use the orthogonal splitting $T_x^{CR}H = C_x \oplus T_x^{CR}M_x$ of Proposition 6.3.11. Identifying $\mathfrak{g}_0$ with its dual by the Killing form, $T_x^{CR}M_x$ can be regarded as the tangent space at ξ of the adjoint orbit $\mathrm{Ad}(G_0)(\xi)$. Since the (nondegenerate) coadjoint symplectic form is defined in this way by the projection of Lie brackets

$[\gamma, \delta]$, where $\gamma, \delta \in \mathfrak{n}^1_{\xi,0} \oplus \theta\mathfrak{n}^1_{\xi,0}$, to the space generated by ξ, we see that $\mathcal{L}_H$ is nondegenerate on $T^{CR}_x M$.

For the part C_x we note that it is the tangent space of an orbit $Z'_G(\xi).\widetilde{A}^1(x)$, where $\widetilde{A}^1 \subset \widetilde{A}$ is the complex subgroup defined by $T_x(\widetilde{A}(x) \cap H) = T_x(\widetilde{A}^1(x))$. The restriction of ρ to this orbit is of the form $\widetilde{\mathcal{E}}(\widetilde{u})$ where $\widetilde{u}$ is strictly convex, but the new base point x plays the role of the old $\widetilde{x}_0$ of this complex homogeneous space equipped with a real form as in our basic situation. Thus this restriction is strictly plurisubharmonic near x. Consequently $\mathcal{L}_H$ is nondegenerate on C_x as well. □

This completes the proof of Theorem 6.3.1, which we stated in the introduction of this section.

7

B-Invariant Hypersurfaces in $\mathcal{M}_Z$

Our main tool for relating the cycle spaces $\mathcal{M}_D$ of G_0 flag domains to the universal domain $\mathcal{U} = \mathcal{U}(G_0)$ is a sort of incidence geometry that pairs cycles in D with certain Schubert varieties of complementary dimension. The resulting incidence varieties are algebraic subvarieties of $\mathcal{M}_Z$ which are invariant under certain special Borel subgroups of G. We begin here with first definitions and background information on these groups, their invariant varieties, and basic aspects of the intersection theory.

Section 1 begins with a summary of the basic facts on actions of Borel subgroups $B \subset G$ on the varieties at hand, in particular on G/Q and G/K. After this general introduction we restrict to Iwasawa Borel subgroups, i.e., those that contain the solvable part A_0N_0 of an Iwasawa decomposition $G_0 = K_0A_0N_0$.

If $\mathcal{O}$ is a B-orbit on Z, then the associated Schubert variety is defined as its closure $S := c\ell(\mathcal{O}) = \mathcal{O}\dot{\cup}Y$. The Schubert varieties of dimension $q := \dim_{\mathbb{C}} C_0$, whose intersection with C_0 is not empty, play a fundamental role throughout this monograph. The key is that the associated incidence varieties

$$H := I_Y := \{C \in \mathcal{M}_Z \mid C \cap Y \neq \emptyset\}$$

are B-invariant analytic hypersurfaces in the complement $\mathcal{M}_Z \setminus \mathcal{M}_D$ (Proposition 7.4.11).

In Section 2 we introduce the notion of the envelope $\mathcal{E}_{\mathcal{H}}$ defined by a B-invariant hypersurface H in $\mathcal{M}_Z$. It is the topological component of the base point in the complement $\mathcal{M}_Z \setminus (\bigcup_{g \in G_0} g(H))$. For later purposes in the hermitian holomorphic case, where $\mathcal{M}_Z$ is the compact dual of the bounded symmetric domain $\mathcal{B} = G_0/K_0$, we show that $\mathcal{E}_{\mathcal{H}} = \mathcal{B}$ (Proposition 7.2.2). In the nonholomorphic cases, where $\mathcal{M}_D$ is contained in an at most finite quotient of the affine symmetric space G/K, it is shown in Section 2 that $\mathcal{E}_{\mathcal{I}} = \mathcal{U}$ (Theorem 7.2.7). One of our main results, in Chapter 11 (Corollary 11.3.2), is that $\mathcal{E}_{\mathcal{H}} = \mathcal{U}$ for every B-invariant hypersurface H in G/K which is not a lift.

Section 3 is devoted to proving the basic facts on Iwasawa–Schubert slices Σ; see Propositions 7.3.7 through 7.3.11. If S is a q-codimensional Schubert variety which has nonempty intersection with C_0, then the associated Schubert slices Σ are defined to be the connected components of the intersection $S \cap D$. For example, it is shown that the intersection $\Sigma \cap C_0$ is transversal and consists of exactly one point z, and $\Sigma = A_0 N_0(z)$. Furthermore, Σ is closed in D and its closure $c\ell(\Sigma)$ in Z has nonempty intersection with every K_0-orbit in $c\ell(D)$.

The main goal of Section 4 is to prove the result on incidence hypersurfaces mentioned above (Proposition 7.4.11). However in Section 4 we also take the opportunity to introduce the basic notions of Barlet's cycle space theory and even to indicate some of the ideas that go into the construction of that cycle space. We also prove the basic properties of the trace transform (Section 7.4B); they could well be useful in semisimple representation theory.

The viewpoint presented here, as well as the consequences and developments in Chapter 9, was initiated in [HS] for the case of $SL(n;\mathbb{R})$, presented with a certain degree of generality in [H], and developed to their present form in [HW3] by use of methods described in Chapter 8.

7.1 Iwasawa–Borel subgroups and their Schubert varieties

7.1A Spherical varieties

Recall that by B we denote a Borel subgroup of a complex reductive Lie group G, i.e., a maximal connected, solvable subgroup. Such is a complex algebraic subgroup of G and any two are conjugate. If $Z = G/Q$ is a projective G-homogeneous manifold, then as a consequence of Borel's fixed point and normalizer theorems, every B has a unique fixed point in Z.

Remark 7.1.1. In this section we will continue to speak of reductive algebraic groups, but it should be noted that the center of G fixes Z pointwise, and consequently in this context it would be sufficient to consider the semisimple case. ◇

Definition 7.1.2. If G is a complex reductive group, then an irreducible algebraic G-variety X is **spherical** if one (hence every) Borel subgroup B of G has an open orbit in X.

The following is a first basic theorem in the subject of spherical G-varieties (see, e.g., [Br]).

Proposition 7.1.3. *If X is a spherical G-variety and B is a Borel subgroup of G, then B has only finitely many orbits in X.*

As abstract manifolds B-orbits are quite simple.

Proposition 7.1.4. *Let S be a connected, complex algebraic group and $H \subset S$ a complex algebraic subgroup. If S is solvable, then the homogeneous space $X = S/H$ is algebraically equivalent to $(\mathbb{C}^*)^m \times \mathbb{C}^n$.*

Proof. Let $S = L \ltimes U$ be a Levi decomposition of S. Here U is the unipotent radical of S and L is a maximal reductive subgroup. Then U is isomorphic to some $\mathbb{C}^n$, and L is isomorphic to $(\mathbb{C}^*)^r$. Thus L is a maximal complex torus in S. If $U = \{1\}$, then S is a complex torus, and the assertion is immediate.

If $U \neq \{1\}$, let Z_U be the identity component of its center and consider the principal bundle

$$X = S/H \to S/Z_U H = (\mathbb{C}^*)^k \times \mathbb{C}^\ell := Y,$$

where the identification of the base with $(\mathbb{C}^*)^k \times (\mathbb{C})^\ell$ follows by the induction assumption. Since the structure group of this bundle is an abelian algebraic group isomorphic to some $(\mathbb{C}^p, +)$ and $H^1(Y; \mathcal{O}^p) = 0$, the bundle $X \to Y$ is algebraically trivial. The assertion follows. □

The two types of homogeneous spaces of complex Lie groups that are relevant for this book, the rational projective manifolds G/Q and the affine homogeneous spaces G/K, are spherical varieties. In the former case this is noted above. In the latter case it follows from the fact that if $G_0 = A_0 N_0 K_0$ is an Iwasawa decomposition and B is a Borel subgroup of G that contains $A_0 N_0$, then BK is open in G.

As pointed out in the Bruhat Lemma 2.1.1, the B-orbits in G/B, and thus also in G/Q, are holomorphically equivalent to complex vector spaces $\mathbb{C}^{\dim \mathcal{O}}$, and, in particular, are topological cells. On the other hand, B-orbits on G/K need not be cells.

Example 7.1.5. Let $G = SL(2; \mathbb{C})$. Then $K \cong \mathbb{C}^*$, and a given Borel subgroup B of G has trivial intersection with a generic conjugate of K. Thus the open B-orbit in G/K is complex algebraically isomorphic to $\mathbb{C}^* \times \mathbb{C}$. ◇

We emphasize the algebraic nature of the setting described just above. Let us formally state the basic properties that we use without further discussion in what follows.

Fact. *Let G be a complex algebraic group, L an algebraic subgroup, and $X = G/L$ the associated algebraic homogeneous space. If H is an algebraic subgroup of G, then every orbit $H(x) \subset X$ is Zariski open in its closure.*

Definition 7.1.6. A B-orbit $\mathcal{O}$ in $Z = G/Q$ is a **Schubert cell**. Its closure $S = c\ell(\mathcal{O})$ is the associated **Schubert variety**.

Notice that the Schubert cells $\mathcal{O} = \mathbb{C}^{m(\mathcal{O})}$ define a CW complex structure on $Z = G/Q$. This is particularly simple, because $\dim_{\mathbb{R}} \mathcal{O} = 2m$ and the boundary of every cell $\mathcal{O}$ is comprised of cells which are of real dimension $\leqq \dim_{\mathbb{R}} \mathcal{O} - 2$. Thus the homology of Z can be described as follows in terms of the set $\mathcal{S}$ of all Schubert varieties, and incidentally this gives an algebraic geometry proof of Proposition 4.3.5, whose proof was topological.

Proposition 7.1.7. *The homology $H_*(Z, \mathbb{Z})$ of a compact algebraic homogeneous manifold $Z = G/Q$ of a reductive group is the free $\mathbb{Z}$-module generated by the set $\mathcal{S}$ of Schubert varieties associated to a Borel subgroup B of G.*

7.1B Schubert varieties related to the G_0-action

Let us return to the basic situation of this monograph. We are studying the action of a simple real form G_0 of the complex semisimple group G on a complex flag manifold $Z = G/Q$. The Iwasawa decompositions of G_0 play a basic role in our considerations. If $G_0 = K_0A_0N_0$ is such a decomposition, we refer to the solvable group A_0N_0 as its **Iwasawa component**.

Following our usual notational convention, A and N are the respective complexifications of A_0 and N_0 in G. They are algebraic subgroups of G, and $AN = A \ltimes N$ is a Levi decomposition of the solvable complex algebraic group AN. Note that $A \cong (\mathbb{C}^*)^r$, i.e., A is an algebraic torus, and that N is a unipotent subgroup of G.

Definition 7.1.8. We refer to a Borel subgroup G which contains some Iwasawa component A_0N_0 of G_0 as an **Iwasawa–Borel subgroup**. If B is such a subgroup and $\mathcal{O} = B(z)$ is a B-orbit in $Z = G/Q$, then its closure $S := c\ell(\mathcal{O})$ in Z will be called an **Iwasawa–Schubert variety**.

The Iwasawa–Borel subgroups can be geometrically described in terms of the G_0-action on G/B. For this we first recall the general fact that the set $\mathrm{Orb}_Z(G_0)$ of G_0-orbits in any $Z = G/Q$ is finite and that there is a unique closed orbit $\gamma_{c\ell} \in \mathrm{Orb}_Z(G_0)$ (see [W2, Theorem 3.3] or Chapter 8 below). If $Z = G/B$, we denote this orbit by $\gamma_{c\ell}^{G/B}$. Recall that $Z = G/B$ parameterizes the set of Borel subgroups of G by $z \mapsto B_z$, because B is its own normalizer in G.

Proposition 7.1.9. *The closed G_0-orbit $\gamma_{c\ell}^{G/B}$ in $Z = G/B$ parameterizes the set of Iwasawa–Borel subgroups in G.*

Proof. Let $G_0 = K_0A_0N_0$ be an Iwasawa decomposition of G_0. The Iwasawa component A_0N_0 is a simply connected real algebraic group; in particular its maximal compact subgroup is $\{1\}$. Therefore any compact A_0N_0-orbit in a space where it is acting algebraically is just a fixed point. Consequently A_0N_0 fixes a point x in $\gamma_{c\ell}^{G/B}$, i.e., $B_x \supset A_0N_0$ is an Iwasawa–Borel subgroup. Since K_0 acts transitively on $\gamma_{c\ell}^{G/B}$ and on the Iwasawa decompositions with compact part K_0, it follows that every point in $\gamma_{c\ell}^{G/B}$ has an Iwasawa–Borel subgroup as its isotropy group.

Conversely, if $B \supset A_0N_0$ for an Iwasawa decomposition $G_0 = K_0A_0N_0$, and $B = B_z$, then $G_0(z)$ is compact, so $G_0(z) = K_0(z) = \gamma_{c\ell}^{G/B}$ and $z \in \gamma_{c\ell}^{G/B}$. □

7.2 Envelope construction

If F is an open relatively compact set in $\mathbb{R}^2$, then one can define its **envelope** $\mathcal{E}(F)$ as the intersection of the half-planes that contain it. For this construction it is enough to consider an appropriate compact family $\mathcal{L}$ of lines L in the complement of F and then

$$\mathcal{E}_{\mathcal{L}}(F) := \left(\mathbb{R}^2 \setminus \bigcup\{L \mid L \in \mathcal{L}\}\right)^0,$$

where the superscript indicates the connected component containing F. By a *compact family* $\{Y_s\}_{s \in S}$ of closed subsets in a Hausdorff topological space X we mean a closed subset $\mathcal{Y}$ of $S \times X$, where S is compact with the following properties. First, the restriction $p : \mathcal{Y} \to S$ of the projection $S \times X \to S$ is surjective, and second, if $s \in S$, then $p^{-1}(s) = Y_s$. Note that the restriction $\mathcal{Y} \to X$ of $S \times X \to X$ is automatically proper.

This type of construction makes very good sense in complex analysis. There the appropriate analog of $\mathbb{R}^2$, where the notion of convexity is available, is that of a Stein manifold (or complex space), which we discussed and defined in Section 4.6D.

For example, suppose that F is a (not necessarily relatively compact) domain in a Stein manifold X and $\mathcal{H}$ is a compact family of complex analytic hypersurfaces H contained in $X \setminus F$.

Since X is smooth, a complex hypersurface in X can be locally defined by defined by one equation. More precisely, there is an open covering $\mathcal{U} = \{U_\alpha\}$ of X with the property that if $H \in \mathcal{H}$ there are holomorphic functions $f_\alpha \in \mathcal{O}(U_\alpha)$ with

$$H \cap U_\alpha = \{x \in U_\alpha \mid f_\alpha(x) = 0\}.$$

We may choose those functions with the additional property that $f_\alpha = f_{\alpha\beta} f_\beta$ on the overlap $U_\alpha \cap U_\beta$ where the $f_{\alpha\beta} : U_\alpha \cap U_\beta \to \mathbb{C}^*$ are holomorphic and nowhere vanishing. We may therefore regard $\{f_{\alpha\beta}\}$ as the set of transition functions for a holomorphic line bundle $\mathbb{L} \to X$ and regard $\{f_\alpha\}$ as defining a holomorphic section $s \in \Gamma(X; \mathbb{L})$. Now define $\mathcal{E}_{\mathcal{H}}(F) = \left(X \setminus \bigcup\{H \mid H \in \mathcal{H}\}\right)^0$.

This construction of $\mathbb{L}$ from H, just above, is the standard construction of a line bundle from a divisor.

Proposition 7.2.1. *The envelope $\mathcal{E}_{\mathcal{H}}(F)$ of a domain F in a Stein manifold X defined by a compact family of complex hypersurfaces $\mathcal{H}$ is Stein.*

Proof. For every point $p \in \mathrm{bd}(\mathcal{E}_{\mathcal{H}}(F))$ there is a complex supporting hypersurface $H \in \mathcal{H}$ which contains p and is contained in the complement of $\mathcal{E}_{\mathcal{H}}(F)$. Let $H = \{x \in H \mid s(x) = 0\}$ for some $s \in \Gamma(X; \mathbb{L})$. Since X is Stein, it follows that $\Gamma(X; \mathbb{L})$ is quite large; in particular there exists $t \in \Gamma(X; \mathbb{L})$ with $t(p) \neq 0$. Now $m = t/s$ is a meromorphic function and its restriction f to $\mathcal{E}_{\mathcal{H}}(F)$ is holomorphic. Further, p is in the polar set of f, and since $t(p) \neq 0$, f is not indeterminate at p. Thus, for $\{p_n\} \subset \mathcal{E}_{\mathcal{H}}(F)$ with $p_n \to p$, $\lim_{n \to \infty} |h(p_n)| = \infty$. □

Our strategy for describing the cycle spaces $\mathcal{M}_D$ in terms of universal domains is to pinch them between certain envelopes which for complex analytic reasons turn out to be equal. As was explained in Chapter 5, the ambient cycle space $\mathcal{M}_Z$ is either a compact hermitian symmetric space $X = G/Q$ or is a finite quotient of the affine symmetric space G/K. Thus we begin our discussion of relevant envelopes by considering these two G-homogeneous spaces.

7.2A Hermitian symmetric spaces

Let G_0 be simple and of hermitian type. Let $X = G/P$ be the compact hermitian symmetric space dual to G_0/K_0. Consider the action of an Iwasawa–Borel subgroup

B on $X = G/P$. Recall that P is a maximal parabolic subgroup of G; consequently the Betti number $b_2(X) = 1$. By Poincaré duality the irreducible components of the complement of the open B-orbit in X generate $H^2(X; \mathbb{Z})$. So there is a unique irreducible B-invariant hypersurface H in X. Define $\mathcal{H} = \{k_0(H) \mid k_0 \in K_0\}$, which is the same as $\{g_0(H) \mid g_0 \in G_0\}$.

If $\mathcal{B} = G_0(x_0)$ is the bounded hermitian symmetric space G_0/K_0 embedded as an open G_0-orbit in X, with $K_0 = \{g \in G_0 \mid g(x_0) = x_0\}$, and if B contains the Iwasawa component of a decomposition $G_0 = A_0N_0K_0$, then $\mathcal{B}$ is contained in the open cell $X \setminus H$. Thus $\mathcal{B}$ is contained in the complement of every hypersurface in $\mathcal{H}$ and we consider the envelope $\mathcal{E}_{\mathcal{H}}(\mathcal{B})$.

Proposition 7.2.2. $\mathcal{E}_{\mathcal{H}}(\mathcal{B}) = \mathcal{B}$.

Proof. Since the complement of every hypersurface $H \in \mathcal{H}$ is just $\mathbb{C}^n$, which of course is Stein, every such H has nonempty intersection with every open G_0-orbit whose (compact) base cycle is positive-dimensional. Therefore $\mathcal{E}_{\mathcal{H}}(\mathcal{B})$ is contained in the complement of the union of the closures of such open orbits.

If $\mathcal{B}$ is not of tube type, then it is the only open G_0-orbit in which the base cycle is just a point. Thus it is enough to consider the tube type case where there are two copies of $\mathcal{B}$ embedded as open orbits in X. Those G_0-orbits can be characterized by the fact that their base cycles are just points. However, in this case we see that $\mathcal{E}_{\mathcal{H}}(\mathcal{B}) = \mathcal{B}$ as well. This follows, e.g., because $\mathrm{bd}(\mathcal{B})$ contains an open dense G_0-orbit γ whose Levi form is not completely degenerate (here we disregard the one-dimensional case). If this piece were part of the boundary of the other copy, then along that boundary the other copy would be pseudoconcave, contrary to $\mathcal{B}$ being Stein. □

7.2B Affine symmetric spaces

Throughout this section $G_0 = K_0A_0N_0$ is an Iwasawa decomposition, B is an Iwasawa–Borel subgroup of G that contains A_0N_0, and we consider the action of B on the (spherical) affine homogeneous space $\Omega = G/K$.

If x_0 is the base point $1K$ in $\Omega := G/K$, then $B(x_0)$ is open and its complement is a B-invariant algebraic set H. Note that G_u/K_0 is a strong deformation retract of Ω and $B(x_0)$ is of the form $\mathbb{C}^k \times (\mathbb{C}^*)^\ell$. Thus, for topological reasons $B(x_0)$ is not equal to the full space Ω. Furthermore, since $B(x_0)$ is Stein, its complement is an analytic hypersurface in Ω. Of course that complement consists of a number of irreducible components.

We now wish to show that the hypersurfaces H, discussed above, lie in the complement of the universal domain $\mathcal{U}$. The following property of plurisubharmonic functions plays a key role in our proof.

Lemma 7.2.3. *Let ρ be a strictly plurisubharmonic function on a n-dimensional complex manifold X. Let M be a real submanifold of X. Suppose that M is contained in the zero set $\{d\rho = 0\}$. Then M is isotropic with respect to the symplectic form $dd^c\rho$. In particular M is totally real and* $\dim_{\mathbb{R}} M \leq n$.

Proof. If $\lambda := d^c\rho$ and $i : M \hookrightarrow X$ is the natural injection, then $\imath^*(\lambda) = 0$, and $i^*(dd^c\rho) = di^*(\lambda) = 0$. □

Proposition 7.2.4. *$\Omega \setminus H$ contains the universal domain $\mathcal{U}$ with base point x_0, defined in* (6.1.7).

Proof. Corollary 6.3.2 gives us a strictly plurisubharmonic G_0-invariant function $\rho : \mathcal{U} \to \mathbb{R}^+$ that is an exhaustion function modulo G_0 of $\mathcal{U}$. Denote $\Sigma := K_0 \exp(i\omega_0)(x_0)$ and note that $\rho|_\Sigma$ is a proper exhaustion.

If $H \cap \mathcal{U} \neq \emptyset$, then $H \cap \Sigma \neq \emptyset$ because H is A_0N_0-invariant. Let $x_1 \in H \cap \Sigma$ such that $\rho(x_1) = \min\{\rho(x) \mid x \in H \cap \Sigma\}$. It follows that $\rho|_{AN(x_1)}$ has a local minimum along $A_0N_0(x_1)$. Thus by Lemma 7.2.3 the orbit $A_0N_0(x_1)$ is a totally real submanifold of $AN(x_1)$. On the other hand $\mathcal{U}$ can be identified with a domain in the tangent bundle of G_0/K_0 (see Proposition 6.1.6) and in that realization every A_0N_0-orbit is a section over G_0/K_0. In particular $\dim_{\mathbb{R}} A_0N_0(x_0) = \dim_{\mathbb{C}} AN(x_1)$. But since $A_0N_0(x_1)$ is totally real, this is only possible when $AN(x_1)$ is open in Ω, contrary to H being a B-invariant proper analytic subset. □

Remark 7.2.5. In the second author's proof [H] of Proposition 7.2.4 only a G_0-invariant strictly plurisubharmonic function without the additional exhaustion condition was used. This is not sufficient, because the minimum might not be achieved. Thus here we use an exhaustion modulo the action of G_0. In the meantime T. Matsuki gave an algebraic proof that holds in greater generality [M4]. ◇

Now let $\mathcal{I} := \{k_0(H) : k_0 \in K_0\}$. As noted earlier, $\mathcal{I} = \{g_0(H) : g_0 \in G_0\}$. Since $H \subset \Omega \setminus \mathcal{U}$ and $\mathcal{U}$ is G_0-invariant, we may consider the envelope $\mathcal{E}_{\mathcal{I}}(\mathcal{U})$. Proposition 7.2.4 can be reformulated as $\mathcal{E}_{\mathcal{I}}(\mathcal{U}) \supset \mathcal{U}$. The reverse inclusion was proved by L. Barchini [Ba]. The following adapts the central part of her argument to the spirit of the present text.

Proposition 7.2.6. $\mathcal{E}_{\mathcal{I}}(\mathcal{U}) \subset \mathcal{U}$.

Proof. Assume that $\mathcal{E}_{\mathcal{I}}(\mathcal{U})$ is not contained in $\mathcal{U}$. Then there exists a sequence $\{z_n\} \subset \mathcal{U} \cap \mathcal{E}_{\mathcal{I}}(\mathcal{U})$ with $z_n \to z \in \mathrm{bd}(\mathcal{U}) \cap \mathcal{E}_{\mathcal{I}}(\mathcal{U})$. From the definition of $\mathcal{U}$ it follows that there exist $\{g_m\} \subset G_0$ and $\{w_m\} \subset \exp(i\omega_0)$ such that $g_m(w_m) = z_m$.

Write $g_m = k_m a_m n_m$ in a $K_0A_0N_0$-decomposition of G_0. Since $\{k_m\}$ is contained in the compact group K_0, we may assume $k_m \to k$, and therefore that $g_m = a_m n_m$.

Since ω_0 is relatively compact in $\mathfrak{a}_0$, we may assume $w_m \to w \in c\ell(\exp(i\omega_0))$. Thus $w_m = s_m x_0$, where $\{s_m\} \subset \exp(i\omega_0)$ and $s_m \to s$. Write $a_m n_n(w_m) = a_m n_m s_m x_0 = \widetilde{a}_m \widetilde{n}_m x_0$, where $\widetilde{a}_m = a_m s_m$ and $\widetilde{n}_m = s_m^{-1} n_m s_m$ are elements of A and N, respectively.

Now $\{z_m\}$ and the limit z are contained in $\mathcal{E}_{\mathcal{I}}(\mathcal{U})$ which is in turn contained in $AN(x_0)$. Furthermore, AN acts freely on this orbit. Thus $\widetilde{a}_m \to \widetilde{a} \in A$ and $\widetilde{n}_m \to \widetilde{n} \in N$ with $\widetilde{a}\widetilde{n}(x_0) = z$. Since $s_m \to s$, it follows that $a_m \to a \in A_0$ and $n_m \to n \in N_0$ with $an(w) = z$. Since $z \notin \mathcal{U}$, it follows that $w \in \mathrm{bd}(\exp(i\omega_0))$, and $z \in \mathcal{E}_{\mathcal{I}}(\mathcal{U})$ implies that $w \in \mathcal{E}_{\mathcal{I}}(\mathcal{U})$.

On the other hand, $w \in \mathrm{bd}(\exp(i\omega_0))$. Hence the isotropy subgroup of G_0 at w is noncompact. But $\mathcal{E}_{\mathcal{I}}(\mathcal{U})$ is Kobayashi hyperbolic (see [H], or see Remark 7.2.9). Therefore the G-action on $\mathcal{E}_{\mathcal{I}}(\mathcal{U})$ is proper (see, e.g., [H] and also Chapter 11) below. Consequently $w \neq \mathcal{E}_{\mathcal{I}}(\mathcal{U})$. That contradicts the assumption $\mathcal{E}_{\mathcal{I}}(\mathcal{U}) \not\subset \mathcal{U}$. □

Combining Propositions 7.2.4 and 7.2.6 we have

Theorem 7.2.7. $\mathcal{E}_{\mathcal{I}}(\mathcal{U}) = \mathcal{U}$.

Remark 7.2.8. This result also follows from the inclusion $\mathcal{U} \subset \mathcal{E}_{\mathcal{H}}(\mathcal{U})$ and the fact that $\mathcal{U}$ is a maximal G_0-invariant, Stein, Kobayashi hyperbolic domain in Ω. See Theorem 11.3.1 below. ◇

Remark 7.2.9. The Kobayashi hyperbolicity of envelopes such as $\mathcal{E}_{\mathcal{I}}(\mathcal{U})$ is a completely general fact, proved below in Chapter 11. We have taken the liberty of using it here. While the concept of hyperbolicity is of importance in our proofs of classification results, it should be emphasized that, except in very special cases, we have essentially no concrete information on the Kobayashi metric. ◇

In Chapter 12 we will see that one can define a cycle space $\mathcal{C}\{\gamma\}$ for every G_0-orbit $\gamma \in \mathrm{Orb}_Z(G_0)$, which is quite analogous to the cycle space $\mathcal{M}_D$ of an open orbit D. See [GM] for the first results on such cycle spaces. Those cycle spaces are closely related to the duality between the K-orbits and G_0-orbits in Z that we will discuss in Chapter 8.

In any flag manifold $Z = G/Q$ the K-orbit dual to the closed G_0-orbit $\gamma_{c\ell}$ is just the open K-orbit κ_{op}. To visualize the geometry, just note that since $\gamma_{c\ell}$ is a G_0-orbit, it is not contained in a proper complex analytic subset of Z. Since K_0 also acts transitively on $\gamma_{c\ell}$, it is immediate that the K-orbit, κ_{op}, of any point on $\gamma_{c\ell}$ is open in Z. Duality in this case is the simple statement that κ_{op} contains $\gamma_{c\ell}$. The dual statement to (5.1.5) is

$$\mathcal{C}'\{\gamma_{c\ell}\} : \text{ topological component of } \kappa_{\mathrm{op}} \text{ in } \{g(\kappa_{\mathrm{op}}) \mid g \in G \text{ and } g(\kappa_{\mathrm{op}}) \supset \gamma_{c\ell}\}.$$

Note that $\mathcal{C}'\{\gamma_{c\ell}\} = G\{\gamma_{c\ell}\}/\check{K}$ where $\check{K} = \{g \in G \mid g(\kappa_{\mathrm{op}}) = \kappa_{\mathrm{op}}\}$ and

$$G\{\gamma_{c\ell}\} : \text{ topological component of 1 in } \{g \in G \mid g(\kappa_{\mathrm{op}}) \supset \gamma_{c\ell}\}.$$

The isotropy $\check{K}$ contains K so we have the projection $G\{\gamma_{c\ell}\}/K \to \mathcal{C}'\{\gamma_{c\ell}\}$. As in the case of $\mathcal{M}_D$ we would like to lift $\mathcal{C}'\{\gamma_{c\ell}\}$ up to $G\{\gamma_{c\ell}\}/K$, which is an open subset of G/K, equipped with a base point x_0, where it can be compared to $\mathcal{E}_{\mathcal{I}}(\mathcal{U})$. Thus we define the cycle space to be

$$\mathcal{C}\{\gamma_{c\ell}\} := G\{\gamma_{c\ell}\}/K. \tag{7.2.10}$$

Proposition 7.2.11. *Suppose that Q is a Borel subgroup of G. Then the cycle space $\mathcal{C}\{\gamma_{c\ell}\}$ of the closed G_0-orbit $\gamma_{c\ell}$ agrees with the Iwasawa envelope $\mathcal{E}_{\mathcal{I}}(\mathcal{U})$. Thus $\mathcal{U} = \mathcal{E}_{\mathcal{I}}(\mathcal{U}) = \mathcal{C}\{\gamma_{c\ell}\}$.*

Proof. Both $\mathcal{E}_{\mathcal{I}}(\mathcal{U})$ and $\mathcal{C}\{\gamma_{c\ell}\}$ are defined as connected components of open sets which contain a fixed base point. Thus it is enough to show that the full sets, i.e., without going to the connected components, are the same. We do this in G before dividing out by the right action of K. At that level the appropriate statement is $\{g \in G : g(\kappa_{\text{op}}) \supset \gamma_{c\ell}\} = \bigcap_{k_0 \in K_0} k_0 ANK$. For this observe that if $B \supset AN$ is regarded as a point in $\gamma_{c\ell}$, then

$$g \in \{g' \in G \mid g'(\kappa_{\text{op}}) \supset \gamma_{c\ell}\} \Leftrightarrow g^{-1}G_0 B \subset KB = KAN \Leftrightarrow g^{-1}G_0 \subset KAN.$$

Thus

$$\begin{aligned}\{g \in G : g(\kappa_{\text{op}}) \supset Z_0\} &= \{g \in G : g^{-1}G_0 \subset KAN\} \\ &= \{g \in G : g^{-1}g_0 \in KAN \ \forall\, g_0 \in G_0\} \\ &= \{g \in G : g_0^{-1}g \in ANK \ \forall\, g_0 \in G_0\} \\ &= \bigcap_{g_0 \in G_0} g_0 ANK = \bigcap_{k_0 \in K_0} k_0 ANK.\end{aligned}$$

This completes the proof. □

Remark 7.2.12. Proposition 7.2.11 holds for any flag manifold $Z = G/Q$; see Theorem 12.5.3 below. The proof given above, the case $Q = B$, was communicated to us by Roger Zierau. ◇

7.3 Schubert intersection properties

Here we return to our considerations of cycles in open G_0-orbits. As before, K_0 is a fixed choice of a maximal compact subgroup of G_0, D is an open G_0-orbit in $Z = G/Q$, and C_0 is the unique complex K_0-orbit contained in D.

Fix an Iwasawa–Borel subgroup B containing the Iwasawa component A_0N_0 of a decomposition $G_0 = K_0A_0N_0$. Denote

(7.3.1)

$$\mathcal{S}_{C_0} := \{B\text{-Schubert varieties } S \mid \dim S + \dim C_0 = \dim D \text{ and } S \cap C_0 \neq \emptyset\}.$$

The B-Schubert varieties generate the homology of Z, and C_0 is topologically nontrivial. Consequently, $\mathcal{S}_{C_0} \neq \emptyset$.

If $S \in \mathcal{S}_{C_0}$, then S is the closure of a B-orbit $\mathcal{O}$, and we express that as $S = \mathcal{O} \dot\cup Y$. In this section we will see that $S \cap D \subset \mathcal{O}$. We will also see that the intersection $S \cap C$ is finite and transversal at each of its points for every $C \in \mathcal{M}_D$. In fact we will see that each connected component Σ of $S \cap D$ contains exactly one point of $S \cap C$ and is a closed A_0N_0-orbit in D.

The connection to B-invariant hypersurfaces in $\mathcal{M}_Z$ is given by incidence geometry: the B-invariant incidence subvariety

$$I_Y := \{C \in \mathcal{M}_Z \mid C \cap Y \neq \emptyset\}$$

of $\mathcal{M}_Z$ is a hypersurface whose components are polar sets of certain rational functions on Ω. Those functions are constructed by the trace-transform method of [BK]. The union H of all hypersurface components of I_Y is contained in $\mathcal{M}_D$ in $\mathcal{M}_Z \setminus \mathcal{M}_D$. Therefore, if $\mathcal{Y} := \{k_0(H) \mid k_0 \in K_0\}$, then the envelope $\mathcal{E}_{\mathcal{Y}}(\mathcal{M}_D)$ is an outer approximation to $\mathcal{M}_D$.

We begin as usual with a fixed Iwasawa decomposition $G_0 = K_0 A_0 N_0$ and a Borel subgroup B of G that contains $A_0 N_0$.

Lemma 7.3.2. *Every A_0N_0-orbit in D has nonempty intersection with C_0. If p is such an intersection point, then*

$$T_p(A_0N_0(p)) + T_p(C_0) = T_pD. \tag{7.3.3}$$

Proof. The first statement follows immediately from the fact that $A_0N_0(C_0) = A_0N_0K_0(p) = G_0(p) = D$. The second statement is again just a consequence of $G_0 = K_0A_0N_0$. □

Corollary 7.3.4. *Recall that $q := \dim_{\mathbb{C}} C_0$. If S is an Iwasawa–Schubert variety with* $\operatorname{codim}_{\mathbb{C}} S \geqq q + 1$, *then $S \cap D = \emptyset$.*

Proof. If $p \in S \cap D$, then $\dim_{\mathbb{R}} T_p(A_0N_0(p)) + \dim_{\mathbb{R}} T_p(C_0) \leqq 2(\dim_{\mathbb{C}} S + \dim_{\mathbb{C}} C_0) < \dim_{\mathbb{R}} D$, in contradiction to Lemma 7.3.2. □

Lemma 7.3.2 has a number of other useful consequences.

Proposition 7.3.5. *If $S \in \mathcal{S}_{C_0}$ and $p \in S \cap C_0$, then $A_0N_0(p)$ is open in S and closed in D.*

Proof. Since $\dim_{\mathbb{C}} S + \dim_{\mathbb{C}} C_0 = \dim_{\mathbb{C}} D$, the sum of the spaces in (7.3.3) is a direct sum and $\dim_{\mathbb{R}} A_0N_0(p) = \dim_{\mathbb{R}} S$. Thus $A_0N_0(p)$ is open in S. If this orbit were not closed in D, then it would have an A_0N_0-orbit of lower dimension on its boundary. But this would be contrary to Lemma 7.3.2, because its real codimension would be smaller than the real dimension of C_0. □

Definition 7.3.6. An orbit $\Sigma := A_0N_0(p)$ of a point $p \in S \cap C_0$ is called a **Schubert slice**.

Let $S = \mathcal{O} \dot{\cup} Y$ be the closure of a B-orbit. Suppose that $S \in \mathcal{S}_{C_0}$. Since $T_p\Sigma \oplus T_pC_0 = T_pD$, we have the following observation.

Proposition 7.3.7. *The intersection $S \cap C$ is finite and contained in $\mathcal{O}$, and at each of its points it is transversal in D. Express $S \cap C = \{z_1, \dots z_d\}$. Then the $\Sigma_j := A_0N_0(z_j)$ are the associated Schubert slices, and $S \cap D = \mathcal{O} \cap D$ is the disjoint union of the Σ_j.*

Proof. Transversality, and thus finiteness of the intersection, follow from the above direct sum decomposition of T_pD and the fact that the Schubert slices are open in S.

Since the components of $S \cap D$ are A_0N_0-invariant, every A_0N_0-orbit in D has nonempty intersection with C_0, and for dimension reasons such orbits must be open in $S \cap D$. It is therefore immediate that such a component is a Schubert slice. □

We wish to understand phenomena at the boundary of D. For this the following is a basic tool.

Proposition 7.3.8. *The closure* $c\ell(\Sigma)$ *in* $c\ell(D)$ *of a Schubert slice* Σ *satisfies* $K_0.c\ell(\Sigma) = c\ell(D)$. *In particular,* $\text{bd}(\Sigma)$ *has nonempty intersection with every* G_0*-orbit in* $\text{bd}(D)$.

Proof. Since K_0 is compact, $K_0.c\ell(\Sigma)$ is compact; in particular it is closed in $c\ell(D)$. Note $K_0.\Sigma = K_0A_0N_0(p) = G_0(p) = D$ for the $p \in C_0 \cap S$ that is the base point for Σ. Thus $K_0.c\ell(\Sigma)$ is dense in $c\ell(D)$ and equality follows. □

We now turn to the proof of the following basic fact.

Proposition 7.3.9. *The intersection* $C_0 \cap \Sigma$ *of the base cycle with any Schubert slice consists of exactly one point.*

The proof uses the Retraction Theorem 8.3.1 of Chapter 8, which holds for any G_0-orbit in Z. We state it here in our present context of open orbits. Compare [W2, Theorem 5.4].

Proposition 7.3.10. *The base cycle* C_0 *is a strong deformation retract of the open orbit* D*. In particular,* D *is simply connected.*

Thus the G_0-isotropy at points of $C_0 \cap S$ splits according to the Iwasawa decomposition.

Proposition 7.3.11. *If* $S \in \mathcal{S}_{C_0}$ *and* $z \in S \cap C_0$*, then the mapping*

$$\alpha : (K_0 \cap Q_z) \times (A_0N_0 \cap Q_z) \to (G_0 \cap Q_z), \text{ given by } \alpha(k_0, a_0n_0) = k_0a_0n_0,$$

is a diffeomorphism.

Proof of Proposition 7.3.11. The Iwasawa decomposition $G_0 = K_0A_0N_0$ defines a diffeomorphism $K_0 \times A_0N_0 \to G_0$ that restricts to α, and consequently α is a diffeomorphism onto its image.

For surjectivity note that $\dim(K_0 \cap Q_z) + \dim(A_0N_0 \cap Q_z) = \dim(G_0 \cap Q_z)$ because $\dim K_0 + \dim(A_0N_0) = \dim G_0$ and $\dim K_0(z) + \dim A_0N_0(z) = \dim G_0(z)$. Thus the image $\text{Image}(\alpha)$ is open in $G_0 \cap Q_z$. Observe that the compact group $K_0 \cap Q_z$ acts freely and locally transitively on $(G_0 \cap Q_z)/(A_0N_0 \cap Q_z)$. Consequently, $\text{Im}(\alpha)$ is also closed in $G_0 \cap Q_z$. We will see (Theorem 8.3.1) that C_0 is a strong deformation retract of D, and therefore $\pi_1(D) = \{1\}$. Thus $G_0 \cap Q_z$ is connected and $\text{Image}(\alpha) = G_0 \cap Q_z$. □

Proof of Proposition 7.3.9. For $z \in \Sigma \cap C_0$ suppose that $\Sigma = (A_0N_0)(z)$ intersects C_0 in a (possibly additional) point z'. Since $z' \in C_0$, there exists $k_0 \in K_0$ with $k_0(z') = z$. But $z' = (a_0n_0)(z)$ for some $a_0n_0 \in A_0N_0$ as well. So $k_0a_0n_0 \in (G_0 \cap Q_z)$, and it follows from Proposition 7.3.11 above that $k_0 \in (K_0 \cap Q_z)$ and therefore $z' = z$. □

As a consequence of Proposition 7.3.9, we obtain precise information about the intersection $C \cap S$ for any $C \in \mathcal{M}_D$.

Theorem 7.3.12. *Let $S \in \mathcal{S}_{C_0}$ and define d to be the topological intersection number $[S].[C_0]$. Then $S \cap D$ is a disjoint union $\Sigma_1 \dot{\cup} \ldots \dot{\cup} \Sigma_d$ of d Schubert slices Σ_j. Furthermore, for each j and every cycle $C \in \mathcal{M}_D$ the intersection $C \cap \Sigma_j$ is transversal and consists of a single point.*

Proof. The intersection $S \cap C_0 = \{z_1, \ldots, z_d\}$ is transversal at each of its points, and by Proposition 7.3.9 each of the Schubert slices $\Sigma_j = (A_0 N_0)(z_j)$ satisfies $\Sigma_j \cap C_0 = \{z_j\}$. It follows that there are exactly d distinct Schubert slices. They are disjoint because two distinct $A_0 N_0$-orbits must be disjoint.

If $C \in \mathcal{M}_D$ and $\dim C \cap S > 0$, then $C \cap S$ could not be contained in the Stein manifold $\mathcal{O}$. Thus $C \cap Y \neq \emptyset$, contrary to the fact that $Y \subset (Z \setminus D)$. Thus $C \cap S = C \cap \mathcal{O}$ is finite.

Now $\mathcal{O}$ is the disjoint union of the Σ_j, and the Σ_j all are closed submanifolds of D. Let J be the set of $C \in \mathcal{M}_\mathcal{D}$ such that $C \cap \Sigma_j$ is nonempty and transversal for all j. Clearly J is an open subset of $\mathcal{M}_\mathcal{D}$ and we replace it by its component which contains C_0. Now $[S].[C] = d$ for all $C \in \mathcal{M}_\mathcal{D}$, and $C \cap \mathcal{O}$ is finite and contains at least d points for all $C \in J$. If $C \in J$ it follows that $|C \cap \Sigma_j| = 1$ for all j. If C_1 were a boundary point of J in $\mathcal{M}_\mathcal{D}$, then $C_1 \cap \Sigma_j$ would still consist of one point for all j, but at least one of these intersections, $C_i \cap \Sigma_{j_0}$, would not be transversal. However, the condition of nontransversal intersection with Σ_{j_0} defines an analytic set A in $\mathcal{M}_\mathcal{D}$, and every cycle C in $\mathcal{M}_\mathcal{D}$ which is near C_1 intersects Σ_{j_0} in at least two points. Since we can choose C to be in J, this is a contradiction. Consequently J is closed in $\mathcal{M}_\mathcal{D}$, and $J = \mathcal{M}_\mathcal{D}$ as desired. □

7.4 Trace transform

Here we begin with a brief discussion of the construction and basic properties of the cycle space $\mathcal{C}_q(X)$ of q-dimensional cycles in a complex space X.

After introducing the trace transform and proving certain of its general properties, we turn to the setting of this book and discuss the trace transform defined by Iwasawa–Schubert varieties in $Z = G/Q$. The resulting polar sets of trace transformed meromorphic functions, or equivalently the associated incidence divisors, are of basic importance for our concept of Schubert domain, defined here, and the resulting description of the cycle spaces $\mathcal{M}_D$ (and later the more general cycle spaces $\mathcal{C}\{\gamma\}$).

Throughout this work we have tried to minimize the technical difficulties involved with singular complex spaces. In fact, for most of our considerations it is enough to stay in the realm of complex manifolds. However, this is not possible when dealing with cycle spaces. The complex spaces in this setting are, however, *reduced*. This means that their local models are zero sets $A = \{f_1 = \cdots = f_k = 0\}$ of finitely many holomorphic functions on open subsets U in some $\mathbb{C}^n$. Local holomorphic functions then are just restrictions of holomorphic functions on open subsets of U. In many contexts it is sufficient to consider *normal* complex spaces. This means that Riemann's theorem on removable singularities holds: If X is the complex space, Y is an analytic subset and $f \in \mathcal{O}(X \setminus Y)$ is locally bounded near Y, then f extends to a holomorphic function on X. See [GR1] and [GR2] for precise definitions and the basic theory.

7.4A Introduction to Barlet cycle spaces

Our brief discussion of cycle spaces (in the sense of D. Barlet) begins at the level of points.

If X is a normal complex space, then the set of unordered k-tuples of points in X is denoted by $\mathrm{Sym}^k(X)$. It can be identified with the quotient of the k-fold product $X \times \cdots \times X$ by the standard action of the symmetric group $\mathfrak{S}_k$.

$\mathrm{Sym}^k(X)$ carries the quotient topology for $\pi : X \times \cdots \times X \to \mathrm{Sym}^k(X)$. A function f on an open subset U of $\mathrm{Sym}^k(X)$ is said to be holomorphic if and only if $f \circ \pi$ is holomorphic on $\pi^{-1}(U)$. In this way, starting with a normal complex space, it follows [Car] that $\mathrm{Sym}^k(X)$ is a normal complex space and π is holomorphic.

Let us write a point in $\mathrm{Sym}^k(X)$ as a sum $c = n_1 x_1 + \cdots + n_m x_m$, where the coefficients n_j are positive integers and the x_j are 0-dimensional analytic subsets of X, each consisting of one point which we also denote by x_j. In this notation $\mathrm{Sym}^k(X)$ is the **cycle space** of all 0-dimensional cycles c.

An n-dimensional cycle in an arbitrary complex space X is a formal sum $C = n_1 C_1 + \cdots + n_m C_m$, where the coefficients are positive integers, and the C_j are irreducible n-dimensional compact analytic subsets of X. Its support $|C|$ is defined to be the union of the varieties C_j. The space of all n-dimensional cycles in X is denoted by $\mathcal{C}_n(X)$.

The space $\mathcal{C}_n(X)$ has a canonical complex structure. Near a cycle C it is defined by setting up local charts that are called **scales**. An essential step for this is to set up a local analytic set as an **analytic cover** of, for example, a polydisk.

For this, let A be an analytic subset of pure dimension n in an open subset W of $\mathbb{C}^{n+p}$. For simplicity it may be assumed that the origin 0 is the base point in A. Then there are linear subspaces $\mathbb{C}^n$ and $\mathbb{C}^p$ and polydisks $U \subset \mathbb{C}^n$ and $B \subset \mathbb{C}^p$ such that $c\ell(U) \times c\ell(B) \subset W$ and $A \cap (c\ell(U) \times \mathrm{bd}(B)) = \emptyset$.

Replacing A by $A \cap (U \times B)$, we consider the projection $\pi : A \to U$ onto the first factor. This is an analytic cover representation of A near the base point. It is a finite-fibered, proper, surjective holomorphic map. Outside a nowhere dense analytic subset $E \subset U$ it is a k-sheeted unramified covering map. That defines a holomorphic map $U \setminus E \to \mathrm{Sym}^k(B)$ which extends holomorphically, by Riemann's theorem on removable singularities, to a holomorphic map $\psi : U \to \mathrm{Sym}^k(B)$. In a naïve way we may think of ψ as a local coordinate representation of a piece A of a cycle, and moving ψ in $\mathrm{Hol}(U, \mathrm{Sym}^k(B))$ gives us candidates for local charts in nearby cycles.

Definition 7.4.1. A **scale** in X is a triple $E = (U, B, f)$ such that $U \subset \mathbb{C}^n$ and $B \subset \mathbb{C}^p$ are relatively compact polydisks, and $f : X_E \to \mathbb{C}^{n+p}$ is an embedding of an open subset X_E of X into an open neighborhood of $c\ell(U) \times c\ell(B)$ in $\mathbb{C}^{n+p}$. If $C \in \mathcal{C}_n(X)$ is a cycle and the **analytic cover condition**

$$f(|C| \cap X_E) \cap (c\ell(U) \times \mathrm{bd}(B)) = \emptyset$$

is fulfilled, then one says that E is **adapted** to C.

If E is a scale which is adapted to an n-cycle C, then C induces a ramified cover of degree $\deg_E C = k_E$ onto U and its associated map $U \to \mathrm{Sym}^{k_E}(B)$.

The following definition gives the correct condition for gluing local ramified covering representations in $\mathrm{Hol}(U, \mathrm{Sym}^k(B))$.

Definition 7.4.2. Let S be a complex space and let $\{C_s\}_{s \in S}$ be a family of n-cycles in X.

1. The family $\{C_s\}$ is called **analytic** if, for every $s_0 \in S$ and every scale $E = (U, B, f)$ adapted to C_{s_0}, there exists an open neighborhood S_E of s_0 in S such that
   a) E is adapted to C_s for all $s \in S_E$,
   b) $deg_E C_s = deg_E C_{s_0}$ for all $s \in S_E$, and
   c) the map $g_E : S_E \times U \to \mathrm{Sym}^{k_E}(B)$, defined by the condition that $g_E(s, \cdot)$ is the holomorphic map induced by C_s, is holomorphic.
2. The family is called an **analytic family of cycles** if for every $s_0 \in S$ and every open neighborhood W of $|C_{s_0}|$ there exists an open neighborhood S' of s_0 such that $|C_s| \subset W$ for all $s \in S'$.

The graph $\mathfrak{X}_S$ of a family $\{C_s\}$ of n-cycles is defined by the natural incidence relation:

$$\mathfrak{X}_S := \{(s, x) \in S \times X : x \in |C_s|\}.$$

It is an analytic subset of $S \times X$. The maps $p : \mathfrak{X}_S \to S$ and $\pi : \mathfrak{X}_S \to \mathcal{C}_n(X)$ are defined by the respective projections. The topological condition for $\{C_s\}$ to be an analytic family of cycles is that π be proper.

The following basic existence theorem is due to D. Barlet [B1].

Theorem 7.4.3. *The set $\mathcal{C}_n(X)$ carries a complex structure such that the following conditions are fulfilled:*

(1) *The family $\{C_s\}_{s \in \mathcal{C}_n(X)}$ is an analytic family of compact n-cycles.*
(2) *The maps p and π are holomorphic and π is proper.*
(3) *If $\{C_s\}_{s \in S}$ is an analytic family of n-cycles, the map $S \to \mathcal{C}_n(X)$, $s \mapsto C_s$, is holomorphic.*

7.4B Basic properties of the trace transform

Here we introduce the trace transform in its simplest version and prove those of its properties that are needed for our work. These are just the tip of the iceberg of a much deeper subject which has been developed by D. Barlet and his coworkers; see, e.g., [BKa], [BK], [BM1], [BM2], and [BM3].

For our applications it is enough to consider a complex manifold Z, a closed analytic set S in Z of pure codimension q and a closed analytic subset Y of S of pure codimension 1. Let $\Gamma(S, \mathcal{O}(*Y))$ denote the space of meromorphic functions on S with poles contained in Y. The incidence variety I_Y is defined by $I_Y := \{C \in \mathcal{C}_q(Z) : C \cap Y \neq \emptyset\}$, and the trace transform is a canonically defined linear map $\mathrm{Tr} : \Gamma(S, \mathcal{O}(*Y)) \to \Gamma(\mathcal{C}_q(Z), \mathcal{O}(*I_Y))$. For the definition of Tr we consider the set

$$\Xi := \{C \in \mathcal{C}_q(Z) \mid C \cap Y = \emptyset \text{ and } C \cap T \text{ is finite}\}.$$

Lemma 7.4.4. *Ξ is Zariski open in $\mathcal{C}_q(Z)$.*

Proof. It is sufficient for us to show that each of the conditions (i) $C \cap Y \neq \emptyset$ and (ii) $\dim C \cap Y \geqq 1$ defines a closed analytic subset of $\mathcal{C}_q(Z)$.

Consider the graph $\mathfrak{X} \subset Z \times \mathcal{C}_q(Z)$ of the universal family of q-cycles parameterized by $\mathcal{C}_q(Z)$ and let $p : \mathfrak{X} \to Z$ and $\pi : \mathfrak{X} \to \mathcal{C}_q(Z)$ be the projections. Then C satisfies the first condition if and only if $C \in \pi(p^{-1}(Y))$. Since π is proper, $\pi(p^{-1}(Y))$ is closed and therefore the first condition defines a closed analytic subset.

Observe that C satisfies the second condition if and only if the restriction of π to $p^{-1}(Y)$ has a positive-dimensional fiber over the point C in $\mathcal{C}_q(Z)$. Again since π is proper, this is a closed analytic subset. Now Ξ is Zariski open in $\mathcal{C}_q(Z)$. □

For $f \in \Gamma(S, \mathcal{O}(*))$ define the function $\mathrm{Tr}(f) : \Xi \to \mathbb{C}$ by

$$\mathrm{Tr}(f)(C) = \sum_{z_j \in C \cap T} f(z_j)$$

(counting multiplicities). The following is the main result required for our applications. Also see [HS, Appendix].

Proposition 7.4.5. *Let X be a closed irreducible subspace of $\mathcal{C}_q(Z)$ such that $X \cap \Xi \neq \emptyset$. In particular $X \cap \Xi$ is a dense, Zariski open set in X. Then $f \mapsto \mathrm{Tr}(f)$ is holomorphic on $X \cap \Xi$ and extends meromorphically to X.*

Proof. The essential point is to observe that $f \mapsto \mathrm{Tr}(f)$ is the composition of two maps. The first is the intersection map

$$I : \Gamma \cap \Xi \to \mathrm{Sym}^k(S \setminus Y)$$

given by $I(C) = C \cap S$ (in the cycle sense), where $k \in \mathbb{N}$ depends on X. Moreover, the map I extends to a meromorphic map $\widehat{I} : X \to \mathrm{Sym}^k(T)$, because

$$H = \{(C, (t_1 \dots t_k), t) \in X \times \mathrm{Sym}^k(T) \times T \mid t \in C \text{ and } \exists i \text{ such that } t_i = t\}$$

is a closed, analytic set in $X \times \mathrm{Sym}^k(T) \times S$ which is proper over X.

The second map is $f \mapsto \mathrm{trace}(f)$, where $\mathrm{trace}(f) : \mathrm{Sym}^k(T \setminus Y) \to \mathbb{C}$ is defined by $\mathrm{trace}(f)(t_1 \dots t_k) = \sum_{j=1}^k f(t_j)$. Of course $\mathrm{trace}(f)$ is holomorphic. We now prove that it extends meromorphically to the full space $\mathrm{Sym}^k(T)$.

Let $(t_1^0 \dots t_k^0) \in \mathrm{Sym}^k(T)$ such that some of the t_i^0 belong to Y. Choose g holomorphic near $\{t_1^0\} \cup \dots \cup \{t_k^0\}$ such that $F := g \times f$ is holomorphic; for example take $g \equiv 1$ near t_i^0 if $t_i^0 \neq Y$. Let $\varphi(t_1 \dots t_k) = \prod_{j=1}^k g(t_j)$ for $(t_1 \dots t_k)$ near $(t_1^0 \dots t_k^0)$. Then

$$(\varphi \times \mathrm{trace}(f))(t_1 \dots t_k) = \sum_{j=1}^{k} g(t_1) \cdot \dots \cdot \widehat{g(t_j)} \cdot \dots \cdot g(t_k) \cdot F(t_j)$$

is holomorphic on $\mathrm{Sym}^k(T)$ near $(t_1^0 \dots t_k^0)$ by the standard theorem on symmetric functions. Hence trace(f) is meromorphic on $\mathrm{Sym}^k(T)$.

Now $\mathrm{Tr}(f)$ is meromorphic as the composition trace$(f) \circ \widehat{I}$ of two meromorphic maps. □

Using the above result it is shown in [BK] that $\mathrm{Tr}(f)$ is holomorphic on $X \setminus I_Y$. In our applications this is clear because, if $S \cap Y$ is positive-dimensional, then $C \cap Y \neq \emptyset$. In any case, as stated above, if we replace I_Y by $X \cap I_Y$, then $\mathrm{Tr} : \Gamma(S, \mathcal{O}(*Y)) \to \Gamma(X, \mathcal{O}(*I_Y))$.

We would also like to remark on the connection to the Andreotti–Norguet transform (see [AN] and [BM1]). For this let Ω^q be the sheaf of holomorphic q-forms and $H^q(Z \setminus Y; \Omega^q)$ the associated Dolbeault cohomology space. A class $\xi = [\alpha]$ in $H^q(Z \setminus Y; \Omega^q)$ defines a holomorphic function $AN(\xi)$ on $\mathcal{C}_q(Z \setminus Y)$ by $AN(\xi)(C) = \int_C \omega$.

Now a function $f \in \Gamma(S, \mathcal{O}(*Y))$ defines a Dolbeault class ξ in a very natural way. For this, let c_S be the fundamental class of S (see [B1]). This is the class of the integration current $[S]$, and for f as above we may regard fc_S as an element of $H^q(Z \setminus Y; \Omega^q)$. It then follows that $\mathrm{Tr}(f) = AN(fc_S)$; see [BM1], [HS].

The above discussion would be of no interest if $\Gamma(S, \mathcal{O}(*Y))$ contained only the constant functions. Hence, we now restrict to a situation where we are guaranteed many nonconstant meromorphic functions. Let us say that Y **has an ample Cartier structure in** S if S is compact and there is a holomorphic embedding of S in a projective space $\mathbb{P}(V)$ so that Y is the intersection of S with a projective hyperplane H.

Theorem 7.4.6. *Suppose that Y has an ample Cartier structure, and let X be is a closed irreducible subset of the cycle space $\mathcal{C}_q(Z)$. Let $\{C_n\}$ be a sequence of cycles in $X \cap \Xi$ such that there exist points $p_n \in C_n \cap S$ with the property that $p_n \to p \in Y$. Then, after going to a subsequence and renaming the coordinates, there exists $f \in \Gamma(S, \mathcal{O}(*Y))$ such that $\lim |Tr(f)(C_n)| = \infty$. In particular, $I_Y = \{C \in X \mid C \cap Y \neq \emptyset\}$ is an analytic hypersurface.*

Proof. Embed S in a projective space $\mathbb{CP}_m$ so that Y is the intersection with a hyperplane H, and regard the complement as being $\mathbb{C}^m$ with coordinates $(z_1, \dots, z_m)$. Since $p_n \to p \in Y$, by going to a subsequence, we may assume that $z_1(p_n) \to \infty$.

Furthermore, it may be assumed that the number of points in $C_n \cap S$ is constant, say k. We write this intersection as $\{p_n^1, \dots, p_n^k\}$, where $p_n = p_n^1$. Thus for each n we have the k-tuple $v_n = (z_1(p_n^1), \dots, z_1(p_n^k)) \in \mathbb{C}^k$.

Now let s_n be the point in $\mathrm{Sym}^k(\mathbb{C})$ associated to v_n. Since v_n is a divergent sequence in $\mathbb{C}^k$, it follows that s_n diverges in $\mathrm{Sym}^k(\mathbb{C})$, and consequently there is a regular function f on $\mathrm{Sym}^k(\mathbb{C})$ with $\limsup |f(s_n)| = \infty$.

We regard f as a regular function on $\mathbb{C}^k$ that is invariant under the permutation group $\mathfrak{S}_k$. The ring of $\mathfrak{S}_k$-invariant functions on $\mathbb{C}^k$ is the polynomial algebra generated by the Newton polynomials $N_0, \dots, N_k$. Thus, there is a Newton polynomial N_ℓ such that $\limsup |N_\ell(v_n)| = \infty$, i.e., $\limsup \left|\sum_j z_1(p_n^j)^\ell\right| = \infty$. Hence we let $P(z_1, \dots, z_k) = z_1^\ell$, and it follows that $\limsup |\operatorname{Tr}(P)(C_n)| = \infty$. □

7.4C Schubert trace transform

Here we apply Theorem 7.4.6 to the case where S is a Schubert variety.

Recall our standard situation, where $Z = G/Q$ is the given homogeneous projective manifold, G_0 is a noncompact real form of G, B is an Iwasawa–Borel subgroup and $S = \mathcal{O} \dot{\cup} Y$ is a Schubert variety which is the closure of some B-orbit $\mathcal{O}$. In order to apply Theorem 7.4.6, we determine a B-equivariant embedding $\varphi : S \to \mathbb{P}(V)$ into a certain projective space.

For this, let $\mathbb{L} \to Z$ be a very ample bundle. Without loss of generality we may assume that G is simply connected, and therefore $\mathbb{L}$ is a G-bundle. The embedding $\varphi : Z \to \mathbb{P}(V)$, where $V = \Gamma(Z, \mathbb{L})^*$, is G-equivariant. Restricting φ to S we have

Proposition 7.4.7. *There exists a B-equivariant embedding $\varphi : S \to \mathbb{P}(V)$ into a projective space.*

In this situation the following observation is extremely useful.

Lemma 7.4.8. *All orbits of a connected unipotent group U of affine transformations of $\mathbb{C}^n$ are closed.*

Proof. The proof goes by induction on n, where the beginning step is straightforward.

We regard such an action on $\mathbb{C}^n$ as an action on $\mathbb{P}_n$ with homogeneous coordinates $[z_0 : \cdots : z_n]$ so that the hyperplane $\{z_0 = 0\}$ is stabilized, and its complement is the $\mathbb{C}^n$ in question. Equip the latter with affine coordinates $(w_1, \dots, w_n)$, where $w_j = z_j/z_0$. We may also assume that the action is in triangular form with respect to these coordinates, i.e.,

$$(w_1, \dots, w_n) \mapsto (w_1 + a_{10}, w_2 + a_{21}w_1 + a_{20}, \dots, w_n + a_{nn-1}w_{n-1} + \cdots + a_{n0}).$$

Thus the projection $\pi : \mathbb{C}^n \to \mathbb{C}^{n-1}$, $(w_1, \dots, w_n) \mapsto (w_1, \dots w_{n-1})$, is U-equivariant with respect to the obvious affine action of U on $\mathbb{C}^{n-1}$. By the induction assumption we may assume that every U-orbit in $\mathbb{C}^{n-1}$ is closed.

Now let $p \in \mathbb{C}^n$ be given, define $q := \pi(p)$, and consider the equivariant map $\pi : U(p) \to U(q)$ of orbits. Note that the closure $c\ell(U(p))$ is mapped onto the closed orbit $U(q)$. Thus any orbit on the boundary of $U(p)$ would necessarily have at least the same dimension as that of $U(q)$. Consequently, if $U(p)$ and $U(q)$ have the same dimension, then $U(p)$ is closed.

We may therefore suppose that the map $\pi : U(p) \to U(q)$ has one-dimensional fiber. Due to the fact that the group is unipotent, this fiber is isomorphic to $\mathbb{C}$. Hence, such fibers equal the corresponding fibers of the full projection $\pi : \mathbb{C}^n \to \mathbb{C}^{n-1}$. Thus $U(p)$ is the full π-preimage of the closed set $U(q)$ and is consequently closed. □

Corollary 7.4.9. *Let U be a unipotent algebraic group acting algebraically on a projective space $\mathbb{P}(V)$. If $S = \mathcal{O} \cup Y$ is the closure on a U-orbit $\mathcal{O}$ which is not contained in a proper projective subspace, then Y is the intersection of S with a U-invariant linear hyperplane H.*

Proof. Since U is unipotent, it possesses an invariant hyperplane H. Clearly $H \cap S$ is contained in the complement Y of the open orbit $\mathcal{O}$. Now U acts as a group of affine transformations on the complement of H in $\mathbb{P}(V)$. Hence, it follows from the above Proposition that the open U-orbit in S is closed in $\mathbb{P}(V) \setminus H$. Thus it is closed in $S \setminus H$, and the desired result follows. □

Corollary 7.4.10. *If $S = \mathcal{O} \cup Y$ is a Schubert variety in Z of an Iwasawa–Borel subgroup B of G, then Y has an ample Cartier structure. In particular, if S is q-codimensional and X is an irreducible closed subspace of $\mathcal{C}_q(Z)$, then the conclusions of Theorem* 7.4.6 *hold.*

Proof. By Proposition 7.4.7 we may regard S as being B-equivariantly embedded in some projective space $\mathbb{P}(V)$. We replace $\mathbb{P}(V)$ by the smallest projective subspace which contains S. Thus the assumptions of Corollary 7.4.9 are fulfilled for the maximal unipotent subgroup U of B, because U acts transitively on $\mathcal{O}$. □

As a first consequence of this corollary we obtain an outer approximation for $\mathcal{M}_D$ in $\mathcal{M}_Z$. For this we first recall our setup.

The domain D is an open G_0-orbit in $Z = G/Q$. It contains the base cycle C_0, which is determined by a choice of maximal compact subgroup K_0 in G_0. We consider an Iwasawa–Borel subgroup B, i.e., a Borel subgroup of G that contains the A_0N_0 of an Iwasawa decomposition $G_0 = K_0A_0N_0$.

If S is a B-Schubert variety in $\mathcal{S}_{C_0}$, in other words S has codimension q and $S \cap C_0 \neq \emptyset$, then we may apply the above corollary and reach the conclusion of Theorem 7.4.6 for the situation where X is the closure of $\mathcal{M}_Z$ in $\mathcal{C}_q(Z)$.

To minimize notation, replace I_Y by its intersection H with $\mathcal{M}_Z$. This is clearly nonempty because there are elements $g \in G$ that move the base point of D to a point in Y. Furthermore, since $\operatorname{codim}_{\mathbb{C}} Y = q + 1$, we know from Corollary 7.3.4 that $Y \cap D = \emptyset$. Now the following is immediate.

Proposition 7.4.11. *For every $S \in \mathcal{S}_{C_0}$ the incidence variety H is a B-invariant analytic hypersurface in the complement of $\mathcal{M}_D$ in $\mathcal{M}_Z$.*

Let $\mathcal{H}$ denote the family $\{k(H)\}_{k \in K_0}$. Proposition 7.4.11 implies that the envelope $\mathcal{E}_{\mathcal{H}}$ contains $\mathcal{M}_D$. Of course every $S \in \mathcal{S}_{C_0}$ yields such an envelope.

Definition 7.4.12. The connected component containing $\mathcal{M}_D$ in the intersection of the envelopes $\mathcal{E}_{\mathcal{H}}$, $S \in \mathcal{S}_{C_0}$, is called the **Schubert domain** $\mathcal{S}_D$.

As a consequence of Proposition 7.4.11, the fact (Proposition 7.2.1) that envelopes are Stein, and the facts that intersections and connected components of Stein domains are Stein, we have the following result.

Corollary 7.4.13. *The Schubert domain $\mathcal{S}_D$ is a G_0-invariant Stein domain in $\mathcal{M}_Z$ that contains the cycle space $\mathcal{M}_D$.*

We regard $\mathcal{S}_D$ as the best outer approximation of $\mathcal{M}_D$ which can be constructed by Schubert intersection theory. It is one of the main results of this book that, in fact, $\mathcal{M}_D = \mathcal{S}_D$. See Corollary 9.2.6 and Theorem 11.3.7 below.

8

Orbit Duality via Momentum Geometry

The starting point for considerations of the cycle space $\mathcal{M}_D$ is that there is a unique complex K_0-orbit C_0 in every open G_0-orbit D in $Z = G/Q$. Equivalently, C_0 is the unique closed K-orbit in D. Here we shall show that this reflects a duality between the set $\mathrm{Orb}_Z(K)$ of K-orbits and the set $\mathrm{Orb}_Z(G_0)$ of G_0-orbits in Z: there is a bijective correspondence $\gamma \leftrightarrow \kappa$ between $\mathrm{Orb}_Z(G_0)$ and $\mathrm{Orb}_Z(K)$ such that $\gamma \in \mathrm{Orb}_Z(G_0)$ is in correspondence with $\kappa \in \mathrm{Orb}_Z(K)$ if and only if $\gamma \cap \kappa$ is a K_0-orbit.

In this generality the duality was first proved by combinatorial methods [M1], [M2]. More recently, it was derived as a by-product of understanding certain aspects of the K_0-momentum geometry of Z [MUV], [BL]. Here we adapt the latter viewpoint, following the treatment in [BL]. The main result can be formulated as follows.

Let ω be a G_u-invariant Kähler form on Z, and let $\mu_{K_0} : Z \to \mathfrak{k}_0^*$ be the associated K_0-moment map. Define $E := \|\mu_{K_0}\|^2$ to be the energy function computed with respect to a K_0-invariant Killing norm, and ∇E its gradient field computed with respect to the associated Kähler metric. Then the critical set $C = \{z \in Z \mid \nabla E = 0\}$ is a finite union of K_0-orbits κ_0. Let $\mathrm{Crit}_Z(K_0)$ be the set of all such K_0-orbits κ_0. If $z \in \kappa_0$, define $\kappa(\kappa_0) := K(z)$ and $\gamma(\kappa_0) := G_0(z)$. The basic duality theorem states that these maps are bijective correspondences:

$$\mathrm{Orb}_Z(K) \cong \mathrm{Crit}_Z(K_0) \cong \mathrm{Orb}_Z(G_0).$$

In fact, $\kappa = \kappa(\kappa_0)$ and $\gamma = \gamma(\kappa_0)$ are dual to each other if and only if $\kappa \cap \gamma = \kappa_0$. Furthermore, the ($K_0$-equivariant) gradient flow of ∇E is tangent to all K- and G_0-orbits. It is hyperbolic in the sense that if (κ, γ) is a dual pair with $\kappa \cap \gamma = \kappa_0$, then the flow of ∇E realizes κ_0 as a strong deformation retract of γ and the flow of $-\nabla E$ realizes κ_0 as a strong deformation retract of κ. See Theorem 8.3.1.

This chapter is organized as follows. Certain standard information concerning G_u-invariant Kähler structures on Z is summarized in Section 1. In Section 2 the energy function E is computed in terms of the Lie algebra structures at hand. This yields a precise Lie algebraic condition for a point $z \in Z$ to be a critical point of ∇E: the reductive part of the isotropy algebra $\mathfrak{q}_z$ (which is the complexification of $(\mathfrak{g}_u)_z$)

contains a Cartan subalgebra which is both σ-invariant and τ-invariant (Theorem 8.2.5). This is used to give an exact description of the Hessian of E at any such critical point (Theorem 8.2.13). In Section 3 the duality theorem is then proved as a consequence of Bott's results on Morse theory (Theorems 8.3.1 and 8.3.2). Finally, in Section 4 we prove the ordering principle; it says that if (γ_1, κ_1) and (γ_2, κ_2) are dual pairs, then γ_1 is contained in the closure of γ_2 if and only if κ_2 is contained in the closure of κ_1.

8.1 Coadjoint orbits

8.1A Symplectic structures and moment maps

Let L be a connected Lie group acting via Ad^* on the dual space $\mathfrak{l}^*$ of its Lie algebra. If $\xi \in \mathfrak{l}$ let $\widehat{\xi}$ denote the associated vector field on $\mathfrak{l}^*$. Recall that if $\alpha \in \mathfrak{l}^*$, then

$$\omega_\alpha(\widehat{\zeta}_\alpha, \widehat{\eta}_\alpha) := \alpha([\zeta, \eta])$$

defines a nondegenerate antisymmetric bilinear form on the tangent space $T_\alpha \mathcal{O}$ of the orbit $\mathcal{O} := \mathrm{Ad}^*(L)(\alpha)$ which is invariant under the action of the isotropy group L_α. The resulting differential form $\omega_\mathcal{O}$ on $\mathcal{O}$ is closed and is therefore an L-invariant symplectic form.

Suppose that L acts as a group of symplectic diffeomorphisms on a symplectic manifold (M, ω), and that $\mu : M \to \mathfrak{l}^*$ is a smooth equivariant map. Then μ is called a **moment map** for the action of L if, for every $\zeta \in \mathfrak{l}$, the function $\mu_\zeta := \zeta \circ \mu$ is a Hamiltonian of the field $\widehat{\zeta}_M$, i.e., $d\mu_\zeta = i_{\widehat{\zeta}_M}\omega$. In the case of a coadjoint orbit $(\mathcal{O}, \omega_\mathcal{O})$ the inclusion $id_\mathcal{O} \hookrightarrow \mathfrak{l}^*$ is a moment map. Furthermore, if $\widetilde{L}$ is a Lie subgroup of L, then a moment map $\mu_{\widetilde{L}} : \mathcal{O} \to \widetilde{\mathfrak{l}}^*$ is given by the canonical projection $\mathfrak{l}^* \to \widetilde{\mathfrak{l}}^*$.

If L is semisimple with Killing form $\langle\, ,\, \rangle$ we identify $\mathfrak{l}$ with $\mathfrak{l}^*$ by $\xi \mapsto \langle \xi, \cdot \rangle$. In this way the invariant symplectic form on the orbit $\mathcal{O} := \mathrm{Ad}(L)(\xi)$ can be written as

$$\omega_\mathcal{O}(\xi)(\widehat{\zeta}(\xi), \widehat{\eta}(\xi)) = \langle \xi, [\zeta, \eta] \rangle.$$

The moment map $\mu : \mathcal{O} \to \mathfrak{l}$ is unique in the case of semisimple groups and, again, is just the inclusion $\mathcal{O} \hookrightarrow \mathfrak{l}$.

8.1B Compatibility of the complex structure

Now we restrict to our principal case of interest, the case where Z is a complex flag manifold G/Q. Thus Z is a compact complex homogeneous G-manifold. We suppose that the parabolic subgroup Q corresponds to the base point $z_0 \in Z$. As usual let G_u be a compact real from of G which is likewise acting transitively on Z. Recall that σ denotes the complex conjugation which defines G_u.

Let L_u denote the isotropy subgroup of G_u at the base point z_0. It is a compact real form of the Levi factor (reductive component) of Q. If ξ is a generic element of the center $\mathfrak{z}(\mathfrak{l}_u)$, then L_u is its centralizer in G_u. Thus we may regard Z as the adjoint orbit $\mathcal{O} := \mathrm{Ad}(G_u)(\xi)$ equipped with the invariant symplectic form $\omega_{\mathcal{O}}$. This identifies ξ with the base point $z_0 \in G_u/L_u = G/Q$. We now show that with these identifications, $\omega_{\mathcal{O}}$ is compatible with the complex structure of Z.

The inclusion $\mathfrak{g}_u \hookrightarrow \mathfrak{g}$ defines an isomorphism $\mathfrak{g}_u/\mathfrak{l}_u \to \mathfrak{g}/\mathfrak{q}$ of real vector spaces. That isomorphism defines an L_u-invariant complex vector space structure J on $\mathfrak{g}_u/\mathfrak{l}_u$. Compatibility means that

$$\langle \xi, [J\zeta, J\eta]\rangle = \langle \xi, [\zeta, \eta]\rangle \text{ for all } \zeta, \eta \in \mathfrak{g}_u.$$

To show this let T_u be a maximal torus in L_u, i.e., $T := T_u^{\mathbb{C}}$ is a Cartan subgroup of $L := L_u^{\mathbb{C}}$. Choose a positive root system Σ^+ such that $\mathfrak{q} = \mathfrak{r}_- + \mathfrak{l}$, where $\mathfrak{r}_- = \sum \mathfrak{g}_{-\alpha}$ as α runs over the set of positive roots not involved in $\mathfrak{l}$. Also, let $\mathfrak{r}_+ = \sum \mathfrak{g}_{+\alpha}$, where α runs over the positive roots not involved in $\mathfrak{l}$. Identify $\mathfrak{r}_+$ with the holomorphic tangent space $\mathfrak{g}/\mathfrak{q}$ and define

$$I : \mathfrak{r}_+ = \mathfrak{g}/\mathfrak{q} \to \mathfrak{g}_u/\mathfrak{l}_u \text{ by } \zeta \mapsto \tfrac{1}{2}(\zeta + \sigma(\zeta)) \operatorname{mod}(\mathfrak{l}_u).$$

It is immediate that I is the projection onto the image of the inclusion $\mathfrak{g}_u \hookrightarrow \mathfrak{g}$, and therefore the complex structure J is given by

$$J(\zeta + \sigma(\zeta)) := i(\zeta - \sigma(\zeta)).$$

In what follows we often let $\langle , \rangle$ denote the complex bilinear extension of the Killing form of $\mathfrak{g}_u$ to $\mathfrak{g}$. For example, if ξ is the base point chosen as above, then $b(\zeta, \eta) := \langle \xi, [\zeta, \eta]\rangle$ defines an antisymmetric $\mathbb{C}$-bilinear form for which $\mathfrak{r}_+$ and $\mathfrak{r}_- = \sigma(\mathfrak{r}_+)$ are isotropic subspaces. Thus, for example,

$$(8.1.1) \quad \langle \xi, [\zeta + \sigma(\zeta), \eta + \sigma(\eta)]\rangle = \langle \xi, [\zeta, \sigma(\eta)] + [\sigma(\zeta), \eta]\rangle = 2\langle \xi, [\zeta, \sigma(\eta)]\rangle.$$

Proposition 8.1.2. *The complex structure J preserves the coadjoint symplectic form $\omega_{\mathcal{O}}$.*

Proof. Since the last expression in (8.1.1) is left unchanged if both ζ and η are multiplied by i, the result is immediate. □

Now ω is nondegenerate, but of course we must choose ξ more carefully in order to insure that it is positive definite. For this it is useful to note that the associated hermitian metric is given by

$$(8.1.3) \qquad h(\zeta, \eta) = \omega(i\zeta, \eta) + i\omega(\zeta, \eta) = 2\langle i\xi, [\zeta, \sigma(\eta)]\rangle$$

for $\zeta, \eta \in \mathfrak{g}$ (projected to $\mathfrak{g}/\mathfrak{q} \cong \mathfrak{r}_+$).

Proposition 8.1.4. *For every root space $\mathfrak{g}_\alpha$ in $\mathfrak{r}_+$ choose $0 \neq \zeta_\alpha \in \mathfrak{g}_\alpha$ and define $iH_\alpha := [\zeta_\alpha, \sigma(\zeta_\alpha)]$. Let $P = \{\nu \in \mathfrak{t}_u \mid \langle \nu, H_\alpha\rangle < 0$ whenever $\mathfrak{g}_\alpha \in \mathfrak{r}_+\}$. Then P is a nonempty open cone, and if $\xi \in P$, the associated form ω is kählerian.*

Proof. Let $\rho = \frac{1}{2}\sum_{\alpha\in\Sigma^+}\alpha$. Then $\rho(H_\alpha) > 0$, in other words, $\langle \rho, H_\alpha\rangle > 0$, for every $\alpha \in \Sigma^+$. Thus $-\rho \in P$ and so P is not empty. It is open because it is defined by strict inequalities. From (8.1.3), the definition of P implies that $h(\zeta_\alpha, \zeta_\alpha) > 0$ for all such root spaces (see (8.1.3)). Since the root decomposition of $\mathfrak{r}_+$ is h-orthogonal, the desired result is immediate. □

8.2 The K_0-energy function

Recall our standard setup: τ is complex conjugation of G over G_0, θ is the Cartan involution of G_0 with fixed point set K_0, and $\sigma = \tau\theta = \theta\tau$ is complex conjugation of G over its compact real form G_u. As usual θ also denotes the holomorphic extension of θ to G.

Equip $\mathfrak{g}_u$ with its standard inner product structure based on the positive definite hermitian inner product $(\xi, \eta) := -\langle \xi, \sigma(\eta) \rangle$ on $\mathfrak{g}$. Then $\mathfrak{g}_u = \mathfrak{k}_0 + i\mathfrak{s}_0$ is orthogonal. We split the G_u-moment map, $id_Z = \mu_{G_u} = \mu_{\mathfrak{k}_0} + \mu_{i\mathfrak{s}_0}$, where $\mu_{\mathfrak{k}_0}$ and $\mu_{i\mathfrak{s}_0}$ are the projections on the respective components.

As was observed above, $\mu_{\mathfrak{k}_0}$ is indeed a moment map for the K_0-action on Z. Note that since Z is a single orbit $\mathrm{Ad}(G_u)(\xi)$, $\|id_Z\|^2$ is a constant. We normalize that constant to be 1. Thus

$$1 = \|\mu_{\mathfrak{k}_0}\|^2 + \|\mu_{i\mathfrak{s}_0}\|^2.$$

8.2A Basics on the gradient field

Here we explicitly compute the gradient of the energy function $E := \|\mu_{\mathfrak{k}_0}\|^2$ with respect to the metric defined by the Kähler form ω. For this it is convenient to think in general terms of a moment map $\mu : M \to \mathfrak{g}_u^*$. Take $\{\xi_1^*, \ldots, \xi_m^*\}$ to be an orthonormal basis of $\mathfrak{g}_u^*$. Since $\mu_\xi = \xi \circ \mu : M \to \mathfrak{g}_\mu^*$ (by definition) it follows that $\mu = \sum \mu_{\xi_i} \xi_i^*$. Note that $d\mu_{\xi_i} = \iota_{\xi_i}\omega$ (interior product). Therefore

$$\|\mu\|^2 = \sum \mu_{\xi_i}^2 \text{ and}$$

$$d\|\mu\|^2 = 2\sum \mu_{\xi_i} d\mu_{\xi_i} = 2\sum \mu_{\xi_i} i_{\widehat{\xi_i}}\omega = 2\sum \mu_{\xi_i} Re\, h(J\widehat{\xi_i}, \cdot).$$

The following is a translation into our concrete setting.

Proposition 8.2.1. *If $\{\xi_1, \ldots, \xi_k\}$ is an orthonormal basis of $\mathfrak{k}_0$ and $\{\xi_{k+1}, \ldots, \xi_m\}$ is an orthonormal basis of $i\mathfrak{s}_0$, then*

$$\nabla E = \nabla\|\mu_{\mathfrak{k}_0}\|^2 = 2\sum_{j=1}^{k} \mu_{\xi_j} J\widehat{\xi_j} = -\nabla\|\mu_{i\mathfrak{s}_0}\|^2 = -2\sum_{j=k+1}^{m} \mu_{\xi_j} J\widehat{\xi_j}.$$

Corollary 8.2.2. *The gradient field ∇E is tangent to both the K- and G_0-orbits in Z.*

8.2B Characterization of the critical points

Let $C = \{z \in Z : \nabla E(z) = 0\}$, the ($K_0$-invariant) set of critical points of the energy function.

If $g \in G_u$ and $z = g(z_0) \in Z$, then $\mathrm{Ad}(g)$ sends our choices of $\mathfrak{l}_u$, $\mathfrak{l}$, $\mathfrak{r}_-$ and $\mathfrak{r}_+$ at z_0 over to the corresponding spaces at z. The results are independent of choice of g that sends z_0 to z. We will commit an abuse of notation, identifying the objects at z_0 with their images at z, thus speaking of the decomposition

$$\mathfrak{g} = \mathfrak{q} + \mathfrak{r}_+ = \mathfrak{r}_- + \mathfrak{l} + \mathfrak{r}_+$$

without reference to z. We will also refer to $\mathrm{Ad}(g)\xi$ as ξ when we are working at $z = g(z_0)$.

The gradient field ∇E is computed as follows. First, since $\mu_{\mathfrak{k}_0}$ is defined by projection,

$$\mu_{\mathfrak{k}_0}(z) = \tfrac{1}{2}(id + \tau)\mu_{G_u}(z) = \tfrac{1}{2}(\xi + \tau(\xi)).$$

Thus

$$E(z) = -\tfrac{1}{4}\langle \xi + \tau(\xi), \xi + \tau(\xi)\rangle = \tfrac{1}{2}(1 - \langle \xi, \tau(\xi)\rangle).$$

Proposition 8.2.3. *If $\zeta \in \mathfrak{g}_u$, then $\widehat{\zeta}(E) = -\langle \zeta, [\xi, \tau\xi]\rangle$.*

Proof. First, note that

$$\widehat{\zeta}(\langle \xi, \tau(\xi)\rangle) = \frac{d}{dt}\Big|_{t=0}\langle \mathrm{Ad}(\exp(\zeta t))(\xi), \tau(\mathrm{Ad}(\exp(\zeta t))(\xi))\rangle.$$

Thus,

$$\begin{aligned}\widehat{\zeta}(E) &= -\tfrac{1}{2}\widehat{\zeta}\langle(\xi, \tau(\xi))\rangle = -\tfrac{1}{2}\big(\langle[\zeta, \xi], \tau(\xi)\rangle + \langle \xi, \tau([\zeta, \xi])\rangle\big)\\ &= -\langle \zeta - \tau(\zeta), [\xi, \tau(\xi)]\rangle = -\langle \zeta, [\xi, \tau(\xi)]\rangle,\end{aligned}$$

as asserted. □

Proposition 8.2.3 tells us that $\nabla E = 0$ just where $[\xi, \tau(\xi)] = 0$. Write $\xi = \xi^{\tau} + \xi^{-\tau}$ according to the decomposition $\mathfrak{g}_u = \mathfrak{k}_0 + i\mathfrak{s}_0$. This yields the following

Corollary 8.2.4. $C = \{z \in Z \mid [\xi^{\tau}, \xi^{-\tau}] = 0\}$.

Using this characterization of the set of critical points, we now prove the basic fact from which all further results follow by elementary calculation.

Theorem 8.2.5. *A point $z \in Z$ is critical for ∇E, i.e., $z \in C$, if and only if the reductive part $\mathfrak{l}_z$ of the isotropy algebra $\mathfrak{q}_z$ has a Cartan subalgebra that is both τ- and σ-invariant.*

Before turning to the proof, we note that $\mathfrak{l} = \mathfrak{l}_{z_0}$ has Cartan subalgebras that are τ- and σ-invariant.

Lemma 8.2.6. *Let $\mathfrak{l}$ be a complex semisimple Lie algebra with involutions σ and τ as above. Then there exists a Cartan subalgebra $\mathfrak{t}$ in $\mathfrak{l}$ which is both σ- and τ-invariant.*

Proof. If $\mathfrak{g}_0 = \mathfrak{k}_0 + \mathfrak{s}_0$ is the Cartan decomposition, $\mathfrak{a}_0$ is a maximal abelian subspace of $\mathfrak{s}_0$, and $\mathfrak{t}_0$ is a maximal torus in the centralizer $\mathfrak{m}_0$ of $\mathfrak{a}_0$ in $\mathfrak{k}_0$, then the complexification $\mathfrak{t}$ of $\mathfrak{t}_0 + \mathfrak{a}_0$ does the job. □

Proof of Theorem 8.2.5. Suppose $z \in C$, i.e., that ξ^{τ} and $\xi^{-\tau}$ commute. Since ξ^{τ} and $\xi^{-\tau}$ are semisimple, they are contained in a Cartan subalgebra $\mathfrak{h}$ of $\mathfrak{l}$. Thus $Z_G(\xi^{\tau}, \xi^{-\tau}) := Z_G(\xi^{\tau}) \cap Z_G(\xi^{-\tau})$ is connected, reductive, and σ- and τ-invariant. Note that $\mathfrak{h}$ is a Cartan subalgebra of $\mathfrak{g}$ because $\mathfrak{l}$ has full rank in $\mathfrak{g}$.

Conversely, let $\mathfrak{h}$ be a τ-invariant Cartan subalgebra of $\mathfrak{h}$. Since $\xi \in \mathfrak{z}(\mathfrak{l})$, it follows that $\xi \in \mathfrak{h}$, and since $\tau(\xi) \in \mathfrak{h}$, it follows that ξ and $\tau(\xi)$ commute. □

8.2C Tangent spaces at the critical points

Here we compute the tangent spaces $T_z(K(z))$ and $T_z(G_0(z))$ at a critical point $z \in C$. As usual

$$\mathfrak{g} = \mathfrak{q} + \mathfrak{r}_+ = \sigma(\mathfrak{r}_+) + \mathfrak{l} + \mathfrak{r}_+ \text{ and } \mathfrak{r}_- = \sigma(\mathfrak{r}_+) \text{ at } z.$$

Our discussions of roots in this section refer to a τ- and σ-invariant Cartan subalgebra $\mathfrak{h}$ of $\mathfrak{l}$, as in Lemma 8.2.6.

The set of roots $\Sigma(\mathfrak{r}_+, \mathfrak{h})$ which are involved in $\mathfrak{r}_+$ is the disjoint union of three sets of roots. Those and the corresponding subspaces of $\mathfrak{r}_+$ are as follows:

$$\text{(i)}\quad \Sigma_1 = \{\alpha \in \Sigma(\mathfrak{r}_+, \mathfrak{h}) \mid \tau^*(\alpha) \in \Sigma(\mathfrak{r}_+, \mathfrak{h})\} \text{ and } \mathfrak{r}_1 = \sum\nolimits_{\alpha \in \Sigma_1} \mathfrak{g}_\alpha,$$

$$\text{(ii)}\quad \Sigma_2 = \{\alpha \in \Sigma(\mathfrak{r}_+, \mathfrak{h}) \mid \tau^*(\alpha) \in \Sigma(\mathfrak{r}_-, \mathfrak{h})\} \text{ and } \mathfrak{r}_2 = \sum\nolimits_{\alpha \in \Sigma_2} \mathfrak{g}_\alpha,$$

$$\text{(iii)}\quad \Sigma_3 = \{\alpha \in \Sigma(\mathfrak{r}_+, \mathfrak{h}) \mid \tau^*(\alpha) \in \Sigma(\mathfrak{l}, \mathfrak{h})\} \quad \text{and } \mathfrak{r}_3 = \sum\nolimits_{\alpha \in \Sigma_3} \mathfrak{g}_\alpha.$$

Now $\mathfrak{r}_+ = \mathfrak{r}_1 + \mathfrak{r}_2 + \mathfrak{r}_3$, vector space direct sum, where

$$\begin{aligned} \mathfrak{r}_1 &= \mathfrak{r}_+ \cap \tau(\mathfrak{r}_+) = \mathfrak{r}_+^\tau + \mathfrak{r}_+^{-\tau}, \\ \mathfrak{r}_2 &= \mathfrak{r}_+ \cap \tau(\mathfrak{r}_-) = \mathfrak{r}_+ \cap \theta(\mathfrak{r}_+) = \mathfrak{r}_+^\theta + \mathfrak{r}_+^{-\theta}, \quad \text{and} \\ \mathfrak{r}_3 &= \mathfrak{r}_+ \cap \tau(\mathfrak{l}). \end{aligned}$$

Proposition 8.2.7. *Let $z \in Z$ be a critical point of the gradient field ∇E, i.e., $z \in C$. Identify $T_z(Z)$ with $\mathfrak{r}_+$. Then*

$$T_z(G_0(z)) = \mathfrak{r}_+^\tau + \mathfrak{r}_2 + \mathfrak{r}_3,\ T_z(K(z)) = \mathfrak{r}_1 + \mathfrak{r}_+^\theta + \mathfrak{r}_3,\ \textit{and}\ T_z(K_0(z)) = \mathfrak{r}_+^\tau + \mathfrak{r}_+^\theta + \mathfrak{r}_3.$$

Proof. Let $\mathfrak{m} := \mathfrak{l} + \tau(\mathfrak{l})$, $\mathfrak{u}_1 := \sigma(\mathfrak{r}_1) + \mathfrak{r}_1$, and $\mathfrak{u}_2 := \sigma(\mathfrak{r}_2) + \mathfrak{r}_2$. Since the decomposition $\mathfrak{g} = \mathfrak{u}_1 + \mathfrak{u}_2 + \mathfrak{m}$ is τ- and σ-invariant, it is a matter of projecting the fixed point spaces of τ, σ and (the complex linear involution) θ in these components into $\mathfrak{r}_+$.

Let us begin with $\mathfrak{m}$. Its projection into $\mathfrak{r}_+$ is $\mathfrak{r}_3 = \mathfrak{r}_+ \cap \tau(\mathfrak{l})$. Observe that if $\zeta \in \mathfrak{r}_3$, then $(id + \sigma)(id + \tau)(\zeta)$ is in $\mathfrak{m}$, is τ- and σ-fixed, and projects to a multiple of ζ. Thus $T_z(K_0(z)) \supset \mathfrak{r}_3$, and consequently the same holds for $T_z(G_0(z))$ and $T_z(K(z))$.

Since the decomposition $\mathfrak{u}_1 = \sigma(\mathfrak{r}_1) + \mathfrak{r}_1$ is τ-invariant, the projection of $\mathfrak{u}_1^\tau$ is just $\mathfrak{r}_1^\tau = \mathfrak{r}_+^\tau$. For $\zeta \in \mathfrak{r}_+^\tau$ observe that $(id + \sigma)(id + \tau)(\zeta)$ projects to 2ζ. Consequently, $\mathfrak{r}_+^\tau \subset T_z(K_0(z))$. Since the intersection of $T_z(G_0(z))$ with $\mathfrak{r}_1$ has already been shown to be $\mathfrak{r}_+^\tau$, the same holds for $T_z(K_0(z))$. Finally, $\mathfrak{r}_1$ is the complexification of $\mathfrak{r}_+^\tau$ and thus the $\mathfrak{r}_1$ is contained in the complex vector space $T_z(K(z))$.

To complete the task, consider $\mathfrak{u}_2$. Its decomposition $\mathfrak{u}_2 = \tau(\mathfrak{r}_2) + \mathfrak{r}_2$ is θ-invariant. Hence, the projection of $\mathfrak{u}_2^\theta$ is $\mathfrak{r}_2^\theta = \mathfrak{r}_+^\theta$. Computing as above, if $\zeta \in \mathfrak{r}_+^\theta$, then $(id + \sigma)(id + \tau)(\zeta)$ projects to 2ζ, and therefore the intersections of both $T_z(K(z))$ and $T_z(K_0(z))$ with $\mathfrak{r}_2$ are $\mathfrak{r}_+^\theta$. Finally, $\mathfrak{u}_2^\tau \cap \mathfrak{r}_2 = \{0\}$. So by dimension it projects surjectively, in other words, $T_z(G_0(z)) \supset \mathfrak{r}_2$. □

8.2D The Hessian of the energy function

Using Proposition 8.2.3, we are able to directly compute the Hessian Hess(E) of E at a critical point $z \in C$.

Lemma 8.2.8. *If $\zeta, \eta \in \mathfrak{g}_u$ and $z \in C$ then, at z,*

$$\mathrm{Hess}(E)(\widehat{\zeta}, \widehat{\eta}) = \widehat{\eta}\widehat{\zeta}(E) = \tfrac{1}{2}\langle[\zeta - \tau(\zeta), \tau(\xi)], [\eta - \tau(\eta), \xi]\rangle.$$

Proof. From the next to last expression for $\widehat{\zeta}(E)$ in the argument of Proposition 8.2.3 we have

$$\begin{aligned} 2\widehat{\eta}(\widehat{\zeta}(E)) &= -\widehat{\eta}\langle\zeta - \tau(\zeta), [\xi, \tau(\xi)]\rangle \\ &= -\tfrac{d}{dt}\Big|_{t=0}\langle\zeta - \tau(\zeta), [\mathrm{Ad}(\exp(t\eta))(\xi), \tau(\mathrm{Ad}(\exp(t\eta))(\xi))]\rangle. \end{aligned}$$

Standard manipulations, and the fact that $\nabla E(z) = 0$ and thus $[\xi, \tau(\xi)] = 0$, lead us to

$$\begin{aligned} 2\,\mathrm{Hess}(E)(\widehat{\zeta}, \widehat{\eta}) &= -\langle\zeta - \tau(\zeta), [[\eta, \xi], \tau(\xi)] + [\xi, \tau([\eta, \xi])]\rangle \\ &= -\langle\zeta - \tau(\zeta), [[\eta - \tau(\eta), \xi], \tau(\xi)]\rangle \\ &= \langle[\zeta - \tau(\zeta), \tau(\xi)], [\eta - \tau(\eta), \xi]\rangle, \end{aligned}$$

as asserted. □

This formula becomes quite simple when expressed in terms of root vectors.

Proposition 8.2.9. *Let $z \in C$. Identify $T_z(Z)$ with $\mathfrak{r}_+$, as usual. Let $\zeta_\alpha \in \mathfrak{g}_\alpha$ and $\zeta_\beta \in \mathfrak{g}_\beta$ be root vectors in $\mathfrak{r}_+$. Let ζ'_α and ζ'_β denote the real vector fields on Z defined by the respective projections $\frac{1}{2}(\zeta_\alpha + \sigma(\zeta_\alpha))$ and $\frac{1}{2}(\zeta_\beta + \sigma(\zeta_\beta))$ to $\mathfrak{g}_u$. Set $c(\alpha) = -\frac{1}{2}\alpha(\xi)\alpha(\tau(\xi))$. Then*

$$\mathrm{Hess}(E)(\zeta'_\alpha, \zeta'_\beta) = c(\alpha)\langle(\zeta_\alpha, \sigma(\zeta_\beta) - \tau(\zeta_\beta) - \theta(\zeta_\beta)\rangle \text{ at } z. \tag{8.2.10}$$

Proof. If γ is a root, denote $V_\gamma := (\zeta_\gamma + \sigma(\zeta_\gamma) - \tau(\zeta_\gamma + \sigma(\zeta_\gamma))$. Then Lemma 8.2.8 gives us

$$8\,\mathrm{Hess}(E)(\zeta'_\alpha, \zeta'_\beta) = \langle[V_\alpha, \tau(\xi)], [V_\beta, \xi]\rangle.$$

Using $\sigma(\xi) = \xi$, $\sigma(V_\alpha) = V_\alpha$, $\tau(V_\alpha) = -V_\alpha$, $\tau\sigma = \theta$, and $\langle\mathfrak{r}_+, \mathfrak{r}_+\rangle = 0$, we compute

$$\begin{aligned} \langle[V_\alpha, \tau(\xi)], [V_\beta, \xi]\rangle &= 2\langle[\zeta_\alpha, \tau(\xi)], [V_\beta, \xi]\rangle + 2\langle[\zeta_\alpha, \xi], [V_\beta, \tau(\xi)]\rangle \\ &= -2\alpha(\tau(\xi))\langle\zeta_\alpha, [V_\beta, \tau(\xi)]\rangle - 2\alpha(\xi)\langle\zeta_\alpha, [V_\beta, \tau(\xi)]\rangle \\ &= -4\alpha(\xi)\alpha(\tau(\xi))\langle\zeta, V_\beta\rangle. \end{aligned}$$

Thus $\mathrm{Hess}(E)(\zeta'_\alpha, \zeta'_\beta) = c(\alpha)\langle\zeta_\alpha, \sigma(\zeta_\beta) - \tau(\zeta_\beta) - \theta(\zeta_\beta)\rangle$, as asserted. □

Corollary 8.2.11. *If $z \in C$, then the decomposition*

$$T_z Z \cong \mathfrak{r}_1 + \mathfrak{r}_2 + \mathfrak{r}_3$$

is Hess(E)*-orthogonal.*

Proof. This follows immediately from the above formula (8.2.10), the fact that the decomposition $\mathfrak{g} = \mathfrak{u}_1 + \mathfrak{u}_2 + \mathfrak{m}$ is invariant with respect to all involutions at hand, and that the root spaces $\mathfrak{g}_\alpha$ and $\mathfrak{g}_\beta$ are Killing form orthogonal unless $\beta = -\alpha$. □

Now let us compute $\mathrm{Hess}(E)|_{\mathfrak{r}_i}$, $i = 1, 2$.

Lemma 8.2.12. *Root spaces $\mathfrak{g}_\alpha$ and $\mathfrak{g}_\beta$ in $\mathfrak{r}_1$ are* Hess(E)*-orthogonal unless $\beta = \alpha$ or $\beta = \tau(\alpha)$. In $\mathfrak{r}_2$ they are* Hess(E)*-orthogonal unless $\beta = \alpha$ or $\beta = \theta(\alpha)$.*

Proof. This follows immediately from formula (8.2.10), because, for example in the case of $\mathfrak{r}_1$, if $\tau(\alpha) \neq \beta \neq \alpha$, then no term in $\sigma(\zeta_\beta) - \tau(\zeta_\beta) - \theta(\zeta_\beta)$ is in $\mathfrak{g}_{-\alpha}$. □

Finally, let us compute the Hess(E) norms $\|\cdot\|^2_{\mathrm{Hess}}$ in the root spaces in $\mathfrak{r}_1$ and $\mathfrak{r}_2$. For $\mathfrak{g}_\alpha \subset \mathfrak{r}_1$ and $\zeta_\alpha \in \mathfrak{g}_\alpha \setminus \{0\}$

$$\mathrm{Hess}(E)(\zeta'_\alpha, \zeta'_\alpha) = c(\alpha)\langle \zeta_\alpha, \sigma(\zeta_\alpha) - \theta(\zeta_\alpha)\rangle$$

and

$$\mathrm{Hess}(E)(\zeta'_\alpha, \tau(\zeta_\alpha)') = c(\alpha)\langle \zeta_\alpha, \theta(\zeta_\alpha) - \sigma(\zeta_\alpha)\rangle.$$

Also note that $\mathrm{Hess}(E)(\zeta'_\alpha, \zeta'_\alpha) = \mathrm{Hess}(E)(\tau(\zeta_\alpha)', \tau(\zeta_\alpha)')$.

Suppose $\tau(\zeta_\alpha) \notin \mathfrak{g}_\alpha$. If $\zeta := a\zeta_\alpha + b\tau(\zeta_\alpha)$ for $a, b \in \mathbb{C}$, then

$$\|a\zeta'_\alpha + b\tau(\zeta_\alpha)'\|^2_{\mathrm{Hess}} = c(\alpha)|a - b|^2\langle \zeta_\alpha, \sigma(\zeta_\alpha)\rangle.$$

Now suppose $\tau(\zeta_\alpha) \in \mathfrak{g}_\alpha$. Since $\mathfrak{r}^\tau$ is contained in the tangent space $T_z(K_0(z))$, which is in the degeneracy of Hess(E), we may assume that $\tau(\zeta_\alpha) = -\zeta_\alpha$; then Hess$(E)$-norm is just twice that above.

In both of the cases above the norm square has a factor $|a - b|$. Its vanishing defines a line $\{t(\zeta_\alpha + \tau(\zeta_\alpha))\}$ in the complex vector space $\mathfrak{r}_+$. That line is either contained in $\mathfrak{r}_+$ or has intersection 0 with $\mathfrak{r}_+$. But if t is real, then $t(\zeta_\alpha + \tau(\zeta_\alpha))$ is in the fixed point set of τ, which has intersection 0 with $\mathfrak{r}_+$. Thus Hess(E) is positive definite on $\mathfrak{r}_1$.

In case $\mathfrak{g}_\alpha$ is in $\mathfrak{r}_2$ and $\zeta = a\zeta_\alpha + b\theta(\zeta_\alpha)$, we give a similar argument. If $\tau(\zeta_\alpha) \notin \mathfrak{g}_{-\alpha}$, then

$$\mathrm{Hess}(E)(\zeta'_\alpha, \zeta'_\alpha) = c(\alpha)\langle \zeta_\alpha, \sigma(\zeta_\alpha)\rangle$$

and

$$\mathrm{Hess}(E)(\zeta'_\alpha, \theta(\zeta_\alpha)') = -c(\alpha)\langle \zeta_\alpha, \sigma(\zeta_\alpha)\rangle.$$

If $\tau(\zeta_\alpha) \in \mathfrak{g}_{-\alpha}$, then $\theta(\zeta_\alpha) = \pm\zeta_\alpha$, and since $\mathfrak{r}^\theta$ is contained in $T_z(K_0(z))$, which is in the degeneracy of Hess(E), we may assume that $\theta(\zeta_\alpha) = -\zeta_\alpha$. Again the Hess$(E)$-norm is just twice that above.

We have $c(\alpha) < 0$ because $\mathfrak{g}_\alpha$ is in $\mathfrak{r}_2$. Consequently

$$\|\zeta'\|^2_{\mathrm{Hess}} < 0 \text{ unless } a = b.$$

But by the same argument as for the $\mathfrak{r}_1$ case above, if $a = b$, then $\zeta \notin \mathfrak{r}_2$.

These calculations can be summarized as follows.

Theorem 8.2.13. *Let $\kappa \in \mathrm{Orb}_Z(K)$ and $\gamma \in \mathrm{Orb}_Z(G_0)$, and suppose that $z \in \kappa \cap \gamma$ is a critical point of the energy function E. Then the degeneracy of the Hessian of the energy function E at z is the tangent space $T_z\kappa_0$ at z of $\kappa_0 = K_0(z)$. The induced forms on the quotients $T_z\kappa/T_z\kappa_0$ and $T_z\gamma/T_z\kappa_0$ are, respectively, positive and negative definite.*

8.3 Duality

Here we apply elementary aspects of Bott–Morse theory to complete the proofs of the duality and retraction statements which were mentioned in the introduction of this chapter.

As we have seen in the previous sections, the energy function $E : Z \to \mathbb{R}^{\geqq 0}$ has the following properties:

- $C = \{z \in Z : \nabla E(z) = 0\}$ consists of finitely many K_0-orbits κ_0.
- For $z \in \kappa_0 \subset C$ the orbits $\kappa = K(z)$ and $\gamma = G_0(z)$ satisfy

$$T_z\kappa + T_z\gamma = T_z Z \text{ and } T_z\kappa \cap T_z\gamma = T_z\kappa_0.$$

- The degeneracy of the Hessian of E is $T_z\kappa_0$ and the induced form on $T_z\kappa/T_z\kappa_0$ (respectively, $T_z\gamma/T_z\kappa_0$) is positive (respectively, negative) definite.

In particular, E is a Bott–Morse function (see, e.g., [AB]).

The Bott–Morse Lemma then yields a smooth embedding of a tubular neighborhood T_{κ_0} of the normal bundle of κ_0 in Z, such that ∇E is tangent to its fibers, and such that every base point $z \in \kappa_0$ of each such fiber $(T_{\kappa_0})_z$ an isolated critical point of $\nabla E|_{(T_{\kappa_0})_z}$. This has strong consequences for the limiting behavior of ∇E.

We formulate these for $t \to +\infty$ and $\gamma \in \mathrm{Orb}_Z(G_0)$. The analogous results hold for $t \to -\infty$ and $\kappa \in \mathrm{Orb}_Z(K)$.

Let $\{e(t)\}_{t\in\mathbb{R}}$ denote the one-parameter group for ∇E. If $z \in (T_{\kappa_0})_{z_0}$, then classical Morse theory applies. For example, if there exists a sequence $t_n \to \infty$ so that $\{e(t_n)(z)\} \subset T_{\kappa_0}$, then

$$\lim_{t\to\infty} e(t)(z) = z_0.$$

In this case, $z \in \gamma$, where γ is the G_0-orbit of Theorem 8.2.13.

Theorem 8.3.1. *For every $z \in Z$ the following hold:*

1. $\lim_{t\to\infty} e(t)(z) =: \pi(z)$ *exists and is in C.*
2. *Every $\gamma \in \mathrm{Orb}_Z(G_0)$ contains a unique critical orbit $\kappa_0 \subset C$. The mapping $\pi : \gamma \to \kappa_0$, $z \mapsto \lim_{t\to\infty} e(t)(z)$, is well defined, smooth, and K_0-equivariant.*
3. *The action mapping $\mathbb{R} \times \gamma \to \gamma$, $z \mapsto e(t)(z)$, extended to $\mathbb{R} \cup \{\infty\}$ by the limiting process, realizes κ_0 as a K_0-equivariant strong deformation retract of γ.*

All of the above hold if γ is replaced by $\kappa \in \mathrm{Orb}_Z(K)$ and $e(t)$ by $e(-t)$.

Proof. For Assertion 1 observe that if $z \in \gamma$, then $e(t).z \in \gamma$ for all $t \in \mathbb{R}$. If not, then let t_0 be the smallest (in absolute value) time such that $e(t_0)(z)$ is in some other orbit $\widetilde{\gamma}$ in $\mathrm{bd}(\gamma)$. Since ∇E is tangent to all G_0-orbits, it follows that $e(t) \in \widetilde{\gamma}$ for all t near t_0, contrary to the minimality assumption.

Now let $t_n \to \infty$ with $z_n := e(t_n)(z) \to z_0$. Of course, z_0 is contained in some $\kappa_0 \subset C$. Therefore, for n sufficiently large, $z_n \in T_{\kappa_0}$. Thus the local observations above show that $z_0 \in \gamma = G_0(z)$, and $z_0 = \lim_{t\to\infty} e(t)(z)$ is well defined.

For Assertion 2 we remark that if $e(t)(z)$ is in some T_{κ_0}, then $e(t)(U) \subset T_{\kappa_0}$ for U a sufficiently small neighborhood of z. Thus

$$A(\kappa_0) := \{z \in \gamma : \lim_{t\to\infty} e(t)(z) \in \kappa_0\}$$

is open in γ. Since γ is connected, it follows that $A(\kappa_0) = \gamma$, where κ_0 is the unique critical K_0-orbit in γ. Thus the fact that $\pi|_{(T_{\kappa_0}\cap\gamma)}$ is well defined and smooth implies the global result.

Assertion 3 follows immediately from Assertion 2.

The same argument applies to the situation where $e(t)$ is replaced by $e(-t)$ and γ by $\kappa \in \mathrm{Orb}_Z(K)$. □

For $z \in \kappa_0 \subset C$ define $\kappa(\kappa_0) := K(z)$ and $\gamma(\kappa_0) := G_0(z)$. Then Assertion 2 can be reformulated by stating that $\kappa : C \to \mathrm{Orb}_Z(K)$ and $\gamma : C \to \mathrm{Orb}_Z(G_0)$ are bijections. These in turn establish a bijection

$$\mathrm{Orb}_Z(K) \cong \mathrm{Orb}_Z(G_0).$$

If κ and γ correspond to each other in this bijection, then we say that (κ, γ) is a **dual pair**. We now collect some useful characterizations of dual pairs.

Theorem 8.3.2. *The following are equivalent:*

1. $\kappa \cap \gamma$ *consists of exactly one* K_0*-orbit* κ_0.
2. $\kappa \cap \gamma$ *contains an isolated* K_0*-orbit* κ_0.
3. (κ, γ) *is a dual pair.*
4. $\kappa \cap \gamma$ *is compact.*

Proof. Since ∇E is tangent to both the G_0- and K-orbits in Z, if κ_0 is isolated in $\kappa \cap \gamma$, it is contained in C, i.e., (κ, γ) is a dual pair. Thus (1) ⇒ (2) ⇒ (3).

For the remaining implications note that if the intersection $\kappa \in \mathrm{Orb}_Z(K)$ and $\gamma \in \mathrm{Orb}_Z(G_0)$ contains a point $z \notin C$, then the full noncompact orbit $e(t).z$ is contained in this intersection. Now, e.g., $\lim_{t\to\infty} e(t).z$ is in some orbit $\widetilde{\kappa}$ on the boundary of κ, because the critical set in κ is repulsive with respect to this flow. Thus, if $\kappa \cap \gamma$ contains a noncritical point, then it is noncompact.

Reversing this, if $\kappa \cap \gamma$ is compact, then it is contained in C and therefore consists of a unique critical orbit κ_0, i.e., (4) ⇒ (1).

Finally, assume that (κ, γ) is a dual pair. If the intersection $\kappa \cap \gamma$ were not compact, it would contain a point $z \notin C$. For example, this would imply that

$$\lim_{t\to-\infty} e(t).z = z_0 \in \kappa_0 \subset \gamma.$$

But this is contrary to $\lim_{t\to-\infty} e(t).z$ being in some orbit $\widetilde{\gamma}$ in $\mathrm{bd}(\gamma)$. Therefore (3) $\Rightarrow$ (4), and the proof is complete. □

Corollary 8.3.3. *Suppose that (κ, γ) is not a dual pair and that $\kappa \cap \gamma \neq \emptyset$. Then for every $w \in \kappa \cap \gamma$ the gradient ∇E is nowhere tangent to $K_0.w$, and for ε sufficiently small, the action map*

$$(-\varepsilon, \varepsilon) \times K_0.w \to \kappa \cap \gamma, \quad by\ (t, z) \mapsto e(t)(z),$$

is a diffeomorphism onto its image M, which is a locally closed submanifold of $\kappa \cap \gamma$.

Proof. If ∇E were everywhere tangent to $K_0.w$, then $e(t)$ would not flow to the critical orbit. Since $e(t)$ is K_0-equivariant, it follows that ∇E is nowhere tangent to $K_0.w$ and the desired result is immediate. □

8.4 Orbit ordering

In this brief subsection, we complete Chapter 8 with a proof that duality reverses the partial ordering on orbits that is defined by orbit closure.

If $\gamma_1, \gamma_2 \in \mathrm{Orb}_Z(G_0)$, then we write $\gamma_1 < \gamma_2$ whenever $\gamma_2 \subset c\ell(\gamma_1) \setminus \gamma_1$. Similarly, $\kappa_1 < \kappa_2$ means that $\kappa_2 \subset c\ell(\kappa_1) \setminus \kappa_1$ for $\kappa_1, \kappa_2 \in \mathrm{Orb}_Z(K)$.

Theorem 8.4.1. *If (γ_1, κ_1) and (γ_2, κ_2) are dual pairs, then $\gamma_1 < \gamma_2$ if and only if $\kappa_2 < \kappa_1$.*

The proof of Theorem 8.4.1 requires some technical preparation beyond the basic results of this section on gradient flows. For this, let (γ, κ) be a dual pair and suppose $p_0 \in \gamma \cap \kappa$. Since

$$T_{p_0}\kappa + T_{p_0}\gamma = T_{p_0}Z,$$

it follows that there exists a local submanifold $\Delta \subset G_0$ containing 1 such that (i) the action map $\Delta \to Z$, $g \mapsto g(p_0)$, is a diffeomorphism of Δ onto its image $\Delta(p_0)$, and (ii) $\Delta(p_0)$ is transversal to κ in Z. This transversality says

$$T_{p_0}\kappa \oplus T_{p_0}\Delta(p_0) = T_{p_0}Z. \tag{8.4.2}$$

In the next section we will show that Δ can be chosen so that $\Delta(p_0)$ has very special properties, but here any such submanifold Δ, $1 \in \Delta \subset G_0$ and satisfying (i) and (ii) above, suffices.

Lemma 8.4.3. *If U is a sufficiently small neighborhood of p_0 in κ, and Δ is chosen sufficiently small, then the action map*

$$\alpha : \Delta \times U \to Z$$

is a diffeomorphism of $\Delta \times U$ onto an open neighborhood V of p_0 in Z.

Proof. It follows immediately from the transversality (8.4.2) that the differential of α at $(1, p_0)$ is an isomorphism onto $T_{p_0}Z$. □

Corollary 8.4.4. *If Δ and U are chosen as above and $\Delta \times U$ is identified with its α-image $\Delta(U)$, then every orbit $\tilde{\gamma} \in \mathrm{Orb}_Z(G_0)$ that meets $\Delta \times U$ also meets U. In particular, then $\tilde{\gamma} \cap \kappa \neq \emptyset$.*

Proof. If $p = (g_0, u) \in \tilde{\gamma} \cap (\Delta \times U)$, then $g_0^{-1}(p) = (1, \tilde{u}) \in \tilde{\gamma} \cap U$. □

Proof of Theorem 8.4.1. Suppose that $\gamma_1 < \gamma_2$, i.e., $\gamma_2 \subset c\ell(\gamma_1) \backslash \gamma_1$. An application of the above corollary to $\gamma := \gamma_2$, $\kappa := \kappa_2$, and $\tilde{\gamma} := \gamma_1$ shows that $\kappa_2 \cap \gamma_1 \neq \emptyset$.

Now use the gradient flow φ_t. It realizes $\gamma_1 \cap \kappa_1$ as a strong deformation retract of γ_1. Applying it to $p \in \kappa_2 \cap \gamma_1$ it follows that $\lim_{t \to \infty} \varphi_t(p) \in \kappa_1$. But $c\ell(\kappa_2)$ is invariant by this flow, and therefore it follows that $\kappa_1 \subset c\ell(\kappa_2) \setminus \kappa_2$, i.e., $\kappa_2 < \kappa_1$. □

9

Schubert Slices in the Context of Duality

With the basic results on duality available (see Chapter 8), it is possible to prove the existence of Schubert slices for arbitrary G_0-orbits γ in Z. As a consequence one observes that every point in the boundary $\mathrm{bd}(D)$ of an open G_0-orbit in Z is contained in a $(q+1)$-codimensional Schubert variety which in turn is contained in the complement of D.

The existence of these boundary supporting varieties is sufficient for completing the detailed description of $\mathcal{M}_D$ in the case where G_0 is of hermitian type, thus showing that $\mathcal{M}_D$ is $\mathcal{B}$ or $\overline{\mathcal{B}}$ for domains of holomorphic type, and is $\mathcal{M} = \mathcal{B} \times \overline{\mathcal{B}} = \mathcal{U}$ otherwise.

Even if G_0 is not of hermitian type, this also implies the inclusions

(9.0.1) $$\mathcal{U} \subset \mathcal{M}_D \subset \mathcal{E}_{\mathcal{H}}$$

for every incidence hypersurface H which is defined by a q-codimensional Schubert variety which has nonempty intersection with C_0. Since the Schubert domain S_D is defined to be the connected component containing the base point of the intersection of all such $\mathcal{E}_{\mathcal{H}}$, it is immediate that $\mathcal{M}_D \subset S_D$. In fact another more delicate construction shows that $\mathcal{M}_D = S_D$. In particular this shows that $\mathcal{M}_D$ is a Stein domain, a result whose proof for the measurable case (by completely different methods) was indicated in Section 5.3.

The above summarizes the main results of the present chapter. We now outline its organization. The existence and basic properties of Schubert slices for arbitrary G_0-orbits (Theorem 9.1.1) is proved at the outset of Section 1. This yields the existence of the boundary supporting varieties mentioned above (see Proposition 9.1.4).

The incidence variety I_S in $\mathcal{M}_Z$ of a boundary supporting variety S may not be a hypersurface, i.e., it may have higher codimension. However, since every boundary point of $\mathcal{M}_D$ is contained in such an S, it follows that the boundary of $\mathcal{M}_D$ is contained in the complement of the intersection of the open B-orbits as B ranges over all Iwasawa–Borel subgroups of G. In particular, $\mathcal{U} \subset \mathcal{M}_D$. This is the new ingredient which leads to the inclusions (9.0.1); see Theorem 9.1.5.

The description of $\mathcal{M}_D$ in the hermitian case (Propositions 9.1.7 and 9.1.8) is an immediate consequence of this result together with Theorem 5.4.7.

Finally, Section 2 is devoted to a somewhat refined construction which, given a point C in the boundary of $\mathcal{M}_D$, provides a B-invariant incidence hypersurface which contains C and which is contained in the complement of $\mathcal{M}_D$ in $\mathcal{M}_Z$; see Corollary 9.2.4. This proves that $\mathcal{M}_D = \mathcal{S}_D$. It should be noted that the completely different methods of Chapters 10 and 11 show that $\mathcal{E}_\mathcal{H} = \mathcal{U}$. Therefore the inclusions (9.0.1) imply that $\mathcal{U} = \mathcal{M}_D = \mathcal{S}_D$, unless of course D is of hermitian holomorphic type. This is an essentially stronger result than $\mathcal{M}_D = \mathcal{S}_D$ which is proved here. However, since the method of proof of Corollary 9.2.4 is constructive, it might be useful for applications.

9.1 Schubert slices in arbitrary G_0-orbits

As above let $\mathrm{Orb}_Z(G_0)$ and $\mathrm{Orb}_Z(K)$ be the sets of orbits of G_0 and K in a given homogeneous rational manifold $Z = G/Q$. For an Iwasawa decomposition $G_0 = K_0A_0N_0$ we consider a Borel subgroup B of G which contains A_0N_0. We let $\mathcal{S}_\kappa$ denote the set of B-Schubert varieties S of complementary dimension to κ in Z such that $S \cap c\ell(\kappa) \neq \emptyset$. Since the B-Schubert varieties generate the integral homology $H_*(Z, \mathbb{Z})$, it follows that $\mathcal{S}_\kappa$ is nonempty. As usual $S = \mathcal{O} \cup Y$ denotes the decomposition of a Schubert variety into its open B-orbit $\mathcal{O}$ and its boundary Y.

9.1A Existence theorem

This section is devoted to the following existence theorem for Schubert slices Σ in any G_0-orbit. Many of the proofs are essentially the same as those for the case of an open G_0-orbit.

Theorem 9.1.1. *If $(\gamma, \kappa) \in \mathrm{Orb}(G_0) \times \mathrm{Orb}(K)$ is a dual pair and $S \in \mathcal{S}_\kappa$, then the following hold.*

1. *The intersection $S \cap c\ell(\kappa)$ is finite and contained in $\gamma \cap \kappa$. If $w \in S \cap \kappa$, then $(AN)(w) = B(w) = \mathcal{O}$, in particular the intersection $S \cap c\ell(\kappa)$ takes place at smooth points of both varieties. Furthermore, this intersection is transversal at each of its points:*
$$T_w S \oplus T_w \kappa = T_w Z.$$
2. *The orbit $\Sigma = \Sigma(\gamma, S, w) = (A_0N_0)(w)$ is open in S and is closed in γ with $\Sigma \cap \kappa = \{w\}$.*
3. *The map $K_0 \times c\ell(\Sigma) \to c\ell(\gamma)$, given by $(k_0, z) \mapsto k_0(z)$, is surjective.*

Proof of Assertion 1. Let $w \in S \cap c\ell(\kappa)$. Then $T_w(AN(w)) + T_w(K(w)) = T_w(Z)$ because $\mathfrak{g} = \mathfrak{a} + \mathfrak{n} + \mathfrak{k}$. As $w \in S = c\ell(\mathcal{O})$ and $AN \subset B$, we have $\dim AN(w) \leqq \dim B(w) \leqq \dim \mathcal{O} = \dim S$. Furthermore $w \in c\ell(\kappa)$. Thus $\dim K(w) \leqq \dim \kappa$.

If w were not in κ, this inequality would be strict, in violation of the above additivity of the dimensions of the tangent spaces. Thus $w \in \kappa$ and $T_w(Z) = T_w(S) + T_w(\kappa)$. Since $\dim S + \dim \kappa = \dim Z$, this sum is direct, i.e., $T_w(Z) =$

$T_w(S) \oplus T_w(\kappa)$. In particular the intersection is transversal at each of its points and therefore finite. It also follows that $\dim AN(w) = \dim S$. Thus $AN(w)$ is open in S. Now if $B \cap K =: B_K$, then $B = B_K AN$, and since B_K stabilizes both S and κ, it follows that $B_K \subset B_w$ and consequently $AN(w) = B(w) = \mathcal{O}$.

We have already seen that $K(w)$ is open in κ, forcing $K(w) = \kappa$. It only remains to show that $S \cap \kappa$ is contained in the dual γ of κ. For this let $\widehat{\gamma} = G_0(w)$. Since $\Sigma = (A_0 N_0)(w)$ is transversal to κ in Z, it is also transversal to $K_0(w)$ in $\widehat{\gamma}$. If $\widehat{\gamma}$ were not dual to κ, then $\widehat{\gamma} \cap \kappa \supset M$ as in Corollary 8.3.3, and thus $(A_0 N_0(w)) \cap M$ would be positive-dimensional. This is of course contrary to $A_0 N_0(w)$ being transversal to κ at w, and it follows that $\widehat{\gamma} = \gamma$ is dual to κ. □

Proof of Assertion 2. $G_0 = K_0 A_0 N_0$ implies $T_w(K_0(w)) + T_w(A_0 N_0(w)) = T_w \gamma$. Furthermore, $T_w(K(w)) \oplus T_w(AN(w)) = T_w Z$. Also, by duality, we have $\dim(K_w) + \dim \gamma = \dim Z - \dim K_0(w)$. Hence $\dim AN(w) + \dim K_0(w) = \dim \gamma$ and $\dim A_0 N_0(w) = \dim AN(w)$ is forced. Therefore, $\Sigma = A_0 N_0(w)$ is open in S.

If Σ were not closed in γ, there would be an orbit of lower dimension on its boundary. But since $G_0 = K_0 A_0 N_0$, every $A_0 N_0$-orbit in γ meets $K_0(w)$, and if p is such an intersection point, then $T_p(\gamma) = T_p(K(p)) + T_p(A_0 N_0(p))$. This would be impossible, because $A_0 N_0(p)$ has smaller dimension than the orbit $A_0 N_0(w)$, which is transversal to $K_0(w) = K_0(p)$ in γ.

Finally, we must show that $\Sigma \cap \kappa = \{w\}$. The proof of this exactly follows the line of the proof of Proposition 7.3.9. As in that proof we show that the natural map $\alpha : (K_0 \cap Q_w) \times (A_0 N_0 \cap Q_w) \to (G_0 \cap Q_w)$, given by $\alpha(k_0, a_0 n_0) = k_0 a_0 n_0$ and the image is open in $G_0 \cap Q_w$.

Since $K_0(w)$ is a strong deformation retract of γ, by Theorem 8.3.1. Thus $(G_0 \cap Q_w)/(K_0 \cap Q_w)$ is connected; in other words, every component of $G_0 \cap Q_w$ contains a component of $K_0 \cap Q_w$, and it follows that α is surjective. The proof of the one point intersection property then continues exactly as in the proof for Proposition 7.3.9. □

Proof of Assertion 3. Since K_0 is compact, the image $K_0(c\ell(\gamma))$ is closed in $c\ell(\gamma)$. Since $\gamma = K_0(\Sigma)$, this image is clearly dense in $c\ell(\gamma)$. □

9.1B Supporting Schubert varieties at points of bd(D)

We now return to the setting of an open G_0-orbit D in Z with q-dimensional base cycle C_0. Given a point $p \in \mathrm{bd}(D)$ we use Schubert slices at generic points of $\mathrm{bd}(D)$ to construct a $(q+1)$-codimensional Iwasawa–Schubert variety S_{bd} with $p \in S_{\mathrm{bd}}$ and $S_{\mathrm{bd}} \subset Z \setminus D$.

Although this confirms a sort of complex analytic q-convexity of $\mathrm{bd}(D)$, it should be underlined that, since S_{bd} contains a Schubert slice Σ that is contained in $\mathrm{bd}(D)$, it is not clear how Σ fits into the Levi geometry of the given boundary orbit. In particular, in the case where the boundary component at hand is a nondegenerate, mixed Levi signature hypersurface, the tangent bundle of Σ could be Levi isotropic.

We will say that a point $p \in \mathrm{bd}(D)$ is **generic**, and write $p \in \mathrm{bd}_{\mathrm{gen}}(D)$ if $\gamma_p = G_0(p)$ is open in $\mathrm{bd}(D)$. It is clear that $\mathrm{bd}_{\mathrm{gen}}(D)$ is open and dense in $\mathrm{bd}(D)$.

For $p \in \mathrm{bd}_{\mathrm{gen}}(D)$ the orbit $\gamma = \gamma_p$ need not be a real hypersurface in Z. For example, $G_0 = SL(n+1;\mathbb{R})$ has exactly two orbits in $\mathbb{P}_n(\mathbb{C})$, an open orbit and its complement $\mathbb{P}_n(\mathbb{R})$.

Lemma 9.1.2. *For $p \in \mathrm{bd}(D)_{\mathrm{gen}}$, $\gamma = G_0(p)$, and κ dual to γ, it follows that $\kappa \cap D \neq \emptyset$. Furthermore, if C_0 is the base cycle in D, then $q = \dim C_0 < \dim \kappa$.*

Proof. The property $\kappa \cap D \neq \emptyset$ follows from the fact that γ is open in $\mathrm{bd}(D)$, and from the transversality of the intersection $\kappa \cap \gamma$ in Z which follows from the fact that (κ, γ) is a dual pair.

For the dimension estimate, note that it follows from the K_0-equivariant retraction property of Theorem 8.3.1 that C_0 is dimension-theoretically a minimal K_0-orbit in D, e.g., the K_0-orbits in $\kappa \cap D$ are at least of its dimension. As κ is not compact, now $\dim \kappa > \dim C_0$. □

The above lemma is also an immediate consequence of Theorem 8.4.1.

We will also make use of the following basic fact about Schubert varieties.

Lemma 9.1.3. *Let B be a Borel subgroup of G and S a k-dimensional B-Schubert variety in Z. Choose $\ell \in \mathbb{N}$ so that $\dim Z \geqq \ell \geqq k$. Then there exists a B-Schubert variety S' with $\dim S' = \ell$ and $S' \supset S$.*

Proof. We may assume that $S \neq Z$. Let $\mathcal{O}$ be the open B-orbit in S and let $\mathcal{O}'$ be a B-orbit of minimal dimension among those orbits with $c\ell(\mathcal{O}') \supsetneq \mathcal{O}$.

For $p \in \mathcal{O}$ it follows that $c\ell(\mathcal{O}') \setminus \mathcal{O} = \mathcal{O}'$ near p. Since $\mathcal{O}'$ is affine, it then follows that $\dim \mathcal{O}' = \dim \mathcal{O} + 1$. Applying this argument recursively, we find Schubert varieties $S' := c\ell(\mathcal{O}')$ of every intermediate dimension ℓ. □

We now come to the main result on supporting Schubert varieties.

Proposition 9.1.4. *Let D be an open G_0-orbit on Z. Fix a boundary point $p \in \mathrm{bd}(D)$. Then there exist an Iwasawa decomposition $G_0 = K_0A_0N_0$, an Iwasawa–Borel subgroup $B \supset A_0N_0$, and a B-Schubert variety S_{bd}, such that*

(1) $p \in S_{\mathrm{bd}} \subset Z \setminus D$,
(2) $\mathrm{codim}_Z\ S_{\mathrm{bd}} = q + 1$, *and*
(3) *the incidence variety* $I_{S_{\mathrm{bd}}} := \{C \in \mathcal{M}_Z : C \cap S_{\mathrm{bd}} \neq \emptyset\}$ *is contained in* $\mathcal{M}_Z \setminus \mathcal{M}_D$.

Proof. Let $p \in \mathrm{bd}(D)_{\mathrm{gen}}$, let $\gamma = G_0(p)$, and let κ be dual to γ. First consider the case where $p \in \gamma \cap \kappa$. From Lemma 9.1.2, $\mathrm{codim}\, S \geqq q + 1$ for every $S \in \mathcal{S}_\kappa$.

Now, given $S \in \mathcal{S}_\kappa$ and $p \in S \cap \kappa$ as above, let S_{bd} be a $(q+1)$-codimensional Schubert variety containing S; see Lemma 9.1.3. Suppose that S_{bd} meets C_0. Since $\dim_{\mathbb{C}} C_0 = q$ and $\mathrm{codim}_{\mathbb{C}}\ S_{\mathrm{bd}} = q + 1$, and every A_0N_0-orbit in D has nonempty intersection with C_0, there would be a point of intersection $z \in S_{\mathrm{bd}} \cap C_0$. But $A_0N_0(z) \subset S_{\mathrm{bd}}$, and so it would follow that

$$q = \dim_{\mathbb{C}} C_0 \geqq \mathrm{codim}_{\mathbb{C}}\, A_0N_0(z) \geqq \mathrm{codim}_{\mathbb{C}}\, S_{\mathrm{bd}} = q + 1.$$

This contradiction implies that S_{bd} does not meet D. On the other hand, using Theorem 9.1.1, it meets every G_0-orbit in $c\ell(\gamma)$. Thus, by conjugating appropriately, we have the desired result for any point in the closure of γ. Since γ was chosen to be an arbitrary orbit in $\mathrm{bd}_{\mathrm{gen}}(D)$ which is dense, the result follows for every point of $\mathrm{bd}(D)$. Furthermore, $S_{bd} \subset Z$ and consequently $I_{S_{bd}}$ is contained in the complement of $\mathcal{M}_D$ in $\mathcal{M}_Z$. □

In view of Proposition 9.1.4 we are now able to pinch $\mathcal{M}_D$ between a universal domain and the Schubert domain $\mathcal{S}_D$. In order to state this result in as uniform a way as possible, we introduce some notation.

Recall that there are two possibilities for $\mathcal{M}_D$, as described in Chapter 5. In the hermitian holomorphic case it is the bounded symmetric domain $\mathcal{B} = G_0/K_0$ (or its complex conjugate structure $\overline{\mathcal{B}}$) inside the compact dual symmetric flag manifold $\mathcal{M}_Z = G/KS_-$. In the general case it is a domain in the affine homogeneous space $\mathcal{M}_Z = G/\breve{K}$ where $\pi : G/K \to G/\breve{K}$ is a finite cover. In the hermitian holomorphic case the base point z_0 corresponding to C_0 has been chosen to be the identity coset (in other words the origin) in $\mathcal{B}$, and in the general case it is the identity coset $1\breve{K} \in G/\breve{K}$. Here, in the general case, we write $\breve{K}$ instead of the previously used J to emphasize the relation with K.

The universal domain $\mathcal{U}$ was defined (6.1.7) as a subdomain of G/K, but here we consider its image $\breve{\mathcal{U}}$ in $G/\breve{K}$. In the hermitian holomorphic case we let $\breve{\mathcal{U}} := \mathcal{B}$. In this way we can directly compare $\breve{\mathcal{U}}$ to the cycle spaces $\mathcal{M}_D$ and their various envelopes.

At the level of G/K it was shown in Chapter 7 that $\mathcal{U}$ agrees with the Iwasawa-envelope $\mathcal{E}_{\mathcal{I}}(\mathcal{U})$; see Proposition 7.2.11. This carries over to exactly the same statement in $G/\breve{K}$. Thus, unless we are in the hermitian holomorphic setting, we know that $\breve{\mathcal{U}} = \mathcal{E}_{\mathcal{I}}(\breve{\mathcal{U}}) \subset \mathcal{S}_D$. In the holomorphic hermitian setting we have already shown that $\breve{\mathcal{U}}$ agrees with any of its envelopes; see Proposition 7.2.2. Thus $\breve{\mathcal{U}} = \mathcal{S}_D$ in that case.

Theorem 9.1.5. *If D is an open G_0-orbit in a flag manifold $Z = G/Q$, then $\breve{\mathcal{U}} \subset \mathcal{M}_D \subset \mathcal{S}_D$.*

Proof. Proposition 9.1.4 shows that every boundary point of $\mathcal{M}_D$ contains a point of an incidence variety $I_{S_{bd}}$, which is invariant by some Iwasawa–Borel subgroup B of G. Such points are contained in B-invariant hypersurfaces and therefore are in the complement of $\mathcal{E}_{\mathcal{I}}(\breve{\mathcal{U}})$. Thus $\mathrm{bd}(\mathcal{M}_D)$ is contained in the complement of $\mathcal{E}_{\mathcal{I}}(\breve{\mathcal{U}}) = \breve{\mathcal{U}}$ and consequently $\breve{\mathcal{U}} \subset \mathcal{M}_D$.

Corollary 7.4.13 implies $\mathcal{M}_D \subset \mathcal{S}_D$. □

Remark 9.1.6. As we will see below, Theorem 9.1.5 combines with Theorem 5.4.7 to give a detailed classification in the hermitian case. In the non-hermitian case, methods involving Kobayashi hyperbolicity show that $\breve{\mathcal{U}} = \mathcal{S}_D$; see Chapters 11 and 12. Furthermore, the projection $\pi : \mathcal{U} \to \breve{\mathcal{U}}$ is biholomorphic; see Lemma 11.3.4. Thus in the non-hermitian case we have the classification $\mathcal{U} \cong \breve{\mathcal{U}} = \mathcal{M}_D = \mathcal{S}_D$. ◇

9.1C Classification in the hermitian case

Let us begin by giving yet another proof of the classification theorem in the hermitian holomorphic case, this time in the spirit of Schubert intersection theory.

Proposition 9.1.7. *In the hermitian holomorphic case $\mathcal{M}_D$ is $\mathcal{B}$ or $\overline{\mathcal{B}}$.*

Proof. Since $\check{\mathcal{U}} = \mathcal{B}$ in this case, Theorem 9.1.5 shows that $\mathcal{B} \subset \mathcal{M}_D \subset \mathcal{S}_D$. Now $\mathcal{S}_D$ is the intersection of all envelopes defined by the Schubert varieties $S \in \mathcal{S}_{C_0}$. Thus it is enough to show that one such envelope $\mathcal{E}_\mathcal{Y}(\mathcal{B})$ agrees with $\mathcal{B}$.

Here the complement of the open B-orbit in G/P is an irreducible hypersurface H which contains *every* proper B-Schubert variety, and it follows that H contains the B-Schubert slices in every G_0-orbit. In particular H has nonempty intersection with every G_0-orbit γ in $\mathrm{bd}(\mathcal{B})$. Thus it is clear that $\mathcal{E}_\mathcal{Y}(\mathcal{B}) = \mathcal{B}$. □

Now let us turn to the nonholomorphic hermitian case. By Theorem 5.4.7 we already know that $\mathcal{M}_D \subset \check{\mathcal{U}}$. Thus the inclusion in Theorem 9.1.5 does the job.

Proposition 9.1.8. *In the nonholomorphic hermitian case $\mathcal{M}_D = \check{\mathcal{U}} \cong \mathcal{U} \cong \mathcal{B} \times \overline{\mathcal{B}}$.*

Proof. As mentioned above, Theorem 5.4.7, together with Theorem 9.1.5, yields $\mathcal{M}_D = \check{\mathcal{U}}$. As noted earlier, $\pi|_\mathcal{U} : \mathcal{U} \to \check{U}$ is biholomorphic. Finally, the equivalence of $\mathcal{U}$ and $\mathcal{B} \times \overline{\mathcal{B}}$ is the content of Proposition 6.1.9. □

9.2 Supporting hypersurfaces at the boundary of $\mathcal{M}_D$

As mentioned above, in Chapter 10, we use indirect methods to prove that $\mathcal{U}$ agrees with the Schubert domain $\mathcal{S}_D$; see Chapter 11. By definition every point of the boundary of $\mathcal{S}_D$ contains a hypersurface which is contained in the polar set of a rational function on $\mathcal{M}_Z$ which is regular on $\mathcal{M}_D$. In this section we explicitly construct these polar sets and functions, and thereby show that $\mathcal{M}_D = \mathcal{S}_D$.

Precisely speaking, given $C \in \mathrm{bd}(\mathcal{M}_D)$, we determine a point $p \in C \cap \mathrm{bd}(D)$ and a Schubert variety $S \in \mathcal{S}_{C_0}$ such that $S = \mathcal{O} \dot\cup Y$ and $p \in Y$. Of course $\mathrm{codim}\, Y = q+1$ and $Y \subset Z \setminus D$ as above. Thus we may apply the results of Section 7.4, in particular Proposition 7.4.11, to obtain a B-invariant incidence hypersurface that contains the boundary point C and is contained in the complement $\mathcal{M}_Z \setminus \mathcal{M}_D$.

Here, given $p \in C \cap \mathrm{bd}(D)$, we consider Iwasawa–Schubert varieties $S = \mathcal{O} \dot\cup E$ of minimal possible dimension that satisfy the conditions

(9.2.1)
(1) $p \in E$,
(2) $S \cap D \neq \emptyset$, and
(3) $Z \setminus D$ contains every irreducible component of E that contains p.

Notation:

A_E is the union of all the irreducible components of E contained in $Z \setminus D$, *and*
B_E is the union of the remaining components of E.

So of course $E = A_E \cup B_E$. Note that by starting with the Schubert variety $S_0 := S_{\mathrm{bd}}$ as in the Proposition 9.1.4, and by considering a chain $S_0 \subset S_1 \subset \ldots$ with $\dim S_{i+1} = \dim S_i + 1$, we eventually come to a Schubert variety $S = S_k$ with the properties (9.2.1). Given p, the Schubert variety S may not be unique, but $\dim S = n - q + \delta \geqq n - q$ for some $\delta = \delta_S \geqq 0$.

The following proposition gives a constructive method for determining an B-invariant incidence hypersurface that contains C and is itself contained in the complement $\mathcal{M}_Z \setminus \mathcal{M}_D$. Starting with S that satisfies (9.2.1), we apply the following result.

Proposition 9.2.2. *If $\delta > 0$, then $C \cap A_E \cap B_E \neq \emptyset$.*

Applying Proposition 9.2.2, if $\delta > 0$, we take a point $p_1 \in C \cap A_E \cap B_E$ and replace S by a B_E-component S_1 of E that contains p_1. Possibly there are components of $E_1 := S_1 \setminus \mathcal{O}_1$ that contain p_1 and also have nonempty intersection with D. If this is the case, we replace S_1 by any such component. Since this S_1 still has nonempty intersection with the Iwasawa–Borel invariant A_E, at least some of the components of its E_1 do not intersect in this way. Continuing in this way, we eventually determine an S_1 that satisfies the conditions (9.2.1) at p_1. The procedure stops because Schubert varieties of dimension less than $n - q$ have empty intersection with D.

Corollary 9.2.3. *If S_0 satisfies the conditions* (9.2.1) *at p_0, there exist $p_1 \in \mathrm{bd}(D)$ and a Schubert subvariety $S_1 \subset S_0$ that satisfies these conditions at p_1 and has dimension $n - q$.*

Proof. We recursively apply the procedure indicated above until $\delta = 0$. □

Corollary 9.2.4. *If $C \in \mathrm{bd}(\mathcal{M}_D)$, then there exists an Iwasawa–Schubert variety $S \in \mathcal{S}_{C_0}$ of dimension $n - q$ such that $Y := S \setminus \mathcal{O}$ meets C.*

Corollary 9.2.5. *Let $C \in \mathrm{bd}(\mathcal{M}_D)$. Then there exist an Iwasawa–Borel subgroup B of G and a B-Schubert variety $S = \mathcal{O}\dot{\cup} Y$ in $\mathcal{S}_{C_0}$ such that the incidence variety I_Y is an analytic hypersurface and $C \subset I_Y \subset (\mathcal{M}_Z \setminus \mathcal{M}_D)$.*

Proof. This follows immediately from Proposition 7.4.11. □

This can be equivalently formulated as follows.

Corollary 9.2.6. $\mathcal{M}_D = \mathcal{S}_D$.

Let us now turn to certain technical preparations for the proof of Proposition 9.2.2.

Recall the basic maps $\nu : \mathfrak{X}_Z \to \mathcal{M}_Z$ and $\mu : \mathfrak{X}_Z \to Z$. Define $\mathfrak{X}_S$ to be the μ-preimage of S in $\mathfrak{X}_Z$. Since ν is proper and $\mathfrak{X}_S$ is closed, the restriction $\nu|_{\mathfrak{X}_S}$ is likewise proper. Furthermore, since $[S].[C_0] = d > 0$ and $\mathcal{M}_Z$ is connected, every cycle parameterized by $\mathcal{M}_Z$ has nonempty intersection with S, and consequently $\nu|_{\mathfrak{X}_S} : \mathfrak{X}_S \to \mathfrak{X}_Z$ is surjective. Finally, the $\nu|_{\mathfrak{X}_S}$-preimage of a cycle C can be identified with $C \cap S$.

All orbits of the Iwasawa–Borel group that defined S are transversal to the base cycle C_0 in the sense that the sum of the tangent spaces at an intersection point is that of the ambient space Z. In particular, $C_0 \cap S$ has pure dimension with $\dim C_0 \cap S = \delta$. Thus the generic cycle in $\mathcal{M}_Z$ has this property as well.

Choose a one-dimensional (local) disk Δ in $\mathcal{M}_Z$ with C corresponding to its origin, such that $I_z := \nu^{-1}(z)$ is δ-dimensional for $z \neq 0$. Define $\mathfrak{Z}$ to be the closure of $\nu^{-1}(\Delta \setminus \{0\})$ in $\mathfrak{X}_S$. The map $\nu_{\mathfrak{Z}} := \nu|_{\mathfrak{Z}} : \mathfrak{Z} \to \Delta$ is proper and its fibers are purely δ-dimensional.

In what follows we use the standard moving lemma of intersection theory and argue using a desingularization $\widetilde{\pi} : \widetilde{S} \to S$, where only points E are blown up. Let $\widetilde{E}$, $\widetilde{A}$ and $\widetilde{B}$ denote the corresponding $\widetilde{\pi}$-preimages. By taking Δ in generic position, we may assume that for $z \neq 0$ no component of I_z is contained in E. Hence we may lift the family $\mathfrak{Z} \to \Delta$ in $\mathfrak{X}_S$ to a family $\widetilde{\mathfrak{Z}} \to \Delta$ in $\mathfrak{X}_{\widetilde{S}}$ of δ-dimensional varieties such that $\widetilde{\mathfrak{Z}} \to \mathfrak{Z}$ is finite to one outside of the fiber over $0 \in \Delta$. Let $\widetilde{I}_z$ denote the fiber of $\widetilde{\mathfrak{Z}} \to \Delta$ at $z \in \Delta$, and shrink $\widetilde{\mathfrak{Z}}$ so that $\widetilde{I} := \widetilde{I}_0$ is connected. Since $\widetilde{I}_z \cap \widetilde{A} = \emptyset$ for $z \neq 0$, it follows that the intersection class $[\widetilde{I}].[\widetilde{A}]$ in the homology of $\widetilde{S}$ is zero.

An irreducible component of $\widetilde{I}$ is one of the following types: it intersects $\widetilde{A}$ but not $\widetilde{B}$, or it intersects both $\widetilde{A}$ and $\widetilde{B}$, or it intersects $\widetilde{B}$ but not $\widetilde{A}$. Write $\widetilde{I} = \widetilde{I}_A \cup \widetilde{I_{AB}} \cup \widetilde{I}_B$ correspondingly.

Lemma 9.2.7. $\widetilde{I_{AB}} \neq \emptyset$.

Proof. Since $\widetilde{I}_A \cup \widetilde{I_{AB}} \neq \emptyset$, it is enough to consider the case where $\widetilde{I}_A \neq \emptyset$. Let H be a hyperplane section in Z with $H \cap S = E$ (see Proposition 7.4.9), and put H in a continuous family H_t of hyperplanes with $H_0 = H$ such that $H_t \cap I_A$ is $(\delta - 1)$-dimensional for $t \neq 0$ and such that the lift $\widetilde{E}_t$ of $E_t := H_t \cap S$ contains no irreducible component of $\widetilde{I}_A$. In particular, $\widetilde{E}_t.\widetilde{I}_A \neq 0$ for $t \neq 0$. Since $\widetilde{I}_A.\widetilde{A} = \widetilde{I}_A.\widetilde{E} = \widetilde{I}_A.\widetilde{E}_t$, it follows that $\widetilde{I}_A.\widetilde{A} \neq 0$. But $0 = \widetilde{I}.\widetilde{A} = \widetilde{I}_A.\widetilde{A} + \widetilde{I_{AB}}.\widetilde{A}$ and therefore $\widetilde{I_{AB}} \neq \emptyset$. □

Proof of Proposition 9.2.2. We first consider the case $\delta \geqq 2$. Since $\widetilde{I_{AB}} \neq \emptyset$, it follows that some irreducible component I' of I has nonempty intersection with both A and B. Of course $I' \cap E = (I' \cap A) \cup (I' \cap B)$. But E is the support of a hyperplane section, and since $\dim I' \geqq 2$, it follows that $(I' \cap E)$ is connected. In particular $(I' \cap A)$ meets $(I' \cap B)$. Therefore $I' \cap A \cap B \neq \emptyset$, and consequently $C \cap A \cap B \neq \emptyset$.

Now suppose that $\delta = 1$, i.e., that $\widetilde{I}$ is one-dimensional. Since $\widetilde{I}.\widetilde{A} = 0$, the (nonempty) intersection $\widetilde{I} \cap \widetilde{A}$ is not discrete. We will show that some component of $\widetilde{I_{AB}}$ is contained in $\widetilde{A}$. It will follow immediately that $C \cap A \cap B \neq \emptyset$.

For this we assume to the contrary that every component of $\widetilde{I}$ which is contained in $\widetilde{A}$ is in $\widetilde{I}_A$. We decompose $\widetilde{I} = \widetilde{I}_1 \cup \widetilde{I}_2$, where $\widetilde{I}_1$ consists of those components of $\widetilde{I}$ which are contained in $\widetilde{A}$ and $\widetilde{I}_2$ of those which have discrete or empty intersection with $\widetilde{A}$.

Now $\widetilde{I}_1.\widetilde{A} = \widetilde{I}_1.\widetilde{E}$. Choosing H_t as above, we have $\widetilde{I}_1.\widetilde{E} = \widetilde{I}_1.\widetilde{E}_t \geqq 0$ for $t \neq 0$. If $\widetilde{I}_2 \neq \widetilde{I}_B$, then $\widetilde{I}_2.\widetilde{A} > 0$. This would contradict $0 = \widetilde{I}.\widetilde{A} = \widetilde{I}_1\widetilde{A} + \widetilde{I}_2.\widetilde{A}$. Thus $\widetilde{I}_2 = \widetilde{I}_B$ and $\widetilde{I}_1 = \widetilde{I}_A$. But $\widetilde{I}_A$ and $\widetilde{I}_B$ are disjoint, contrary to $\widetilde{I}$ being connected. Thus it follows that $\widetilde{I_{AB}}$ does indeed contain a component that is contained in $\widetilde{A}$. The proof is complete. □

10

Analysis of the Boundary of $\mathcal{U}$

Let us review our progress toward a detailed description of the cycle spaces $\mathcal{M}_D$. In the previous chapter the hermitian case was completely handled: in the hermitian holomorphic case $\mathcal{M}_D$ is the associated bounded symmetric domain $\mathcal{B}$ (or $\overline{\mathcal{B}}$); otherwise $\mathcal{M}_D$ is naturally identified with $\mathcal{B} \times \overline{\mathcal{B}}$. See Propositions 9.1.7 and 9.1.8. Furthermore, in general we have the pinching $\breve{\mathcal{U}} \subset \mathcal{M}_D \subset \mathcal{S}_D$. Except in the hermitian holomorphic case the natural projection $G/K \to G/\breve{K} = \mathcal{M}_Z$ identifies $\breve{\mathcal{U}}$ with the universal domain $\mathcal{U}$. Since $\mathcal{S}_D$ is defined as the intersection of envelopes defined by certain B-invariant hypersurfaces, we may also think of $\mathcal{S}_D$ as contained in G/K. Thus, if we are not in the hermitian holomorphic case, we have the inclusions $\mathcal{U} \subset \mathcal{M}_D \subset \mathcal{S}_D$ of G_0-invariant Stein domains in G/K.

In Chapter 11 we will complete the discussion by showing that domains of the type $\mathcal{S}_D$ are Kobayashi hyperbolic and that the only G_0-invariant, Stein, Kobayashi hyperbolic domain containing $\mathcal{U}$ is $\mathcal{U}$ itself. Consequently, either $\mathcal{U} = \mathcal{M}_D = \mathcal{S}_D$ or D is hermitian holomorphic and the above mentioned results apply.

The proof that $\mathcal{U}$ is maximal with respect to the properties of being G_0-invariant, Stein and Kobayashi hyperbolic makes strong use of the results in the present chapter. The original research for these two chapters was carried out in [FH]. Before outlining these in detail, we recall our notation.

As usual G_0 is a simple, noncompact real form of the complex semisimple group G. It is defined by the antiholomorphic involution τ at both the Lie algebra and group level. We follow here the convention of using the same notation for involutions at the group and Lie algebra levels. The Cartan involution $\theta : G_0 \to G_0$ which defines the compact form K_0 of G_0 extends uniquely to a holomorphic involution of G, which we also denote by θ. If G is simply connected, then the fixed point set $G^\theta = K = (K_0)^{\mathbb{C}}$ is connected; in the general case we only have $K \subset G^\theta \subset N_G(K)$. If not otherwise stated, we assume in this chapter that G is a *simply connected* complex semisimple Lie group. The holomorphic involution θ commutes with the complex conjugation τ of G over G_0, and $\sigma := \tau\theta = \theta\tau$ defines another involution of G, it is complex conjugation of G over a certain compact real form G_u and thus is a Cartan involution of G.

Note that σ is the antiholomorphic extension of θ. As before $\mathfrak{g}_0 = \mathfrak{k}_0 + \mathfrak{s}_0$ is the Cartan decomposition of $\mathfrak{g}_0$ into (± 1)-eigenspaces of θ: $\mathfrak{k}_0 = \mathfrak{g}_0^\theta$ and $\mathfrak{s}_0 = \mathfrak{g}_0^{-\theta}$. Similarly the holomorphic extension θ defines the complexified Cartan decomposition $\mathfrak{g} = \mathfrak{k} + \mathfrak{s} := \mathfrak{g}^\theta + \mathfrak{g}^{-\theta}$.

One goal of this chapter is to describe aspects of the invariant theory of the G_0-action on the complex symmetric space G/K which are relevant for our understanding of this action on $\mathrm{bd}(\mathcal{U})$. In particular we characterize the closed orbits and, given a nonclosed orbit, we show how to reach a point in the closed orbit in its closure by a simple limiting procedure.

For this and virtually every consideration in this chapter the restricted root decomposition of $\mathfrak{g}$ is of central importance. This is reviewed in Section 10.1, where basic principles of actions of real forms are also summarized. There we also discuss in some detail the group $\mathrm{Aut}_{\mathbb{R}}(\mathfrak{g})$ of automorphisms of the underlying real Lie group structure of $\mathfrak{g}$.

In Section 2 the basic G_0-equivariant *linearization* map

$$\eta : G/K \to \mathrm{Aut}_{\mathbb{R}}(\mathfrak{g}), \ g \mapsto \tau\,\mathrm{Ad}(g)\theta\,\mathrm{Ad}(g^{-1})$$

is introduced and studied in detail. For example, it is shown that $G_0(x)$ is closed in G/K if and only if $\eta(x)$ is semisimple (Proposition 10.2.7). The Jordan decomposition $\psi = s_\psi u_\psi$ of $\psi \in \mathrm{Aut}_{\mathbb{R}}(\mathfrak{g})$ is of fundamental importance. Lifting to $\mathfrak{g}$, it is shown in Lemma 10.2.9 that there is a unique $E \in \mathfrak{g}^{s_\psi}$ so that $u_\psi = \mathrm{Ad}(\exp(E))$. These ad-nilpotent elements are later used to construct $\mathfrak{sl}_2$-triples which are useful for determining the closed orbit in the boundary of a given G_0-orbit.

A semisimple element $s \in \mathrm{Aut}_{\mathbb{R}}(\mathfrak{g})$ can be further decomposed as a product $s = s_{\mathrm{ell}} s_{\mathrm{hyp}}$, where the eigenvalues of s_{ell} (respectively, s_{hyp}) are in the unit circle S^1 (respectively, in $\mathbb{R}^{>0}$). We say that an element $x \in G/K$ is elliptic if $\eta(x) = s = s_{\mathrm{ell}}$ is a semisimple element of elliptic type. The main result of Section 3 is that the set of elliptic elements is just $G_0.\exp(i\mathfrak{a}_0)(x_0)$, where x_0 is the base point corresponding to K in G/K Proposition 10.3.1 As a consequence the closed orbits in $\mathrm{bd}(\mathcal{U})$ are those orbits in the set

$$(\exp i\mathfrak{a}_0)(x_0) \cap \mathrm{c}\ell(\mathcal{U}) = \mathrm{c}\ell((\exp i\omega_0)(x_0)) = \exp(\mathrm{c}\ell(i\omega_0))(x_0),$$

where as usual ω_0 denotes the polyhedral domain in $\mathfrak{a}_0$ which defines $\mathcal{U}$ by $\mathcal{U} := G_0.\exp(i\omega_0)(x_0)$.

The Luna Slice Theorem is the key tool for understanding the G_0-action in a neighborhood of a closed orbit. After explaining this in Section 4, we proceed to compute its first main ingredient, namely the (reductive) isotropy group of the G_0-action at an arbitrary point in $\exp(i\mathfrak{a}_0)$ (Proposition 10.4.7). In particular, this gives us the basic information required in Section 6 for computations in a slice neighborhood of a closed orbit in $\mathrm{bd}(\mathcal{U})$.

One main application of our boundary analysis is, given $z \in \mathrm{bd}(\mathcal{U})$, we determine an $\mathfrak{sl}_2$ triple in $\mathfrak{g}_0$ so that the position with respect to $\mathcal{U}$ of the orbit $S(z)$ of the associated complex group gives us important information on G_0-invariant domains that contain $\mathcal{U}$. The isotropy group S_z is either a complex torus in S or its normalizer, i.e., a 2:1

extension of a torus. In the former case this is the two-dimensional affine quadric which has $\mathbb{P}_1(\mathbb{C}) \times \mathbb{P}_1(\mathbb{C})$ as its unique S-equivariant compactification. We have discussed this two-dimensional example in Chapter 6, but we do so again here in Section 5 in the context of the $\mathfrak{sl}_2$ triples of the present chapter.

The orbit $S(z)$ gives satisfactory information only at generic points of $\mathrm{bd}(\mathcal{U})$. In fact the example of $G_0 = SL(5;\mathbb{R})$ shows that our method fails for at least certain nongeneric boundary points (see Section 7).

By definition, generic boundary points are those points z so that the closed orbit in $G_0(z)$ is of the form $G_0(z_0)$, where $z_0 = \exp(i\xi)$ for $\xi \in \mathrm{bd}(\omega_0)$ which is in only one root hyperplane, i.e., in a maximal-dimensional face of the polyhedron ω_0.

Our first main goal in Section 6 is to show that the set $\mathrm{bd}^{\mathrm{gen}}(\mathcal{U})$ of generic boundary points is open and dense in $\mathrm{bd}(\mathcal{U})$. In fact we show that its complement has codimension $\geqq 1$ (Corollary 10.6.4). This result is proved by showing that going from generic to nongeneric fibers of the (real) categorical quotient has the desired semicontinuity. For this we make strong use of our detailed information on the nilpotent cone of the representation in the Luna slice (see Lemma 10.6.3) and implement a basic theorem of Kostant and Rallis [KR].

Having shown that the generic points are indeed open and dense in the boundary, we then prove the desired result about the intersection $S(z) \cap \mathcal{U}$, namely that it is the universal domain $\mathcal{U}_S$ in the affine quadric S/K_S or its 2:1 quotient (Theorem 10.6.9). This is one of the key results for the final step in the description of the cycle space $\mathcal{M}_D$ which is given in the next chapter.

10.1 Preparation

In this preparatory section we collect material which is used in the sequel.

10.1A Restricted roots

We now summarize some basic information on restricted root systems and decompositions. All is completely standard, and can be found with complete proofs in most books on symmetric spaces (e.g., [Hel]) or on the representation theory of semisimple Lie groups (e.g., [War]), but it is convenient for nonspecialists in Lie theory to have it collected here.

As before, θ is a Cartan involution of the real semisimple Lie algebra $\mathfrak{g}_0$ and $\mathfrak{g}_0 = \mathfrak{k}_0 + \mathfrak{s}_0$ is the corresponding Cartan decomposition. So the Killing form $\langle \cdot, \cdot \rangle$ is negative definite on $\mathfrak{k}_0$ and positive definite on $\mathfrak{s}_0$, and thus $h(\xi, \zeta) := -\langle \xi, \theta(\zeta) \rangle$ is positive definite on $\mathfrak{g}_0$. Fix an abelian subalgebra $\mathfrak{a}_0$ of $\mathfrak{g}_0$ that is maximal among those contained in $\mathfrak{s}_0$. Then $\mathfrak{a}_0$ is contained in a θ-stable Cartan subalgebra $\mathfrak{h}_0 = (\mathfrak{h}_0 \cap \mathfrak{k}_0) + \mathfrak{a}_0$ of $\mathfrak{g}_0$. Restricting roots $\alpha \in \Sigma(\mathfrak{g}, \mathfrak{h})$ from $\mathfrak{h}_0$ to $\mathfrak{a}_0$ the root space decomposition $\mathfrak{g} = \mathfrak{h} + \sum_{\alpha \in \Sigma(\mathfrak{g}, \mathfrak{h})} \mathfrak{g}_\alpha$ leads to a decomposition

$$(10.1.2) \qquad \mathfrak{g}_0 = \mathfrak{z}_{\mathfrak{g}_0}(\mathfrak{a}_0) + \sum\nolimits_{\lambda \in \Sigma(\mathfrak{g}_0, \mathfrak{a}_0)} \mathfrak{g}_0^\lambda = (\mathfrak{m}_0 + \mathfrak{a}_0) + \sum\nolimits_{\lambda \in \Sigma(\mathfrak{g}_0, \mathfrak{a}_0)} \mathfrak{g}_0^\lambda.$$

Here $\mathfrak{z}_{\mathfrak{g}_0}(\mathfrak{a}_0)$ is the $\mathfrak{g}_0$-centralizer of $\mathfrak{a}_0$. Since $\mathfrak{a}_0$ is θ-stable and is maximal abelian in $\mathfrak{s}_0$ that centralizer has form $\mathfrak{m}_0 + \mathfrak{a}_0$ where $\mathfrak{m}_0 = \mathfrak{z}_{\mathfrak{g}_0}(\mathfrak{a}_0) \cap \mathfrak{k}_0$, and $\Sigma(\mathfrak{g}_0, \mathfrak{a}_0)$ consists of the nonzero $\alpha|_{\mathfrak{a}_0}$ for $\alpha \in \Sigma(\mathfrak{g}, \mathfrak{h})$. Those restrictions are the **restricted roots** or **$\mathfrak{a}_0$-roots**. Every $\alpha \in \Sigma(\mathfrak{g}, \mathfrak{h})$ takes real values on $\mathfrak{a}_0$, so $\mathrm{ad}(\mathfrak{a}_0)$ is diagonalizable over $\mathbb{R}$; here the $\mathfrak{g}_0^\lambda$ for $\lambda \neq 0$ (we write the λ as superscript to avoid cumbersome notation) are the joint eigenspaces on $\mathfrak{g}_0$. They are called **restricted root spaces** or **$\mathfrak{a}_0$-root spaces**. The set $\Sigma(\mathfrak{g}_0, \mathfrak{a}_0)$ is called the **restricted root system** or **$\mathfrak{a}_0$-root system** of $\mathfrak{g}_0$.

The restricted root system $\Sigma(\mathfrak{g}_0, \mathfrak{a}_0)$ inherits many properties from the root system $\Sigma(\mathfrak{g}, \mathfrak{h})$. For example, with a slight bit of redundancy,

- using $\theta|_{\mathfrak{a}_0} = -\mathrm{Id}$, if $\lambda \in \Sigma(\mathfrak{g}_0, \mathfrak{a}_0)$, then $-\lambda \in \Sigma(\mathfrak{g}_0, \mathfrak{a}_0)$ and $\theta(\mathfrak{g}_0^\lambda) = \mathfrak{g}_0^{-\lambda}$,
- the Killing form is nondegenerate on each of $\mathfrak{m}_0$, $\mathfrak{a}_0$, and the $\mathfrak{g}_0^\lambda + \mathfrak{g}_0^{-\lambda}$ for $\lambda \in \Sigma(\mathfrak{g}_0, \mathfrak{a}_0)$, and those spaces are mutually orthogonal,
- the Killing form is zero on each $\mathfrak{g}_0^\lambda$ and pairs $\mathfrak{g}_0^\lambda$ with $\mathfrak{g}_0^{-\lambda}$, and
- if $\lambda, \mu, \lambda + \mu \in \Sigma(\mathfrak{g}_0, \mathfrak{a}_0)$ then $[\mathfrak{g}_0^\lambda, \mathfrak{g}_0^\mu] \subset \mathfrak{g}_0^{\lambda+\mu}$.

Since the Killing form is positive definite on $\mathfrak{a}_0$ it carries over to a positive definite form (also denoted $\langle \cdot, \cdot \rangle$) on $\mathfrak{a}_0^*$. For computation it will be convenient to have the notation

if $\lambda \in \mathfrak{a}_0^*$, then $H_\lambda \in \mathfrak{a}_0$ defined by $\langle \xi, H_\lambda \rangle = \lambda(\xi)$,
if $\lambda \in \mathfrak{a}_0^*$, then $h_\lambda \in \mathfrak{a}_0$ defined by $h_\lambda := 2H_\lambda / \langle H_\lambda, H_\lambda \rangle$,
if $\lambda, \mu \in \mathfrak{a}_0^*$, then $\langle \lambda | \mu \rangle := \frac{2\langle \lambda, \mu \rangle}{\langle \mu, \mu \rangle} = \lambda(h_\mu)$,
if $\lambda, \mu \in \Sigma(\mathfrak{g}_0, \mathfrak{a}_0)$, then the reflection $s_\lambda : \nu \mapsto \nu - \langle \nu | \lambda \rangle \lambda$ preserves $\Sigma(\mathfrak{g}_0, \mathfrak{a}_0)$.

If $\lambda \in \Sigma(\mathfrak{g}_0, \mathfrak{a}_0)$, then H_λ is called the **coroot** for λ and h_λ is the **normalized coroot**. The group $W(\mathfrak{g}_0, \mathfrak{a}_0)$ generated by the s_λ, $\lambda \in \Sigma(\mathfrak{g}_0, \mathfrak{a}_0)$, is the **small Weyl group**, and those s_λ are the **root reflections** or **Weyl reflections**. The small Weyl group can also be characterized as $N_{G_0}(\mathfrak{a}_0)/Z_{G_0}(\mathfrak{a}_0)$. As in the complex case it acts simply transitively on the set of all positive root systems.

The restricted root system is an example of abstract root system. For this and for use in Part IV we recall the definition.

Definition 10.1.3. An **abstract root system** is a finite subset Σ of a real vector space V that is furnished with a (positive definite) scalar product $\langle \ , \ \rangle$ such that

RS1. Σ generates V as a vector space,
RS2. Σ is preserved by all reflections $s_\lambda : v \mapsto v - 2\frac{\langle v, \lambda \rangle}{\langle \lambda, \lambda \rangle} \cdot \lambda$ with $\lambda \in \Sigma$, and
RS3. if $\lambda, \mu \in \Sigma$, then the numbers $2\frac{\langle \mu, \lambda \rangle}{\langle \lambda, \lambda \rangle}$ are integers.

It follows that if $\lambda \in \Sigma$ and also a multiple $c\lambda \in \Sigma$, then $c \in \{\pm\frac{1}{2}, \pm 1, \pm 2\}$.

RRS. *If* $\Sigma \subset (V, \langle \ , \ \rangle)$ *satisfies* RS1, RS2, *and* RS3, *and if* $\lambda, c\lambda \in \Sigma$ *implies* $c = \pm 1$, *then* Σ *is called a* **reduced** *root system.*

To every nonreduced root system Σ one can associate a reduced root system Σ^{reduced} either by removing all longest elements in Σ or by removing all shortest elements.

As in Section 1.1, a **positive root system** in an abstract root system Σ is a subset $\Sigma^+ = \Sigma^+(\mathfrak{g}, \mathfrak{h})$ such that (i) Σ is the disjoint union of Σ^+ and $-\Sigma^+$, and (ii) if $\lambda, \mu \in \Sigma^+$ and $\lambda + \mu \in \Sigma$, then $\lambda + \mu \in \Sigma^+$. The **Weyl chambers** are the topological components of $V \setminus \left(\bigcup_{\lambda \in \Sigma} \lambda^\perp\right)$ as before. Weyl chambers are on one-to-one correspondence $C \leftrightarrow \Sigma^+$ with the set of all positive root systems, by $C = \{\xi \in V \mid \lambda(\xi) > 0 \text{ for all } \lambda \in \Sigma^+\}$. Thus one can also specify positive root systems as subsets of Σ on one side of a hyperplane that does not meet Σ.

Also as in Section 1.1, a positive abstract root system Σ^+ specifies the **simple subsystem** $\Psi = \{\psi \in \Sigma^+ \mid \psi \neq \lambda + \mu \text{ for any } \lambda, \mu \in \Sigma^+\}$, in other words the indecomposable elements in Σ^+.

A restricted root system is a (possibly nonreduced) root system. We indicate the proof because it depends on a notion that we need for the remainder of this chapter.

A triple $\{E^+, H, E^-\} \subset \mathfrak{g}_0$ is called an $\mathfrak{sl}_2$ **triple** if

$$[H, E^+] = 2E^+, \quad [H, E^-] = -2E^-, \quad \text{and} \quad [E^+, E^-] = H.$$

It will be convenient to refer to E^+, H, and E^- as the nil-positive, hyperbolic and nil-negative elements, respectively. Of course they generate a subalgebra isomorphic to $\mathfrak{sl}(2; \mathbb{R})$. A basic property is the following.

Lemma 10.1.3. *If $0 \neq E' \in \mathfrak{g}_0^\lambda$, then there is a multiple $E = cE'$ such that $\{E, -\theta(E), [E, -\theta(E)]\}$ is an $\mathfrak{sl}_2$ triple.*

An $\mathfrak{sl}_2$ triple $\{E^+, H, E^-\}$ is ***θ-adapted*** *if $E^- = -\theta(E^+)$.*

The proof of Lemma 10.1.3 is a matter of careful normalization. One can use the result to prove the following.

Proposition 10.1.4. *The set $\Sigma(\mathfrak{g}_0, \mathfrak{a}_0)$ of restricted roots is a (possibly nonreduced) root system.*

Proof. The set $\Sigma(\mathfrak{g}_0, \mathfrak{a}_0)$ spans $\mathfrak{a}_0^*$, since otherwise the subalgebra $\{\zeta \in \mathfrak{a}_0 : \lambda(\zeta) = 0 \text{ for all } \lambda \in \Sigma\}$ would be nonzero and central in $\mathfrak{g}_0$, contradicting the assumption that $\mathfrak{g}_0$ is semisimple. That proves RS1.

RS2 and RS3 are direct consequences of Lemma 10.1.3, as follows. If $\lambda \in \Sigma(\mathfrak{g}_0, \mathfrak{a}_0)$ the resulting θ-adapted $\mathfrak{sl}_2$ triple has form $\{E^{+\lambda}, E^{-\lambda}, h_\lambda\}$, where $E^{\pm\lambda} \in \mathfrak{g}_0^{\pm\lambda}$ and h_λ is the normalized coroot. Now the adjoint action of the corresponding $\mathfrak{sl}(2; \mathbb{R})$, call it $\mathfrak{g}_0[\lambda]$, shows that every λ-string $\{\mu + k\lambda \in \Sigma(\mathfrak{g}_0, \mathfrak{a}_0) \mid k \in \mathbb{Z}\}$ is connected, so $s_\lambda(\mu) \in \Sigma(\mathfrak{g}_0, \mathfrak{a}_0)$, which is RS2. This is the same argument as for ordinary complex root systems. The integrality RS3 follows from the fact that h_λ has integral eigenvalues $\{-d, -d+2, \ldots, d-2, d\}$ in an irreducible $\mathfrak{g}_0[\lambda]$-module of dimension $d + 1$. □

Remark 10.1.5. As already mentioned, comparing the restricted roots (and the corresponding root spaces) with the full root system and root spaces, the major differences are that (1) the restricted root system may be nonreduced and (2) $\dim \mathfrak{g}_0^\lambda$ may be bigger than one. The list of *simple* real forms $\mathfrak{g}_0$ for which $\Sigma(\mathfrak{g}_0, \mathfrak{a}_0)$ is nonreduced is

$$\mathfrak{su}(p,q), |p-q| \geqq 2; \quad \mathfrak{sp}(p,q), |p-q| \geqq 1; \quad \mathfrak{so}^*(4r+2); \quad \mathfrak{e}_{6(-14)}; \quad \mathfrak{f}_{4(-20)}.$$

Here the numbers in parentheses on $\mathfrak{e}_6$ and $\mathfrak{f}_4$ are $\dim \mathfrak{s}_0 - \dim \mathfrak{k}_0$; they determine the real form at hand. ◇

The obvious $[\mathfrak{g}_0^\lambda, \mathfrak{g}_0^\mu] \subset \mathfrak{g}_0^{\lambda+\mu}$ can be sharpened, using θ-adapted $\mathfrak{sl}_2$ triples, to prove the following.

Lemma 10.1.6. *Suppose that λ, μ, $\lambda+\mu \in \Sigma(\mathfrak{g}_0, \mathfrak{a}_0)$.*

1. *If $0 \neq E^\lambda \in \mathfrak{g}_0^\lambda$ and $0 \neq E^\mu \in \mathfrak{g}_0^\mu$, then either $[E^\lambda, \mathfrak{g}_0^\mu] = \mathfrak{g}_0^{\lambda+\mu}$ or $[\mathfrak{g}_0^\lambda, E^\mu] = \mathfrak{g}_0^{\lambda+\mu}$. In particular, $[\mathfrak{g}_0^\lambda, \mathfrak{g}_0^\mu] = \mathfrak{g}_0^{\lambda+\mu}$.*
2. *If $\lambda \in \Sigma(\mathfrak{g}_0, \mathfrak{a}_0)$, then $[\mathfrak{g}_0^\lambda, \mathfrak{g}_0^{-\lambda}] \subset \mathbb{R} h_\lambda \oplus \mathfrak{m}_0 \subset \mathfrak{a}_0 \oplus \mathfrak{m}_0$.*

Proof. The integrality condition RS3 implies that $\lambda(h_\mu)\mu(h_\lambda) \in \{0,1,2,3\}$ for nonproportional roots. It is also clear for $\lambda = \pm\mu$. The remaining case is $\mu = \pm 2\lambda$. In that case $\lambda(h_{\pm 2\lambda}) = \pm 1$ and $\pm 2\lambda(h_\lambda) = \pm 4$. Hence, either λ and μ are orthogonal or at least one of the integers $\lambda(h_\mu)$, $\mu(h_\lambda)$ is equal ± 1. In the latter case assume $\lambda(h_\mu) = \pm 1$.

Given $\lambda, \mu \in \Sigma(\mathfrak{g}_0, \mathfrak{a}_0)$, we may now assume that $\lambda(h_\mu) \in \{-1, 0, 1\}$. Consider the θ-adapted $\mathfrak{sl}_2$ triple $\{E^\mu, E^{-\mu}, h_\mu\}$. The corresponding subalgebra $\mathfrak{g}_0[\mu]$ stabilizes the subspace $W := \sum \mathfrak{g}_0^{\lambda+j\mu}$, in other words W is a finite-dimensional $\mathfrak{g}_0[\mu]$-module. The $\mathfrak{g}_0^{\lambda+j\mu}$ are the h_μ-eigenspaces with eigenvalues $(\lambda+\mu)(h_\mu) + 2j$. By choice of λ and μ the eigenvalue $\lambda(h_\mu) \in \{1,2,3\}$. The elementary representation theory of $\mathfrak{g}_0[\mu]$ now tells us that $\mathrm{ad}(E^\mu) : \mathfrak{g}_0^\lambda \to \mathfrak{g}_0^{\lambda+\mu}$ is surjective.

We have just proved assertion 1. Assertion 2 is obvious and we omit its proof. □

10.1B The group $\mathrm{Aut}_{\mathbb{R}}(\mathfrak{g})$

Whenever $\mathfrak{g}$ is a complex Lie algebra we define $\mathrm{Aut}_{\mathbb{R}}(\mathfrak{g})$ denote the group of Lie algebra automorphisms of the underlying real Lie algebra structure of $\mathfrak{g}$. In our case $\mathfrak{g}$ is semisimple so the adjoint representation maps $\mathfrak{g}$ isomorphically onto the Lie algebra of $\mathrm{Aut}_{\mathbb{R}}(\mathfrak{g})$. Thus $\mathrm{Aut}(\mathfrak{g})$ is an open subgroup of $\mathrm{Aut}_{\mathbb{R}}(\mathfrak{g})$. Since G is connected and semisimple, $\mathrm{Ad}(G)$ is the identity component of $\mathrm{Aut}_{\mathbb{R}}(\mathfrak{g})$. The Cartan involution $\sigma = \tau\theta$ of G (whose fixed point set is the compact real form G_u) carries over to $\mathrm{Aut}(\mathfrak{g})$ as the Cartan involution

$$\widehat{\sigma} : \mathrm{Aut}_{\mathbb{R}}(\mathfrak{g}) \to \mathrm{Aut}_{\mathbb{R}}(\mathfrak{g}) \text{ by } \widehat{\sigma}(\gamma) = \sigma\gamma\sigma^{-1}.$$

If γ is a linear transformation of $\mathfrak{g}$ let γ^* denote its adjoint relative to the positive definite $h(\xi,\eta) := -\langle \xi, \theta\eta \rangle$ on the real vector space structure of $\mathfrak{g}$. One checks that $\widehat{\sigma} : \gamma \mapsto (\gamma^*)^{-1}$, and the fixed point set

$$K_{\mathrm{Aut}} := \{\gamma \in \mathrm{Aut}(\mathfrak{g}) \mid \gamma^* = \gamma^{-1}\} = \mathrm{Aut} \cap O(\mathfrak{g}, h) \tag{10.1.7}$$

is a maximal compact subgroup of $\mathrm{Aut}_{\mathbb{R}}(\mathfrak{g})$. Then K_{Aut} meets every topological component of $\mathrm{Aut}_{\mathbb{R}}(\mathfrak{g})$.

Proposition 10.1.7. *Suppose that $\mathfrak{g}$ is a complex simple Lie algebra. Then every element $\varphi \in \mathrm{Aut}_{\mathbb{R}}(\mathfrak{g})$ is either a holomorphic or an antiholomorphic linear map. In particular, if ν is any antiholomorphic automorphism of $\mathfrak{g}$, e.g., complex conjugation over a real form, then $\mathrm{Aut}_{\mathbb{R}}(\mathfrak{g}) = \mathrm{Aut}(\mathfrak{g}) \dot{\cup} \nu\, \mathrm{Aut}(\mathfrak{g})$.*

Proof. Let J denote the complex structure on $\mathfrak{g}$. Then $\mathfrak{g}^{\mathbb{C}} = \mathfrak{g}^{1,0} \oplus \mathfrak{g}^{0,1}$, the direct sum of ideals. Here $\mathfrak{g}^{1,0}$ is the holomorphic tangent space, the $(+i)$-eigenspace of $J^{\mathbb{C}}$, and $\mathfrak{g}^{0,1}$ is the antiholomorphic tangent space, the $(-i)$-eigenspace of $J^{\mathbb{C}}$. These ideals are unique, so either $\varphi^{\mathbb{C}}(\mathfrak{g}^{1,0}) = \mathfrak{g}^{1,0}$ and φ is holomorphic, or $\varphi^{\mathbb{C}}(\mathfrak{g}^{1,0}) = \mathfrak{g}^{0,1}$ and φ is antiholomorphic. □

10.1C Varieties defined over $\mathbb{R}$

A complex linear algebraic group $H \subset GL(V)$ is said to be **defined over** $\mathbb{R}$ if it is the complexification of a real linear algebraic group. That means that there is a **real structure**, i.e., an antiholomorphic $\mathbb{R}$-linear involution $\mu : V \to V$ such that H is stable under the induced involutive automorphism $\widehat{\mu} : \psi \mapsto \mu\psi\mu^{-1}$ of $GL(V)$. This definition is equivalent to the definition which says that H has a real structure if the defining ideals of H have a generating system consisting of polynomials with real coefficients. The subset $H_{\mathbb{R}} = \{h \in H \mid \widehat{\mu}(h) = h\}$ is called the **set of real points**. For example, G_0, G_u are the sets of real points $G_{\mathbb{R}}$ of G with respect to the real structures τ and σ, respectively.

Similarly, subvariety Y of a complex vector space W is defined over $\mathbb{R}$ if it is the complexification of a real affine algebraic variety, in other words if it has a real structure, i.e., if it is stable under a holomorphic linear involution $\nu : W \to W$. Again, this is equivalent to the condition that the defining ideals are generated by polynomials with real coefficients, and then $Y_{\mathbb{R}} = \{y \in Y \mid \nu(y) = y\}$ is called the set of real points.

A complex algebraic action $H \times Y \to Y$ is said to be defined over $\mathbb{R}$ if H, Y are defined over $\mathbb{R}$ as above and $\nu(h(y)) = \widehat{\mu}(h)(\nu(y))$ for all $h \in H$, $y \in Y$.

Many actions have nonclosed orbits. A basic example is the conjugation action of H on $Y = H$: The orbits here are the conjugacy classes. These are nonclosed (for reductive H) if and only if the corresponding elements have nontrivial unipotent part in its Jordan decomposition. The set of *closed* orbits however can be parameterized in a very convenient way (we discuss this in more detail in the next paragraph). We conclude this paragraph by giving a relation between closed H and $H_{\mathbb{R}}$ orbits. The proof of the following proposition can be found in [BoHC, Proposition 2.3] or [Br, Proposition 5.3].

Proposition 10.1.8. *If $H \times Y \to Y$ is an algebraic action defined over $\mathbb{R}$, then the intersection $H \cdot y \cap Y_{\mathbb{R}}$ is either empty or is a finite union $\bigcup_{l \leqq j \leqq \ell} (H_{\mathbb{R}})^0 \cdot y_j$ of the* *orbits of the real form* $(H_{\mathbb{R}})^0$. *The orbits* $(H_{\mathbb{R}})^0 \cdot y_j$, $y_j \in Y_{\mathbb{R}}$, *are closed if and only if the* $H \cdot y_j$ *are closed.*

10.1D Real and complex quotients

If there are nonclosed H-orbits in Y, then, furnished with the quotient topology, the orbit space Y/H is not Hausdorff. In such a situation it is often appropriate to consider an invariant-theoretic quotient. An example of such is the **categorical quotient** which behaves well with respect to many natural functors and is by definition Hausdorff. Let us sketch its basic properties

In the complex affine category a variety Y is completely determined by its (finitely generated) algebra of global regular algebraic functions $\mathcal{O}(Y)$ as the (maximal) spectrum $\mathrm{Spec}^{\mathrm{m}}(\mathcal{O}(Y))$.

If the (complex) algebraic group H is reductive, the subalgebra $\mathcal{O}(Y)^H$ of H-invariant functions is finitely generated (see, e.g., [Kra]). The categorical quotient $Y/\!\!/H$ is the affine variety $\mathrm{Spec}^{\mathrm{m}}(\mathcal{O}(Y))$.

Let $\pi : Y \to Y/\!\!/H$ be the canonical quotient map which is an affine algebraic map. If Y is a normal space, the quotient $Y/\!\!/H$ is likewise normal. However, if Y is smooth, the quotient may nevertheless be singular. The defining property of this quotient is that in each fiber of π there is precisely one closed H-orbit. In particular, the closure of every H-orbit contains precisely one closed H-orbit. It should be noted that this property fails for nonaffine actions. For example, for the action of main concern in this monograph, namely $K = (K_0)^{\mathbb{C}}$ acting on flag manifolds G/Q, there are several closed K-orbits, i.e., the base cycles in the respective open G_0 orbits, which are all contained in the closure of the unique *open* K-orbit in G/Q.

For real affine algebraic actions the situation with respect to closed orbits is more complicated than that in the complex case. Let us consider for example a real reductive group H with global Cartan decomposition $H = K_0 \cdot \exp \mathfrak{s}_0$, and suppose that it is acting algebraically on a real affine variety Y. As in the complex situation, the underlying set of the real categorical quotient $Y/\!\!/H$ parameterizes the *closed* orbits. The real categorical quotient is a Hausdorff space, the quotient map $\mathrm{p} : Y \to Y/\!\!/H$ is continuous and by definition every fiber of p contains precisely one closed H-orbit. Unlike the complex case, one can only furnish $Y/\!\!/H$ with a structure of an *semialgebraic* set [BCR], which in general is not algebraic. The simplest example in this context is the action $\mathbb{Z}_2 \times \mathbb{R} \to \mathbb{R}$ by change of sign. The quotient is the *semi*-algebraic set $\mathbb{R}/\!\!/\mathbb{Z}_2 = \mathbb{R}/\mathbb{Z}_2 = [0, \infty)$.

In many situations the real affine action arises as the restriction to the real points of a complex affine action which is defined over $\mathbb{R}$. Let $H \times Y \to Y$ be such an action and $Y_{\mathbb{R}} \subset Y$ be the real affine subvariety of real points. It is instructive to compare the sets $(Y/\!\!/H)_{\mathbb{R}}$, $\pi(Y_{\mathbb{R}})$ and $Y_{\mathbb{R}}/\!\!/H_{\mathbb{R}}$, which in general are all different. The following diagram explains the situation:

$$\begin{array}{ccccc} Y_{\mathbb{R}} & & Y_{\mathbb{R}} & & Y \\ \downarrow \mathrm{p} & & \downarrow \pi & & \downarrow \pi \\ Y_{\mathbb{R}}/\!\!/H_{\mathbb{R}} & \xrightarrow{P} & \pi(Y_{\mathbb{R}}) \subseteq (Y/\!\!/H)_{\mathbb{R}} & \subset & Y/\!\!/H \end{array}.$$

The map $P : Y_{\mathbb{R}}/\!\!/H_{\mathbb{R}} \to \pi(Y_{\mathbb{R}})$ is a finite semialgebraic map and $\pi(Y_{\mathbb{R}})$, by a theorem of Tarski and Seidenberg, is a semialgebraic subset of $(Y/\!\!/H)_{\mathbb{R}}$.

10.2 Linearization and the Jordan decomposition

10.2A Reduction to an action by conjugation

One of the objectives of this chapter is to give a description of the G_0-orbit structure in the boundary $\mathrm{bd}(\mathcal{U})$ of the universal domain $\mathcal{U} = G_0 \cdot \exp i\omega_0(x_0)$, considered as an open subset of the smooth complex affine space $\Omega = G/K$. Here, x_0 denotes the base point $1\,K \in G/K$.

Recall that ω_0 is the convex polyhedron in $\mathfrak{a}_0$, a maximal abelian subalgebra in $\mathfrak{s}_0$ given by $\omega_0 := \{\xi \in \mathfrak{a}_0 : |\alpha(\xi)| < \frac{\pi}{2} \text{ for all } \alpha \in \Sigma(\mathfrak{g}_0, \mathfrak{a}_0)\}$, where $\Sigma(\mathfrak{g}_0, \mathfrak{a}_0)$ denotes the system of restricted roots. By definition the subset $\exp i\omega_0(x_0)$ meets every G_0-orbit in $\mathcal{U}$.

The structure of the general G_0-orbits in G/K is more difficult to describe. To put this in perspective, let us recall the G_0-equivariant polar coordinate mapping $\Pi : G_0 \times_{K_0} \mathfrak{s}_0 \to G/K$ of Chapter 6. The G_0-action on $G_0 \times_{K_0} \mathfrak{s}_0$, the total space, is quite simple, because (1) it is proper, in particular, every orbit is closed, (2) $[1, i\mathfrak{a}_0] = \mathfrak{a}_0$ meets every G_0-orbit and (3) the intersection of a G_0-orbit with $\mathfrak{a}_0$ coincides with an orbit of the (small) Weyl group $W(\mathfrak{g}_0, \mathfrak{a}_0)$.

While $\mathcal{U}_C = \Pi^{-1}(\mathcal{U}) \xrightarrow{\Pi} \mathcal{U}$ is a diffeomorphism, the property of Π being a (local) diffeomorphism breaks down at the boundary points of $\mathcal{U}_C$. The restriction $\Pi : \mathrm{bd}(\mathcal{U}_C) \to \mathrm{bd}(\mathcal{U})$ is neither injective nor surjective; in particular, as we will see in Corollary 10.6.4, the image $\Pi(\mathrm{bd}(\mathcal{U}_C))$ is of positive codimension in $\mathrm{bd}(\mathcal{U})$. Further, it should be noted that there are nonclosed orbits in G/K, in fact most orbits in $\mathrm{bd}(\mathcal{U})$ are not closed, and the closure of the set $\mathcal{U} = G_0 \cdot \exp i\mathfrak{a}_0(x_0)$ has a nonempty complement in G/K.

Although the group action $G_0 \times G/K \to G/K$ and the corresponding orbit structure seems at a first glance to be quite complicated, they can be well understood by replacing the action by a certain action by conjugation. This is a well-known technique (see [dCP], [BoHC], [HeSch]). For instance the action $K \times G/K \to G/K$ is equivalent to a conjugation action using the map $\eta : G \to G$, $g \mapsto \theta(g)g^{-1}$, where θ is the defining involution of the symmetric subgroup K. A simple check shows that η factors over G/K and induces a K-equivariant map $\eta : G/H \to G$ where K acts on G by conjugation.

Sometimes it is more convenient to work in $\mathrm{Ad}(G) = \mathrm{Int}(\mathfrak{g}) \subset GL(\mathfrak{g})$ rather than in G. In that case the map η is given by $g \mapsto \theta \cdot \mathrm{Ad}(g) \cdot \theta \cdot \mathrm{Ad}(g^{-1})$. In our situation the real form $G_0 = G^{\tau}$ is acting on G/K and thus η should be modified as follows (see [M2]):

(10.2.1) $$\eta : G \to \mathrm{Aut}_{\mathbb{R}}(\mathfrak{g}),\ g \longmapsto \tau \cdot \mathrm{Ad}(g) \cdot \theta \cdot \mathrm{Ad}(g^{-1}).$$

Note that the image $\mathrm{Image}(\eta) \subset \sigma\, \mathrm{Aut}_{\mathbb{R}}(\mathfrak{g})^0$.

If $\mu : \mathfrak{g} \to \mathfrak{g}$ is a linear map, we write ${}^g\mu := \mathrm{Ad}(g)\cdot\mu\cdot\mathrm{Ad}(g^{-1})$. In this notation, $\eta(g) = \tau \cdot {}^g\theta$. Further, we also write $g \cdot \varphi := {}^g\varphi = \mathrm{Ad}(g)\cdot\varphi\cdot\mathrm{Ad}(g^{-1})$ when we wish to stress that the action is by conjugation on $\mathrm{Aut}_{\mathbb{R}}(\mathfrak{g})$.

The proofs of the following statements are elementary and are omitted.

Lemma 10.2.2. *Let $\eta : G \to \mathrm{Aut}_{\mathbb{R}}(\mathfrak{g})$ as defined in* (10.2.1).

(i) *The map η factors through G/K.*
(ii) *The induced map $\eta : G/K \to \mathrm{Aut}_{\mathbb{R}}(\mathfrak{g})$ is G_0-equivariant with respect to the action by left translations on G/K and by conjugation in $\mathrm{Aut}_{\mathbb{R}}(\mathfrak{g})$.*

For later purposes we state some simple properties of η with respect to various involutions. Given an involution $\nu \in \mathrm{Aut}_{\mathbb{R}}(\mathfrak{g})$, $\widehat{\nu}(\psi) = \nu\cdot\psi\cdot\nu^{-1}$ is the corresponding involution in $\mathrm{Aut}(\mathrm{Aut}_{\mathbb{R}}(\mathfrak{g}))$.

Lemma 10.2.3. *Let μ and ν be involutions of G. Then*

(i) $\widehat{\tau}(\eta(x)) = \tau\cdot\eta(x)\cdot\tau = \eta(x)^{-1}$, *and*
(ii) *if μ, ν commute, then for every involution ζ that commutes with both μ and ν, $\widehat{\zeta}(\eta(x)) = \eta(\zeta(x))$.*

10.2B Geometric properties of η

Here we give more detailed information on the fibers of $\eta : G/K \to \mathrm{Aut}_{\mathbb{R}}(\mathfrak{g})$ and the image $\mathrm{Im}(\eta)$ in the real reductive group $\mathrm{Aut}_{\mathbb{R}}(\mathfrak{g})$.

Lemma 10.2.4. *The image $\mathrm{Im}(\eta)$ is Zariski closed in $\mathrm{Aut}_{\mathbb{R}}(\mathfrak{g})$.*

Proof. The proof relies on the following well-known fact concerning conjugacy classes (see [Hu2, Proposition 18.2] or [BoHC, Proposition 10.1]).

Proposition 10.2.5. *Let H be a closed algebraic subgroup of $GL(V)$ and let $s \in GL(V)$ be a semisimple element which normalizes H. Regard H as acting on $GL(V)$ by conjugation. Then the orbit $H \cdot s := \mathrm{Ad}(H)(s)$ is closed in $GL(V)$.*

Note that $\mathrm{Im}(\eta)$ is closed if and only if $\{\mathrm{Ad}(g)\theta\,\mathrm{Ad}(g^{-1}) \mid g \in G\}$ is closed in $GL(\mathfrak{g})$. Apply Proposition 10.2.5 with $s = \theta$ and $H = \mathrm{Ad}(G)$. □

Lemma 10.2.6 ($N_G(K)$-invariance). *The map η factors through a G_0-equivariant embedding*

$$G/N_G(K) \hookrightarrow \mathrm{Aut}_{\mathbb{R}}(\mathfrak{g});$$

in other words, $\eta(x) = \eta(y)$ if and only if $y = xg^{-1}$ for some $g \in N_G(K)$.

Proof. We may write $y = xg^{-1}$ for some $g \in G$. Thus it must be shown that $\eta(x) = \eta(xg^{-1})$ if and only if $g \in N_G$. But $\eta(x) = \eta(xg^{-1})$ is equivalent to $\mathrm{Ad}(g)\theta = \theta\,\mathrm{Ad}(g)$, which in turn is equivalent to the fact that $\mathrm{Ad}(g)$ stabilizes the complexified Cartan decomposition $\mathfrak{g} = \mathfrak{k} + \mathfrak{s}$. If $\mathrm{Ad}(g)$ stabilizes $\mathfrak{k} + \mathfrak{s}$, then $\mathrm{Ad}(g)(\mathfrak{k}) = \mathfrak{k}$, i.e., $g \in N_G(K)$. On the other hand, if $g \in N_G(K)$, then $\mathrm{Ad}(g)(\mathfrak{s}) = \mathfrak{s}$ because $\mathfrak{s}$ is the orthogonal complement of $\mathfrak{k}$ with respect to the Killing form of $\mathfrak{g}$. □

We note that $N_G(K)/K$ is a finite abelian group. See [Fe1] for the classification.

Summary: η induces a G_0-equivariant map $\eta : G/K \to \mathrm{Im}(\eta) \subset \mathrm{Aut}_{\mathbb{R}}(\mathfrak{g})$ which is a finite abelian covering. An orbit $G_0{\cdot}x$ in G/K is closed if and only if the G_0-conjugacy class of $\eta(x)$ is closed in $\mathrm{Aut}_{\mathbb{R}}(\mathfrak{g})$. Hence, for our purposes it suffices to understand the conjugation action $G_0 \times \mathrm{Im}(\eta) \to \mathrm{Im}(\eta)$.

We have already pointed out that $G \cdot \eta(x)$ $(= \mathrm{Ad}(\mathrm{Ad}(G))(\eta(x)))$ is closed if and only if $\eta(x)$ is semisimple. The same statement remains true if we replace G by the real form G_0.

Proposition 10.2.7. *A G_0-orbit $G_0{\cdot}x$ in G/K is closed if and only if the element $\eta(x) \in \mathrm{Aut}_{\mathbb{R}}(\mathfrak{g})$ is semisimple.*

Proof. If $\eta(x)$ is semisimple, $G{\cdot}\eta(x)$ is closed by Proposition 10.2.5. The orbit $G{\cdot}\eta(x)$ is the homogeneous space $G/Z_G(\eta(x))$. Note that as a consequence of Lemma 10.2.3(i) the isotropy subgroup $Z_G(\eta(x))$ is a τ-stable. The "if" part of the proposition follows from the fact that $G_0{\cdot}\eta(x)$ is closed in $G{\cdot}\eta(x)$. This is true under broader conditions. More precisely, we have the following.

Lemma 10.2.8. *Let $\tau : G \to G$ be an arbitrary involutive automorphism of a semisimple Lie group, G_0 an open subgroup of G^τ and $H \subset G$ an arbitrary closed τ-stable subgroup. Then $G_0{\cdot}x_0$ is closed in G/H with $x_0 = 1\,H$.*

Proof of the lemma. From the action of τ on the tangent space of G/H at x_0, G_0 acts locally transitively on the fixed point set of τ, which itself is closed. Thus every G_0 orbit in that fixed point set is already closed. □

In order to prove the "only if" part we will show that if $\eta(x)$ is not semisimple, then one can construct a one-parameter subgroup $\{\gamma(t)\} \subset G_0$ such that the closure of $\{\gamma(t)\eta(x)\gamma(t)^{-1}\}$ belongs to an orbit $G_0.\eta(y)$ with semisimple $\eta(y)$. This is a classical idea and of course the key tool here is the Jordan decomposition. We recall the basic facts in the next subsection.

10.2C Jordan decomposition

Classically one knows that every element $z \in GL_{\mathbb{R}}(\mathfrak{g})$ has a unique multiplicative Jordan decomposition $z = s_z u_z = u_z s_z$ with a semisimple s_z and a unipotent u_z. Let $\mathrm{End}^{nil}_{\mathbb{R}}(\mathfrak{g})$ denote the set of nilpotent elements in $\mathrm{End}_{\mathbb{R}}(\mathfrak{g})$ and $GL^{uni}_{\mathbb{R}}(\mathfrak{g})$ the set of unipotent elements in $GL_{\mathbb{R}}(\mathfrak{g})$. The exponential map defines a bijection between $\mathrm{End}^{nil}_{\mathbb{R}}(\mathfrak{g})$ and $GL^{uni}_{\mathbb{R}}(\mathfrak{g})$. The reverse map, $\log : GL^{uni}_{\mathbb{R}}(\mathfrak{g}) \to \mathrm{End}_{\mathbb{R}}(\mathfrak{g})$ is given explicitly by the formula $u \mapsto \sum_{n=1}^{\infty}(-1)^{n-1}(u - I)^n/n$, where the sum contains only finitely many nonzero terms.

We begin here by describing the particular properties of s_ψ and u_ψ for ψ an automorphism of $\mathfrak{g}$. Since $\mathrm{Aut}_{\mathbb{R}}(\mathfrak{g})$ is an algebraic subgroup, the semisimple and unipotent parts of an element $\psi \in \mathrm{Aut}_{\mathbb{R}}(\mathfrak{g})$ also belong to $\mathrm{Aut}_{\mathbb{R}}(\mathfrak{g})$. Let $\mathfrak{g}^\psi$ denote the subalgebra of elements fixed by ψ.

Lemma 10.2.9. *Let $s_\psi \cdot u_\psi$ be the Jordan decomposition of $\psi \in \mathrm{Aut}_{\mathbb{R}}(\mathfrak{g})$. Then there is a unique* ad-*nilpotent element $E \in \mathfrak{g}$ with $u_\psi = \mathrm{Ad}(\exp E)$ and $E \in \mathfrak{g}^\psi$.*

Proof. Define $N := \log u_\psi \in \mathrm{End}(\mathfrak{g})$ and note that $\exp : \mathbb{R}N \to GL_{\mathbb{R}}(\mathfrak{g})$ is polynomial. Since $\exp \mathbb{Z}N \subset \mathrm{Aut}_{\mathbb{R}}(\mathfrak{g})$ now $\exp \mathbb{R}N \subset \mathrm{Aut}^0(\mathfrak{g})$ and $N \in \mathrm{ad}(\mathfrak{g})$, $N = \mathrm{ad}(E)$ for some $E \in \mathfrak{g}$. Since $s_\psi \cdot \mathrm{Ad}(\exp E) = \mathrm{Ad}(\exp E) \cdot s_\psi$ and $\psi \cdot \mathrm{Ad}(\exp E) = \mathrm{Ad}(\exp E) \cdot \psi$, we conclude that $E \in \mathfrak{g}^\psi$. □

Now we are in the position to finish the following proof.

Proof of "only if" in Proposition 10.2.7. Assume that $\eta = \eta(x)$ is not semisimple. Let $\eta = s_\eta \cdot \mathrm{Ad}(\exp E) = s_\eta \cdot u_\eta$, Jordan decomposition with $E \in \mathfrak{g}^{\eta(x)} \setminus 0$, by Lemma 10.2.9. The subalgebras $\mathfrak{g}^\eta$ and $\mathfrak{g}^{s_\eta}$ carry more structure: given two involutions $\mu, \nu \in \mathrm{Aut}_{\mathbb{R}}(\mathfrak{g})$, a direct computation shows that the subalgebra of fixed points of $\mu \cdot \nu$ is given by $\mathfrak{g}^{\mu \cdot \nu} = \mathfrak{g}^{\nu \cdot \mu} = \mathfrak{g}^\mu \cap \mathfrak{g}^\nu \oplus \mathfrak{g}^{-\mu} \cap \mathfrak{g}^{-\nu}$. In our situation, i.e., for $\eta(x) = \tau \cdot {}^x\theta$, we have $\tau\eta(x)\tau = \eta(x)^{-1}$. Consequently, $\tau s_\eta \tau = s_\eta^{-1}$ and $\tau\,\mathrm{Ad}(\exp(E))\tau = \mathrm{Ad}(\exp(-E))$. This shows that

$$\mathfrak{g}^{s_\eta} = (\mathfrak{g}^{s_\eta} \cap \mathfrak{g}^\tau) + (\mathfrak{g}^{s_\eta} \cap \mathfrak{g}^{-\tau}) \tag{10.2.10}$$

is reductive and τ-stable. Further, since $s_\eta \cdot u_\eta|_{\mathfrak{g}^\eta} = \mathrm{Id}$, we also have $s_\eta|_{\mathfrak{g}^\eta} = u_\eta|_{\mathfrak{g}^\eta} = \mathrm{Id}$, and consequently $\mathfrak{g}^\eta \subset \mathfrak{g}^{s_\eta}$. We have proved the following. □

Observation 10.2.11. *If $E = E_{\eta(x)}$ is the nilpotent element in the Jordan decomposition of $\eta(x)$ as in Lemma* 10.2.9*, then $E \in (\mathfrak{g}^{-\tau} \cap \mathfrak{g}^{-({}^x\theta)}) \subset (\mathfrak{g}^{s_\eta} \cap \mathfrak{g}^{-\tau})$.*

Next we construct a one-parameter subgroup $\{\gamma(t)\} \subset \mathrm{Aut}_{\mathbb{R}}(\mathfrak{g})$ with the property $\lim_{t\to-\infty} \gamma(t)(s_\eta u_\eta)\gamma(t)^{-1} \in G_0.s_\eta$. The key ingredient here is the Jacobson–Morozov theorem. We state it in the form most suitable for us. Recall that a triple $\{E, F, H\} \subset \mathfrak{g}$ is called an $\mathfrak{sl}_2$-triple if $[H, E] = 2E$, $[H, F] = -2F$ and $[E, F] = H$, in other words if $\{E, F, H\}$ spans a Lie algebra isomorphic to $\mathfrak{sl}_2(\mathbb{K})$. An obvious modification of the proof in [KR, Proposition 4], where the statement was proved for complex symmetric pairs, yields the following.

Proposition 10.2.12. *Let $\mathfrak{l}$ be a real reductive Lie algebra and $E \in \mathfrak{l}$ a nonzero* ad-*nilpotent element. Then there exist $F, H \in \mathfrak{l}$ such that $\{E, F, H\}$ is an $\mathfrak{sl}_2$-triple. If $\mu : \mathfrak{l} \to \mathfrak{l}$ is an involutive automorphism and $E \in \mathfrak{l}^{-\mu}$, then one can choose $F \in \mathfrak{l}^{-\mu}$ and $H \in \mathfrak{l}^\mu$.*

As already remarked, $\mathfrak{g}^{s_\eta}$ is a τ-stable reductive subalgebra such that the nilpotent element $E = E_{\eta(x)}$ lies in $\mathfrak{g}^{s_\eta} \cap \mathfrak{g}^{-\tau}$. The above proposition gives us the elements $H \in \mathfrak{g}^{s_\eta} \cap \mathfrak{g}^{\tau}$ and $F \in \mathfrak{g}^{s_\eta} \cap \mathfrak{g}^{-\tau}$ such that E, H, F is a $\mathfrak{sl}_2$-triple. Define $\gamma(t) := \exp tH$. Then

$$\begin{aligned}\gamma(t)\eta(x)\gamma(t)^{-1} &= \gamma(t)(s_\eta u_\eta)\gamma(t)^{-1} \\ &= (\gamma(t)s_\eta\gamma(t)^{-1})(\gamma(t)u_\eta\gamma(t)^{-1}) \\ &= s_\eta(\gamma(t)\operatorname{Ad}(\exp E)\gamma(t)^{-1}) = s_\eta \operatorname{Ad}(\exp(e^{2t}E) \overset{t\to-\infty}{\longrightarrow} s_\eta.\end{aligned}$$

We have proved that if $\eta(x) = s_\eta u_\eta$ with $u_\eta \neq 1$, then $G_0.s_\eta$ is contained in the closure of $G_0.\eta(x)$. The proof of Proposition 10.2.7 is now complete. □

Although $\eta : G \to \operatorname{Aut}_{\mathbb{R}}(\mathfrak{g})$ is not a group homomorphism we can lift the Jordan decomposition of $\eta(x)$ to the group G as follows.

Proposition 10.2.13 (Lifting of the Jordan decomposition).

(i) *Given $g \in G$, let $\eta(g) = s_\eta u_\eta = s_\eta \cdot \operatorname{Ad}(\exp E)$ be the Jordan decomposition in $\operatorname{Aut}_{\mathbb{R}}$, where $E \in \mathfrak{g}^{-\tau} \cap \mathfrak{g}^{-(g\theta)}$ is the corresponding nilpotent element as in Observation 10.2.11. Then $\eta(\exp(\frac{1}{2}E)\cdot g) = s_\eta$.*

(ii) *Let $g \in G$ such that $G_0 \cdot gx_0$ is closed in G/K, i.e., $\eta(g)$ is semisimple. Let $0 \neq E \in \mathfrak{g}^{-\tau} \cap \mathfrak{g}^{-g\theta}$ be ad-nilpotent and define $y := \exp(-\frac{1}{2}E)\cdot gx_0$. Then $\eta(y)$ has Jordan decomposition $s_y u_y$ with $s_y = \eta(g)$ and $u_y = \operatorname{Ad}(\exp E)$. In particular, $G_0 \cdot \exp(-\frac{1}{2}E)\cdot gx_0$ is not closed.*

Proof. We evaluate $\eta(\exp(\frac{1}{2}E)g)$. Since $\tau(E) = -E$ and $\operatorname{Ad}(\exp tE)$ commutes with s_η for every t,

$$\begin{aligned}\eta(\exp(\tfrac{1}{2}E)g) &= \tau\operatorname{Ad}(\exp(\tfrac{1}{2}E))\cdot\operatorname{Ad}(g)\cdot\theta\cdot\operatorname{Ad}(g^{-1})\cdot\operatorname{Ad}(\exp(-\tfrac{1}{2}E)) \\ &= \operatorname{Ad}(\exp(-\tfrac{1}{2}E))\cdot\tau\cdot\operatorname{Ad}(g)\cdot\theta\cdot\operatorname{Ad}(g^{-1})\cdot\operatorname{Ad}(\exp(-\tfrac{1}{2}E)) \\ &= \operatorname{Ad}(\exp(-\tfrac{1}{2}E))\cdot s_\eta u_\eta \cdot\operatorname{Ad}(\exp(-\tfrac{1}{2}E)) \\ &= s_\eta\cdot\operatorname{Ad}(\exp(-\tfrac{1}{2}E))\cdot\operatorname{Ad}(\exp(E))\cdot\operatorname{Ad}(\exp(-\tfrac{1}{2}E)) = s_\eta.\end{aligned}$$

A similar computation proves the claim for $E \in \mathfrak{g}^{-\tau} \cap \mathfrak{g}^{-g\theta}$ as in (ii). □

10.3 Characterization of closed orbits

Polar decomposition of the eigenvalues of a semisimple element $s \in GL_{\mathbb{R}}(V)$ gives $s = s_{\text{ell}}s_{\text{hyp}} = s_{\text{hyp}}s_{\text{ell}}$, where s_{ell} is elliptic (eigenvalues all of absolute value 1) and s_{hyp} is hyperbolic (eigenvalues all positive real). If $g \in GL_{\mathbb{R}}(V)$ that refines its Jordan decomposition $g = su$ to $g = s_{\text{ell}}s_{\text{hyp}}u$, where all three terms commute with each other. If $g \in \operatorname{Aut}_{\mathbb{R}}(\mathfrak{g})$ then all three terms belong to $\operatorname{Aut}_{\mathbb{R}}(\mathfrak{g})$. With $\eta : G/K \to \operatorname{Aut}_{\mathbb{R}}(\mathfrak{g})$ in mind, we define the set $(G/K)_{\text{ell}} \subset G/K$ of **η-elliptic** elements as the preimage of the set of elliptic semisimple elements in $\mathrm{Aut}_{\mathbb{R}}(\mathfrak{g})$. As for any reductive linear group, an element of $\mathrm{Aut}_{\mathbb{R}}(\mathfrak{g})$ is elliptic semisimple if and only if it belongs to some compact subgroup.

The main observation of this section is that every closed G_0-orbit in $c\ell(\mathcal{U})$ is contained in the set $(G/K)_{\mathrm{ell}}$, (see Proposition 10.3.2). Before we give the precise statement, we characterize the subset of η-elliptic elements in G/K.

Proposition 10.3.1 (Elliptic elements). *The set* $(G/K)_{\mathrm{ell}}$ *of* η*-elliptic elements coincides with* $G_0 \cdot \exp(i\mathfrak{a}_0)(x_0)$.

Proof. The elements of $G_0.\eta(g) = \{\mathrm{Ad}(h)\eta(g)\,\mathrm{Ad}(h)^{-1} \mid h \in G_0\}$ all have the same eigenvalues so $(G/K)_{\mathrm{ell}}$ is G_0-invariant. Thus $\exp(i\mathfrak{a}_0)(x_0) \subset (G/K)_{\mathrm{ell}}$ will imply $G_0 \cdot \exp(i\mathfrak{a}_0)(x_0) \subset (G/K)_{\mathrm{ell}}$. To prove this inclusion note that, for $\xi \in \mathfrak{a}_0$, we have

$$\begin{aligned}\eta(\exp(i\xi(x_0))) &= \tau\,\mathrm{Ad}(\exp(i\xi))\theta\,\mathrm{Ad}(\exp(-i\xi))\\ &= \tau\theta\,\mathrm{Ad}(\exp(-2i\xi)) = \sigma\,\mathrm{Ad}(\exp(-2i\xi)).\end{aligned}$$

Since $\exp i\mathfrak{a}_0 \subset G_u$, $\mathrm{Ad}(\exp -2i\xi)$ commutes with σ, so $\eta(\exp i\xi(x_0))$ commutes with σ. By (10.1.7) we conclude that $\eta(\exp(i\mathfrak{a}_0)\cdot x_0)$ is contained in a maximal compact subgroup of $\mathrm{Aut}_{\mathbb{R}}(\mathfrak{g})$, i.e., every element in $\eta(\exp(i\mathfrak{a}_0)\cdot x_0)$ is elliptic. We have proved $G_0 \cdot \exp(i\mathfrak{a}_0)(x_0) \subset (G/K)_{\mathrm{ell}}$.

For the opposite inclusion, let $\eta(g)$ be elliptic semisimple. Then it commutes with a Cartan involution σ'' of $\mathfrak{g}$.

Claim. There exists $h \in G_0$ such that $\eta(hg)$ commutes with the original Cartan involution σ.

Proof of the Claim. If σ'' does not commute with τ, we replace it by another Cartan involution σ' which does. For define $\kappa(\xi, \zeta) = -\langle \xi, \sigma''\zeta\rangle$ where we work on the underlying real Lie algebra structure ${}_R\mathfrak{g}$ of $\mathfrak{g}$. Then κ is a positive definite symmetric bilinear form on ${}_R\mathfrak{g}$. Now compute $\kappa(\tau\sigma''\xi, \zeta) = -\langle\tau\sigma''\xi, \sigma''\zeta\rangle = -\langle\sigma''\xi, \tau\sigma''\zeta\rangle = -\langle\tau\sigma''\zeta, \sigma''\xi\rangle = \kappa(\tau\sigma''\zeta, \xi)$. This shows that $\tau\sigma''$ is diagonalizable over $\mathbb{R}$ as a linear transformation of the real vector space ${}_R\mathfrak{g}$. Thus $\nu := (\tau\sigma'')^2$ lies on a one-parameter subgroup $t \mapsto \nu^t$ of $\mathrm{Aut}(\mathfrak{g})$. Define $\sigma' = \nu^{\frac{1}{4}}\sigma''\nu^{-\frac{1}{4}}$. Following [Hel, Chapter III, §7], σ' commutes with τ. By direct calculation one verifies that ν, and thus the ν^t, commute with $\eta(g)$. It follows that σ' commutes with $\eta(g)$.

In the above argument, $\nu^t \in \mathrm{Ad}(G_0)$ because ν commutes with τ. So the one-parameter subgroup $t \mapsto \nu^t$ is of the form $t \mapsto \exp(t\alpha)$ for some $\alpha \in \mathfrak{g}_0$.

Both σ' and our original σ commute with τ. So we have a Cartan involution θ' of $\mathfrak{g}_0$ with $\sigma' = \tau\theta'$, and θ' is of the form $\mathrm{Ad}(\mathrm{Ad}(h_1))\theta$ for some $h_1 \in G_0$. Thus $\sigma' = \mathrm{Ad}(h_1)\sigma\,\mathrm{Ad}(h_1^{-1})$.

In analogy to the above construction of ν, one argues that $\mu := \sigma\sigma'\sigma\sigma'$ lies on a one-parameter subgroup $t \mapsto \mu^t$ of $\mathrm{Aut}(\mathfrak{g})$, and $\mu^{\frac{1}{4}}\sigma'\mu^{-\frac{1}{4}}$ commutes with σ. Both are Cartan involutions of $\mathfrak{g}$, so they are equal, $\sigma = \mu^{\frac{1}{4}}\sigma'\mu^{-\frac{1}{4}}$. Again, μ commutes with τ and it follows that $t \mapsto \mu^t$ is of the form $t \mapsto \exp(t\beta)$ for some $\beta \in \mathfrak{g}_0$. Now

$h := \exp(\frac{1}{4}\beta) \in G_0$ has the property that $\eta(hg)$ commutes with σ. The Claim is proved. □

Replacing g by hg we may now assume that $\eta(g)$ and σ commute. Now we will adjust g so that it lies in $G_u = G^\sigma$. With respect to the global Cartan decomposition of G defined by σ, write $g = u\exp(\zeta)$ with $u \in G_u$ and $\sigma(\zeta) = -\zeta$. We show that in fact $\exp(\zeta) \in K$. Since σ commutes with τ and $\mathrm{Ad}(u)$ and $\sigma\,\mathrm{Ad}(\exp\zeta) = \mathrm{Ad}(\exp(-\zeta))\sigma$, we have

$$\begin{aligned}\sigma\eta(g) &= \sigma\cdot(\tau\,\mathrm{Ad}(u)\,\mathrm{Ad}(\exp(\zeta))\theta\,\mathrm{Ad}(\exp(-\zeta))\,\mathrm{Ad}(u^{-1}))\\ &= \tau\,\mathrm{Ad}(u)\,\mathrm{Ad}(\exp(-\zeta))\theta\,\mathrm{Ad}(\exp(\zeta))\,\mathrm{Ad}(u^{-1})\cdot\sigma.\end{aligned}$$

On the other hand

$$\sigma\eta(g) = \eta(g)\sigma = \tau\,\mathrm{Ad}(u)\,\mathrm{Ad}(\exp(\zeta))\theta\,\mathrm{Ad}(\exp(-\zeta))\,\mathrm{Ad}(u^{-1})\cdot\sigma.$$

Combining these two equations, we obtain

$$\mathrm{Ad}(\exp(\zeta))\cdot\theta\cdot\mathrm{Ad}(\exp(-\zeta)) = \mathrm{Ad}(\exp(-\zeta))\cdot\theta\cdot\mathrm{Ad}(\exp(\zeta)),$$

so $\mathrm{Ad}(\exp(2\zeta))$ commutes with θ. Since the restriction $\mathrm{Ad} : \exp(i\mathfrak{g}_u) \to \mathrm{Aut}(\mathfrak{g})$ is injective, it follows that $\theta(\exp(\zeta)) = \exp(\zeta)$, i.e., $\exp(\zeta) \in K$. Replacing g by $g\exp(-\zeta)$, it follows that $g(x_0) = g\exp(-\zeta)(x_0)$; hence, we may assume that $g \in G_u$.

Since $G_u = K_0\cdot\exp(i\mathfrak{a}_0)\cdot K_0$, we may assume that $g \in K_0\exp(i\mathfrak{a}_0)$ and then translate it by left multiplication by an element of K_0 to reach the following conclusion: If $g \in G$ is such that $\eta(g)$ is elliptic, then there exists $h \in G_0$ and $l \in K$ with $hgl \in \exp i\mathfrak{a}$. In other words, there is $h \in G_0$ with $hg(x_0) \in \exp(i\mathfrak{a}_0)(x_0)$. This proves the inclusion $(G/K)_{\mathrm{ell}} \subset G_0\cdot\exp(i\mathfrak{a}_0)(x_0)$. □

The boundary $\mathrm{bd}(\mathcal{U})$ in G/K contains nonclosed orbits. However, since G/K is a real affine variety, every orbit $G_0(y)$ has unique closed G_0-orbit in its closure. The closed orbits in $\mathrm{bd}(\mathcal{U})$ can be described as follows.

Proposition 10.3.2. *Every closed orbit $G_0(y)$ in $\mathrm{bd}(\mathcal{U})$ meets $\exp\mathrm{bd}(i\omega_0)(x_0)$ and every orbit $G_0(x)$ with $x \in \exp\mathrm{bd}(i\omega_0)(x_0)$ is closed. Furthermore,*

$$(\exp i\mathfrak{a}_0(x_0)) \cap c\ell(\mathcal{U}) = c\ell(\exp i\omega_0(x_0)) = \exp(c\ell(i\omega_0))(x_0).$$

Proof. Let $G_0(y)$ be a closed orbit in $\mathrm{bd}(\mathcal{U})$. Choose sequences $\{\xi_n\}$ in ω_0 and $\{g_n\}$ in G_0 so that $x_n := g_n\exp(i\xi_n)(x_0) \to y$. Consider the corresponding sequences $\eta(x_n) \to \eta(y)$. Since the set of eigenvalues $\mathrm{Spec}(\varphi) \in \mathbb{C}^n/\mathfrak{S}_n$ varies continuously in φ, ellipticity of the $\eta(x_n)$ implies ellipticity for $\eta(y)$. By Proposition 10.3.1 we have $y \in G_0\cdot\exp(i\zeta)(x_0)$ for some $\zeta \in \mathfrak{a}_0$. By continuity, $\zeta \in \mathrm{bd}(\omega_0)$, and we have proved $\mathrm{bd}(\mathcal{U}) \cap (\exp(i\mathfrak{a}_0)(x_0)) \subset \exp(i\,\mathrm{bd}(\omega_0))(x_0)$. The opposite inclusion is trivial. It follows that $\mathrm{bd}(\mathcal{U}) \cap (\exp i\mathfrak{a}_0(x_0)) = \exp(i\,\mathrm{bd}(\omega_0))(x_0)$. Now $(\exp(i\mathfrak{a}_0)(x_0)) \cap c\ell(\mathcal{U}) = \exp(c\ell(i\omega_0))(x_0)$. □

10.4 The slice theorem and related isotropy computations

An important tool for the investigation of nonclosed orbits in $\mathrm{bd}(\mathcal{U})$ is (a real version of) the Luna slice theorem. Roughly speaking, given an algebraic action of a reductive group G on a smooth affine variety Y, a G-invariant neighborhood of a closed G-orbit $G(y) \cong G/H$ is G-equivariantly isomorphic to a neighborhood of the zero section of the homogeneous bundle $G \times_H V$, where $V \subset T_y Y$ is a H-stable subspace, complementary to $T_y(G(y))$, and $H \times T_y Y \to T_y Y$ is the linear isotropy action. Here is the precise statement.

Theorem 10.4.1 (Slice Theorem). *Let $H \times Y \to Y$ be an affine algebraic action defined over $\mathbb{R}$ with H reductive and Y smooth. Let $y \in Y_{\mathbb{R}}$ such that $H(y)$ is closed and let L denote the isotropy group at y. Then there is an L- stable open (in the Hausdorff topology) neighborhood $U \subset T_y Y / T_y(G(y))$ and an L- equivariant isomorphism $e : U \to e(U) \subset Y$ such that*

$$H \times_L U \longrightarrow Y \qquad [h, v] \longmapsto h(e(v))$$

is an H-equivariant isomorphism, defined over $\mathbb{R}$, onto the image which is an open H-stable neighborhood of $H(y)$. In particular, the above map defines an $H_{\mathbb{R}}$-equivariant isomorphism of $H_{\mathbb{R}} \times_{L_{\mathbb{R}}} U_{\mathbb{R}}$ onto an open $H_{\mathbb{R}}$-invariant neighborhood of $H_{\mathbb{R}}(y)$ in $Y_{\mathbb{R}}$.

See [Br] for proofs and further details.

We apply the above theorem to a closed orbit $G_0(ax_0) \subset G/K = \Omega$ for $a = \exp(i\xi)$ and $\xi \in \mathfrak{a}_0$. In the above notation, $Y = G/K \times G/K$, $H = \mathrm{diag}(G) \subset G \times G$ and the real structures are $\widehat{\tau} : Y \to Y$, $(y_1, y_2) \mapsto (\tau(y_2), \tau(y_1))$, and $\tau : H \to H$. We identify Ω with $Y_{\mathbb{R}} = \{(gK, \tau(g)K) \mid g \in G\}$. Before going further we must derive explicit information concerning the isotropy representation of $(G_0)_{ax_0}$ on $T_{ax_0}(G/K)$.

10.4A Geometry of real orbits in G/K

Starting with a general situation, let $K \subset G$ denote a complex symmetric subgroup which is the fixed point set of a holomorphic involution $\theta : G \to G$. As usual x_0 denotes the base point $1K$ in G/K. Given $x = gx_0 \in G/K$ let $\mathfrak{g}_x$ denote the Lie algebra of the isotropy group gKg^{-1}. We then have the identification $T_x(G/K) = \mathfrak{g}/\mathfrak{g}_x = \mathfrak{g}^{-({}^g\theta)}$ where ${}^g\theta = \mathrm{Ad}(g)\cdot\theta\cdot\mathrm{Ad}(g^{-1})$.

The real form $G_0 = G^\tau \subset G$ acts on $\Omega := G/K$ by left translations. Consider an orbit $M := G_0(x)$ in Ω. Since G_0 acts by biholomorphic transformations, the complex structure on G/K induces a Cauchy–Riemann structure on M. In this connection see Section 6.3. The first generalities in this situation can be stated as follows. Let $T_y^{CR}M$ denote the Cauchy–Riemann tangent space $T_y M \cap iT_y M$ of M at the point y.

Lemma 10.4.2. *Let G be a complex Lie group, H a complex subgroup and $X = G/H$ the corresponding complex manifold. Let G_0 be a real form of G defined by the* *involution τ. Consider the action by left translations $G_0 \times X \to X$. Every orbit $M := G_0 \cdot y$ (which is a locally closed submanifold) carries a natural G_0-invariant Cauchy–Riemann structure. The corresponding subspaces of $T_y X = \mathfrak{g}/\mathfrak{g}_y$ are given as follows:*

$$\begin{array}{ccccccc} T_y^{CR} M & \hookrightarrow & T_y M & \hookrightarrow & T_y X & & y = gx_0 \\ \| & & \| & & \| & & \\ \dfrac{\mathfrak{g}_y + \tau\mathfrak{g}_y}{\mathfrak{g}_y} & \hookrightarrow & \dfrac{\mathfrak{g}_0 + \mathfrak{g}_y}{\mathfrak{g}_y} & \hookrightarrow & \dfrac{\mathfrak{g}}{\mathfrak{g}_y}. & & \end{array}$$

In our particular situation $H = K$ is a symmetric subgroup, and we have the identification $T_{gx_0}(G/K) = \mathfrak{g}^{-(^g\theta)} = \mathrm{Ad}(g)(\mathfrak{s})$. Let $\mathrm{pr} : \mathfrak{g} \to \mathrm{Ad}(g)(\mathfrak{s})$ denote the linear projection. In order to describe the image of $\mathrm{pr} : \mathfrak{g}_0 \to \mathfrak{g}^{-(^g\theta)}$, needed for a description of the subspaces $T_x M$ and $T_x^{CR} M$, it is necessary to give a more explicit description of $\mathrm{Ad}(g)$. Further, in view of the above Slice Theorem, the G_0-orbit structure in a neighborhood of a closed orbit $M = G_0(x)$ is determined by the isotropy action of $G_0 \cap G_x$ on $T_x\Omega/T_x M$.

Here we need only describe $\mathrm{Ad}(g)$ for $g \in \exp i\mathfrak{a}_0$. For $x \in \exp(i\mathfrak{a}_0)(x_0)$ we also describe the normal space $T_x\Omega/T_x M$ and the isotropy Lie algebra $(\mathfrak{g}_0)_x$ in terms of certain root spaces. The computation of $\mathrm{Ad}(\exp i\xi)$ for $\xi \in \mathfrak{a}_0$ will be given with respect to some appropriately chosen basis of $\mathfrak{g}_0$ which is related to pairs $\mathfrak{g}_0^{\lambda}, \mathfrak{g}_0^{-\lambda}$ of restricted root spaces.

10.4B $\mathrm{Ad}(\exp(i\xi))$ and related subspaces

Fix a Cartan subalgebra $\mathfrak{t}$ of $\mathfrak{m}$, so $\mathfrak{h} := \mathfrak{t}+\mathfrak{a}$ is a Cartan subalgebra of $\mathfrak{g}$. We emphasize our convention that $\mathfrak{a}$-root spaces are written with superscript while $\mathfrak{h}$-root spaces are written with subscript. Thus, given a restricted root $\lambda \in \Sigma(\mathfrak{g}_0, \mathfrak{a}_0)$, the $\mathfrak{a}$-root and $\mathfrak{a}_0$-root spaces are

$$\mathfrak{g}_0^{\lambda} = \mathfrak{g}_0 \cap \mathfrak{g}^{\lambda} \text{ where } \mathfrak{g}^{\lambda} := \sum\nolimits_{\{\alpha\in\Sigma(\mathfrak{g},\mathfrak{h}) \mid \alpha|_{\mathfrak{a}}=\lambda\}} \mathfrak{g}_{\alpha}, \tag{10.4.3}$$

and the resulting restricted root decompositions are

$$\mathfrak{g}_0 = \mathfrak{m}_0 + \mathfrak{a}_0 + \sum\nolimits_{\Sigma(\mathfrak{g}_0,\mathfrak{a}_0)} \mathfrak{g}_0^{\lambda} \text{ and } \mathfrak{g} = \mathfrak{m} + \mathfrak{a} + \sum\nolimits_{\Sigma(\mathfrak{g}_0,\mathfrak{a}_0)} \mathfrak{g}^{\lambda}. \tag{10.4.4}$$

If $\lambda \in \Sigma(\mathfrak{g}_0, \mathfrak{a}_0)$ then $\theta(\mathfrak{g}^{\lambda}) = \sigma(\mathfrak{g}^{\lambda}) = \mathfrak{g}^{-\lambda}$ and $\tau(\mathfrak{g}^{\lambda}) = \mathfrak{g}^{\lambda}$. So while the root space summands in (10.4.4) are preserved by τ they are pairwise interchanged by σ and θ. For this reason it will sometimes be convenient to have the notation

$$\mathfrak{g}_0^{[\lambda]} = \mathfrak{g}_0 \cap \mathfrak{g}^{[\lambda]} \text{ where } \mathfrak{g}^{[\lambda]} = \mathfrak{g}^{\lambda} + \mathfrak{g}^{-\lambda}$$

for the restricted root space parts of (10.4.4) preserved by all of τ, θ and σ.

Lemma 10.1.3 tells us that, for every pair $\{\lambda, -\lambda\} \in \Sigma(\mathfrak{g}_0, \mathfrak{a}_0)$ we can select bases $\{E_1^{\lambda}, \ldots, E_k^{\lambda}\}$ of $\mathfrak{g}_0^{\lambda}$ and $\{E_1^{-\lambda}, \ldots, E_k^{-\lambda}\}$ of $\mathfrak{g}_0^{-\lambda}$ such that the $\{E_j^{\lambda}, E_j^{-\lambda}, h_{\lambda}\}$ are θ-adapted $\mathfrak{sl}_2$-triples. Then we have

$$X_j^\lambda := E_j^\lambda - E_j^{-\lambda} = E^\lambda + \theta E^\lambda \in \mathfrak{g}_0^\theta \text{ and}$$
$$Y_j^\lambda := E_j^\lambda + E_j^{-\lambda} = E^\lambda - \theta E^\lambda \in \mathfrak{g}_0^{-\theta}. \tag{10.4.5}$$

They satisfy the identities

$$[X_j^\lambda, Y_j^\lambda] = 2h_\lambda \qquad [h_\lambda, X_j^\mu] = \mu(h_\lambda)Y_j^\mu \qquad [h_\lambda, Y_j^\mu] = \mu(h_\lambda)X_j^\mu. \tag{10.4.6}$$

Each of the complex 2-planes $\langle\langle X_j^\lambda, Y_j^\lambda \rangle\rangle_{\mathbb{C}}$ is stable under the action of $\mathrm{Ad}(\exp i\mathfrak{a}_0)$. The proof of the following proposition makes extensive use of this basis.

We underline the fact that most of the spaces in the following proposition should be viewed as real subspaces of the underlying real vector space structure of $\mathfrak{g}$.

Proposition 10.4.7. *Let $a = \exp(i\xi) \in \exp(i\mathfrak{a}_0)$, $\Omega = G/K$ and $M = G_0(ax_0)$. Select a positive system $\Sigma^+(\mathfrak{g}_0, \mathfrak{a}_0)$. In the direct sums λ runs through elements of $\Sigma^+(\mathfrak{g}_0, \mathfrak{a}_0)$ which satisfy the given (in)equalities.*

$$T_{ax_0}\Omega = \mathfrak{a}_0 + i\mathfrak{a}_0 + \sum_{\lambda(\xi)\in\mathbb{Z}\pi} (\mathfrak{g}^{[\lambda]})^{-\theta} + \sum_{\lambda(\xi)\in\frac{\pi}{2}+\mathbb{Z}\pi} (\mathfrak{g}^{[\lambda]})^{\theta} + \sum_{\lambda(\xi)\notin\mathbb{Z}\frac{\pi}{2}} \mathrm{Ad}(a)((\mathfrak{g}^{[\lambda]})^{-\theta})$$
$$T_{ax_0}M = \mathfrak{a}_0 + \sum_{\lambda(\xi)\in\mathbb{Z}\pi} (\mathfrak{g}_0^{[\lambda]})^{-\theta} + \sum_{\lambda(\xi)\in\frac{\pi}{2}+\mathbb{Z}\pi} (\mathfrak{g}_0^{[\lambda]})^{\theta} + \sum_{\lambda(\xi)\notin\mathbb{Z}\frac{\pi}{2}} \mathrm{Ad}(a)((\mathfrak{g}^{[\lambda]})^{-\theta})$$
$$T_{ax_0}^{CR}M = \sum_{\lambda(\xi)\notin\mathbb{Z}\frac{\pi}{2}} \mathrm{Ad}(a)((\mathfrak{g}^{[\lambda]})^{-\theta}).$$

Further, the real reductive subalgebra $\mathfrak{g}^{\eta(a)} = (\mathfrak{g}_0 \cap \mathfrak{g}^{^a\theta}) + (i\mathfrak{g}_0 \cap \mathfrak{g}^{-(^a\theta)})$ defines a real symmetric pair for which the isotropy algebra $^a\mathfrak{h} := (\mathfrak{g}_0 \cap \mathfrak{g}^{^a\theta}) = (\mathfrak{g}_0)_{ax_0}$ and the tangent component $^a\mathfrak{t} := (i\mathfrak{g}_0 \cap \mathfrak{g}^{-(^a\theta)})$ is a $(\mathfrak{g}_0)_{ax_0}$-stable complement to $T_{ax_0}M$. In terms of root spaces we have

$$^a\mathfrak{h} = \mathfrak{g}_0 \cap \mathfrak{g}^{(^a\theta)} = (\mathfrak{g}_0)_{ax_0} = \mathfrak{m}_0 + \sum_{\lambda(\xi)\in\mathbb{Z}\pi} (\mathfrak{g}_0^{[\lambda]})^{\theta} + \sum_{\lambda(\xi)\in\frac{\pi}{2}+\mathbb{Z}\pi} (\mathfrak{g}_0^{[\lambda]})^{-\theta} \tag{10.4.8}$$

and

$$^a\mathfrak{t} = i\mathfrak{g}_0 \cap \mathfrak{g}^{-(^a\theta)} \cong (T_{ax_0}M)^\perp \cong i\mathfrak{a}_0 + \sum_{\lambda(\xi)\in\mathbb{Z}\pi} i(\mathfrak{g}_0^{[\lambda]})^{-\theta} + \sum_{\lambda(\xi)\in\frac{\pi}{2}+\mathbb{Z}\pi} i(\mathfrak{g}_0^{[\lambda]})^{\theta}. \tag{10.4.9}$$

Proof. Let $a := \exp i\xi$ for $\xi \in \mathfrak{a}_0$. We first compute $\mathrm{Ad}(a)$ on $\langle\langle X_j^\lambda, Y_j^\lambda \rangle\rangle_{\mathbb{C}}$ with respect to the basis of (10.4.5). The matrix presentation $\mathrm{Mat}(\mathrm{Ad}(\exp i\xi))$ with respect to $\{X_j^\lambda, Y_j^\lambda\}$ is of the following shape:

$$\mathrm{Mat}(\mathrm{Ad}(\exp i\xi)) = \begin{pmatrix} \cosh\lambda(i\xi) & \sinh\lambda(i\xi) \\ \sinh\lambda(i\xi) & \cosh\lambda(i\xi) \end{pmatrix} = \begin{pmatrix} \cos\lambda(\xi) & i\sin\lambda(\xi) \\ i\sin\lambda(\xi) & \cos\lambda(\xi) \end{pmatrix}. \tag{10.4.10}$$

For $\lambda(\xi) \in \frac{\pi}{2} + \mathbb{Z}\pi$ or $\lambda(\xi) \in \mathbb{Z}\pi$ the above matrices are particularly simple. Next, we compute the restrictions of $^a\theta$ and $\eta(a)$ to $\langle\langle X_j^\lambda, Y_j^\lambda \rangle\rangle_{\mathbb{C}}$ with respect to the real basis $\{X_j^\lambda, iX_j^\lambda, Y_j^\lambda, iY_j^\lambda\}$:

$$\operatorname{Mat}({}^{a}\theta) = \begin{pmatrix} \cos 2\lambda(\xi) & 0 & 0 & \sin 2\lambda(\xi) \\ 0 & \cos 2\lambda(\xi) & -\sin 2\lambda(\xi) & 0 \\ 0 & -\sin 2\lambda(\xi) & -\cos 2\lambda(\xi) & 0 \\ \sin 2\lambda(\xi) & 0 & 0 & -\cos 2\lambda(\xi) \end{pmatrix},$$

$$(10.4.11) \qquad \operatorname{Mat}(\eta(a)) = \begin{pmatrix} \cos 2\lambda(\xi) & 0 & 0 & \sin 2\lambda(\xi) \\ 0 & -\cos 2\lambda(\xi) & \sin 2\lambda(\xi) & 0 \\ 0 & -\sin 2\lambda(\xi) & -\cos 2\lambda(\xi) & 0 \\ -\sin 2\lambda(\xi) & 0 & 0 & \cos 2\lambda(\xi) \end{pmatrix}.$$

We are now in the position to compute the various subspaces in Lemma 10.4.2 in terms of root spaces. In order to determine the projection

$$\mathfrak{g}_0 \hookrightarrow \mathfrak{g} = \operatorname{Ad}(a)(\mathfrak{k}) + \operatorname{Ad}(a)(\mathfrak{s}) \to \operatorname{Ad}(a)(\mathfrak{s}),$$

we first determine $\operatorname{Ad}(a^{-1})(\mathfrak{g}_0) \subset \mathfrak{g}$ and its projection onto $\mathfrak{s}$. For this see

$$\begin{array}{ccccc} \mathfrak{g}_0 & \hookrightarrow & \mathfrak{g} = \mathfrak{g}^{{}^{a}\theta} + \mathfrak{g}^{-({}^{a}\theta)} & \longrightarrow & \mathfrak{g}^{-({}^{a}\theta)} \supset \operatorname{pr}(\mathfrak{g}_0) = T_{ax_0}M \\ \operatorname{Ad}(a)\uparrow & & & & \operatorname{Ad}_a \uparrow \\ \operatorname{Ad}(a^{-1})\mathfrak{g}_0 & \hookrightarrow & \mathfrak{g} = \mathfrak{g}^{\theta} + \mathfrak{g}^{-\theta} & \longrightarrow & \mathfrak{g}^{-\theta} = \mathfrak{s}. \end{array}$$

We need only to compute the projections of $\operatorname{Ad}(a^{-1})\mathfrak{g}_0^{[\lambda]} \hookrightarrow \mathfrak{g}^{[\lambda]} \to (\mathfrak{g}^{[\lambda]})^{-\theta}$. Recall that the basis (10.4.5) respects the decomposition $(\mathfrak{g}^{[\lambda]})^{\theta} + (\mathfrak{g}^{[\lambda]})^{-\theta}$ and we have $\mathfrak{g}^{[\lambda]} = \langle\!\langle X_j^{\lambda} : 1 \leqq j \leqq k \rangle\!\rangle + \langle\!\langle Y_j^{\lambda} : 1 \leqq j \leqq k \rangle\!\rangle$. A straightforward computation using (10.4.10) shows that $\operatorname{Ad}(a^{-1})\mathfrak{g}_0^{[\lambda]} \to (\mathfrak{g}^{[\lambda]})^{-\theta}$ is not surjective only if $\lambda(\xi) \in \frac{\pi}{2}\mathbb{Z}$. More precisely, we have the following diagram:

$$\begin{array}{ccccccc} \operatorname{pr}(\operatorname{Ad}(a^{-1})\mathfrak{g}_0) = \mathfrak{a}_0 + & \sum\limits_{\lambda(\xi)=\mathbb{Z}\pi} (\mathfrak{g}_0^{[\lambda]})^{-\theta} & + & \sum\limits_{\lambda(\xi)=\frac{\pi}{2}+\mathbb{Z}\pi} i(\mathfrak{g}_0^{[\lambda]})^{-\theta} & + & \sum\limits_{\lambda(\xi)\neq\mathbb{Z}\frac{\pi}{2}} (\mathfrak{g}^{[\lambda]})^{-\theta} \\ \big\downarrow \operatorname{Ad}(a) & \big\downarrow \begin{pmatrix} \pm 1 & \\ & \pm 1 \end{pmatrix} & & \big\downarrow \begin{pmatrix} & \pm i \\ \pm i & \end{pmatrix} & & \\ T_{ax_0}M \quad = \mathfrak{a}_0 + & \sum\limits_{\lambda(\xi)=\mathbb{Z}\pi} (\mathfrak{g}_0^{[\lambda]})^{-\theta} & + & \sum\limits_{\lambda(\xi)=\frac{\pi}{2}+\mathbb{Z}\pi} (\mathfrak{g}_0^{[\lambda]})^{\theta} & + & \sum\limits_{\lambda(\xi)\neq\mathbb{Z}\frac{\pi}{2}} \operatorname{Ad}(a)((\mathfrak{g}^{[\lambda]})^{-\theta}). \end{array}$$

The matrices next to the vertical arrows give the matrix description of $\operatorname{Ad}(a)$ with respect to X_j^{λ}, Y_j^{λ}. This proves the formulas in the first part of the proposition.

Using (10.4.11), a similar computation gives the decomposition

$$\mathfrak{g}^{\eta(a)} = (\mathfrak{g}_0 \cap \mathfrak{g}^{{}^{a}\theta}) + (i\mathfrak{g}_0 \cap \mathfrak{g}^{-({}^{a}\theta)})$$

of the reductive subalgebra $\mathfrak{g}^{\eta(a)}$, as desired. Comparing the decompositions of $i\mathfrak{g}_0 \cap \mathfrak{g}^{-{}^{a}\theta}$ and $T_{ax_0}M$ into the generalized root spaces, we see that these two subspaces are complementary in $T_{ax_0}\Omega$ and stable under the real isotropy group. This proves the second part of the statement. □

From the above description concerning the isotropy Lie algebra we see that the isotropy groups $(G_0)_{\exp(i\xi)(x_0)}$ are compact for $\xi \in \omega_0$, and are equal to $M_0 = Z_{K_0}(\mathfrak{a}_0)$ for generic points. The isotropy group becomes noncompact for $\xi \in \operatorname{bd}(\omega_0)$, and the above proposition gives the precise description of the corresponding Lie algebra.

10.5 Example: Two-dimensional affine quadric

We illustrate the concepts introduced in this chapter with the example of the two-dimensional quadric. This example appears at various points in this monograph, for example Section 6.1C, and it is useful to look at it from several viewpoints.

Let $V \cong \mathbb{C}^2$, $G = SL(V) = SL(2;\mathbb{C})$ and $G_0 = SU(1,1)$. Thus V has a positive definite scalar product $\langle\cdot,\ \cdot\rangle$, we fix an orthonormal basis $\{v_+, v_-\}$, G_0 is the special unitary group of the hermitian form with matrix $\left(\begin{smallmatrix}1 & 0\\ 0 & -1\end{smallmatrix}\right)$, and $K \cong \mathbb{C}^*$ is the subgroup of G that preserves each of the subspaces $V_\pm = \mathbb{C}v_\pm$. We will study the orbit structure for the action of G_0 on G/K.

10.5A Matrix realizations

The matrix realizations of our groups and their Lie algebras, relative to the basis $\{v_+, v_-\}$ of V, are

$$
\begin{aligned}
\mathrm{Mat}(G_0) &= \left\{\begin{pmatrix} z & \overline{w}\\ w & \overline{z}\end{pmatrix}\middle|\, |z|^2 - |w|^2 = 1\right\},\\
\mathrm{Mat}(\mathfrak{g}_0) &= \left\{\begin{pmatrix} i\kappa & \overline{w}\\ w & -i\kappa\end{pmatrix}\middle|\, \begin{matrix}\kappa\in\mathbb{R}\\ w\in\mathbb{C}\end{matrix}\right\} = \left\{\begin{pmatrix} i\kappa & 0\\ 0 & -i\kappa\end{pmatrix}\right\}\oplus\left\{\begin{pmatrix} 0 & \overline{w}\\ w & 0\end{pmatrix}\right\},\\
\mathrm{Mat}(K) &= \left\{\begin{pmatrix} d & 0\\ 0 & d^{-1}\end{pmatrix}\middle|\, d\in\mathbb{C}^*\right\},\\
\mathrm{Mat}(K_0) &= \left\{\begin{pmatrix} e^{i\kappa} & \\ & e^{-i\kappa}\end{pmatrix}\middle|\, \kappa\in\mathbb{R}\right\}.
\end{aligned}
\tag{10.5.1}
$$

In terms of matrices, the involutions θ, τ, σ are given as follows. The Cartan involution σ of G, complex conjugation over its compact real form $G_u = SU(2)$ defined by $\langle\cdot,\ \cdot\rangle$, is $A \mapsto -A^*$ on $\mathfrak{g}_u$, $A \mapsto (A^*)^{-1}$ on G_u. We usually write $I_{p,q}$ for $\left(\begin{smallmatrix} I_p & 0\\ 0 & -I_q\end{smallmatrix}\right)$ where I_r is the $r\times r$ identity matrix, so $I_{1,1} = \left(\begin{smallmatrix}1 & 0\\ 0 & -1\end{smallmatrix}\right)$. Now

$$
\begin{aligned}
\mathrm{Mat}(\sigma(X)) &= -\overline{\mathrm{Mat}(X)}^t =: {}^\theta\,\mathrm{Mat}(X),\\
\mathrm{Mat}(\theta(X)) &= I_{1,1}\cdot\mathrm{Mat}(X)\cdot I_{1,1},\\
\mathrm{Mat}(\tau(X)) &= I_{1,1}\cdot{}^\theta\,\mathrm{Mat}(X)\cdot I_{1,1}.
\end{aligned}
\tag{10.5.2}
$$

Select a maximal abelian subalgebra $\mathfrak{a}_0 \subset \mathfrak{s}_0$, say $\mathfrak{a}_0 = \mathbb{R}\left(\begin{smallmatrix}0 & -i\\ i & 0\end{smallmatrix}\right)$. Then $\mathfrak{a}_0$ is a split Cartan subalgebra in $\mathfrak{g}_0 = \mathfrak{su}(1,1) \cong \mathfrak{sl}_2(\mathbb{R})$, and the corresponding roots are $\Sigma(\mathfrak{g}_0, \mathfrak{a}_0) = \{\pm\gamma\}$ with $\gamma\left(\begin{smallmatrix}0 & -ia\\ ia & 0\end{smallmatrix}\right) = 2a$. We give the root decomposition in terms of a θ-adapted $\mathfrak{sl}_2$-triple $\{E^-, E^+, H\} \subset \mathfrak{g}_0$ with $H = h_\gamma$. Of course $\mathfrak{g}_0 = \mathfrak{a}_0 + \mathfrak{g}_0^\gamma + \mathfrak{g}_0^{-\gamma}$, and therefore

$$
\mathfrak{g}_0 = \mathbb{R}\begin{pmatrix}0 & -i\\ i & 0\end{pmatrix} + \mathbb{R}\cdot\frac{1}{2}\begin{pmatrix}-i & 1\\ 1 & i\end{pmatrix} + \mathbb{R}\cdot\frac{1}{2}\begin{pmatrix}i & 1\\ 1 & -i\end{pmatrix} \cong \mathbb{R}h_\gamma + \mathbb{R}E^+ + \mathbb{R}E^-.
$$

For later purposes, we give a decomposition with θ-stable summands:

$$\mathfrak{g}_0 = \mathfrak{a}_0 + (\mathfrak{g}_0^{[\gamma]})^{\theta} + (\mathfrak{g}_0^{[\gamma]})^{-\theta}$$

and thus

$$\begin{aligned}\mathfrak{g}_0 &= \mathbb{R}\begin{pmatrix} 0 & -i \\ i & 0 \end{pmatrix} + \mathbb{R}\begin{pmatrix} -i & 0 \\ 0 & i \end{pmatrix} + \mathbb{R}\begin{pmatrix} 0 & 1 \\ 1 & 0 \end{pmatrix} \\ &= \mathbb{R}h_\gamma + \mathbb{R}(E^+ - E^-) + \mathbb{R}(E^+ + E^-).\end{aligned}$$

Also note

$$\begin{aligned}&\exp(t\cdot ih_\gamma) = \exp\begin{pmatrix} 0 & t \\ -t & 0 \end{pmatrix} = \begin{pmatrix} \cos t & \sin t \\ -\sin t & \cos t \end{pmatrix} \in \exp i\mathfrak{a}_0, \\ &\exp(\pm\tfrac{\pi i}{4}h_\gamma) = \tfrac{\sqrt{2}}{2}\begin{pmatrix} 1 & \pm 1 \\ \mp 1 & 1 \end{pmatrix}, \\ &\mathrm{Mat}({}^{a}\theta(X)) = \mathrm{R}_2 \cdot \mathrm{Mat}(X) \cdot \mathrm{R}_2 \ \text{ where } a = \exp(\pm\tfrac{\pi i}{4}h_\gamma).\end{aligned} \tag{10.5.3}$$

Note that along the subset $\exp \mathbb{R}\, ih_\gamma$ the G_0-isotropy is positive-dimensional precisely at the points $\exp(\frac{ih_\gamma}{4}\mathbb{Z})$.

10.5B Natural compactifications

The two-dimensional homogeneous space $\Omega := G/K$ can be realized as an open orbit in $\mathbb{P}^1(\mathbb{C}) \times \mathbb{P}^1(\mathbb{C})$ and $G/N_G(K)$ as an open orbit in $\mathbb{P}^2$. These realizations are the unique smooth G-equivariant compactifications of the respective homogeneous spaces. The 2:1 covering $G/K \to G/N(K)$ can be extended to a branched covering $\mathbb{P}^1(\mathbb{C}) \times \mathbb{P}^1(\mathbb{C}) \to \mathbb{P}^2(\mathbb{C})$.

In order to define the an appropriate action on the compactifications, we consider $\mathbb{P}(V) \times \mathbb{P}(V)$ as a flag manifold of $\widehat{G} := SL(2;\mathbb{C}) \times SL(2;\mathbb{C}) = G \times G$. Let $\varrho : \mathbb{P}(V) \times \mathbb{P}(V) \hookrightarrow \mathbb{P}(V \otimes V) = \mathbb{P}(S^2V \oplus \wedge^2 V)$ be the $\widehat{G}$-equivariant embedding given by $([v],[w]) \mapsto [v \otimes w]$. Note that $V \otimes V$ is an irreducible $SL(2;\mathbb{C}) \times SL(2;\mathbb{C})$-module. Consider G as the diagonal subgroup of $G \times G$. As a G-module, $V \otimes V = S^2V \oplus \wedge^2 V$. These two irreducible G-submodules are given explicitly by the embeddings

$$\begin{aligned}&\jmath_S : S^2V \to V \otimes V \text{ is the linear extension of } v\cdot w \mapsto \frac{v \otimes w + w \otimes v}{2}, \\ &\jmath_\wedge : \wedge^2 V \to V \otimes V \text{ is the linear extension of } v \wedge w \mapsto \frac{v \otimes w - w \otimes v}{2}.\end{aligned} \tag{10.5.4}$$

If π_S and $\pi_\wedge$ denote the respective projections of $V \otimes V$ to $S^2(V)$ and $\wedge^2(V)$ then $\pi_S \circ \jmath_S = \mathrm{Id}$ and $\pi_\wedge \circ \jmath_\wedge = \mathrm{Id}$.

Any nonzero symplectic form on V, for example $\omega := v_+^* \wedge v_-^*$, determines G. Also, ω defines $V \cong V^*$, identifying $V \otimes V$ with $\mathrm{End}\, V$. The image of $\varrho : \mathbb{P}(V) \times \mathbb{P}(V) \hookrightarrow \mathbb{P}(V \otimes V)$ is the quadric $\{[A] \in \mathbb{P}(\mathrm{End}\, V) : \det(A) = 0\}$.

The point $\{p\} = [\wedge^2 V] \in \mathbb{P}(V \otimes V)$ does not belong to $\varrho(\mathbb{P}_1 \times \mathbb{P}_1)$. Consider the meromorphic projection $\pi : \mathbb{P}(V \otimes V) \to \mathbb{P}(S^2 V)$ with $\{p\}$ as center, i.e., π blows up the point $\{p\}$ but is holomórphic on $\mathbb{P}(V \otimes V) \setminus \{p\}$. This map defines a holomorphic map $\pi : \mathbb{P}_1 \times \mathbb{P}_1 \to \mathbb{P}(S^2 V)$, which is a branched 2:1 covering (see the following diagram).

$$\begin{array}{ccccccc} \operatorname{diag}\mathbb{P}_1 & \hookrightarrow & \mathbb{P}_1 \times \mathbb{P}_1 & = & \mathbb{P}_1 \times \mathbb{P}_1 & \hookrightarrow & \mathbb{P}(V \otimes V) \\ \downarrow & & \downarrow & & \downarrow & & \downarrow \pi \\ C & \hookrightarrow & (\mathbb{P}_1 \times \mathbb{P}_1)/\mathbb{Z}_2 & \longmapsto & \mathbb{P}(S^2 V) & = & \mathbb{P}(S^2 V). \end{array}$$

The branch locus is formed by all points in $\varrho(\mathbb{P}_1 \times \mathbb{P}_1)$, at which the projection lines $\ell \ni \{p\}$ in $\mathbb{P}(V \otimes V)$ are tangent to $\mathbb{P}_1 \times \mathbb{P}_1$.

This map can also be described as the quotient of $\mathbb{P}_1 \times \mathbb{P}_1$ by the $\mathbb{Z}_2$ action, which is given by interchanging the factors in $\mathbb{P}_1 \times \mathbb{P}_1$.

10.6 $\mathfrak{sl}_2$ models at generic points of bd($\mathcal{U}$)

In this section we determine certain two-dimensional $\mathfrak{sl}_2$ models at generic points of the boundary bd$(\mathcal{U})$. These are closed subsets of G/K isomorphic to one of the two affine quadrics, the quotient of $S \cong SL(2; \mathbb{C})$ by $K_S \cong \mathbb{C}^*$ or by $N_G(K_S)$. Given a "generic" point y of bd$(\mathcal{U})$, we construct an $\mathfrak{sl}_2$-triple such that the orbit $S(y)$ of the associated complex group is one of the two homogeneous spaces mentioned above. We refer to this as a SL_2-model, because $S(y) = S/K_S$ is an affine homogeneous space of the three-dimensional group S and the intersection $S(y) \cap \mathcal{U} = \mathcal{U}_S$ is its associated universal domain.

Our first step here is to give a precise definition of the notion *generic boundary point* and to show that the set of such points is open and dense in bd$(\mathcal{U})$. Then, given a generic point, we determine the appropriate $\mathfrak{sl}_2$-triple and show that S has the properties indicated above.

10.6A Generic boundary points

In order to define generic points in the boundary bd$(\mathcal{U})$ we start with the boundary of the polytope $\omega_0 \subset \mathfrak{a}_0$. The generic points in bd ω_0 are defined as follows:
(10.6.1)

$$\mathrm{bd}^{\mathrm{gen}}(\omega_0) := \left\{ \zeta \in \mathfrak{a}_0 \;\middle|\; \begin{array}{l} \text{there exists precisely one } \lambda \in \Sigma(\mathfrak{g}_0, \mathfrak{a}_0) \text{ such that} \\ \lambda(\zeta) = \pi/2, \text{ and } \mu(\zeta) \notin \frac{\pi}{2}\mathbb{Z} \text{ for all } \mu \in \Sigma \setminus \{\pm\lambda\} \end{array} \right\}.$$

This set is stable under the (small) Weyl group $W(\mathfrak{a}_0)$. If $y \in$ bd$(\mathcal{U})$ we know that the closure of $G_0(y)$ contains exactly one closed G_0-orbit. The point y is called **generic** if that closed orbit is $G_0 \cdot \exp(i\zeta)(x_0)$ for some $\zeta \in \mathrm{bd}^{\mathrm{gen}}(\omega_0)$. We write $\mathrm{bd}^{\mathrm{gen}}(\mathcal{U})$ for the set of generic boundary points.

Clearly $\mathrm{bd}^{\mathrm{gen}}(\omega_0)$ is open and dense in bd(ω_0). However, it is not at all obvious that $\mathrm{bd}^{\mathrm{gen}}(\mathcal{U})$ is dense in bd$(\mathcal{U})$. We show that the complement of $\mathrm{bd}^{\mathrm{gen}}(\mathcal{U})$ in bd$(\mathcal{U})$ is of dimension less than dim bd$(\mathcal{U})$; that justifies the name "generic." As before, $a = \exp i\xi$ with $\xi \in \mathfrak{a}_0$. A key tool for our description of all nonclosed orbits in a neighborhood of a closed one is the Slice Theorem 10.4.1, together with the explicit description of the isotropy representation of $(G_0)_{ax_0}$ on the complement $T^\perp$ of $T_{ax_0}(G_0 \cdot ax_0)$ in the tangent space of G/K as worked out in Proposition 10.4.7. In (10.4.9) we analyzed that complement as a real subspace ${}^a\mathfrak{t}$ of $\mathfrak{g}$ stable under conjugation by the real isotropy group ${}^aH := G_0 \cap aKa^{-1}$. Also, we gave a complete description of the Lie algebra ${}^a\mathfrak{h}$ of ${}^aH$ in (10.4.8). Those descriptions are augmented in (10.6.2) and the preceding discussion.

Due to the Slice Theorem, the description of the orbit structure in a neighborhood of a closed orbit $G_0 \cdot \exp(i\xi)(x_0)$, $\xi \in \mathrm{bd}(\omega_0)$, can be deduced from the orbit structure by the action of the isotropy group ${}^aH := G_0 \cap aKa^{-1}$ on ${}^a\mathfrak{t}$. We are especially interested in the description of the nonclosed orbits. Every such orbit $G_0(y)$ which contains an orbit $G_0(ax_0)$ in its closure intersects the cone ${}^a\mathcal{N} \subset {}^a\mathfrak{t}$ of ad-nilpotent elements in ${}^a\mathfrak{t}$. It is therefore necessary to estimate the size of ${}^a\mathcal{N}$.

Let $p : \Omega \to \Omega /\!/ G_0$ be the categorical quotient. Then a fiber $p^{-1}(p(ax_0))$ is the union of all orbits which have $G_0(ax_0)$ in their closures. The generic points $\mathrm{bd}^{\mathrm{gen}}(\mathcal{U})$ are those points which belong to the preimage of the subset $p(\exp(i\,\mathrm{bd}^{\mathrm{gen}}(\omega_0))(x_0)) \subset \Omega /\!/ G_0$. Note that $\mathrm{bd}(\omega_0) \setminus \mathrm{bd}^{\mathrm{gen}}(\omega_0)$ is a closed 1-codimensional subset in the piecewise linear boundary $\mathrm{bd}(\omega_0)$. The same is true of the image $p(\exp\big(i(\mathrm{bd}(\omega_0) \setminus \mathrm{bd}^{\mathrm{gen}}(\omega_0))\big)(x_0))$ in the categorical quotient. The main difficulty is to show that the fibers over the remaining points, that is, the points of $p(\exp\big(i\,\mathrm{bd}(\omega_0) \setminus \mathrm{bd}^{\mathrm{gen}}(\omega_0)\big)(x_0))$ are no bigger than the fibers over $p(\exp(i\,\mathrm{bd}^{\mathrm{gen}}(\omega_0))(x_0))$.

In general the fibers of a quotient $V \to V /\!/ H$ need not all have the same dimension. A simple counterexample is the linear action of $H = SL(2;\mathbb{C})$ on the $(d+1)$-dimensional space $V = \mathbb{C}[X_1, X_2]_d$ of homogeneous polynomials of degree $d \geqq 5$ where the fiber through $0 \in V$ is dimension-theoretically bigger than the generic fiber.

One key point in our considerations is that the actions arise as isotropy actions on symmetric pairs. Recall ${}^a\mathfrak{h} = (\mathfrak{g}_0 \cap \mathfrak{g}^{{}^a\theta})$ from (10.4.8) and ${}^a\mathfrak{t} = (i\mathfrak{g}_0 \cap \mathfrak{g}^{-({}^a\theta)})$ from (10.4.9), and the result $\mathfrak{g}^{\eta(a)} = {}^a\mathfrak{h} + {}^a\mathfrak{t}$ from Proposition 10.4.7. Since a belongs to $G_u = G^\sigma$, it follows that $\sigma\eta(a) = \eta(a)\sigma$ (see Lemma 10.2.3) and $\sigma \cdot {}^a\theta = {}^a\theta \cdot \sigma$. Now the reductive subalgebra $\mathfrak{g}^{\eta(a)}$ and the subspaces ${}^a\mathfrak{h}, {}^a\mathfrak{t}$ are stable under σ. Thus also the semisimple part $(\mathfrak{g}^{\eta(a)})^{ss} = [\mathfrak{g}^{\eta(a)}, \mathfrak{g}^{\eta(a)}]$ and the center of $\mathfrak{g}^{\eta(a)}$, are stable under σ, ${}^a\theta$ and τ. At any rate, we have

$$\mathfrak{g}^{\eta(a)} = {}^a\mathfrak{h} \cap \mathfrak{g}_u + {}^a\mathfrak{h} \cap i\mathfrak{g}_u + {}^a\mathfrak{t} \cap \mathfrak{g}_u + {}^a\mathfrak{t} \cap i\mathfrak{g}_u. \tag{10.6.2}$$

We sometimes drop the upper index "a" if it is clear from the context to which $a = \exp i\xi$ we refer.

Recall that ${}^a\mathcal{N} \subset {}^a\mathfrak{t}$ denotes the cone of ad-nilpotent elements (nil cone). It is the 0-fiber in the categorical quotient ${}^a\mathfrak{t} \to {}^a\mathfrak{t} /\!/ (G_0 \cap aKa^{-1})$.

Lemma 10.6.3. *Let $a = \exp i\xi$ with $\xi \in \mathfrak{a}_0$ and $\mathfrak{g}^{\eta(a)} = {}^a\mathfrak{h} + {}^a\mathfrak{t}$ as above. Then $i\mathfrak{a}_0$ is a maximal toral subalgebra in ${}^a\mathfrak{t}$. Further,*

(i) *the codimension of the nil cone ${}^a\mathcal{N}$ in ${}^a\mathfrak{t}$ is greater or equal to* $\dim_{\mathbb{R}} \mathfrak{a}_0$;

(ii) *if* $a \in \exp i\mathrm{bd}^{\mathrm{gen}}(\omega_0)$ *then* $\operatorname{codim}_{{}^a\mathfrak{q}} {}^a\mathcal{N} = \dim_{\mathbb{R}} \mathfrak{a}_0$.

Proof. For the dimension considerations we may deal only with the connected component of ${}^aH$ at the identity. If there were an abelian subalgebra $\mathfrak{t} \supsetneq i\mathfrak{a}_0$ then, according to Proposition 10.4.7, there would be a nonzero element

$$\zeta \in \mathfrak{t} \cap \sum_{\lambda(\xi)=\mathbb{Z}\pi} i(\mathfrak{g}_0^{[\lambda]})^{-\theta} + \sum_{\lambda(\xi)=\frac{\pi}{2}+\mathbb{Z}\pi} i(\mathfrak{g}_0^{[\lambda]})^{\theta}$$

which is absurd since $i\mathfrak{a}_0$ acts nontrivially on every subspace $i(\mathfrak{g}_0^{[\lambda]})^{\pm\theta}$.

To simplify the notation during the proof, we write ${}^a\mathfrak{l}$ for $\mathfrak{g}^{\eta(a)}$ and then drop the upper index "a" on ${}^a\mathfrak{l}$, ${}^a\mathfrak{h}$ and ${}^a\mathfrak{r}$. This is possible because $a = \exp(i\xi)$ does not change during the proof.

In order to estimate the dimension of ${}^a\mathcal{N} = \mathcal{N}$, we consider the complexification $\mathfrak{l}^{\mathbb{C}} = \mathfrak{h}^{\mathbb{C}} + \mathfrak{r}^{\mathbb{C}}$ of $\mathfrak{l}$; since $\mathfrak{l}$ is totally real in $\mathfrak{g}$, $\mathfrak{l}^{\mathbb{C}}$ can be viewed as a subalgebra of $\mathfrak{g}$). As shown above, the subalgebra $\mathfrak{a} = \mathfrak{a}_0 + i\mathfrak{a}_0$ is a maximal toral subalgebra in $\mathfrak{r}^{\mathbb{C}}$. The key fact now is that, due to a theorem of Kostant and Rallis [KR, Theorem 3], the dimension of the complex nil cone $\widehat{\mathcal{N}} \subset \mathfrak{r}^{\mathbb{C}}$ coincides with the dimension of a generic closed orbit, and the codimensions of these two sets in $\mathfrak{r}^{\mathbb{C}}$ is $\dim \mathfrak{a}$. Now, $\mathcal{N} = \widehat{\mathcal{N}} \cap \mathfrak{r}$, and since ${}^a\mathcal{N}$ is totally real in $\widehat{\mathcal{N}}$, we have $\dim_{\mathbb{R}} \mathcal{N} \leqq \dim_{\mathbb{C}} \widehat{\mathcal{N}}$. Note that this inequality may be proper: if $\mathfrak{l} = \mathfrak{h} + \mathfrak{r}$ is riemannian symmetric, then $\widehat{\mathcal{N}} \cap \mathfrak{r} = \{0\}$.

Next we show that if $\mathfrak{l}$ comes from a generic $a \in \exp(i\,\mathrm{bd})^{\mathrm{gen}}(\omega_0)$, then $\dim_{\mathbb{R}} \mathcal{N} = \dim_{\mathbb{C}} \widehat{\mathcal{N}}$. Since $a \in \exp i\,\mathrm{bd}^{\mathrm{gen}}(\omega_0)$,

$$\begin{aligned} \mathfrak{l} &= \mathfrak{m}_0 + (\mathfrak{g}_0^{[\lambda]})^{-\theta} + \{\zeta \in i\mathfrak{a}_0 \mid \lambda(\zeta) = 0\} + \mathbb{R}ih_\lambda + i(\mathfrak{g}_0^{[\lambda]})^{\theta} \\ \text{and } \mathfrak{r} &= \{\zeta \in i\mathfrak{a}_0 \mid \lambda(\zeta) = 0\} + \mathbb{R}ih_\lambda + i(\mathfrak{g}_0^{[\lambda]})^{\theta}. \end{aligned}$$

Here λ is the restricted root with $\lambda(\xi) = \pi/2$. Note further that the subalgebra $\{\zeta \in i\mathfrak{a}_0 \mid \lambda(\zeta) = 0\}$ lies in the center of $\mathfrak{l}$. We claim that the real nil cone

$$\mathcal{N} \subset \mathbb{R}ih_\lambda + i(\mathfrak{g}_0^{[\lambda]})^{\theta} = \mathfrak{r}^{ss} \subset \mathfrak{m}_0 + (\mathfrak{g}_0^{[\lambda]})^{-\theta} + \mathbb{R}ih_\lambda \oplus i(\mathfrak{g}_0^{[\lambda]})^{\theta} =: \mathfrak{l}^{ss}$$

has codimension 1 in $\mathfrak{r}^{ss}$. To see this, note that, since $(\mathfrak{l}^{ss}, \mathfrak{h} \cap \mathfrak{l}^{ss})$ is a rank-one symmetric pair, the ring of invariant functions is generated by a single quadratic function, the real Killing form $\kappa(\zeta) := \langle \zeta, \zeta \rangle$. Since $\mathcal{N} = \kappa^{-1}(0)$ and this quadratic form has negative (on $\mathbb{R}ih_\lambda$) and positive (on $i(\mathfrak{g}_0^{[\lambda]})^{\theta}$) eigenvalues, $\kappa^{-1}(0)$ is a real hypersurface in $\mathfrak{r}^{ss}$ and the codimension of $\mathcal{N}$ in $\mathfrak{r}$ is $\dim_{\mathbb{R}} \mathfrak{a}_0$. □

Corollary 10.6.4. *The set of nongeneric boundary points,* $\mathrm{bd}(\mathcal{U}) \setminus \mathrm{bd}^{\mathrm{gen}}(\mathcal{U})$, *is of codimension at least* 1 *in* $\mathrm{bd}(\mathcal{U})$

Proof. We only need to compare codimensions (in $\Omega = G/K$) of the fibers over $p(\mathrm{bd}(\mathcal{U}))$ in the categorical quotient $p : \Omega \longrightarrow \Omega /\!/ G_0$ Making use of the preceding lemma we have $\operatorname{codim} p^{-1}(p(z)) = \dim \mathfrak{a}_0$ for $z \in \exp(i\,\mathrm{bd}^{\mathrm{gen}}(\omega))(x_0)$, while

$$\operatorname{codim} p^{-1}(p(z)) \geqq \dim \mathfrak{a}_0 \quad \text{for } z \in (\exp(i\,\mathrm{bd}(\omega_0))(x_0) \setminus \exp(i\,\mathrm{bd}^{\mathrm{gen}}(\omega))(x_0)).$$

The result now follows, because $\mathrm{bd}(\omega_0) \setminus \mathrm{bd}^{\mathrm{gen}}(\omega_0)$ is of codimension 1. □

10.6B Construction of the three-dimensional subgroups

We construct two-dimensional quadrics at points $\mathrm{bd}^{\mathrm{gen}}(\mathcal{U})$ as orbits of appropriately chosen subgroups $S \subset G$ which are isomorphic to $SL(2;\mathbb{C})$ or $PSL(2;\mathbb{C})$. Every such orbit $S(y)$ is isomorphic to one of the quadrics discussed in Section 10.5.

As shown in Proposition 10.2.13 and Observation 10.2.11, every nonclosed G_0-orbit whose closure contains $G_0 \exp(i\xi)(x_0)$, for some $\xi \in \mathfrak{a}_0$, passes through a point $\exp E \cdot \exp(i\xi)(x_0)$ such that $E \in \mathfrak{g}^{-({}^a\theta)} \cap i\mathfrak{g}_0 \subset \mathfrak{g}^{\eta(a)} = \mathfrak{g}^{\eta(\exp(i\xi))}$ is nilpotent. The theorem of Jacobson–Morozov (see Proposition 10.2.12) provides us with an $\mathfrak{sl}_2$-triple $\{E = E^+, E^-, H\}$ that generates a subalgebra of $\mathfrak{g}$ isomorphic to $\mathfrak{sl}(2;\mathbb{C})$. Since $E^+ \in {}^a\mathfrak{t}$, we may assume that $E^\pm \in {}^a\mathfrak{t}$ and $H \in {}^a\mathfrak{h}$. This is roughly how we get S, but we still need to modify E appropriately.

Let $\mathfrak{g}^{\eta(a)} = \mathfrak{g}^{\eta(\exp(i\xi))}$ be the totally real subalgebra of $\mathfrak{g}$ determined by $\xi \in \mathrm{bd}(\omega_0) \subset \mathfrak{a}_0$, and $\mathfrak{g}^{\eta(a)} = {}^a\mathfrak{h} + {}^a\mathfrak{t}$ the decomposition of the infinitesimal symmetric space, given by τ or ${}^a\theta$. Since $\exp(i\xi)$ belongs to the maximal compact subgroup $G^\sigma = G_u$ of G, statement (ii) of Lemma 10.2.3 implies that this decomposition is stable under σ. Recall that ${}^aH$ is the isotropy group $G_0 \cap aKa^{-1}$. The fundamental facts in this context are contained in the following

Proposition 10.6.5 (Perfect $\mathfrak{sl}_2$-triples). *Let $E \in {}^a\mathfrak{t}$ be* ad*-nilpotent element and let $\{E = E^+, E^-, H\}$ be an $\mathfrak{sl}_2$-triple as above.*

(i) *There exists an element $h \in {}^aH^0$ such that*

$$\{E_1^+, E_1^-, H_1\} := \{\mathrm{Ad}(h)E^+, \mathrm{Ad}(h)E^-, \mathrm{Ad}(h)H\}$$

is σ-adapted, i.e., $\sigma(H_1) = -H_1$ and $\sigma(E_1^+) = -E_1^-$.

(ii) *Recall that $i\mathfrak{a}_0 \subset {}^a\mathfrak{t}$. Given a σ-adapted triple as in* (i)*, $E_1^+ - E_1^-$ is an elliptic semisimple element, and there exists $k \in K_0 \cap {}^aH$ such that the $\mathfrak{sl}_2$-triple*

$$\{E_2^+, E_2^-, H_2\} := \{\mathrm{Ad}(k)E_1^+, \mathrm{Ad}(k)E_1^-, \mathrm{Ad}(k)H_1\}$$

has the property that $E_2^+ - E_2^- \in i\mathfrak{a}_0$.

Proof. The first statement is known [Sek, Lemma 1.4]. For the second statement recall the decomposition (10.6.2) and note that (a) $i\mathfrak{a}_0 \subset \mathfrak{r} \cap \mathfrak{g}_u$ is a maximal toral subalgebra, (b) $E_1^+ - E_1^- \in \mathfrak{r} \cap \mathfrak{g}_u$ since $\{E_1^+, E_1^-, H_1\}$ is σ-adapted, and (c) $E_1^+ - E_1^-$ is semisimple elliptic in $\mathfrak{g}$ because it is semisimple elliptic in the semisimple subalgebra spanned by $\{E_1^+, E_1^-, H_1\}$. Here we have dropped the "a" notation. The claim follows now from the fact that any two maximal toral subalgebras in $\mathfrak{r} \cap \mathfrak{g}_u$ are conjugate. □

Now select a closed orbit in the boundary of $\mathcal{U}$. Such an orbit contains a point of the following shape (see Proposition 10.3.2):

$$\exp i\xi(x_0) \equiv a(x_0), \quad \xi \in \mathrm{bd}(\omega_0) \subset \mathfrak{a}_0.$$

Lemma 10.6.6. *Every closed orbit* $G_0 \cdot \exp i\xi(x_0) = G_0(ax_0)$ *with* $\xi \in \mathrm{bd}(\omega_0)$ *is contained in the closure of some nonclosed orbit. Let* $\mathfrak{g}^{\eta(\exp i\xi)} = {}^a\mathfrak{h} + {}^a\mathfrak{t}$ *as in Proposition* 10.4.7. *If* $\xi \in \mathrm{bd}(\omega_0)$, *there is a nonzero nilpotent element in* ${}^a\mathfrak{t}$.

Proof. In view of Proposition 10.2.13 the first statement follows from the second. Write $\mathfrak{g}^{\eta(a)}$ for $\mathfrak{g}^{\eta(\exp i\xi)}$. If $\xi \in \mathrm{bd}(\omega_0)$, then, for at least one restricted root, we have $\lambda(\xi) = \pi/2$. Using the basis of (10.4.5), we see that the real span of $\{Y^\lambda, ih_\lambda, iX^\lambda\}$ is a θ-stable subalgebra $\langle\!\langle Y^\lambda, ih_\lambda, iX^\lambda \rangle\!\rangle_{\mathbb{R}} \subset \mathfrak{g}^{\eta(a)}$ with $ih_\lambda, iX^\lambda \in {}^a\mathfrak{t}$. Since $\langle\!\langle Y^\lambda, ih_\lambda, iX^\lambda \rangle\!\rangle_{\mathbb{R}} \cong \mathfrak{sl}(2;\mathbb{R})$ it contains nilpotent elements; some of them, for example, $E = ih_\lambda \pm iX^\lambda$, are contained in ${}^a\mathfrak{t}$. Such elements are ad-nilpotent in $\mathfrak{g}$.□

Given a closed orbit $G_0 \cdot \exp i\xi(x_0)$ in $\mathrm{bd}(\mathcal{U})$, one can select a $\mathfrak{sl}_2$-triple $\{E_2^+, E_2^- i, H_2\}$ with the properties specified in the second part of the Proposition 10.6.5 and such that $G_0(y) = G_0 \cdot \exp(E_2^+)(\exp i\xi)(x_0)$ is a nonclosed orbit with $G_0 \cdot \exp i\xi(x_0)$ in its closure. Define $e^+ := iE_2^+$, $h := H_2$ and $e^- := -iE_2^-$. They span an $\mathfrak{sl}_2$-triple whose real span is contained in $\mathfrak{g}_0$. The element $E_2^+ - E_2^- = -i(e^+ + e^-)$ considered as a vector in $i\mathfrak{a}_0$ with the base point $i\xi \in \mathrm{bd}(i\omega_0)$ may or may not point into the interior of $i\omega_0$. Observe, however, that if necessary one can always adjust $E_2^+ - E_2^-$ by an element $w \in N_{{}^aH}(\mathfrak{a}_0)$ such that $\mathrm{Ad}(w)(E_2^+ - E_2^-)$ points toward the interior of $i\omega_0$.

We now define the group S to be the analytic subgroup of G with Lie algebra $\langle\!\langle E_2^+, E_2^-, H_2 \rangle\!\rangle_{\mathbb{C}} = \langle\!\langle e^+, e^-, h \rangle\!\rangle_{\mathbb{C}} \cong \mathfrak{sl}(2;\mathbb{C})$.

Finally we compute the isotropy algebra for S at ax_0. By construction of the triple $\{E_2^+, E_2^-, H_2\}$, we have $\langle\!\langle E_2^+, H_2, E_2^- \rangle\!\rangle_{\mathbb{C}} \cap \mathfrak{g}^{a\theta} = \mathbb{C}H_2$. This is a Cartan subalgebra in the Lie algebra of S. We conclude that

$$(10.6.7) \qquad \text{either } S(ax_0) \cong SL(2;\mathbb{C})/\mathbb{C}^* \quad \text{or} \quad S(ax_0) \cong SL(2;\mathbb{C})/N(\mathbb{C}^*),$$

where $\mathbb{C}^*$ is the Cartan subgroup with Lie algebra $\mathbb{C}H_2$ and $N(\mathbb{C}^*)$ is its normalizer. In other words, the orbit $S(ax_0)$ is either a simply connected two-dimensional affine quadric or a 2:1 quotient of such a quadric as described in Section 10.5. We refer to such an orbit $S(ax_0)$ as a **basic slice**.

10.6C Intersections of the SL_2-models with $\mathcal{U}$

In this subsection we investigate the intersections of the S-orbits determined above with the universal domain $\mathcal{U} = \mathcal{U}_{G_0} \subset G/K$. This makes use of a real form $S_0 = S \cap G_0$ of S. Since $SL(2;\mathbb{C})/\mathbb{C}^*$ and $SL(2(;C)/N(\mathbb{C}^*)$ are (minimal-dimensional) analogs of G/K and $G/\breve{K}$, we have the universal domain $\mathcal{U}_{S_0}$ in $S/(S \cap K) \cong SL(2;\mathbb{C})/\mathbb{C}^*$ or in $S/(S \cap \breve{H}) \cong SL(2;\mathbb{C})/N(\mathbb{C}^*)$. The main result here is that if $y \in \mathrm{bd}^{\mathrm{gen}}(\mathcal{U})$ then $S(y) \cap \mathcal{U}$ coincides with a universal domain $\mathcal{U}_{S_0}$. We also give an example of $z \in \mathrm{bd}(\mathcal{U}) \setminus \mathrm{bd}^{\mathrm{gen}}(\mathcal{U})$ for which $S(z) \cap \mathcal{U} \neq \mathcal{U}_{S_0}$.

If $y \in \mathrm{bd}^{\mathrm{gen}}(\mathcal{U})$, then the closed orbit in $c\ell(G_0(y))$ has form $G_0 \cdot \exp i\xi(x_0)$ with $\xi \in \mathrm{bd}^{\mathrm{gen}}(\omega_0)$. Then the structure of $\mathfrak{g}^{\eta(\exp i\xi)}$ is particularly simple, because there is precisely one restricted root $\lambda \in \Sigma(\mathfrak{g}_0, \mathfrak{a}_0)$ with $\lambda(\xi) = \pi/2$, and if $\mu \in \Sigma(\mathfrak{g}_0, \mathfrak{a}_0) \setminus \{\pm\lambda\}$ then $\mu(\xi) \notin \frac{\pi}{2}\mathbb{Z}$. According to Proposition 10.4.7,

$$\mathfrak{g}^{\eta(a)} = {}^a\mathfrak{h} + {}^a\mathfrak{r} = (\mathfrak{m}_0 + (\mathfrak{g}_0^{[\lambda]})^{-\theta}) + (i\mathfrak{a}_0 + i(\mathfrak{g}_0^{[\lambda]})^{\theta}).$$

The key point here is that if we select a perfect $\mathfrak{sl}_2$-triple (as explained in Proposition 10.6.5 and the following paragraph) with $E^+ \in {}^a\mathfrak{r}$, then the hyperbolic element H is conjugate in $\langle\langle iE^+, H - iE^-\rangle\rangle$ to the normalized coroot h_λ. Here is the argument.

The Lie algebra of S is spanned by $\{e^+, h, e^-\} := \{iE^+, H, -iE^-\}$ and is contained in the symmetric space dual ${}^a\mathfrak{h} + i{}^a\mathfrak{r} = \mathfrak{m}_0 + \mathfrak{a}_0 + \mathfrak{g}_0^{\lambda} + \mathfrak{g}_0^{-\lambda}$ of $\mathfrak{g}^{\eta(a)} = {}^a\mathfrak{h} + {}^a\mathfrak{r}$. By Proposition 10.6.5(ii) we have $e^+ + e^- \in \mathfrak{a}_0$, and a simple check confirms that

$$\left\{\tfrac{1}{2}(h - e^+ + e^-), \quad e^+ + e^-, \quad \tfrac{1}{2}(h + e^+ - e^-)\right\}$$

is another $\mathfrak{sl}_2$-triple in $\langle\langle e^+, h, e^-\rangle\rangle_{\mathbb{R}}$. We claim that $e^+ + e^- = \pm h_\lambda$ where h_λ is the normalized coroot. This is the consequence of the following

Observation 10.6.8. *Let $\tilde{e}, \tilde{h}, \tilde{f}$ be an $\mathfrak{sl}_2$-triple in $\mathfrak{m}_0 + \mathfrak{a}_0 + \mathfrak{g}_0^{\lambda} + \mathfrak{g}_0^{-\lambda}$ with $\tilde{h} \in \mathfrak{a}_0$. Then $\tilde{h} = \pm h_\lambda$.*

Proof of the observation. The nil-positive element $\tilde{e}$ is contained either in $\mathfrak{g}_0^{\lambda}$ or in $\mathfrak{g}_0^{-\lambda}$, and the nil-negative element $\tilde{f}$ is contained in the other one. From Lemma 10.1.6(ii) we conclude that $\tilde{h} \in \mathbb{R}h_\lambda$, $\tilde{h} = c\cdot h_\lambda$. Then

$$2\tilde{e} = [\tilde{h}, \tilde{e}] = [ch_\lambda, \tilde{e}] = \pm c\lambda(h_\lambda)\tilde{e} = \pm 2c \cdot \tilde{e},$$

where the sign depends on whether $\tilde{e} \in \mathfrak{g}_0^{\lambda}$ or $\tilde{e} \in \mathfrak{g}_0^{-\lambda}$. We conclude that $c = \pm 1$ and the observation is proved. □

As noted earlier, we may assume that h_λ, considered as a vector in $\mathfrak{a}_0$ with basis point ξ, points toward ω_0, i.e., that it lies in the half plane $\lambda(\zeta) < \frac{\pi}{2}$. It then follows from the definition of ω_0 that $\{t \in \mathbb{R} \mid th_\lambda + \xi \subset \omega_0\} = (0, \frac{\pi}{2})$. Since the defining set $\omega_{S,0}$ for the universal domain $\mathcal{U}_{S_0}$ is $(-\frac{\pi}{4}, \frac{\pi}{4})h_\lambda$ in the sense that $\mathcal{U}_{S_0} = S_0 \cdot \exp(i\omega_{S,0})(x_0)$, it follows that $S \cdot \exp(i\xi)(x_0) \cap \mathcal{U}_{G_0} = \mathcal{U}_{S_0}$.

Finally also note that the point $\exp(ie^+)(ax_0)$, which determines a nonclosed G_0-orbit in $\mathrm{bd}^{\mathrm{gen}}(\mathcal{U})$, belongs to $S(ax_0)$.

The main result of this section is the following summary of the preceding discussion. Since the universal domain $\mathcal{U}$ in G/K is determined by the real form G_0, we emphasize this in comparison with $\mathcal{U}_{S_0}$ by writing it as $\mathcal{U}_{G_0}$.

Theorem 10.6.9. *Let $\Omega = G/K$, complexification of $\Omega_0 = G_0/K_0$, as usual, and let $\mathcal{U}_{G_0} \subset G/K$ be the universal domain. Every G_0-orbit in $\mathrm{bd}^{\mathrm{gen}}(\mathcal{U}_{G_0})$ has a point of the form $y := \exp(ie)\exp(i\xi)(x_0)$, with e nilpotent, and there is a three-dimensional complex simple Lie subgroup S that contains $\exp(ie)$, in other words $y \in S \cdot \exp(i\xi)(x_0)$. Further,*

(i) *$S \cdot \exp(i\xi)(x_0)$ is isomorphic to an affine two-dimensional quadric, and*
(ii) *the intersection $S \cdot \exp(i\xi)(x_0) \cap \mathcal{U}_{G_0} = \mathcal{U}_{S_0}$.*

10.6D SL_2-models at nongeneric boundary points

We close this chapter with an example which shows that the intersection with $\mathcal{U}$ of an S-orbit at a nongeneric point $z \in \mathrm{bd}(\mathcal{U})$ with $\mathcal{U}$ can be smaller then the universal domain $\mathcal{U}_S$ in $S \cdot z \cong SL(2;\mathbb{C})/\mathbb{C}^*$.

Let $G = SL(5;\mathbb{C})$ and $K = SO(5;\mathbb{C})$, associated to the real form $G_0 = SL(5;\mathbb{R})$. That real form is split: $\mathfrak{a}_0$ is a full Cartan subalgebra of $\mathfrak{g}_0$. We choose $\mathfrak{a}_0$ as the set of trace zero diagonal matrices $\mathrm{diag}(a_1,\ldots,a_5)$ in $\mathfrak{sl}(5;\mathbb{R})$ and use the simple root system $\Psi = \{\psi_1,\psi_2,\psi_3,\psi_4\}$, $\psi_i(\mathrm{diag}(a_1,\ldots,a_5)) = a_i - a_{i+1}$.

Let ξ in $\mathrm{bd}(\omega_0)$ and $a = \exp(i\xi)$ such that $G_0 \cdot \exp i\xi(x_0) = G_0(ax_0)$ is a totally real submanifold of G/K. The real isotropy Lie algebra here is isomorphic to $\mathfrak{so}(2,3)$. The point we have in mind is uniquely determined by the equations

$$\psi_1(\xi) = \psi_3(\xi) = \psi_4(\xi) = 0 \text{ and } \psi_2(\xi) = \tfrac{\pi}{2}. \tag{10.6.10}$$

In terms of matrices in $\mathbb{M}_5(\mathbb{C})$, the involution of the corresponding infinitesimal symmetric space $\mathfrak{g}_0 = \mathfrak{h} + i\mathfrak{r} = \mathfrak{g}_0^{{}^a\theta} \oplus \mathfrak{g}_0^{-{}^a\theta}$ is given as follows (we drop in $\mathfrak{h}$ and $\mathfrak{r}$ the upper index ^{a} with $a = \exp i\xi$):

$$\begin{aligned} \mathrm{Mat}({}^a\theta)(X) =& I_{2,3}(-X^t)I_{2,3} \text{ and } \mathrm{Mat}(\theta)(X) = -X^t \\ &\text{with } I_{2,3} = \mathrm{diag}(1,1,-1,-1,-1). \end{aligned} \tag{10.6.11}$$

The orbit structure in a neighborhood of $G_0 \cdot \exp i\xi(x_0)$ is determined by the isotropy, i.e., the adjoint action of ${}^a\mathfrak{h} \cong \mathfrak{so}(2,3)$ on $i{}^a\mathfrak{r}$. This action can be understood in the following way: If $V \cong \mathbb{R}^n$ and $b_{2,3}$ is the symmetric bilinear form of signature $(2,3)$, defining $\mathfrak{h}$ then as a submodule $\mathfrak{r} \subset S^2V$ has codimension 1 and has complement $\mathbb{R} \cdot b_{23}$. In terms of matrices

$$\mathfrak{r} = \left\{ \begin{pmatrix} a_1 & p_1 & x_1 & x_2 & x_3 \\ p_1 & a_2 & x_4 & x_5 & x_6 \\ -x_1 & -x_4 & a_3 & q_1 & q_3 \\ -x_2 & -x_5 & q_1 & a_4 & q_2 \\ -x_3 & -x_6 & q_3 & q_2 & a_5 \end{pmatrix} \right\} = \mathfrak{a}_0 + \sum_{\lambda(\xi)=\mathbb{Z}\pi} (\mathfrak{g}_0^{[\lambda]})^{-\theta} + \sum_{\lambda(\xi)=\frac{\pi}{2}+\mathbb{Z}\pi} (\mathfrak{g}_0^{[\lambda]})^{\theta}$$

of trace zero. Also see Proposition 10.4.7. Next, we select an appropriate element of the nil cone in $\mathfrak{r}$. To find such an element we decompose $\mathfrak{r}$ into weight spaces with respect to the split component $\mathfrak{a}_{\mathfrak{h}} \subset \mathfrak{h}$ of a Cartan subalgebra of $\mathfrak{h}$. One such split component is

$$\mathfrak{a}_{\mathfrak{h}} = \left\{ \begin{pmatrix} 0 & 0 & 0 & 0 & d_1 \\ 0 & 0 & 0 & d_2 & 0 \\ 0 & 0 & 0 & 0 & 0 \\ 0 & d_2 & 0 & 0 & 0 \\ d_1 & 0 & 0 & 0 & 0 \end{pmatrix} \right\},$$

which is conjugate to the $\mathfrak{a}_0$ in $\mathfrak{h} = \mathfrak{so}(2,3)$ with joint eigenvalues $\pm d_1, \pm d_2, 0$. Since $\mathfrak{r} \subset S^2V$, we conclude that the nonzero weight spaces are one-dimensional and the weights are $\pm 2d_1, \pm 2d_2 \pm (d_1 + d_2), \pm(d_1 - d_2), \pm d_1 \pm d_2, 0$. A simple but tedious computation using MAPLE shows that any element lying in the sum $\mathfrak{r}^{[d_1-d_2]} + \mathfrak{r}^{[d_2]}$ of the $\mathfrak{a}_\mathfrak{h}$-weight spaces. Specifically, the

$$E = \left\{ \left. \begin{pmatrix} 0 & s & s & 0 & 0 \\ s & 0 & 0 & -s & t \\ -s & 0 & 0 & s & t \\ 0 & s & s & 0 & 0 \\ 0 & -t & t & 0 & 0 \end{pmatrix} \right| s \neq 0 \neq t \right\}$$

are nilpotent elements each of which which determines a nilpotent orbit of maximal possible dimension. In our construction the only candidates for S are given by certain perfectly chosen $\mathfrak{sl}_2$-triples. A given E has to be extended to an $\mathfrak{sl}_2$-triple and then be replaced by and H-conjugate by the first element of a θ-adapted normal $\mathfrak{sl}_2$-triple. Here is an example:

$$e' = \begin{pmatrix} 0 & 1 & 1 & 0 & 0 \\ 1 & 0 & 0 & -1 & \sqrt{3} \\ -1 & 0 & 0 & 1 & \sqrt{3} \\ 0 & 1 & 1 & 0 & 0 \\ 0 & -\sqrt{3} & \sqrt{3} & 0 & 0 \end{pmatrix}, \quad f' = \begin{pmatrix} 0 & 1 & -1 & 0 & 0 \\ 1 & 0 & 0 & 1 & -\sqrt{3} \\ 1 & 0 & 0 & 1 & \sqrt{3} \\ 0 & -1 & 1 & 0 & 0 \\ 0 & \sqrt{3} & \sqrt{3} & 0 & 0 \end{pmatrix}, \quad h' = \begin{pmatrix} 0&0&0&4&0 \\ 0&0&2&0&0 \\ 0&2&0&0&0 \\ 4&0&0&0&0 \\ 0&0&0&0&0 \end{pmatrix}.$$

Conjugating this triple to an $\mathfrak{sl}_2$-triple $\{e, f, h\}$ with $e + f \in \mathfrak{a}_0$, as in Proposition 10.6.5(ii), we have

$$e + f = \begin{pmatrix} 4&0&0&0&0 \\ 0&2&0&0&0 \\ 0&0&0&0&0 \\ 0&0&0&2&0 \\ 0&0&0&0&4 \end{pmatrix}.$$

The key point here is that no matter how the hyperbolic element $e + f$ is adjusted by an element of the Weyl group of the real isotropic group, the line through it does not enter ω_0. This means that at least locally the SL_2 model stays outside the universal domain.

11

Invariant Kobayashi-Hyperbolic Stein Domains

Here we come to the final step in the proof of the description $\mathcal{M}_D \cong \mathcal{U}$. This is done by showing that $\mathcal{U}$ is maximal with respect to the properties of being Stein, Kobayashi hyperbolic and invariant with respect to the G_0-action on G/K. In fact, up to a choice of a base point it is the unique maximal domain with these properties.

In Section 1 we introduce the Kobayashi pseudometric and prove that the domains in G/K which arise as envelopes defined by B-invariant hypersurfaces H are hyperbolic except in the case when H is a lift from the compact dual G/P hermitian symmetric space (Theorem 11.1.9). A basic ingredient for this is the fact that if $\{H_s\}_{s \in S}$ is a family of hyperplanes in $\mathbb{P}(V)$ which is parameterized by a connected real analytic set S in $\mathbb{P}(V^*)$ which is not contained in a proper projective linear subspace, then one can choose $2m+1$ hyperplanes from the family ($m := \dim_{\mathbb{C}} \mathbb{P}(V)$) so that the complement of their union is hyperbolic (Corollary 11.1.7).

The bulk of Section 2 is devoted to the two-dimensional case. To describe the main result, we recall that in the previous chapter it was shown that an SL_2-model can be set up at any generic boundary point $z \in \mathrm{bd}(\mathcal{U})$. Up to obvious covers, this means that there is a subgroup $S \cong SL(2;\mathbb{C})$ of G, which is τ-invariant with associated real form $S_0 \cong SL(2;\mathbb{R})$ such that the orbit $S(z) = S/K_S$ is a toy model of our original situation. In particular, its intersection with $\mathcal{U}$ is the universal domain $\mathcal{U}_{S_0}$ associated to the real form S_0.

Concretely, this is just the simplest hermitian situation, where $\mathcal{U}_{S_0} = \mathcal{B} \times \bar{\mathcal{B}}$ is the two-dimensional polydisk embedded in the affine quadric $Q_2 = S/S_K$ in the way presented throughout this monograph

Using the concrete geometry of the situation, in particular using the realization of SL_2-model Q_2 as the complement of the diagonal in $\mathbb{P}_1 \times \mathbb{P}_1$, the envelope of holomorphy of a G_0-invariant domain which contains $\mathcal{U}$ as a proper subset is described (see Corollaries 11.2.2, 11.2.3 and 11.2.4). This yields the following result in Section 3. The domain $\mathcal{U}_{S_0}$ is the maximal G_0-invariant domain of holomorphy containing the base point in S/S_K which is Kobayashi-hyperbolic (Theorem 11.3.1). This in turn yields one of the basic theorems of this monograph which states that, except in the hermitian holomorphic case, if H is an hypersurface in G/K which is invariant by an Iwasawa–Borel subgroup B of G, then its envelope $\mathcal{E}_{\mathcal{H}}$ agrees with $\mathcal{U}$ (Corollary 11.3.2).

After dealing with a technical difficulty that arises when $\mathcal{M}_D$ is contained in a finite quotient of G/K (see Corollary 11.3.6), the complete description of $\mathcal{M}_D$ is given in Theorem 11.3.7, completing the proof that $\mathcal{M}_D = \mathcal{U}$, except of course for the case where D is of hermitian holomorphic type (where $\mathcal{M}_D$ is either the associated bounded domain $\mathcal{B}$ or its complex conjugate).

11.1 Hyperbolicity of domains in G/K

11.1A The Kobayashi pseudometric

We begin by introducing the Kobayashi pseudometric d_X on a connected complex manifold X (see [K] for much more information). This is based on the basic special case of the unit disk $X = \Delta_1(0) = \{z \in \mathbb{C} \mid |z| < 0\}$, where the infinitesimal form of the Poincaré metric is given by

$$ds^2 = \frac{1}{2\pi i} \frac{1}{(1-|z|^2)^2} dz \otimes d\bar{z}.$$

This metric is defined by its invariance with respect to $\mathrm{Aut}_{\mathcal{O}}(X) \cong PSL(2;\mathbb{R})$. Its geodesics are orbits of one-parameter subgroups of $\mathrm{Aut}_{\mathcal{O}}(X)$ and it is complete. It has constant negative Gauss curvature. The unit disk with this metric is the **hyperbolic plane** $\mathbb{H}_2$. Let d_P denote the associated (Poincaré) distance function.

Two points p and q in an arbitrary connected complex manifold X can be connected by a *chain* of hyperbolic planes as follows. Choose a finite sequence of points $\{p_j\}$, $1 \leqq j \leqq n$, in X with $p_1 = p$ and $p_n = q$ such that there exist holomorphic maps $\varphi_j : \mathbb{H}_2 \to X$ with $\varphi_j(z_j) = p$ and $\varphi_j(w_j) = p_{j+1}$ for some $z_j, w_j \in \mathbb{H}_2$. Let c denote this data and define the c-distance between p and q as

$$d_c(p,q) = \sum\nolimits_{1 \leqq j \leqq n} d_P(z_j, w_j).$$

Now let C denote the set of all data sets c for connecting p to q by a chain of hyperbolic planes. The **Kobayashi pseudometric** is defined by

$$d_X(p,q) = \inf\nolimits_{c \in C} d_c(p,q).$$

Then d_X is symmetric and satisfies the triangle inequality, but it can happen that $d_X(p,q) = 0$ with $p \neq q$.

Example 11.1.1. The Kobayashi pseudometric of the complex plane $\mathbb{C}$ is identically zero.

Proof. Let $\Delta_R(0)$ be the disk of radius R in $\mathbb{C}$ and let $\varphi_R : \Delta_1(0) \to \Delta_R(0)$ be the map given by dilation. For R sufficiently large it follows that $p, q \in \Delta_R$, and we define the chain c_R to consist of exactly the one hyperbolic plane given by φ_R. It is immediate that $\lim_{R\to\infty} d_{c_R}(p,q) = 0$. □

If $\varphi : X \to Y$ is a holomorphic map, then, from the definition of the pseudometric,

$$d_Y(\varphi(p), \varphi(q)) \leqq d_X(p, q).$$

One speaks of holomorphic maps as being **distance decreasing**. For example, for points $p, q \in X$ in the image of a holomorphic curve $\varphi : \mathbb{C} \to X$ it follows that $d_X(p, q) = 0$. If there are no such holomorphic curves, then one refers to X as being **Brody hyperbolic**. If the pseudometric is a metric, i.e., if $d(p, q) = 0$ only when $p = q$, then X is said to be **Kobayashi hyperbolic**. The weaker condition of Brody hyperbolicity is equivalent to Kobayashi hyperbolicity in the case where X is compact; see [K] for details.

The uniformization theorem tells us which one-dimensional manifolds X are hyperbolic and which are not. One sees directly that $d_X = d_P$ for $X = \mathbb{H}_2$. Furthermore, if $\widetilde{X} \to X$ is a holomorphic covering map, then $\widetilde{X}$is hyperbolic if and only if X is hyperbolic. Hence, if X has universal holomorphic covering space $\Delta_1(0)$, then it is Kobayashi hyperbolic, and otherwise X is $\mathbb{P}_1(\mathbb{C})$, $\mathbb{C}$, $\mathbb{C}^*$ or a torus, none of which is Kobayashi hyperbolic. In fact in all of these cases $d_X \equiv 0$.

Since polydisks in $\mathbb{C}^n$ are hyperbolic, the distance decreasing property implies the hyperbolicity of bounded domains in $\mathbb{C}^n$.

From the point of view of group actions hyperbolic manifolds X look very much like bounded domains. This is due to the fact that $Aut_{\mathcal{O}}X$ is acting as a group of isometries of d_X. Using the theorem of Meyers and Steenrod, this in turn implies that $\mathrm{Aut}_{\mathcal{O}}\, X$, equipped with the compact-open topology, is a Lie transformation group acting properly on X (see [K]).

Let us note a consequence of this which is relevant for our considerations of G_0-invariant domains in G/K.

Proposition 11.1.2. *A smooth almost effective action of a linear semisimple group G of holomorphic transformations on a hyperbolic manifold is proper.*

Proof. Let $L := \mathrm{Aut}(X)^0$. Since G has only finitely many components and is semisimple, and the action has discrete kernel, G acts with finite kernel. Thus we may assume that G is connected and is contained in L.

Now let R be the radical of L and $L = R \cdot S$ be a Levi–Malcev decomposition, where S is the semisimple subgroup associated to the Lie algebra $\mathfrak{s}$ of the decomposition $\mathfrak{l} = \mathfrak{r} + \mathfrak{s}$ of the Lie algebra of L.

Project L to $L/R = S/(R \cap S) =: \overline{S}$ and then apply the adjoint representation. That defines a Lie morphism $\varphi : L \to \mathrm{Ad}(\overline{S})$. By definition, $Ker(\varphi|_G)$ is a discrete central subgroup, thus finite because G is linear and semisimple. Since semisimple subgroups of linear algebraic groups are themselves linear algebraic, it follows that $\varphi|_G$ has closed image. Since $\varphi|_G$ has finite kernel, G is closed in L. Since the action of L on X is proper, it follows that the G-action is likewise proper. □

The notion of hyperbolicity plays a role in understanding theorems of Picard type. For example, $\mathbb{P}_1(\mathbb{C})$ punctured at three points, e.g., $\mathbb{C} \setminus \{0, 1\}$, is hyperbolic and therefore a nonconstant meromorphic map $f : \mathbb{C} \to \mathbb{P}_1(\mathbb{C})$ can omit at most two values.

In higher dimensions it is in general not a simple matter to judge whether or not a given manifold is hyperbolic. However, interpreting points in $\mathbb{P}_1(\mathbb{C})$ as hyperplanes, the analogous result holds for $\mathbb{P}_m(\mathbb{C})$:

Theorem 11.1.3 ([D]). *The complement of the union of $2m+1$ hyperplanes in general position in $\mathbb{P}_m(\mathbb{C})$ is Kobayashi hyperbolic.*

11.1B Families of hyperplanes

If V is a complex vector space, then $\mathbb{P}(V^*)$ parameterizes the hyperplanes in $\mathbb{P}(V)$, in other words, the 1-codimensional subvarieties of $\mathbb{P}(V)$ defined by linear functions $f \in V^* \setminus \{0\}$. If $\mathbb{L} \to X$ is a holomorphic line bundle over a complex manifold, then a nonzero subspace V of the space $\Gamma(X, \mathbb{L})$ of sections defines a meromorphic map

$$\varphi_V : X \to \mathbb{P}(V^*) \text{ by } x \mapsto H_x := \{s \in V \mid s(x) = 0\}. \tag{11.1.4}$$

Unless $H_x = V$, in which case x may be a point of indeterminacy of φ_V, H_x is a hyperplane and φ_V indeed takes its values in $\mathbb{P}(V^*)$.

Here we consider the above situation for X a G-homogeneous space, $\mathbb{L} \to X$ a G-bundle and V a G-invariant subspace of $\Gamma(X, \mathbb{L})$. In particular the map φ_V is G-equivariant. Since V is assumed to contain at least one nontrivial section and G acts transitively on X, given $x \in X$ there exists $s \in V$ with $s(x) \neq 0$. Thus in our context $\varphi_V : X \to \mathbb{P}(V^*)$ is always holomorphic.

Recalling the envelope construction of Section 7.2, we begin with a hypersurface H in X that is the zero set $\{s = 0\}$ of some section of $\mathbb{L}$, and delete hypersurfaces $g_0(H)$, for $g_0 \in G_0$, from X. Here the point $[s] \in \mathbb{P}(V)$ is regarded as a hyperplane H_s in $\mathbb{P}(V^*)$ and it follows that $\varphi_V^{-1}(H_s) = H$. More precisely, if $\ell \in (V^*)^* = V$ is a linear function defining H_s, then $\varphi_V^*(\ell)$ is a constant multiple of s.

We keep the above setting in mind and consider families of hyperplanes as subsets S of the projective space $\mathbb{P}(V)$ of a given complex vector space V. Such a subset is said to have the **normal crossing property** if for every $k \in \mathbb{N}$ there exist $H_{s_1}, \dots, H_{s_k}$, $s_\alpha \in S$, as follows. If $I = \{i_1, \dots, i_\ell\} \subset \{1, \dots, k\}$ denote $\Lambda_I := \bigcap_{a\in I} H_{s_a}$. The condition is that Λ_I is of codimension $\ell = |I|$. If $|I| > \dim_{\mathbb{C}} V$, then codimension $|I|$ means that $\Lambda_I = \emptyset$.

Proposition 11.1.5. *If S is an irreducible, real analytic subvariety whose projective linear span $\langle S\rangle_{\mathbb{C}}$ is the entire space $\mathbb{P}(V)$, then S has the normal crossing property.*

Proof. Given a set $\{H_{s_1}, \dots, H_{s_k}\}$ of hyperplanes with the normal crossing property and a subset $I \subset \{s_1, \dots, s_k\}$, let

$$\Delta_I := \bigcap\nolimits_{s\in I} H_s,\ \mathcal{H}(I) := \{s \in S \mid H_s \supset \Delta_I\} \text{ and } \mathcal{E} := \bigcup\nolimits_{\Delta_I \neq \emptyset} \mathcal{H}(I).$$

We wish to prove that $S \setminus \mathcal{E} \neq \emptyset$. For this note that each $\mathcal{H}(I)$ is a real analytic subvariety of S, and if $S = \mathcal{E}$, then $S = \mathcal{H}(I)$ for some I with $\Delta_I \neq \emptyset$. However, in such a case $\{H \in \mathbb{P}(V) : H \supset \Delta_I\}$ would be a proper, linear subset of $\mathbb{P}(V)$. Consequently, $S = \mathcal{E}$ would contradict $\langle S\rangle_{\mathbb{C}} = \mathbb{P}(V)$. Therefore there exists $s \in S \setminus \mathcal{E}$, or equivalently, $\{H_{s_1}, \dots, H_{s_k}, H_s\}$ has the normal crossing property. □

If $\dim_{\mathbb{C}} \mathbb{P}(V^*) = m$ and the hyperplanes $H_1, \ldots H_{2m+1}$ satisfy the normal crossing condition, then they are in general position in the sense of Theorem 11.1.3.

Corollary 11.1.6. *If S is an irreducible real analytic subvariety with $\langle S\rangle_{\mathbb{C}} = \mathbb{P}(V)$, then there are hyperplanes $H_{s_1}, \ldots, H_{s_{2m+1}}$, $s_\alpha \in S$, such that*

$$\mathbb{P}(V^*) \setminus \left(\bigcup H_{s_i}\right)$$

is Kobayashi hyperbolic.

Our main application of this result is in the case where S is an orbit of the real form at hand.

Corollary 11.1.7. *Let G be a reductive complex Lie group, G_0 a real form of G, and V^* an irreducible $(m+1)$-dimensional representation space for G. Suppose that S a G_0-orbit in $\mathbb{P}(V)$. Then there exist hyperplanes $H_1, \ldots, H_{2m+1} \in S$ so that $\mathbb{P}(V^*) \setminus \bigcup H_j$ is Kobayashi hyperbolic.*

Proof. From the irreducibility of the representation of G on V^*, it follows that V is likewise irreducible for G_0. Thus $\langle S\rangle_{\mathbb{C}} = \mathbb{P}(V)$. Now we apply Corollary 11.1.6 and the desired result follows. □

11.1C Hyperbolicity of envelopes

As we have seen throughout this part, hypersurfaces H in $\Omega = G/K$, which are invariant under the action of an Iwasawa–Borel subgroup B, play a key role in the study of G_0-invariant domains.

Recall for example that if H is the complement of the open B-orbit in G/K, i.e., the maximal B-invariant hypersurface, then the Iwasawa envelope $\mathcal{E}_{\mathcal{I}}(\mathcal{U})$ is the connected component containing $\mathcal{U}$ of the set which is obtained by removing all translates $g_0(H)$ for $g_0 \in G_0$. Proposition 7.2.11 shows that this envelope agrees with the universal domain $\mathcal{U}$. Therefore we may look at $\mathcal{U}$ in a new light. The same goes for the Schubert domain $\mathcal{S}_D$, defined in Chapter 7 as the intersection of envelopes that arise from Schubert intersection theory.

Here, independent of considerations of cycle spaces, we begin with any hypersurface H in G/K which is defined as above by a given Iwasawa Borel subgroup B. As in the case of any of the envelopes, we remove all translates $g_0(H)$ for $g_0 \in G_0$. This is a compact family, because $G_0(H) = K_0(H)$. Obviously a component of this set contains the Iwasawa envelope and therefore we regard this component as an envelope $\mathcal{E}_{\mathcal{H}}(\mathcal{U})$.

Suppose first that G_0 is hermitian, D is of holomorphic type, and H is the π-preimage of the (unique) B-invariant hypersurface $\check{H}$ in one of the compact duals $G/P = G/P_{\pm}$. Let $\pi : G/K \to G/P$ denote the natural projection. Then the envelope $\mathcal{E}_{\mathcal{H}}(\mathcal{U})$ is the π-preimage of the envelope $\mathcal{E}_{\check{\mathcal{H}}}(\mathcal{B})$. The latter is just $\mathcal{B}$ itself by Proposition 7.2.2. Therefore $\mathcal{E}_{\mathcal{H}}(\mathcal{U}) = \pi^{-1}\mathcal{B}$.

When we are not in the hermitian holomorphic case just described, we will prove that $\mathcal{E}_{\mathcal{H}}(\mathcal{U})$ is Kobayashi hyperbolic, whether G_0 is of hermitian type or not.

To show this we consider the holomorphic line bundle $\mathbb{L} \to \Omega = G/K$ defined by the hypersurface H. In particular H is the zero set of an algebraic section $s \in \Gamma(G/K; \mathbb{L})$. The span $V := \langle G(s) \rangle$ of its G-orbit in $\Gamma(\Omega, \mathbb{L})$ is finite-dimensional. Since s is a B-eigenvector, $b(s) = \chi(b)s$ for some character χ on B, it follows that the action of G on V is an irreducible representation.

Proposition 11.1.8. *Unless G_0 is hermitian, D is of holomorphic type, and H is a lift of the unique B-invariant hypersurface in a compact dual G/P, the map*

$$\varphi_V : G/K \to G/\breve{K} = G(\varphi_V(x_0)) \subset \mathbb{P}(V^*)$$

of (11.1.4) *is a finite covering map.*

Proof. Since φ_V is G-equivariant, its image is an orbit with G-isotropy $\breve{K}$ at the neutral point $\varphi_V(x_0)$. In the non-hermitian case K is dimension-theoretically maximal. Thus in that case $\breve{K}/K$ is automatically finite.

If G_0 is hermitian, then there is only one situation where

$$J := \{g \in G \mid gC_0 = C_0\}$$

is such that J/K is infinite, namely when $J = KS_{\pm} = P$ and φ_V is one of the two canonical maps onto an associated compact hermitian symmetric space $G/P_{\pm}$. Since φ_V defines this map, it follows that H is a lift of the unique B hypersurface. We have excluded that case. □

The main result of this section is now an immediate consequence.

Theorem 11.1.9. *If G_0 is hermitian and H is a lift of the unique B-invariant hypersurface in a compact dual G/P, then $\mathcal{E}_{\mathcal{H}}(\mathcal{U})$ is the preimage under the canonical map $G/K \to G/P$ of the associated bounded symmetric domain $\mathcal{B}$. Otherwise, $\mathcal{E}_{\mathcal{H}}(\mathcal{U})$ is Kobayashi hyperbolic.*

Proof. By Proposition 7.2.2 it follows that the envelope defined by the unique B-invariant hypersurface in a compact hermitian symmetric space is the bounded domain $\mathcal{B}$. Thus it is enough to consider the case where the map φ_V is finite to one.

In this case the envelope $\mathcal{E}_{\mathcal{H}}(\mathcal{U})$ is the connected component containing the neutral point of the analogously defined set $\mathcal{E}_{\breve{\mathcal{H}}}(\breve{\mathcal{U}})$ in the finite quotient $G/\breve{K}$, where $\breve{H}$ lifts to H by φ_V. The latter set is contained in the complement of the union of the hyperplanes $g_0(\breve{H})$, $g_0 \in G_0$, in $\mathbb{P}(V^*)$. This set of hyperplanes is a G_0-orbit S in $\mathbb{P}(V)$.

Since the G-representation on V is irreducible, we may apply Corollary 11.1.7: The complement of finitely many appropriately chosen hyperplanes from S is Kobayashi hyperbolic. Since we have removed all of the hyperplanes which are parameterized by S, it follows that $\mathcal{E}_{\breve{\mathcal{H}}}(\breve{\mathcal{U}})$ is hyperbolic and, since $\mathcal{E}_{\mathcal{H}}(\mathcal{U}) \to \mathcal{E}_{\breve{\mathcal{H}}}(\breve{\mathcal{U}})$ is a surjective holomorphic map with finite fibers, $\mathcal{E}_{\mathcal{H}}(\mathcal{U})$ is also hyperbolic. □

11.2 The maximal invariant Kobayashi-hyperbolic Stein domain in an $\mathfrak{sl}_2$-model

Let us recall the data of an $\mathfrak{sl}_2$-model. The complex group S may be chosen to be $SL(2;\mathbb{C})$ with the real form $S_0 = SL(2;\mathbb{R})$ embedded in S as the subgroup of matrices which have real entries. Let $K_0 = SO(2)$ be a maximal compact subgroup of S_0. To fix the notation, let D_0 and D_∞ be the open S_0-orbits in $\mathbb{P}_1(\mathbb{C})$. Choose $\mathbb{C} \subset \mathbb{P}_1(\mathbb{C}) = \mathbb{C} \cup \{\infty\}$ in such a way that $0 \in D_0$ and $\infty \in D_\infty$ are the K-fixed points.

Let S act diagonally on $Z = \mathbb{P}_1(\mathbb{C}) \times \mathbb{P}_1(\mathbb{C})$. The open orbit Ω, which is the complement of the diagonal $\text{diag}(\mathbb{P}_1(\mathbb{C}))$ in Z, is the affine symmetric space S/K_S.

There are 4 open $S_0 \times S_0$-orbits in $\mathbb{P}_1(\mathbb{C}) \times \mathbb{P}_1(\mathbb{C})$, the bidisks $D_\alpha \times D_\beta$ for any pair (α, β) from $\{0, \infty\}$. As S_0-spaces (via the diagonal action) the domains $D_0 \times D_\infty$ and $D_\infty \times D_0$ are equivariantly biholomorphic; further, they are actually subsets of Ω, and the riemannian symmetric space S_0/K_0 sits in each of them as the unique two-dimensional totally real S_0-orbit $S_0(0,\infty)$ (or $S_0(\infty, 0)$, respectively).

Depending on which of $(0,\infty)$ and $(\infty, 0)$ is chosen as a base point in Ω, either domain can be considered as the universal domain $\mathcal{U}$ associated to the real form S_0 of S. We let

$$\mathcal{U} = D_0 \times D_\infty = S_0.\exp i\omega_0.(0,\infty)$$

with $\omega_0 = (-\frac{\pi}{4}, \frac{\pi}{4})h$, where h is the normalized coroot.

Our main point here is to understand S_0-invariant Stein domains in Ω that properly contain $\mathcal{U}$. The analysis is the same for each of the 4 open $S_0 \times S_0$-orbits, so we may assume that such a domain meets $D_0 \times D_0$.

Let $\text{diag}(D_0)$ denote the diagonal $\{(u,u) \mid u \in D_0\}$ in $D_0 \times D_0$ and observe that $(D_0 \times D_0) \cap \Omega = D_0 \times D_0 \setminus \text{diag}(D_0)$. Furthermore, other than $\text{diag}(D_0)$, all S_0-orbits in $D_0 \times D_0$ are closed real hypersurfaces. For $p \in D_0 \times D_0$ let $\Omega(p)$ be the domain in $D_0 \times D_0$ bounded by $S_0(p)$ and $\text{diag}(D_0)$. We shall show that any function holomorphic in a neighborhood of $S_0(p)$ extends holomorphically to $\Omega(p)$. For this, define $T := \{(-t,t) : 0 \leq t < 1\} \subset D_0 \times D_0$. It is a geometric slice for the action of S_0, i.e., $S_0(T) = D_0 \times D_0$, and every S_0-orbit in $D_0 \times D_0$ meets T in exactly one point. We say that a (one-dimensional) complex curve $C \subset \mathbb{C}^2 \subset Z$ is a **supporting curve** for $\text{bd}(\Omega(p))$ at p if $C \cap c\ell(\Omega(p)) = \{p\}$. Here, $c\ell(\Omega(p))$ denotes the topological closure in $D_0 \times D_0$.

Proposition 11.2.1. *If $p \in D_0 \times D_0 \setminus \text{diag}(D_0)$, there is a supporting curve for* $\text{bd}(\Omega(p))$ *at* p.

Proof. We consider D_0 embedded in $\mathbb{C}$ as the unit disk. It is enough to construct such a curve $C_t \subset \mathbb{C}^2$ at each point $p_t = (-t,t) \in T, t \neq 0$. We define it explicitly as $C_t := \{(-t+z, t+z) : z \in \mathbb{C}\}$.

To prove that $C_t \cap c\ell(\Omega(p_t)) = \{p_t\}$, let d_P be the Poincaré metric of the unit disc D_0 considered as a function $d_P : D_0 \times D_0 \to \mathbb{R}_{\geq 0}$. It is an S_0-invariant function on $D_0 \times D_0$. In fact the values of d_P parameterize the S_0-orbits. We now claim that

$$d_P(-t+z, t+z) \geqq d_P(-t, t) = d_P(p_t) \quad \text{for } z \in \mathbb{C}$$

with $(-t+z, t+z) \in D_0 \times D_0$ with equality only for $z = 0$. In other words we claim that C_t touches $c\ell(\Omega(p_t))$ only at p_t.

To prove the above inequality, it is convenient to compare the Poincaré length of the Euclidean segment $\text{seg}(z-t, z+t)$ in D_0 with the length of $\text{seg}(-t, t)$. Writing the corresponding integral for the length, it is clear without explicit calculation that $d(-t+x, t+x) > d(-t, t)$ for $z = x \in \mathbb{R} \setminus \{0\}$. The same argument shows that $d(-t+x+iy, t+x+iy) > d(-t+x, t+x)$ for all nonzero $y \in \mathbb{R}$. The proposition follows. □

From the above construction it follows that the boundary hypersurfaces $S_0(p)$ are strongly pseudoconvex from the point of view of $\Omega(p)$. Therefore the following is immediate.

Corollary 11.2.2. *For $p \in D_0 \times D_0 \setminus \text{diag}(D_0)$ every function f which is holomorphic on the complement of $\Omega(p)$ in $D_0 \times D_0$ extends holomorphically across the boundary component $S_0(p)$.*

Observe that the set $\text{bd}_{\text{gen}}(D_0 \times D_\infty)$ of generic boundary points, introduced in Section 10.6, consists of the two S_0-orbits:

$$\text{bd}_{\text{gen}}(D_0 \times D_\infty) = (\text{bd}(D_0) \times D_\infty)\,\dot{\cup}\,(D_0 \times \text{bd}(D_\infty)).$$

In the following we assume without loss of generality that $z \in \text{bd}(D_0) \times D_\infty$.

Corollary 11.2.3. *Let $\widehat{\Omega} \subset Q_2 \subset (\mathbb{P}_1(\mathbb{C}) \times \mathbb{P}_1(\mathbb{C}))$ be an S_0-invariant Stein domain that contains $D_0 \times D_\infty$ and a boundary point $z \in \text{bd}(D_0) \times D_\infty$. Then $\big((D_0 \times \mathbb{P}_1(\mathbb{C})) \setminus \text{diag}(\mathbb{P}_1(\mathbb{C})\big) \subset \widehat{\Omega}$.*

Proof. If $\widehat{\Omega}$ did not contain the full domain $(D_0 \times \mathbb{P}_1(\mathbb{C})) \setminus \text{diag}(\mathbb{P}_1(\mathbb{C}))$, it would have a hypersurface orbit $S_0(p)$ in its boundary as in Corollary 11.2.2. The holomorphic extension property in that corollary shows that $\widehat{\Omega}$ is not Stein. □

The same argument proves the following.

Corollary 11.2.4. *An S_0-invariant Stein domain that contains a boundary point z of $\mathcal{U}$ necessarily contains the preimage of D_0 or D_∞ by one of the natural projections of $\mathbb{P}_1(\mathbb{C}) \times \mathbb{P}_1(\mathbb{C}) \setminus \text{diag}(\mathbb{P}_1(\mathbb{C}))$ to a factor $\mathbb{P}_1(\mathbb{C})$.*

Let $\widehat{\Omega}$ be as in Corollary 11.2.3. Then the fibers of $\widehat{\Omega} \to \mathbb{P}_1(\mathbb{C})$, given by restriction of the natural projections of $\mathbb{P}_1(\mathbb{C}) \times \mathbb{P}_1(\mathbb{C}) \setminus \text{diag}(\mathbb{P}_1(\mathbb{C}))$ to a factor $\mathbb{P}_1(\mathbb{C})$, can be parameterized as nonconstant holomorphic curves $f : \mathbb{C} \to \widehat{\Omega}$. In particular, we have the following.

Corollary 11.2.5. *An S_0-invariant Stein domain $\widehat{\Omega}$ in S/K_S which properly contains the universal domain $\mathcal{U}_{S_0}$ is not Brody hyperbolic.*

This can be reformulated as follows.

Corollary 11.2.6. *The universal domain $\mathcal{U}_{S_0}$ is the unique maximal G_0-invariant Kobayashi hyperbolic Stein domain containing the base point $x_0 = (0, \infty)$ in S/K_S.*

Proof. It remains to prove the uniqueness, but in this case given an S_0-invariant domain W, we consider the connected component I_W of $W \cap (\exp(i\omega_0)(x_0))$ containing x_0. If $I_W \neq \exp(i\omega_0)(x_0)$, then $S_0(I_W)$ is a connected component of W and therefore agrees with W. In particular, in that case W is contained in $\mathcal{U}_{S_0}$. Finally, if $I_W = \exp(i\omega_0)(x_0)$, then W contains $\mathcal{U}$ and the result follows from Corollary 11.2.5. □

11.3 Maximality and the characterization of cycle domains

As a consequence of the existence of an SL_2-model (given by Theorem 10.6.9), the hyperbolicity of the relevant envelopes, and the work above in the two-dimensional setting, we are now in a position to prove the main result of this chapter.

Theorem 11.3.1. *The universal domain $\mathcal{U}$ is the unique maximal G_0-invariant Kobayashi hyperbolic Stein domain in G/K that contains the base point x_0.*

Proof. First, suppose that W is a G_0-invariant Kobayashi hyperbolic Stein domain which contains a boundary point z of $\mathcal{U}$. Since the generic boundary points of $\mathcal{U}$ are dense, we may assume that $z \in \mathrm{bd}_{\mathrm{gen}}(\mathcal{U})$. Therefore by Theorem 10.6.9 there exists an $\mathfrak{sl}_2$-model such that $z \in \mathrm{bd}(\mathcal{U}_{S_0}) \subset S/K_S$. Again using the fact that the set of generic boundary points is open and dense we may assume that $z \in \mathrm{bd}_{\mathrm{gen}}(\mathcal{U}_{S_0})$.

Now apply Corollary 11.2.4 to $W \cap S(z)$. It follows that, since W is Stein, so is this intersection, and therefore $W \cap S(z)$ contains the fibers of the projection to one of the factors. Thus it is not hyperbolic (see Corollary 11.2.5) and as a consequence, contrary to assumption, neither is W.

Therefore, $W \subset \mathcal{U}$, and the result follows from the fact that $\mathcal{U}$ can be identified with the Iwasawa envelope $\mathcal{E}_{\mathcal{I}}(\mathcal{U})$, which is Stein by Proposition 7.2.1 and Kobayashi hyperbolic by Theorem 11.1.9 □

Corollary 11.3.2. *If H is a complex hypersurface in G/K which is invariant under an Iwasawa–Borel subgroup B of G and (in the hermitian case) is not a lift of the B-invariant hypersurface in a compact dual symmetric space G/P, then $\mathcal{E}_{\mathcal{H}}(\mathcal{U}) = \mathcal{U}$.*

Proof. Except in the case when H is such a lift, Theorem 11.1.9 says that $\mathcal{E}_{\mathcal{H}}(\mathcal{U})$ is Kobayashi hyperbolic. It is Stein by Proposition 7.2.1, and by definition it is invariant. Also by definition, $\mathcal{E}_{\mathcal{H}}(\mathcal{U}) \supset \mathcal{E}_{\mathcal{I}}(\mathcal{U})$ and we know that $\mathcal{E}_{\mathcal{I}}(\mathcal{U}) = \mathcal{U}$ (Theorem 7.2.7). Thus the result follows from the fact that $\mathcal{U}$ is a maximal G_0-invariant Kobayashi hyperbolic Stein domain in G/K. □

Now let us turn our attention to the cycle domains $\mathcal{M}_D$. We assume that we are not in the holomorphic hermitian case, which has already been handled. Therefore, by Theorem 9.1.5, $\check{\mathcal{U}} \subset \mathcal{M}_D \subset \mathcal{S}_D \subset \mathcal{M}_Z$, where $\check{\mathcal{U}}$ is the image of $\mathcal{U}$ under the projection $\pi : G/K \to G/\check{K} = \mathcal{M}_Z$.

Since $\mathcal{S}_D$ is the intersection of envelopes of the form $\mathcal{E}_{\check{\mathcal{H}}}(\check{\mathcal{U}})$, it follows from Theorem 11.1.9 that it is hyperbolic. Therefore, since the map π is distance decreasing, the connected component of its π-preimage which contains $\mathcal{U}$ is likewise hyperbolic. Since $\mathcal{S}_D$ is Stein and G_0-invariant, this preimage also has these properties. Hence, by Theorem 11.3.1 it agrees with $\mathcal{U}$. Thus the following is immediate.

Proposition 11.3.3. *Recall $\mathcal{M}_Z = G/\check{K}$ and the finite cover $\pi : G/K \to G/\check{K}$. Define $\check{\mathcal{U}} := \pi(\mathcal{U})$. Then $\check{\mathcal{U}} = \mathcal{M}_D = \mathcal{S}_D$.*

We complete our work in this direction by noting that $\pi|_{\mathcal{U}} : \mathcal{U} \to \check{\mathcal{U}}$ is a biholomorphic covering map. The following is the first step in that direction.

Lemma 11.3.4. *The restriction $\pi|_{\mathcal{U}} : \mathcal{U} \to \check{\mathcal{U}}$ is a covering map.*

Proof. Let $\Gamma := \check{K}/K$ act on G/K on the right. We must show that if $\gamma \in \Gamma$, then either $\gamma(\mathcal{U}) = \mathcal{U}$ or $\gamma(\mathcal{U}) \cap \mathcal{U} = \emptyset$. If this were not the case, then an open piece of $\mathrm{bd}(\mathcal{U})$ would be mapped by γ into $\mathcal{U}$. As a consequence of the fact that every closed G_0-orbit in the boundary of $\mathcal{U}$ is in the closure of some other orbit (Lemma 10.6.6), the set of points $z \in \mathrm{bd}(\mathcal{U})$ such that $G(z_0)$ is closed is a nowhere dense subset of $\mathrm{bd}(\mathcal{U})$. Thus γ would map a nonclosed G_0-orbit into $\mathcal{U}$. This is contrary to the fact that all G_0-orbits in $\mathcal{U}$ are closed. Thus either $\gamma(\mathcal{U}) = \mathcal{U}$ or $\gamma(\mathcal{U}) \cap \mathcal{U} = \emptyset$, as desired. □

Proposition 11.3.5. *The domain $\mathcal{U}$ is homeomorphic to a cell. In particular it is connected, simply connected and contractible.*

Proof. Every element of $\mathcal{U}$ has unique expression $\exp(\zeta)\exp(\xi)(x_0)$ with $\zeta \in \mathfrak{s}_0$ and $\xi \in \mathrm{Ad}(K_0)(i\omega_0) \subset i\mathfrak{s}_0$. Here $\mathfrak{s}_0$ represents the (real) tangent space of the symmetric space $G_0(x_0) = G_0/K_0$, $i\mathfrak{s}_0$ represents the (real) tangent space of the compact dual symmetric space $G_u(x_0) = G_u/K_0$, and the map $(\zeta, \xi) \mapsto \exp(\zeta)\exp(\xi)(x_0)$ is a diffeomorphism of the convex open (in $\mathfrak{s}$) cell $\mathfrak{s}_0 \times \mathrm{Ad}(K_0)(i\omega_0)$ onto $\mathcal{U}$. □

Corollary 11.3.6. *The restriction $\pi|_{\mathcal{U}} : \mathcal{U} \to \check{\mathcal{U}}$ is biholomorphic.*

Proof. By Lemma 11.3.4 the restriction $\pi|_{\mathcal{U}} : \mathcal{U} \to \check{\mathcal{U}}$ is an unramified cover with finite structure group, say Γ. Thus $\check{\mathcal{U}} = \mathcal{U}/\Gamma$. If $\gamma \in \Gamma$ then its fixed point set $\mathcal{U}^\gamma$ has the same Euler–Poincaré characteristic ($= 1$) as $\mathcal{U}$ [Bre]. Hence, $\mathcal{U}^\gamma \neq \emptyset$, and thus $\gamma = 1$. Now $\Gamma = \{1\}$, in other words, $\pi|_{\mathcal{U}}$ is biholomorphic. □

We have now given the complete description of $\mathcal{M}_D$ in terms of the universal domain and Schubert incidence geometry.

Theorem 11.3.7. *Let D be an open G_0-orbit in the flag manifold $Z = G/Q$. Let $\mathcal{M}_D$ be the associated cycle space, viewed as a domain in the G-homogeneous space $\mathcal{M}_Z$. In the hermitian holomorphic case, $\mathcal{B} = \check{\mathcal{U}} = \mathcal{M}_D = \mathcal{S}_D$ or $\overline{\mathcal{B}} = \check{\mathcal{U}} = \mathcal{M}_D = \mathcal{S}_D$. In all other cases, $\pi : G/K \to G/\check{K} = \mathcal{M}_Z$ is a finite covering map with*

$$\mathcal{U} = \{gK \in G/K \mid gC_0 \subset D\}^0, \ \check{\mathcal{U}} = \mathcal{M}_D = \mathcal{S}_D,$$

and $\pi|_{\mathcal{U}} : \mathcal{U} \to \check{\mathcal{U}}$ *is biholomorphic. In particular,*

$$\pi : \{gK \in G/K \mid gC_0 \subset D\}^0 \to \mathcal{M}_D$$

is biholomorphic.

Leaving aside the hermitian holomorphic case, we reiterate as follows the conceptual aspects of the proof. Every boundary point $C \in \mathrm{bd}(\mathcal{M}_D)$ in G/K contains a point p which itself is contained in a B-Iwasawa–Schubert variety S in Z which is of complementary dimension to the base cycle and which is contained in the complement of D. Thus C is contained in a B-invariant incidence variety which is in the complement of $\mathcal{M}_D$. Since $\mathcal{U}$ agrees with the Iwasawa envelope $\mathcal{E}_{\mathcal{I}}$, it contains no such varieties and therefore $\mathcal{U} \subset \mathcal{M}_D$.

On the other hand, there are B-Schubert varieties $S = \mathcal{O}\dot{\cup}Y$ which are of complementary dimension to C and which have nonempty (finite) intersection with it. The variety Y is contained in the complement of D and the incidence variety $I_Y =: H$ is a hypersurface in the complement of $\mathcal{M}_D$. Thus the envelope $\mathcal{E}_H$ contains $\mathcal{M}_D$, and in particular it contains $\mathcal{U}$. But the envelope is G_0-invariant, Stein and Kobayashi hyperbolic, and consequently it agrees with $\mathcal{U}$. This then forces the equality $\mathcal{M}_D = \mathcal{U}$. In particular, every boundary point C of $\mathcal{M}_D$ in G/K is contained a translate $k(H)$ for k an appropriate element of K_0.

12

Cycle Spaces of Lower-Dimensional Orbits

As usual $Z = G/Q$ is a homogeneous rational manifold. We work with a real form $G_0 \subset G$ under the assumption that G_0 is simple. Theorem 11.3.7 gave us a precise description of the cycle space $\mathcal{M}_D$, where D is an open G_0-orbit on Z.

Here we discuss the cycle spaces $\mathcal{C}\{\gamma\}$ of arbitrary orbits $\gamma \in \text{Orb}_Z(G_0)$. As in the case of open orbits, for the definition of these cycle spaces one starts with a dual pair (κ, γ) and then considers all transformations $g \in G$ which preserve the defining property of duality, i.e., the compactness of the intersection $g(\kappa) \cap \gamma$. One must be a bit careful with this definition (see Section 1 below), but in the end the cycle space can be regarded as follows. If $q := \dim_{\mathbb{C}} \kappa$, then $\mathcal{C}\{\gamma\}$ is the connected component containing the base cycle $c\ell(\kappa)$ of the space of all cycles $g(c\ell(\kappa)) \in \mathcal{C}_q(Z)$ such that $g \in G$ and $g(c\ell(\kappa)) \cap \gamma =: M_g$ is a smooth compact submanifold of γ. For computational convenience the elements of $\mathcal{C}\{\gamma\}$ are taken to be the transformations g as opposed to the cycles $g(\kappa)$ or the intersections M_g.

The main result of the present chapter, Theorem 12.1.3, was proved originally in [HN] and [N]. It gives the complete description of all such cycle spaces. The methods of proof look very similar to those which were used for the case of an open orbit, but there are a number of points where essentially new ideas are needed.

First, the Schubert varieties $S = \mathcal{O} \cup Y$ are set up as in the case of an open orbit. In particular Y is contained in the complement of a given nonclosed orbit γ. However, it is a priori unclear whether the incidence variety I_Y is contained in the complement of $C\{\gamma\}$. The material in Sections 2 and 3 is aimed at proving that it is indeed contained in that complement, Theorem 12.3.3 being the concluding result. Except for the usual type of hermitian exception, this yields the desired classification result (Theorem 12.4.2) for the case where γ is not closed. It should be underlined that for this the main result of the previous chapter, that $\mathcal{U}$ is a maximal G_0-invariant domain of holomorphy which is Kobayashi hyperbolic, is essential.

The second area where new methods are implemented is that where the group G_0 at hand is hermitian and every incidence hypersurface in G/K, which is invariant under an Iwasawa–Borel subgroup, is a lift from one of the two compact symmetric spaces. In this case it is shown that $c\ell(\kappa)$ is in fact invariant under one of the two parabolics that contain K and, just as in the case of open orbits of holomorphic type, it follows that the cycle space is either the bounded domain $\mathcal{B}$ or its conjugate (Theorem 12.4.4).

Finally, in Section 5 an additional argument must be made in order to handle the case of closed orbits (Theorem 12.5.3).

12.1 Definition of cycle space

For the definition of $\mathcal{C}\{\gamma\}$, recall the basic duality statements. If $\gamma \in \mathrm{Orb}_Z(G_0)$, there is a unique K-orbit $\kappa \in \mathrm{Orb}_Z(K)$ so that $\kappa \cap \gamma$ is nonempty and compact. Then we say that γ and κ are **dual** or that (κ, γ) is a **dual pair**. If (κ, γ) is a dual pair, then the intersection $\kappa \cap \gamma =: M$ is transversal in Z and M is a K_0-orbit $K_0(z_0)$ which is minimal in both κ and γ. As suggested in [GM], motivated by duality, one would like to define $\mathcal{C}\{\gamma\}$ by means of its lift to G and define that lift as the connected component containing the identity of $\{g \in G \mid g(\kappa) \cap \gamma \text{ is nonempty and compact}\}$. At the outset it is not at all clear that this is a reasonable set, e.g., that it is open. Therefore we define $G\{\gamma\}$ to be the component of 1 in the interior of

$$\{g \in G \mid g(\kappa) \cap \gamma \text{ is nonempty and compact}\}. \tag{12.1.1}$$

One of our first observations below is that the identity is indeed an interior point, and that therefore this definition makes sense.

Note that $G\{\gamma\}$ is invariant under the action of K by right multiplication on G. Thus it is convenient, having chosen a fixed base point in G/K, to pass from $G\{\gamma\}$ to the

$$\text{cycle space: } \mathcal{C}\{\gamma\} := G\{\gamma\}/K \subset G/K. \tag{12.1.2}$$

At this level, the results for an open orbit γ_{op} can be formulated as follows. In the holomorphic hermitian case $\mathcal{C}\{\gamma_{\mathrm{op}}\}$ is the preimage in G/K of a bounded symmetric domain $\mathcal{B}$ in a compact dual $G/P := G/P_{\pm}$. Otherwise, $\mathcal{C}\{\gamma_{\mathrm{op}}\}$ agrees with the universal domain $\mathcal{U}$.

In the case of open orbits γ_{op} there is no need to worry about the interior of $G\{\gamma\}$, because $G\{\gamma_{\mathrm{op}}\}$ is obviously open. If we lift the cycle space $\mathcal{M}_D$ to G as we have done here, then for $\gamma_{\mathrm{op}} = D$ we have exactly the same definition as that which was used in the previous sections.

In this chapter we extend the open orbit result to all cycle spaces $\mathcal{C}\{\gamma\}$.

Theorem 12.1.3. *Let G_0 be an arbitrary simple group of noncompact type. If G_0 is hermitian and $c\ell(\kappa)$ is P_+-or P_--invariant, then $\mathcal{C}\{\gamma\}$ is the preimage by $\pi_+ : G/K \to G/P_+$ or by $\pi_- : G/K \to G/P_-$ of the associated bounded symmetric domain. Otherwise, $\mathcal{C}\{\gamma\} = \mathcal{U}$.*

The spirit of the work here is similar to that in the previous chapters. In addition we make use of the following observation of [GM].

Proposition 12.1.4. *If $\gamma \in \mathrm{Orb}_Z(G_0)$, then $\mathcal{C}\{\gamma\} \supset \bigcap_{\gamma_{\mathrm{op}} \subset G/B} \mathcal{C}\{\gamma_{\mathrm{op}}\}$.*

The description of $\mathcal{C}\{\gamma_{\text{op}}\}$ in Theorem 11.3.7 has the following consequence, which was verified in [GM] for most cases.

Corollary 12.1.5. *If $\gamma \in \text{Orb}_Z(G_0)$, then the cycle space $\mathcal{C}\{\gamma\}$ contains the universal domain $\mathcal{U}$.*

Proof. If G_0 is not of hermitian type, then this is immediate, because $\mathcal{C}\{\gamma_{\text{op}}\} = \mathcal{U}$ for any open G_0-orbit in any G-flag manifold (Theorem 11.3.7).

In the hermitian case there are two special open orbits γ_{op}^+ and γ_{op}^- in G/B. They are the ones that project to the bounded symmetric domains $\mathcal{B}$ and $\overline{\mathcal{B}}$. The corresponding cycle spaces are the preimages of $\mathcal{B}$ and $\overline{\mathcal{B}}$ in G/K and therefore $\mathcal{C}\{\gamma_{\text{op}}^+\} \cap \mathcal{C}\{\gamma_{\text{op}}^-\} = \mathcal{B} \times \overline{\mathcal{B}} = \mathcal{U}$. □

12.2 Intersection with Schubert varieties

The results here also strongly rely on the basic properties of Schubert slices from Chapter 9. Let us recall the essential aspects of these results.

Given $\gamma \in \text{Orb}_Z(G_0)$ and an Iwasawa Borel subgroup B, we consider the (topologically defined) set of B-Schubert varieties $\mathcal{S}_\kappa$ which are of complementary dimension to that of the dual orbit κ and have nonempty intersection with $c\ell(\kappa)$. As usual, we decompose $S \in \mathcal{S}_\kappa$ into its open B-orbit $\mathcal{O}$ and $Y := S \setminus \mathcal{O}$.

By Theorem 9.1.1 we know that $S \cap c\ell(\kappa)$ is a finite set $\{z_1, \ldots z_d\}$ in $\kappa \cap \gamma$ which is contained in $\gamma \cap \mathcal{O}$, and $\mathcal{O} \cap \gamma = \dot{\bigcup} \Sigma_j$, where $\Sigma_j = A_0 N_0(z_j)$ are the Schubert slices which are themselves open in $\mathcal{O}$ and closed in γ.

If Σ is a Schubert slice, then in particular $K_0(\text{bd}(\Sigma)) = \text{bd}(\gamma)$. This has the following consequence.

Proposition 12.2.1. *If (κ, γ) is a dual pair, then*

$$c\ell(\kappa) \cap c\ell(\gamma) = \kappa \cap \gamma.$$

Proof. If $c\ell(\kappa) \cap c\ell(\gamma)$ contained a boundary point p of γ, then, by conjugation with an appropriate element of K_0, we could find a Schubert variety S and a Schubert slice Σ in S with $p \in \text{bd}(\Sigma)$. In particular, p would be in the complement of γ in S. This is contrary to the fact that $c\ell(\kappa) \cap S$ is contained in γ.

Similarly, if $p \in \text{bd}(\kappa) \cap \gamma$, then by conjugating appropriately we could choose a Schubert slice S which contains it, and this is contrary to $S \cap c\ell(\kappa)$ being contained in $\kappa \cap \gamma$. □

Note that this shows that if $g \in G$ is sufficiently near the identity, then $g(\kappa) \cap \gamma$ is still compact. Consequently our definition of the cycle space $\mathcal{C}\{\gamma\}$ makes sense and furthermore it is nonempty.

Observe that if γ is closed, then the Schubert variety associated to an Iwasawa–Borel subgroup is just its fixed point. This case must be handled separately (see Section 12.5). Therefore, unless otherwise stated, from now on we assume that the G_0-orbit γ under consideration is not closed.

The main goal of this chapter is to show that Schubert slices can be used to determine whether a sequence of cycles exits $\mathcal{C}\{\gamma\}$. This is precisely formulated as follows.

Proposition 12.2.2. *Let Σ be a Schubert slice and suppose that $\{g_n\}$ is a sequence in $G\{\gamma\}$ such that $g_n \to g$ with $g_n(c\ell(\kappa)) \cap \Sigma = \{p_n\}$ and p_n diverges in Σ. Then $g \notin G\{\gamma\}$.*

The proof requires a bit of preparation.

Lemma 12.2.3. *For all $g \in G\{\gamma\}$ the number of points in the intersection $g(\kappa) \cap \Sigma$ is bounded by the intersection number $[S].[c\ell(\kappa)]$.*

Proof. Since Σ can be regarded as a domain in $\mathcal{O} \cong \mathbb{C}^n$ and $g(\kappa) \cap \gamma$ is compact, it follows from the maximum principle that $g(\kappa) \cap \Sigma$ is finite, and then its cardinality is bounded by the intersection number $[S].[c\ell(\kappa)]$. □

Now let $\mathcal{J}$ be the connected component containing the identity of the interior of

$$\{g \in G : |g(\kappa) \cap \Sigma| = 1 \text{ for all } \Sigma\}.$$

Since $c\ell(\kappa) \cap c\ell(\gamma) = \kappa \cap \gamma$, and $|g(\kappa) \cap \Sigma| = 1$ for all Σ, it follows that $g \in \mathcal{J}$ for g sufficiently close to the identity. Thus $\mathcal{J}$ is well defined and nonempty.

In the definition of $\mathcal{J}$, *for all* Σ means for all choices of the maximal compact group K_0 and all Iwasawa factors A_0N_0, i.e., all Schubert slices which arise by G_0-conjugation of those Σ which are connected components of $S \cap \gamma$ for a fixed $S \in \mathcal{S}_\kappa$.

Lemma 12.2.4. *For all $g \in \mathcal{J}$, the intersection $g(\kappa) \cap \gamma$ is transversal.*

Proof. If d denotes the intersection number $[S].[c\ell(\kappa)]$, then the base cycle $c\ell(\kappa)$ intersects $\mathcal{O}$ in exactly d points. Furthermore, $\mathcal{O} \cap \gamma$ is a disjoint union of Schubert slices $\Sigma_1, \ldots, \Sigma_d$ with one intersection point in each slice. By definition, if $g \in \mathcal{J}$, then $g(c\ell(\kappa))$ intersects each Σ_i in exactly one point as well. If any of such intersection points were not transversal, then it would follow that $|\tilde{g}(c\ell(\kappa)) \cap S| > d$ for $\tilde{g}$ near g in $\mathcal{J}$. □

The following is a consequence of the above lemma.

Corollary 12.2.5. *The intersection $M_g = g(\kappa) \cap \gamma$ is a connected compact manifold for all $g \in \mathcal{J}$.*

Now we want to prove that $G\{\gamma\}$ is contained in $\mathcal{J}$. The crux of the argument is to show that it does not leak out.

Proposition 12.2.6. $\mathrm{bd}(\mathcal{J}) \cap G\{\gamma\} = \emptyset$.

Proof. Suppose to the contrary that $g \in \mathrm{bd}(\mathcal{J}) \cap G\{\gamma\}$. Let $\{g_n\}$ be a sequence in $\mathcal{J} \cap G\{\gamma\}$ with $g_n \to g$. Then by Corollary 12.2.5, $M_n := g_n(\kappa) \cap \gamma$ is a sequence of compact connected manifolds in γ.

Let $\widetilde{M}$ denote the limiting set, $\widetilde{M} := \lim M_n$. It follows that $\widetilde{M}$ is a connected closed subset of $c\ell(\gamma)$. Now, $\widetilde{M} \subset g(c\ell(\kappa)) \cap c\ell(\gamma) =: A \dot{\cup} E$, where $A = A_1 \dot{\cup} A_2$ with $A_1 := g(\mathrm{bd}(\kappa)) \cap c\ell(\gamma)$ and $A_2 := g(\kappa) \cap \mathrm{bd}(\gamma)$, and $E := g(\kappa) \cap \gamma$.

The set A is closed, because A_1 is the intersection of two closed sets, and a sequence in A_2 which converges in Z will either converge to a point of A_2 or A_1.

Since we have assumed that $g \in G\{\gamma\}$, it follows that E is compact. Thus

$$\widetilde{M} = (\widetilde{M} \cap A) \dot{\cup} (\widetilde{M} \cap E)$$

is a decomposition of $\widetilde{M}$ into disjoint open subsets of $\widetilde{M}$. Since $\widetilde{M}$ is connected and $\widetilde{M} \cap A \neq \emptyset$, we conclude that $\widetilde{M} \subset A$.

Consequently, for every relatively compact open neighborhood U of a point $p \in \gamma$, there exists a positive integer $N = N(U)$ such that $g_n(\kappa) \cap U = \emptyset$ for all $n > N$.

Now we have assumed that $g \in G\{\gamma\}$, in particular, that E is nonempty. Hence, for $p \in E$ we can consider an Iwasawa–Schubert variety $S = \mathcal{O} \dot{\cup} Y$ with $p \in \mathcal{O}$. Since E is compact, the complex analytic set $g(c\ell(\kappa)) \cap S$ must contain p as an isolated point. Thus, for Σ a Schubert slice through p, the intersection $g(\kappa) \cap \Sigma$ is isolated at p, and as a consequence, $g_n(\kappa)$ must have nonempty intersection with any open neighborhood $U = U(p)$ of p if n is sufficiently large. This is contrary to the above statement and therefore $g \notin G\{\gamma\}$. □

Corollary 12.2.7. $G\{\gamma\} \subset \mathcal{J}$.

Proof of Proposition 12.2.2. Again we argue by contradiction. Suppose that $g \in G\{\gamma\}$. Then by Corollary 12.2.7, $g(c\ell(\kappa)) \cap \Sigma$ consists of exactly one point, say q. Since p_n diverges in Σ, we may assume that $p_n \to p \in c\ell(\Sigma) \setminus \Sigma$.

Now by definition $G\{\gamma\}$ is open. Thus there exists a small $h \in G$ with hg still in $G\{\gamma\}$ and $hg(\kappa) \cap \Sigma$ containing points near p and q. Thus, $|hg(\kappa) \cap \Sigma| \geqq 2$, in violation of $G\{\gamma\} \subset \mathcal{J}$. □

Corollary 12.2.8. *Let $\{g_n\}$ be a sequence in $G\{\gamma\}$ such that $g_n \to g$ and such that there exist $p_n \in g_n(\kappa) \cap (S \cap \gamma)$ with the property that the sequence $\{p_n\}$ diverges in $\mathcal{O}$. Then $g \notin G\{\gamma\}$.*

Proof. Since $\mathcal{O} \cap \gamma$ is a finite union of Schubert slices $\Sigma_1 \cup \cdots \cup \Sigma_d$, it follows that some Schubert slice contains infinitely many points of the sequence $\{p_n\}$. We may therefore assume that the sequence $\{p_n\}$ is contained in some fixed Schubert slice Σ. This implies that p_n diverges in $c\ell(\Sigma) \setminus \Sigma$, and it follows from Proposition 12.2.2 that $g \notin G\{\gamma\}$. □

12.3 Hypersurfaces in the complement of the cycle space

Given an Iwasawa–Borel subgroup B and an associated Schubert variety $S = \mathcal{O} \dot{\cup} Y$ in $\mathcal{S}_K$, we now show that the incidence variety I_Y is contained in the complement of $\mathcal{C}\{\gamma\}$ in G/K. Here

$$I_Y = \widetilde{I}_Y/K \quad \text{and} \quad \widetilde{I}_Y = \{g \in G \mid g(c\ell(\kappa)) \cap Y \neq \emptyset\}.$$

Let us recall the pertinent result of Chapter 7, specifically Corollary 7.4.10.

Theorem 12.3.1. *If $g \in \widetilde{I}_Y$ and $\{g_n\} \subset (G \setminus \widetilde{I}_Y)$, then there exists $f \in \Gamma(S, \mathcal{O}(*Y))$ so that the meromorphic function $m = \mathrm{Tr}(f)$ satisfies $\lim m(g_n) \to \infty$.*

It follows that I_Y is a B-invariant hypersurface.

If κ is closed, i.e., $\gamma = D$ is open, then every $g \in G\{\gamma\}$ satisfies $g(c\ell(\kappa)) \cap Y = \emptyset$. Therefore, in the case of open orbits it immediately follows that I_Y is contained in the complement of $C\{\gamma\}$. Our goal here is to prove this for all γ.

Let us begin by reviewing our standard notions and introducing several relevant analytic subsets of G.

Throughout, $(\gamma, \kappa) \in \mathrm{Orb}_Z(G_0) \times \mathrm{Orb}_Z(K)$ denotes a dual pair. For an Iwasawa–Schubert variety $S \in \mathcal{S}_\kappa$ defined by an Iwasawa–Borel subgroup B, we have the decomposition $S = \mathcal{O} \dot{\cup} Y$, where $\mathcal{O}$ is the B-orbit in S. The incidence variety $\widetilde{I}_Y$ is an analytic subset of G with the property that if $g \notin \widetilde{I}_Y$, then $g(c\ell(\kappa)) \cap S$ is finite.

Let p_0 be a base point in $\kappa \cap \gamma \cap S$, and

$$U_S := \{g \in G \mid g(p_0) \in S\}.$$

Note that $Q := G_{p_0}$ acts on U_S on the right and realizes it as a Q-principal bundle $\pi : U_S \to S$. In particular, U_S is an irreducible analytic subset of G. For $s \in S$, let $F_s := \pi^{-1}(s)$.

Set $E := U_S \cap \widetilde{I}_Y$ and note that since U_S is irreducible, it is a nowhere dense, analytic subset of U_S. Finally, let $D_S := \{s \in S \mid F_s \subset E\}$. It is a proper analytic subset of S.

By combining the following remark with the results of the previous section, we will show that $\widetilde{I}_Y$ is contained in the complement of $G\{\gamma\}$.

Proposition 12.3.2. *Given a point $g \in U_S$ and a sequence $\{p_n\} \subset S \setminus D_S$ converging to $p = g(p_0) \in Y$, there exists a sequence of transformations $\{g_n\} \subset U_S \setminus E$ with g_n converging to g and $g_n(p_0) = p_n$.*

Proof. Let $\{U_n\}$ be a sequence of open subsets of U_S contracting to g, that is, $U_n \subset U_{n+1}$ for all n and $\bigcap U_n = \{g\}$. Since $\pi : U_S \to S$ is an open mapping, it follows that $V_n := \pi(U_n)$ is a sequence of open neighborhoods of p. Consequently, we can renumber the sequence p_n such that $p_n \in V_n$ for each n. Since the set E is a nowhere dense analytic subset of U_S and $p_n \notin D_S$, it follows that $E \cap (F_{p_n} \cap U_n)$ is nowhere dense in $F_{p_n} \cap U_n$. We can therefore choose $g_n \notin E \cap (F_{p_n} \cap U_n)$ such that $g_n(p_0) = p_n$ and $g_n \to g$. □

By Corollary 12.2.8 we know that the transformations g in Proposition 12.3.2 are not in $G\{\gamma\}$. This is enough to prove the following main result of this section.

Theorem 12.3.3. *For a noncompact G_0-orbit γ with dual K-orbit κ and a Schubert variety $S = \mathcal{O} \cup Y$ in $\mathcal{S}_\kappa$ it follows that the incidence variety $\widetilde{I}_Y$ is contained in the complement of $G\{\gamma\}$ in G.*

Proof. A given $g \in \widetilde{I}_Y$ might not be in U_S. However, for an arbitrarily small neighborhood U of the identity in G, there exists $k \in K$ and $u \in U$ so that $ugk \in U_S \cap \widetilde{I}_Y$. The use of u allows us to assume that $g(\kappa) \cap Y$ is nonempty, and then applying the appropriate k yields $g(p_0) \in Y$. Since both $G\{\gamma\}$ and $\widetilde{I}_Y$ are invariant under the action of K by right-multiplication, it follows from Proposition 12.3.2 and Corollary 12.2.8 that a dense subset of $\widetilde{I}_Y$ is contained in the complement of $G\{\gamma\}$. Since $\widetilde{I}_Y$ is closed and $G\{\gamma\}$ is open, this yields the desired result. □

12.4 Cycle Spaces of nonclosed orbits

Just above we worked in the lift $G\{\gamma\}$ of the cycle space. Now we turn to the cycle space $\mathcal{C}\{\gamma\} = G\{\gamma\}/K \subset G/K$, and compare it to the universal domain $\mathcal{U}$.

Disregarding the origin of hypersurface $H = I_Y$ as an incidence divisor and defining the envelope $\mathcal{E}_{\mathcal{H}}(\mathcal{U})$ as usual, the main result of the previous section implies that we have at least trapped the cycle space between two reasonable domains.

Theorem 12.4.1. *If $\gamma \in \mathrm{Orb}_{G_0} Z$ is noncompact, then $\mathcal{U} \subset \mathcal{C}\{\gamma\} \subset \mathcal{E}_{\mathcal{H}}(\mathcal{U})$.*

Proof. The first inclusion is given in Corollary 12.1.5, and the second follows immediately from the fact that H is contained in the complement of $\mathcal{C}\{\gamma\}$, i.e., from Theorem 12.3.3. □

If $\mathcal{E}_{\mathcal{H}}(\mathcal{U})$ is Kobayashi-hyperbolic, the desired characterization $\mathcal{C}\{\gamma\} = \mathcal{U}$ now follows as a consequence of Theorem 11.3.7 because in that case $\mathcal{U} = \mathcal{E}_{\mathcal{H}}(\mathcal{U})$. This is always the case if G_0 is not of hermitian type. If G_0 is of hermitian type, it is also the case if and only if H is not a lift of the B-invariant hypersurface from either G/P_+ or G/P_-. Precisely speaking, H being a lift, e.g., from G/P_+, means that the $H = \pi_+^{-1}(H_0)$, where $\pi_+ : G/K \to G/P_+$ is the natural projection and H_0 is the (unique) B-invariant hypersurface in G/P_+. Let us formalize this result.

Theorem 12.4.2. *If γ is noncompact and either G_0 is not of hermitian type or is of hermitian type and for some Iwasawa–Borel subgroup B, there is a B-invariant hypersurface H in the complement of $\mathcal{C}\{\gamma\}$ which is not a lift, then $\mathcal{C}\{\gamma\} = \mathcal{U}$.*

Now let us turn to the situation where every H is a lift. In fact, if some H_+ is a lift from G/P_+ and H_- is a lift from G/P_-, then $H = H_+ + H_-$ is not a lift. Thus it is sufficient to handle the situation in the following theorem.

Theorem 12.4.3. *Suppose that if B is an Iwasawa–Borel subgroup of G, then every B-invariant hypersurface in the complement of $\mathcal{C}\{\gamma\}$ is a lift from G/P_+. Then $c\ell(\kappa)$ is P_+-invariant.*

Proof. Assume to the contrary that $c\ell(\kappa)$ is not P_+-invariant. Let $p_0 \in \gamma$ be a base point with $K(p_0) = \kappa$. Since κ is not P_+-invariant, $c\ell(P_+(p_0))$ contains $c\ell(\kappa)$ as a proper subvariety.

Now the intersection $c\ell(\kappa) \cap S \subset \mathcal{O}$ is transversal in Z. Thus, every component of $P_+(p_0) \cap \mathcal{O}$ is positive-dimensional, and since $\mathcal{O} = \mathbb{C}^{m(\mathcal{O})}$, it follows that every such component has at least one point of Y in its closure.

Thus for every arbitrarily small neighborhood U of the identity in G there exists $h \in P_+$ and $g \in U$ such that $gh(p_0) \in Y$. Thus $gh \in \widetilde{I}_Y$ which by assumption is a lift from G/P_+. But in that case $g \in \widetilde{I}_Y$, and in particular $g \notin G\{\gamma\}$. Of course this violates $\mathcal{C}\{\gamma\}$ being an open neighborhood of the identity. □

In summary, the following is the main result of this section.

Theorem 12.4.4. *Suppose that γ is noncompact and that G_0 is of hermitian type. If $c\ell(\kappa)$ is neither P_+ invariant nor P_- invariant, then $\mathcal{C}\{\gamma\} = \mathcal{U}$. If $c\ell(\kappa)$ is P_+ invariant (respectively, P_- invariant), then $\mathcal{C}\{\gamma\} = \overline{\mathcal{B}}$ (respectively, $\mathcal{C}\{\gamma\} = \mathcal{B}$).*

Proof. It is enough to show that if $c\ell(\kappa)$ is P_- invariant, then $\mathcal{C}\{\gamma\} = \pi^{-1}(\mathcal{B})$. But P_- invariance of $c\ell(\kappa)$ is the same as gP_-g^{-1} invariance of $g(c\ell(\kappa))$. In other words, if $c\ell(\kappa)$ is P_- invariant, then $\mathcal{C}\{\gamma\}$ is the π_--preimage of some domain $\breve{\mathcal{B}}$ in G/P_-. Since G_0 acts transitively on $\mathcal{B}$, it is immediate that $\mathcal{B} \subset \breve{\mathcal{B}}$. On the other hand, for any Iwasawa–Borel group B, the unique B-invariant hypersurface H_0 is contained in the complement of $\breve{\mathcal{B}}$. Thus the envelope $\mathcal{E}_{\mathcal{H}_0}$ generated by H_0 is in the complement. But we know from Theorem 7.2.2 that $\mathcal{B} = \mathcal{E}_{\mathcal{H}_0}$. □

It should be underlined that in this result we are discussing the true cycle space $\mathcal{C}\{\gamma\}$, not just $G\{\gamma\}/K$. Of course in the case where we write $\mathcal{C}\{\gamma\} = \mathcal{U}$ the natural realization of $\mathcal{C}\{\gamma\}$ is in the G-orbit $G/\breve{K}$ of $c\ell(\kappa)$ in $C_q(Z)$, and $\breve{K}$ could very well be a finite extension of K. However, from Theorem 11.3.7, the lift of $\mathcal{C}\{\gamma\}$ from $G/\breve{K}$ to G/K is biholomorphic.

12.5 Cycle spaces of closed orbits

Here we consider the case of the closed G_0-orbit $\gamma_{c\ell}$ and its dual κ_{op} which is the open K-orbit in Z. Note that κ_{op} is dense in Z. Duality is the statement that $\kappa_{\mathrm{op}} \supset \gamma_{c\ell}$. Thus we consider the boundary

$$\mathrm{bd}(\kappa_{\mathrm{op}}) = Z \setminus \kappa_{\mathrm{op}} = A_1 \cup \cdots \cup A_m$$

decomposed as a union of its irreducible components.

Each of the components A_j is a closed analytic subset of Z which contains a Zariski open dense K-orbit κ_j, $j = 1, \ldots, m$. The only K-orbit which has such a κ_j in its boundary is the open K-orbit κ_{op}, i.e., $\kappa_{\mathrm{op}} < \kappa_j$ and κ_{op} is the only K-orbit with this property.

Thus, if γ_j denotes the G_0-orbit which is dual to κ_j, then $\gamma_j < \gamma_{c\ell}$ and $\gamma_{c\ell}$ is the only G_0-orbit with this property (Theorem 8.4.1). In other words,

(12.5.1) $$c\ell(\gamma_j) = \gamma_j \dot\cup \gamma_{c\ell}, \quad j = 1, \dots, m.$$

This implies the following basic fact.

Proposition 12.5.2. *If γ_j is dual to the K-orbit κ_j which is defined to be the open K-orbit in the irreducible component A_j in the boundary of the open orbit κ_{op}, then $\mathcal{C}\{\gamma_j\} \supset \mathcal{C}\{\gamma_{c\ell}\}$, $j = 1, \dots, m$.*

Proof. Since γ_j is noncompact, it follows from work in the previous section that every element $g \in \mathrm{bd}(G\{\gamma_j\})$ is contained in the lift $\widetilde{I_Y}$ of some incidence divisor I_Y. In other words, $g(c\ell(\kappa_j)) \cap Y \neq \emptyset$. But Y consists of a single point which is contained in $\gamma_{c\ell}$. Therefore $g \notin \mathcal{C}\{\gamma_{c\ell}\}$. Thus $\mathrm{bd}(\mathcal{C}\{\gamma_j\})$ is contained in the complement of $\mathcal{C}\{\gamma_{c\ell}\}$, and as a result $\mathcal{C}\{\gamma_j\} \supset \mathcal{C}\{\gamma_{c\ell}\}$. □

The following completes our discussion of the cycle spaces $\mathcal{C}\{\gamma_{c\ell}\}$.

Theorem 12.5.3. *The cycle space $\mathcal{C}\{\gamma_{c\ell}\}$ of the closed G_0-orbit in Z is equal to the universal domain $\mathcal{U}$.*

Proof. By Corollary 12.1.5 and Proposition 12.5.2, if γ_j is any G_0-orbit that is dual to a K-orbit κ_j on the boundary of κ_{op}, then

$$\mathcal{U} \subset \mathcal{C}\{\gamma_{c\ell}\} \subset \mathcal{C}\{\gamma_j\}.$$

Thus, if $c\ell(\kappa_j)$ is neither P_+- nor P_--invariant, then $\mathcal{C}\{\gamma_j\} = \mathcal{U}$, and the desired result follows.

If among the boundary orbits κ_j some $c\ell(\kappa_+)$ is P_+-invariant and another $c\ell(\kappa_-)$ is P_--invariant, then $\mathcal{C}\{\gamma_+\} = \overline{\mathcal{B}}$ and $\mathcal{C}\{\gamma_-\} = \mathcal{B}$, and therefore

$$\mathcal{U} \subset \mathcal{C}\{\gamma_{c\ell}\} \subset \mathcal{C}\{\gamma_+\} \cap \mathcal{C}\{\gamma_-\} = \overline{\mathcal{B}} \times \mathcal{B} = \mathcal{U}.$$

Hence, in this case we also have the desired result.

Finally, if every boundary orbit $c\ell(\kappa_j)$ is, e.g., P_--invariant, then the entire boundary $\mathrm{bd}(\kappa_{\mathrm{op}})$ is P_--invariant. Equivalently, κ_{op} is P_--invariant, and thus $\mathcal{C}\{\gamma_{c\ell}\}$ is P_--invariant.

Since P_- is parabolic and G/K is spherical, the P_--orbit of a generic point in G/K is Zariski open. It follows that the complement of $\mathcal{C}\{\gamma_{c\ell}\}$ in G/K, which is clearly nonempty, is contained in a proper analytic set. But this complement is invariant by the real form G_0, in contradiction to the Identity Principle. □

We close this section with remarks on the complex geometry of certain of the $\gamma_{c\ell}$ and κ_{op}.

Recall for this the $(\mathfrak{m}_0\mathfrak{a}_0\mathfrak{n}_0)$ decomposition. Here we have implicitly chosen a Cartan decomposition $\mathfrak{g}_0 = \mathfrak{k}_0 + \mathfrak{s}_0$, $\mathfrak{a}_0$ is as usual an abelian subalgebra which is maximal among those contained in $\mathfrak{s}_0$, and, having chosen a positive chamber, $\mathfrak{n}_0$ is the direct sum of the positive $\mathfrak{a}_0$-root (restricted root) spaces. The subalgebra $\mathfrak{m}_0$ is the centralizer of $\mathfrak{a}_0$ in $\mathfrak{k}_0$.

If $\mathfrak{b}_\mathfrak{m}$ is a Borel subalgebra in the complexification $\mathfrak{m}$, then $\mathfrak{b} := \mathfrak{b}_\mathfrak{m} + \mathfrak{a} + \mathfrak{n}$ is a Borel subalgebra of $\mathfrak{g}$. In particular, $P = MAN$ is a parabolic subgroup of G which contains the Iwasawa–Borel subgroup $B = B_M AN$. We consider the flag manifold $Z = G/P$ which, due to the fact that P is the complexification of $M_0 A_0 N_0$, is defined over the reals.

Let $z_0 \in Z$ be the neutral point corresponding to P and observe that the isotropy group K_{z_0} is just the reductive group M. Hence, the open orbit $\kappa_{\mathrm{op}} := K(z_0) = K/M$ is an affine K-homogeneous space. In particular, all of the boundary components $A_1, \ldots, A_m$ of κ_{op} in Z are of codimension 1, and γ_{op} is a totally real submanifold of Z which is in fact the set of real points

Now let $\widehat{z}_0$ be the neutral point in $\widehat{Z} = G/B$ which corresponds to the Iwasawa–Borel group B defined above. Let $\pi : G/B \to G/P$ be the canonical projection and note that M acts transitively on the fiber $\pi^{-1}(z_0)$. Thus the open orbit $\widehat{\kappa}_{\mathrm{op}} = K.\widehat{z}_0$ is simply the π-preimage of κ_{op}, and in particular $\pi|\widehat{\kappa}_{\mathrm{op}} : \widehat{\kappa}_{\mathrm{op}} \to \kappa_{\mathrm{op}}$ is proper. This is in fact the Remmert reduction of the holomorphically convex manifold $\widehat{\kappa}_{\mathrm{op}}$, which is constructed by identifying points whenever all holomorphic functions take the same value on them.

Let us summarize these remarks.

Proposition 12.5.4. *Every boundary component $\widehat{A}_j$ of the open K-orbit $\widehat{\kappa}_{\mathrm{op}}$ in $\widehat{Z} = G/B$ has codimension 1 and $\widehat{\kappa}_{\mathrm{op}}$ is holomorphically convex. If P is the minimal parabolic containing B, κ_{op} is the open K-orbit in $Z = G/P$, and $\pi : G/B \to G/P$ is the natural map, then $\widehat{\kappa}_{\mathrm{op}}$ is π-saturated and $\pi : \kappa_{\mathrm{op}} \to \tilde{\kappa}_{\mathrm{op}}$ is the Remmert reduction of κ_{op} onto the affine homogeneous space $\tilde{\kappa}_{\mathrm{op}}$.*

Proof. If some component $\widehat{A}$ is of higher codimension and $\widehat{p} \in \widehat{A}$ is a generic point, e.g., is in no other $\widehat{A}_j$, then every function $\widehat{f} \in \mathcal{O}(\widehat{\kappa}_{\mathrm{op}})$ extends holomorphically across $\widehat{A}$ at $\widehat{p}$.

But if $\{\widehat{p}_n\} \subset \widehat{\kappa}_{\mathrm{op}}$ converges to $\widehat{p}$, then $p_n := \pi(\widehat{p}_n)$ diverges in the Stein manifold κ_{op}. Thus there exists $f \in \mathcal{O}(\kappa_{\mathrm{op}})$ with $\lim |f(p_n)| = \infty$. Hence, $f \circ \pi =: \widehat{f}$ certainly cannot be extended across $\widehat{A}$ at $\widehat{p}$.

The same argument shows that $\widehat{\kappa}_{\mathrm{op}}$ is holomorphically convex, i.e., for every divergent sequence in $\widehat{\kappa}_{\mathrm{op}}$ there is a holomorphic function on $\widehat{\kappa}_{\mathrm{op}}$ with $\lim |f(p_n)| = \infty$.

The Remmert reduction is defined at the set-theoretic level by the equivalence relation $p \sim q$ if and only if every $f \in \mathcal{O}(\widehat{\kappa}_{\mathrm{op}})$ satisfies $f(p) = f(q)$. Since the base κ_{op} of $\pi : \widehat{\kappa}_{\mathrm{op}} \to \kappa_{\mathrm{op}}$ is Stein, the fiber is compact (isomorphic to M/B_M) and it is a locally trivial bundle, it is also immediate that the structure sheaf on the base is the direct image sheaf. This is the structural requirement of the Remmert reduction. □

13

Examples

Since the complex projective homogeneous manifolds $Z = G/Q$ of $G = SL(n; \mathbb{C})$ are just the classical flag manifolds, it is possible to discuss the cycle spaces of real forms of G in terms that are perhaps more familiar to complex geometers. Here we outline the Schubert slice method for flag domains of $SL(n; \mathbb{R})$ in this concrete context. For proofs we refer to [HS].

The case $G_0 = SU(2, 1)$ is handled in more detail. In particular, for the most interesting orbit G_0-orbit in the three-dimensional manifold of full flags, it is shown that the Schubert slices Σ are not Stein manifolds. This is contrary to our original hope of deriving the Stein property of $\mathcal{M}_D$ from that property of Σ.

The $SL(3, \mathbb{R})$-domain D in G/B is also particularly interesting. In this case we show that the full cycle space $\mathcal{C}_q(D)$ is essentially larger than $\mathcal{M}_D$.

Here we also give concrete realizations of the universal domains $\mathcal{U}$ for the real forms $SU(p, q)$, $SO(p, q)$, and $\mathrm{Sp}(p, q)$. For $SU(p, q)$ and $\mathrm{Sp}(p, q)$ they take the traditional form $I - W^*W \gg 0$, but the matrix realization for $SO(p, q)$ is somewhat surprising.

This chapter is organized as follows. In Section 1 we introduce the notation for the classical flag manifolds, and state the characterization of the (at most two) open $SL(n; \mathbb{R})$-orbits in terms of the real structure on the underlying vector space (Proposition 13.1.1). The base cycle in this case is the set of flags that are maximally isotropic with respect to the standard complex bilinear form (Proposition 13.1.3).

The relevant Schubert varieties in the case of $SL(n; \mathbb{R})$ are stabilized by parabolic groups having semisimple part $SL(2, \mathbb{R}) \times \cdots \times SL(2, \mathbb{R})$. Therefore the slices Σ are also of a very special type, products $\mathbb{C}^k \times \Delta^\ell$ of complex vector spaces with polydisks (Proposition 13.1.5).

At the end of Section 1 the $SU(2, 1)$-actions on the three possible flag varieties are discussed in detail; in particular, the orbits are described in terms of the signatures of the restricted forms. The Schubert slices for the most interesting open orbit in G/B are shown to project biholomorphically to the complement $\mathbb{P}_2 \setminus (c\ell(\mathcal{B}) \cup L)$ of the union of the closure of the bounded symmetric space $\mathcal{B}$ with one of its complex tangent lines L.

Either of the standard projections $\pi : G/B \to G/P = \mathbb{P}_2$ induces a holomorphic map $\pi_* : \mathcal{C}_q(\tilde{D}) \to \mathcal{C}_q(D)$ of the full cycle space of the open $SL(n;\mathbb{R})$ in G/B to the full cycle space of the open orbit in G/P. It is shown that $\mathcal{C}_q(D) = \mathcal{U}$. However, $\mathcal{C}_q(\tilde{D})$ is much larger. Here we give a precise description of the fiber of π_* (Proposition 13.1.6).

In Section 2 the universal domains of the real forms mentioned above are concretely realized in G-orbits in Grassmannians of q-planes defined over $\mathbb{C}$, $\mathbb{R}$ and $\mathbb{H}$, respectively. The realizations of the symmetric spaces G_0/K_0 in this way are classical (Proposition 13.2.1). The description of their "thickenings" $\mathcal{U}$ requires a computation with restricted roots (Proposition 13.2.4). This results in a concrete matrix domain picture of the respective universal domains (Proposition 13.2.11).

13.1 Cycle spaces of open $SL(n;\mathbb{R})$-orbits

13.1A Open orbits and base cycles

Let $V = \mathbb{C}^n$ and consider a vector $\delta = (d_1, \ldots, d_k)$ of integers for which $0 < d_1 < \cdots < d_k < n$. A **flag** $(\Lambda_1 \subsetneqq \cdots \subsetneqq \Lambda_k)$ with **symbol** or **dimension sequence** δ is an increasing sequence of subspaces of V with $\dim_{\mathbb{C}} \Lambda_j = d_j$. Let Z_δ denote the set of all flags in V with symbol δ, viewed as a homogeneous space G/Q where $G = SL(n;\mathbb{C})$. Then Q is a parabolic subgroup of G and Z_δ is a homogeneous projective variety. Conversely, if $Z = G/Q$ is a homogeneous projective variety, then it is of the form Z_δ for some symbol δ. Here we sketch some information on the cycle spaces $\mathcal{M}_D$ of open orbits D of $G_0 = SL(n;\mathbb{R})$ on such complex flag manifolds Z_δ.

Let γ denote complex conjugation of $V = \mathbb{C}^n$ over $\mathbb{R}^n$. We say that a flag $(\Lambda_1 \subsetneqq \cdots \subsetneqq \Lambda_k)$ is **γ-generic** if $\gamma(\Lambda_i) \cap \Lambda_j$ has minimal possible dimension for all i, j.

Proposition 13.1.1. *An orbit $G_0(z)$ of a point $z \in Z_\delta$ is open if and only if z is a γ-generic flag.*

For the proof of this and other such statements see [HS].

It can be shown that there are either one or two open orbits of $G_0 = SL(n;\mathbb{R})$ on Z_δ. The case of two open orbits occurs when V has even dimension $n = 2m$ and $m \in \delta$. In that case a γ-generic m-dimensional subspace $\Lambda \subset V$ defines an orientation on the space $V_{\mathbb{R}} = \mathbb{R}^n$ of real points, by associating to an ordered basis $\{v_1, \ldots, v_m\}$ of Λ the ordered basis $\{Re\, v_1, Im\, v_1; \ldots; Re\, v_m, Im\, v_m\}$. The orientation defined by the latter depends only on Λ.

Let us be concrete about this matter of orientation. Let $\{e_1, \ldots, e_{2m}\}$ denote the standard basis of $V = \mathbb{C}^{2m}$. Define $\varepsilon_j = e_{2j-1} + \sqrt{-1}e_{2j}$ for $1 \leqq j \leqq m$. In this setting we will say that $\Lambda_m^0 = \mathrm{Span}(\varepsilon_1, \ldots, \varepsilon_m)$ is positively oriented, and that a flag $(\Lambda_1 \subsetneqq \cdots \subsetneqq \Lambda_m \subsetneqq \cdots \subsetneqq \Lambda_k)$ is **positively oriented** if Λ_m defines the same orientation of $V_{\mathbb{R}}$ as Λ_m^0, **negatively oriented** if it defines the other orientation on $V_{\mathbb{R}}$.

Proposition 13.1.2. *If $m = 2n$ and $m \in \delta$, then the space of γ-generic flags in Z_δ has two topological components, the positively oriented flags and the negatively oriented flags, and each is a G_0-orbit. Otherwise, the space of γ-generic flags in Z_δ is the unique open G_0-orbit on Z_δ.*

Let $b : V \times V \to \mathbb{C}$ be the standard complex bilinear form, $b(x, y) = \sum x_i y_i$ in the basis $\{e_i\}$. It defines the complex special orthogonal group $K = SO(n, \mathbb{C})$. A flag $(\Lambda_1 \subsetneqq \cdots \subsetneqq \Lambda_k)$ is **maximally isotropic** (with respect to b) if, for all i and j, either $\Lambda_i^\perp \subset \Lambda_j$ or $\Lambda_j^\perp \subset \Lambda_i$.

Proposition 13.1.3. *The closed K-orbit C_0 inside an open G_0-orbit D in Z_δ is the manifold of maximally isotropic flags in D.*

13.1B Schubert slices

The transversal slices used in [HS] may carry a bit more information than Schubert slices. We exhibit this in the notationally simple case of the full flag manifold $Z = G/B$ in an even-dimensional space $V = \mathbb{C}^{2m}$.

A natural base point for the G_0-orbit of positively oriented flags is

$$\begin{aligned} z_0 =& \big([\varepsilon_1] \subsetneqq [\varepsilon_1 \wedge \varepsilon_2] \subsetneqq \cdots \subsetneqq [\varepsilon_1 \wedge \cdots \wedge \varepsilon_m] \subsetneqq [\varepsilon_1 \wedge \cdots \wedge \varepsilon_m \wedge \gamma(\varepsilon_m)] \\ & \subsetneqq [\varepsilon_1 \wedge \cdots \wedge \varepsilon_m \wedge \gamma(\varepsilon_m) \wedge \gamma(\varepsilon_{m-1})] \subsetneqq \cdots \\ & \subsetneqq [\varepsilon_1 \wedge \cdots \wedge \varepsilon_m \wedge \gamma(\varepsilon_m) \wedge \cdots \wedge \gamma(\varepsilon_2)]\big). \end{aligned}$$

Now consider the parabolic subgroup $P \subset G$ defined as the stabilizer of

$$w = \big(W_2 \subsetneqq W_4 \subsetneqq \cdots \subsetneqq W_{2m-2}\big), \quad \text{where } W_{2j} = [\varepsilon_1 \wedge \gamma(\varepsilon_1) \wedge \cdots \wedge \varepsilon_j \wedge \gamma(\varepsilon_j)].$$

It is shown in [HS] that the closure $S = c\ell(P(z_0))$ is a Schubert variety that has the Iwasawa–Schubert properties which have been used throughout this book.

From our present viewpoint, given the base point z_0, we would probably choose the Borel subgroup B to be one that stabilizes the flag defined by the standard ordered basis $\{e_1, \ldots, e_{2m}\}$.

Of course the closures are the same, $c\ell(P(z_0)) = S = c\ell(B(z_0))$, but one sees the Schubert slices more easily by using P. For example, let P^{ss} denote the semisimple component $SL(2; \mathbb{C}) \times \cdots \times SL(2; \mathbb{C})$ of P. It is the derived group of a reductive part P^r of P defined by the decomposition $V = W_1 \oplus \cdots \oplus W_m$, where $W_j = [\varepsilon_j \wedge \gamma(\varepsilon_j)]$. Let P^{-n} denote the unipotent radical of P. Hence, $P = P^r P^{-n}$ semidirect product, and we can fiber the P-orbit

$$P(z_0) = P/H \to P/P^{-n}H = \mathbb{P}_1(\mathbb{C}) \times \cdots \times \mathbb{P}_1(\mathbb{C}) \tag{13.1.4}$$

with typical fiber $P^{-n}(z_0) \cong \mathbb{C}^k$, where H is the P-stabilizer of z_0. The intersection $P_0 = P_0^r P_0^{-n} := P \cap G_0$ is locally transitive on the intersection $S \cap D$ of the Schubert variety with the open orbit D. In fact, as we know, $A_0 N_0$ already is locally transitive on that intersection. The action of P_0 gives us the following additional information:

1. The real unipotent radical P_0^{-n} is transitive on the fibers of (13.1.4) over the open P_0^r-orbits in the base.
2. The situation is inductive in the sense that the Schubert slices Σ are just the preimages of the open P_0^{-n}-orbits in the P^r-homogeneous manifold (13.1.4).

Finally, we note that in general the Schubert slices for open $SL(n;\mathbb{R})$-orbits have a very special form.

Proposition 13.1.5. *A Schubert slice Σ of an open $SL(n;\mathbb{R})$-orbit in a flag manifold Z_δ is biholomorphic to the product $\mathbb{C}^k \times \Delta^\ell$ of a k-dimensional complex vector space with an ℓ-dimensional polydisk.*

This could very well shed some light on the representations of $SL(n;\mathbb{R})$ on cohomology spaces associated to flag domains.

13.1C Flag domains of $SL(3;\mathbb{R})$

We specialize to $V = \mathbb{C}^3$, with $G = SL(3;\mathbb{C})$ and $G_0 = SL(3;\mathbb{R})$, in order to give more details on the above remarks concerning $SL(n;\mathbb{R})$ in a concrete setting. In particular, we give an example of the general results in Part IV, explicitly showing that the full cycle space of the open G_0-orbit D, in the manifold of full flags, is essentially bigger than $\mathcal{M}_D$.

The group G_0 has two orbits on $\mathbb{P}(V)$, the closed orbit $\mathbb{P}(V_\mathbb{R})$ consisting of real points, and the open orbit $D = \mathbb{P}(V) \setminus \mathbb{P}(V_\mathbb{R})$. In the notation introduced above, a point in $\mathbb{P}(V)$ is just a flag $z = (L)$, where $L \subset V$ is a one-dimensional subspace. The base cycle C_0 is the manifold of isotropic lines. Choosing $K_0 = SO(V_\mathbb{R}) = SO(3)$ we have $C_0 = \{[z_0, z_1, z_2] \in \mathbb{P}(V) \mid z_0^2 + z_1^2 + z_2^2 = 0\}$.

Choose the base point $z_0 = ([\varepsilon_1]) \in C_0$. Here $\varepsilon_1 = e_1 + \sqrt{-1}e_2$ where $\{e_1, e_2, e_3\}$ is the standard basis of V. We use the Borel subgroup B which is the G-stabilizer of the full flag $([e_1] \subsetneqq [e_1 \wedge e_2])$. The closure $S := c\ell(B(z_0))$ of its orbit of the base point is the projective line $\mathbb{P}([e_1 \wedge e_2])$. The intersection $S \cap \mathbb{P}(V_\mathbb{R})$ is the circle

$$S_\mathbb{R} = \{[a\varepsilon_1 + b\gamma(\varepsilon_1)] \in \mathbb{P}(V) \mid a^2 + b^2 = 1\}$$

and the complement Y of the open B-orbit in S is just the fixed point $z_1 = [e_1]$.

The space $\Gamma(S; \mathcal{O}(*Y))$ of meromorphic functions is the domain of definition of the trace transform. It can be realized as the space of polynomials on $\mathbb{C}$ where z_1 is regarded as the point at ∞. We point out that $S \cap D$ has two components: it is the union $S_1 \cup S_2$ of two disks such that $S_\mathbb{R}$ is their common boundary. In fact each of those disks defines a trace transform.

Now we move to the space Z of full flags $(\Lambda_1 \subsetneqq \Lambda_2)$ in V. Of course we have the $\mathbb{P}_1(\mathbb{C})$-bundle $Z \to \mathbb{P}(V)$ given by $(\Lambda_1 \subsetneqq \Lambda_2) \mapsto \Lambda_1$. The open G_0-orbit $\widetilde{D}$ in Z is the set of all γ-generic flags there, and in this case $(\Lambda_1 \subsetneqq \Lambda_2)$ is γ-generic if and only if $V = \gamma(\Lambda_1) + \Lambda_2$. Note that $\mathbb{P}(V)$ is not measurable in the sense of [W2], and that $\widetilde{D}$ is the **measurable model** or **minimal measurable cover** of the open orbit $D = \mathbb{P}(V) \setminus \mathbb{P}(V_\mathbb{R})$ in the sense of flag duality [HW1].

The base cycle $\widetilde{C}_0$ in $\widetilde{D}$ is a section of the restriction to C_0 of the $\mathbb{P}_1(\mathbb{C})$-bundle $Z \to \mathbb{P}(V)$. It is the K_0-orbit of the base point $\widetilde{z}_0 = ([\varepsilon_1] \subsetneqq [\varepsilon_1 \wedge e_3])$. Since dim V is odd, this expression lacks a bit in symmetry.

The appropriate Borel subgroup B is the stabilizer of the flag $([e_1] \subsetneqq [e_1 \wedge e_2])$, as above, and the two-dimensional orbit $\widetilde{\mathcal{O}} := B(\widetilde{z}_0)$ fibers as a $\mathbb{C}$-bundle over $B(z_0)$. The Schubert variety $\widetilde{S} := c\ell(B(\widetilde{z}_0))$ is the preimage of the one-dimensional $S = c\ell(B(z_0))$ under $\pi : Z \to \mathbb{P}(V)$.

The complement $\widetilde{Y} := \widetilde{S} \setminus \widetilde{\mathcal{O}}$ is the union of the fiber $\widetilde{Y}_1 := \pi^{-1}(z_1)$ and a section $\widetilde{Y}_2$ of $\pi : \widetilde{S} \to S$. Finally, $\widetilde{S} \cup \widetilde{D} = \widetilde{S}_1 \cup \widetilde{S}_2$, union of the preimages of the Schubert slices S_1 and S_2 with the section $\widetilde{Y}_2$ removed. Therefore, in this case a Schubert slice is just a product $\Sigma = \mathbb{C} \times \Delta$ of the complex line and a disk.

In order to obtain the full information afforded by B, we must also consider the dual Schubert variety $\widetilde{S}^*$ which is the π^* preimage of the one-dimensional Schubert variety S^* in $\mathbb{P}(V^*)$. Thus, in the case of the flag manifold Z, one has a total of four Schubert slices for the open $SL(3;\mathbb{R})$-orbit $\widetilde{D}$.

Finally let us compute the topological component $\mathcal{C}_1(\widetilde{D})$ of (the one-dimensional base cycle) $\widetilde{C}_0$ in the full cycle space of $\widetilde{D}$. For this note that $\widetilde{D} \to D$ defines a globally G_0-equivariant, locally G-equivariant, holomorphic map $\mathcal{C}_1(\widetilde{D}) \to \mathcal{C}_1(D)$.

Now, any two nondegenerate quadrics in $\mathbb{P}(V)$ differ by an element of $G = SL(3;\mathbb{C})$. Thus, if such a quadric lies in D, then it is $\mathcal{M}_D$. Furthermore, the only degenerations of such smooth quadrics are two projective lines, either in general position or a double line. Since every two-dimensional complex subspace of V contains a one-dimensional real subspace, it follows that every projective line has a real point. Thus no degenerate quadric is in $\mathcal{C}_1(D)$ and it follows that $\mathcal{C}_1(D) = \mathcal{M}_D$ here.

Now we have a holomorphic fibration $\mathcal{C}_1(\widetilde{D}) \to \mathcal{M}_D$, and we describe the fiber F over the base cycle $C_0 \in \mathcal{M}_D$.

Proposition 13.1.6. *The fiber F is the cycle space of the open $SL(2;\mathbb{C})$-orbit acting on $\mathbb{P}_1(\mathbb{C}) \times \mathbb{P}_1(\mathbb{C})$ as a real form of $SL(2;\mathbb{C}) \times SL(2;\mathbb{C})$ by the action $T(z,w) = (Tz, T^*w)$, where $T^* = ({}^t\overline{T})^{-1}$ as usual.*

Proof. The fiber F is $\{\widetilde{C} \in \mathcal{C}_1(\widetilde{D}) \mid \pi(\widetilde{C}) = C_0\}$, where π is the standard projection $Z \to \mathbb{P}(V)$, and C_0 is the orbit of the base point $z_0 \in \mathbb{P}(V)$ under $K = SO(3;\mathbb{C})$. Therefore it is relevant to analyze the action of K on the preimage $X := \pi^{-1}(C_0)$ in Z.

Recall $z_0 = [\varepsilon_1]$ and $\widetilde{C}_0 = K(\widetilde{z}_0)$, where $\widetilde{z}_0 = ([\varepsilon_1] \subsetneqq [\varepsilon_1 \wedge e_1])$. From Witt's theorem, if P_1 and P_2 are two planes in V that contain ε_1 and are b-nondegenerate, then there is an element of K that fixes ε_1 and maps P_1 to P_2. Consequently, the isotropy subgroup K_{z_0} acts transitively on the complement of $\widetilde{z}_0$ in $\pi^{-1}(z_0)$. Since $\pi|_X : X \to C_0$ is K-equivariant, now K has exactly two orbits in X, the base cycle $\widetilde{C}_0$ and its complement.

The classification of almost homogeneous $SL(2;\mathbb{C})$-manifolds (see, for example, [HL]) says that X is biholomorphic to $\mathbb{P}_1(\mathbb{C}) \times \mathbb{P}_1(\mathbb{C})$, where $SL(2;\mathbb{C})$ acts diagonally and $\widetilde{C}_0$ is the diagonal.

Since $K_0 \subset G_0$, it stabilizes both X and D, and thus stabilizes the complement $A := X \setminus \tilde{D}$. The fiber of $\pi|_X : X \to C_0$ is a projective line $\mathbb{P}_1(\mathbb{C})$ and its intersection with D is the complement $\mathbb{C}$ of a single point. The set A consists of those points and forms a C^∞–section of $\pi|_X : X \to C_0$. This has two consequences. First, A is a K_0-orbit on X. Second, $\dim_{\mathbb{R}} A = \dim_{\mathbb{R}} C_0 = 2$. Of course $\dim_{\mathbb{R}} X = 4$, and consequently A is not a real hypersurface in X.

There are only two K_0-orbits in $\mathbb{P}_1(\mathbb{C}) \times \mathbb{P}_1(\mathbb{C})$ that are not real hypersurfaces: the "antidiagonal" $\{([z_0, z_1], [w_0.w_1]) \in \mathbb{P}_1(\mathbb{C}) \times \mathbb{P}_1(\mathbb{C}) \mid z_0\overline{w_0} + z_1\overline{w_1} = 0\}$ and the usual diagonal. Since the base cycle $\tilde{C}_0$ is a complex K_0-orbit in X, it follows that A is the antidiagonal.

Thus the fiber F over C_0 in the full cycle space of $\tilde{D}$ can be identified with the space of cycles in $\mathbb{P}_1(\mathbb{C}) \times \mathbb{P}_1(\mathbb{C})$ which are in the same irreducible component as the diagonal and are contained in the complement of the antidiagonal. This is just the linear cycle space $\mathcal{M}_W$, where W is the open orbit of $G_0 := SL(2; \mathbb{C})$ which acts on $\mathbb{P}_1(\mathbb{C}) \times \mathbb{P}_1(\mathbb{C})$ as a real form of $SL(2; \mathbb{C}) \times SL(2; \mathbb{C})$ in the usual way: $T(z, w) = (T(z), T^*(w))$, where $T^* = ({}^tT)^{-1}$. □

13.2 Cycle spaces for open $SU(p, q; \mathbb{F})$-orbits

We now describe the cycle spaces $\mathcal{M}_D$ corresponding to flag domains of the real orthogonal groups $SO^0(p, q)$, the special unitary groups $SU(p, q)$, and the quaternionic unitary groups $\mathrm{Sp}(p, q)$, in terms of the riemannian symmetric spaces corresponding to those groups. For that purpose we realize the symmetric spaces as real, complex or quaternionic bounded domains; then the group acts by linear fractional transformations.

13.2A Dual Grassmann manifolds as bounded domains

Let $\mathbb{F}$ denote one of the fields $\mathbb{R}$, $\mathbb{C}$ or $\mathbb{H}$, let $0 \leqq p \leqq q$ be integers with $n := p+q \geqq 1$, and write $\mathbb{F}^{p,q}$ for the right vector space $\mathbb{F}^n$ with hermitian form $h(u, v) = \sum_{1 \leqq a \leqq p} u_a \overline{v_a} - \sum_{1 \leqq b \leqq q} u_b \overline{v_b}$. We write $U(p, q; \mathbb{F})$ for the unitary group of $\mathbb{F}^{p,q}$: $U(p, q; \mathbb{R}) = O(p, q)$, $U(p, q; \mathbb{C}) = U(p, q)$ and $U(p, q; \mathbb{H}) = \mathrm{Sp}(p, q)$. In this context, they define our groups as follows.

	$\mathbb{F} = \mathbb{R}$	$\mathbb{F} = \mathbb{C}$	$\mathbb{F} = \mathbb{H}$
G_0	$SU(p, q; \mathbb{R}) = SO^0(p, q)$	$SU(p, q; \mathbb{C}) = SU(p, q)$	$SU(p, q; \mathbb{H}) = \mathrm{Sp}(p, q)$
K_0	$SO(p) \times SO(q)$	$S(U(p) \times U(q)$	$\mathrm{Sp}(p) \times \mathrm{Sp}(q)$
G	$SO(p+q; \mathbb{C})$	$SL(p+q; \mathbb{C})$	$\mathrm{Sp}(p+q; \mathbb{C})$
G_u	$SO(p+q)$	$SU(p+q)$	$\mathrm{Sp}(p+q)$

The Grassmann manifold of q-dimensional linear subspaces of $\mathbb{F}^n$ is the riemannian symmetric space

$$\mathcal{G}(p, q; \mathbb{F}) = SU(p+q; \mathbb{F})/S(U(p, 0; \mathbb{F}) \times U(0, q; \mathbb{F})).$$

In effect it is $G_u/\breve{K}_0$, where $\breve{K}_0$ is the stabilizer of the subspace spanned by the last q basis vectors, in other words, the centralizer of $\begin{pmatrix} I_p & 0 \\ 0 & -I_q \end{pmatrix}$ in G_u. If $\mathbb{F} \neq \mathbb{R}$, then $\breve{K}_0 = K_0$; if $\mathbb{F} = \mathbb{R}$ then K_0 is a subgroup of index 2 in $\breve{K}_0$. In any case, $\mathcal{G}(p, q; \mathbb{F})$ is a compact dual symmetric space of

$$\mathcal{B}(p, q; \mathbb{F}) = SU(p, q; \mathbb{F})/S(U(p, 0; \mathbb{F}) \times U(0, q; \mathbb{F}))^0 = G_)/K_0.$$

If $\mathbb{F}$ is $\mathbb{C}$ or $\mathbb{H}$, then G_u is simply connected and K_0 is connected. Thus $\mathcal{G}(p, q; \mathbb{F})$ is simply connected in these cases. On the other hand $\mathcal{G}(p, q; \mathbb{R})$ has a 2-sheeted riemannian covering space consisting of all *oriented* q-dimensional linear subspaces of $\mathbb{R}^n$. When $\mathbb{F} = \mathbb{R}$, we set aside the trivial case $p = q = 1$, where $\mathcal{G}(p, q; \mathbb{R})$ is the real projective line.

We recall the following more or less standard result in order to establish notation. Here we view $\mathbb{F}^n$ as consisting of column vectors so that our unitary groups act on it by matrix multiplication from the left, as usual.

Proposition 13.2.1. *Let $\{e_a\}$ be the standard basis of $\mathbb{F}^n$ with respect to which we wrote the hermitian form h. Let $x_0 = \text{Span}\{e_{p+1}, \dots, e_{p+q}\} \in \mathcal{G}(p, q; \mathbb{F})$, base point. Then*

$$\mathcal{B}(p, q; \mathbb{F}) = SU(p, q; \mathbb{F})(x_0)$$

is the space of negative definite q-dimensional subspaces of $\mathbb{F}^{p,q}$.

If $x \in \mathcal{B}(p, q; \mathbb{F})$, then it has a (unique) basis $\{u_1, \dots, u_q\}$ of the form $u_b = (\sum_{1 \leq a \leq p} w_{a,b} e_a) + e_{p+b}$, where the hermitian matrix $(h(u_b, u_c))$ is negative definite. In other words, x is the column span of a (unique) matrix $\begin{pmatrix} W \\ I_q \end{pmatrix}$, where $W = (w_{a,b})$ is $p \times q$ and $I_p - WW^* \gg 0$. Thus we identify $\mathcal{B}(p, q; \mathbb{F})$ with the bounded domain $\{W \in \mathbb{F}^{p \times q} \mid I_p - WW^* \gg 0\}$. In this identification x_0 is identified with the $p \times q$ matrix 0, and the action of $G_0 = SU(p, q; \mathbb{F})$ on $\mathcal{B}(p, q; \mathbb{F})$ is given by linear fractional transformations,

$$\begin{pmatrix} A & B \\ C & D \end{pmatrix} : W \mapsto (AW + B)(CW + D)^{-1},$$

where A is $p \times p$, B is $p \times q$, C is $q \times p$, D is $q \times q$, and $CW + D$ is invertible.

Proof. The hermitian form variation on Witt's theorem shows that $U(p, q; \mathbb{F})$ acts transitively on $\mathcal{B}(p, q; \mathbb{F})$. Hence, as asserted, $\mathcal{B}(p, q; \mathbb{F}) = SU(p, q; \mathbb{F})(x_0)$ is the space of all negative definite q-dimensional linear subspaces of $\mathbb{F}^{p,q}$. If $x \in \mathcal{B}(p, q; \mathbb{F})$ now $x \cap \text{Span}\{e_1, \dots, e_p\} = 0$, so the orthogonal projection of $\mathbb{F}^{p,q}$ onto x_0 with kernel $\text{Span}\{e_1, \dots, e_p\} = 0$ is an isomorphism. Let $\{u_1, \dots, u_q\}$ be the basis of x that projects onto the basis $\{e_{p+1}, \dots, e_{p+q}\}$ of x_0. Negative definiteness of x translates to negative definiteness for the hermitian matrix $(h(u_b, u_c))$, which in turn translates to $I_p - WW^* \gg 0$.

$G_0 = SU(p, q; \mathbb{F})$ acts on the column span x of $\begin{pmatrix} W \\ I_q \end{pmatrix}$ by matrix multiplication, $\begin{pmatrix} A & B \\ C & D \end{pmatrix}\begin{pmatrix} W \\ I_q \end{pmatrix} = \begin{pmatrix} AW+B \\ CW+D \end{pmatrix}$. Here $CW + D$ is nonsingular because that column span is negative definite, and $\begin{pmatrix} AW+B \\ CW+D \end{pmatrix}$ has the same column span as $\begin{pmatrix} (AW+B)(CW+D)^{-1} \\ I_q \end{pmatrix}$. This completes the proof. □

In the case $\mathbb{F} = \mathbb{R}$, the correspondence $x \mapsto W$ of Proposition 13.2.1 is also defined on the Grassmannian of oriented negative definite q-dimensional subspaces of $\mathbb{R}^{p+q}$, but the correspondence ignores orientation, and it is convenient for us to work with the nonoriented Grassmannian.

Proposition 13.2.1 describes $\Omega_0 = G_0(x_0)$ inside $\Omega = G/K$. Now we describe $\Omega_u = G_u(x_0)$ inside Ω.

We start by complexifying the "matrix space" of Proposition 13.2.1. The universal domain will sit naturally inside that complexification. For notational clarity we tensor with a $\mathbb{C}$ whose square root of -1 we denote by ℓ. This is to avoid confusion with pure imaginary elements of $\mathbb{F}$. The following lemma is obvious.

Lemma 13.2.2. *By abuse of notation, write h for the $\mathbb{C}$-hermitian extension of h from $\mathbb{F}^{p,q}$ to its complexification $\mathbb{F}^{p,q}_{\mathbb{C}}$. Let*

$$\mathbb{F}^{p,q}_u = \mathrm{Span}_{\mathbb{F}}\{\ell e_1, \dots, \ell e_p, e_{p+1}, \dots, e_{p+q}\}.$$

By further abuse of notation let h also denote the restriction of the hermitian form from $\mathbb{F}^{p,q}_{\mathbb{C}}$ to $\mathbb{F}^{p,q}_u$. Then $\ell e_a \mapsto e_a$ for $1 \leqq a \leqq p$ and $e_b \mapsto e_b$ for $1 \leqq b \leqq q$ defines an isometry of $\mathbb{F}^{p,q}_u$ onto $\mathbb{F}^{p,q}$.

Lemma 13.2.2 sets up notation for the following obvious remark.

Lemma 13.2.3. *The orbit $G_u(x_0) \cong \mathcal{G}(p, q; \mathbb{F})$ consists of all q-dimensional $\mathbb{F}$-subspaces of $\mathbb{F}^{p,q}_u$.*

When $\mathbb{F} \neq \mathbb{R}$, where K_0 is the isotropy subgroup of G_u at x_0, Lemma 13.2.3 identifies $\Omega_u \subset \Omega$ geometrically as $\mathcal{G}(p, q; \mathbb{F})$. When $\mathbb{F} = \mathbb{R}$, it identifies Ω_u geometrically as the 2-sheeted universal covering space of $\mathcal{G}(p, q; \mathbb{F})$.

13.2B Restricted root analysis of $\frac{1}{2}\Omega_u$

We can now give a complete description of $\frac{1}{2}\Omega_u$. We will describe it in the Grassmannian $\mathcal{G}(p, q; \mathbb{F})$ and see explicitly that it is a cell. As a consequence its lift to $\Omega_u \subset \Omega$ is bijective, and the description applies there as well.

Proposition 13.2.4. *The domain $\frac{1}{2}\Omega_u \subset \Omega_u$ is given as follows:*

1. *Suppose that $\mathbb{F}$ is $\mathbb{C}$ or $\mathbb{H}$. Then $\frac{1}{2}\Omega_u$ consists of all negative definite q-dimensional $\mathbb{F}$-subspaces of $\mathbb{F}^{p,q}_u$. Those are the subspaces that are column spans of matrices $\begin{pmatrix} \ell W \\ I_q \end{pmatrix}$, where $W \in \mathbb{F}^{p\times q}$ such that WW^* has eigenvalue array $\{\tan^2(a_1), \dots, \tan^2(a_p)\}$ with all a_i real and $|a_i| < \pi/4$ for $1 \leqq i \leqq p$. The latter condition can be formulated as $I_p - WW^* \gg 0$.*
2. *Suppose that $\mathbb{F} = \mathbb{R}$. Then $\frac{1}{2}\Omega_u$ consists of all q-dimensional subspaces of $\mathbb{R}^{p,q}_u$ that are column spans of matrices $\begin{pmatrix} \ell W \\ I_q \end{pmatrix}$, where $W \in \mathbb{R}^{p\times q}$ such that WW^* has eigenvalue array $\{\tan^2(a_1), \dots, \tan^2(a_p)\}$ with all a_i real and*

Case $p \geqq 2$: $|a_i| + |a_j| < \pi/2$ *for* $1 \leqq i < j \leqq p$. *That condition can be formulated as* $I_{p(p-1)/2} - \frac{2}{\pi} \bigwedge^2 \left(\arctan\left(\sqrt{WW^*}\right)\right) \gg 0$.

Case $p = 1$: $|a_1| < \pi/2$. *Then W is a row vector and the condition just says that the square length* $||W||^2 < \infty$ *where* $||W||^2 := WW^*$.

Proof. The isotropy subgroup G_u at x_0 is the group $\breve{K}_0 = S(U(p,0;\mathbb{F}) \times U(0,q;\mathbb{F}))$, which is the centralizer in G_u of $\left(\begin{smallmatrix} I_p & 0 \\ 0 & -I_q \end{smallmatrix}\right)$. It has identity component $K_0 = G_u \cap G_0$, and we have decompositions $G_0 = K_0 A_0 K_0$ and $G_u = K_0 A_u K_0$ where $\mathfrak{a}_u = \ell \mathfrak{a}_0$. In matrices,
(13.2.5)

$$A_0 = \text{ all } a_A := \begin{pmatrix} \cosh(A) & \sinh(A) & 0 \\ \sinh(A) & \cosh(A) & 0 \\ 0 & 0 & I \end{pmatrix} \text{ and } A_u = \text{ all } a_{\ell A} := \begin{pmatrix} \cosh(\ell A) & \sinh(\ell A) & 0 \\ \sinh(\ell A) & \cosh(\ell A) & 0 \\ 0 & 0 & I \end{pmatrix},$$

where $A = \operatorname{diag}\{a_1, \ldots, a_p\}$ is a real diagonal $p \times p$ matrix. But $\cosh(\ell a) = \cos(a)$ and $\sinh(\ell a) = \ell \sin(a)$. Hence $a_{\ell A} = \begin{pmatrix} \cos(A) & \ell \sin(A) & 0 \\ \ell \sin(A) & \cos(A) & 0 \\ 0 & 0 & I \end{pmatrix}$, and therefore $a_{\ell A}(x_0)$ is given by

$$\operatorname{Span}_{\mathbb{F}}\{\sin(a_1)\ell e_1 + \cos(a_1)e_{p+1}, \ldots, \sin(a_p)\ell e_p + \cos(a_p)e_{2p}, e_{2p+1}, \ldots, e_{p+q}\}.$$

This exhibits $A_u(x_0)$ as a product of circles $\sin(a)\ell e_b + \cos(a)e_{p+b}$ in $\mathbb{F}_u^{p,q}$.

Now, using Proposition 6.1.4, we run through the various cases to explicitly see $\frac{1}{2}\Omega_u$ in terms of restricted roots.

Case $\mathbb{F} = \mathbb{H}$. Here $G_0 = \mathrm{Sp}(p,q)$, where $1 \leqq p \leqq q$. Let $n = p + q$. We view G_0 as a group of quaternionic matrices, and we use the Cartan subalgebra

$$\mathfrak{h} : \text{ all } \xi = \xi(a_1, \ldots, a_n) := \begin{pmatrix} \operatorname{diag}\{a_{p+1}, \ldots, a_{2p}\} & \operatorname{diag}\{a_1, \ldots, a_p\} & 0 \\ \operatorname{diag}\{a_1, \ldots, a_p\} & \operatorname{diag}\{a_{p+1}, \ldots, a_{2p}\} & 0 \\ 0 & 0 & \operatorname{diag}\{a_{2p+1}, \ldots, a_n\} \end{pmatrix}$$

with $a_i \in \mathbb{C}$. Here $\mathfrak{a}$ consists of the $\eta = \eta(a_1, \ldots, a_p) := \xi(a_1, \ldots, a_p, 0, \ldots, 0)$. Its real form $\mathfrak{a}_0 = \{\eta(a_1, \ldots, a_p) \mid a_i \in \mathbb{R}\}$, and $\mathfrak{a}_u = \ell \mathfrak{a}_0$.

As usual the $\varepsilon_i : \mathfrak{h} \to \mathbb{C}$ are the eigenvalues of the corresponding $2n \times 2n$ matrices. The short roots are the $\pm\varepsilon_i \pm \varepsilon_j$, $1 \leqq i < j \leqq n$, and the long roots are the $\pm 2\varepsilon_i$, $1 \leqq i \leqq n$. For the $\mathfrak{a}$-roots we set $\varepsilon_k = 0$ for $p < k \leqq n$. Now the matrix $a_{\ell A}$ of (13.2.5) is just $\exp(\eta(\ell a_1, \ldots, \ell a_p))$, where $A = \operatorname{diag}\{a_1, \ldots, a_p\}$, $a_i \in \mathbb{R}$. The values of the $\mathfrak{a}$-roots on $\eta(\ell a_1, \ldots, \ell a_p) = \log a_{\ell A}$ are the $\pm 2\ell a_i$ and $\pm \ell a_i$, $1 \leqq i \leqq p$, and the $\pm \ell a_i \pm \ell a_j$, $1 \leqq i < j \leqq p$. Thus we have

$$(13.2.6) \quad |\alpha(\eta)| < \tfrac{\pi}{2} \text{ for all } \mathfrak{a}\text{-roots } \alpha \Leftrightarrow \text{ each } |2\varepsilon_i(\eta)| < \tfrac{\pi}{2} \Leftrightarrow \text{ each } |a_i| < \tfrac{\pi}{4}.$$

It is immediate from (13.2.6) that $A_u(x_0) \cap \frac{1}{2}\Omega_u$ is given by the condition that, for each index i, a_i belongs to the component of 0 in the range $|\sin(a_i)| < |\cos(a_i)|$; so $\frac{1}{2}\Omega_u$ consists of the negative definite q-dimensional $\mathbb{H}$-subspaces of $\mathbb{H}_u^{p,q}$.

Case $\mathbb{F} = \mathbb{C}$. Here $G_0 = SU(p,q)$, where $1 \leqq p \leqq q$. Let $n = p + q$. We view G_0 as a group of complex matrices, and we use the Cartan subalgebra $\mathfrak{h}$ of all matrices

$$\xi = \xi(a_1, \dots, a_n) := \begin{pmatrix} \mathrm{diag}\{a_{p+1},\dots,a_{2p}\} & \mathrm{diag}\{a_1,\dots,a_p\} & 0 \\ \mathrm{diag}\{a_1,\dots,a_p\} & \mathrm{diag}\{a_{p+1},\dots,a_{2p}\} & 0 \\ 0 & 0 & \mathrm{diag}\{a_{2p+1},\dots,a_n\} \end{pmatrix}$$

of trace zero with complex entries. Here $\mathfrak{a}$ consists of the $\eta = \eta(a_1, \dots, a_p) := \xi(a_1, \dots, a_p, 0, \dots, 0)$. Its real form $\mathfrak{a}_0 = \{\eta(a_1, \dots, a_p) \mid a_i \in \mathbb{R}\}$, and $\mathfrak{a}_u = \ell\mathfrak{a}_0$.

As usual the $\varepsilon_i : \mathfrak{h} \to \mathbb{C}$ are the joint eigenvalues. The roots are the $\pm(\varepsilon_i - \varepsilon_j)$, $1 \leqq i < j \leqq n$, and for the $\mathfrak{a}$-roots we set $\varepsilon_k = 0$ for $2p < k \leqq n$ and $\varepsilon_i = \varepsilon_{p+i}$ for $1 \leqq i \leqq n$. Again the matrix $a_{\ell A}$ of (13.2.5) is just $\exp(\eta(\ell a_1, \dots, \ell a_p))$, where $A = \mathrm{diag}\{a_1, \dots, a_p\}$, $a_i \in \mathbb{R}$, and values of the $\mathfrak{a}$-roots on $\eta(\ell a_1, \dots, \ell a_p) = \log a_{\ell A}$ are the $\pm 2\ell a_i$ and $\pm \ell a_i$, $1 \leqq i \leqq p$, and the $\pm(\ell a_i - \ell a_j)$, $1 \leqq i < j \leqq p$. Thus again we have

(13.2.7) $\quad |\alpha(\eta)| < \frac{\pi}{2}$ for all $\mathfrak{a}$-roots $\alpha \Leftrightarrow$ each $|2\varepsilon_i(\eta)| < \frac{\pi}{2} \Leftrightarrow$ each $|a_i| < \frac{\pi}{4}$,

and it is immediate from (13.2.7) that $A_u(x_0) \cap \frac{1}{2}\Omega_u$ is given by the condition that, for each i, a_i belongs to the component of 0 in the range $|\sin(a_i)| < |\cos(a_i)|$. So $\frac{1}{2}\Omega_u$ consists of the negative definite q-dimensional $\mathbb{C}$-subspaces of $\mathbb{C}_u^{p,q}$.

Case $\mathbb{F} = \mathbb{R}$. Here $G_0 = SO(p, q)$, where $1 \leqq p \leqq q$. Let $n = p + q$. We view G_0 as a group of real matrices, and we use the Cartan subalgebra $\mathfrak{h}$ of all

$$\xi = \xi(a_1, \dots, a_m) := \begin{pmatrix} 0 & \mathrm{diag}\{a_1,\dots,a_p\} & 0 \\ \mathrm{diag}\{a_1,\dots,a_p\} & 0 & 0 \\ 0 & 0 & \mathrm{diag}\{r(a_{p+1}),\dots,r(a_m),(0)\} \end{pmatrix},$$

where n is $2m$ or $2m+1$, the $a_i \in \mathbb{C}$, $r(a) = \left(\begin{smallmatrix} 0 & a \\ -a & 0 \end{smallmatrix}\right)$, and the (0) is present just when n is odd. Again, $\mathfrak{a}$ consists of the $\eta = \eta(a_1, \dots, a_p) := \xi(a_1, \dots, a_p, 0, \dots, 0)$. Its real form $\mathfrak{a}_0 = \{\eta(a_1, \dots, a_p) \mid a_i \in \mathbb{R}\}$, and $\mathfrak{a}_u = \ell\mathfrak{a}_0$. As before, the $\varepsilon_i : \mathfrak{h} \to \mathbb{C}$ are the eigenvalues. The roots are

- if $n = 2m + 1$: the $\pm\varepsilon_i \pm \varepsilon_j$, $1 \leqq i < j \leqq m$, and the $\pm\varepsilon_i$, $1 \leqq i \leqq m$;
- if $n = 2m$: the $\pm\varepsilon_i \pm \varepsilon_j$, $1 \leqq i < j \leqq m$.

As before, the matrix $a_{\ell A}$ of (13.2.5) is just $\exp(\eta(\ell a_1, \dots, \ell a_p))$, where $A = \mathrm{diag}\{a_1, \dots, a_p\}$, $a_i \in \mathbb{R}$, and values of the $\mathfrak{a}$-roots on $\eta(\ell a_1, \dots, \ell a_p) = \log a_{\ell A}$ are as follows:

- if $n = 2m + 1$ and $p \geqq 2$: the $\pm\ell a_i \pm \ell a_j$ for $1 \leqq i < j \leqq p$ and the $\pm a_i$ for $1 \leqq i \leqq p$, corresponding to the $\mathfrak{a}$-root system $\{\pm\varepsilon_i \pm \varepsilon_j, \pm\varepsilon_k\}_{1 \leqq i < j \leqq p, 1 \leqq k \leqq p}$;
- if $n = 2m$ and $m - 1 \geqq p \geqq 2$: the $\pm\ell a_i \pm \ell a_j$ for $1 \leqq i < j \leqq p$ and the $\pm a_i$ for $1 \leqq i \leqq p$; then $\{\pm\varepsilon_i \pm \varepsilon_j, \pm\varepsilon_k\}_{1 \leqq i < j \leqq p, 1 \leqq k \leqq p}$ is the corresponding $\mathfrak{a}$-root system;
- if $n = 2m$ and $m = p \geqq 2$: the $\pm\ell a_i \pm \ell a_j$ for $1 \leqq i < j \leqq p$, corresponding to the $\mathfrak{a}$-root system $\{\pm\varepsilon_i \pm \varepsilon_j\}_{1 \leqq i < j \leqq p}$; and
- if $p = 1$: $\pm a_1$, corresponding to the $\mathfrak{a}$-root system $\{\pm\varepsilon_1\}$.

So if $p \geqq 2$, we have $|\alpha(\eta)| < \frac{\pi}{2}$ for all $\mathfrak{a}$-roots α if and only if

(13.2.8) $\quad |\varepsilon_i(\eta)| + |\varepsilon_j(\eta)| < \frac{\pi}{2}$ for $i < j \Leftrightarrow |a_i| + |a_j| < \frac{\pi}{2}$ for $i < j$, and if $p = 1$, we have

$$(13.2.9) \qquad |\alpha(\eta)| < \tfrac{\pi}{2} \text{ for all } \mathfrak{a}\text{-roots } \alpha \Leftrightarrow |\varepsilon_1(\eta)| < \tfrac{\pi}{2} \Leftrightarrow |a_i| < \tfrac{\pi}{2}.$$

Suppose that $p \geqq 2$. Then $A_u(x_0) \cap \frac{1}{2}\Omega_u$ is specified by the condition that, for $1 \leqq i < j \leqq p$, $|a_i| + |a_j| < \frac{\pi}{2}$. Thus $\frac{1}{2}\Omega_u$ consists of the oriented column span of matrices $\begin{pmatrix} \ell W \\ I_p \end{pmatrix}$ such that WW^* is diagonalizable real eigenvalues t_i^2, $1 \leqq i \leqq p$, such that $|t_i| + |t_j| < \frac{\pi}{2}$ for $i < j$. Here $t_i = \tan a_i$. Thus we arrive at the description of $\frac{1}{2}\Omega_u$ claimed in Proposition 13.2.4.

Suppose that $p = 1$. Then $A_u(x_0) \cap \frac{1}{2}\Omega_u$ is specified by the condition that $|a_1| < \frac{\pi}{2}$. Thus $\frac{1}{2}\Omega_u$ consists of the oriented column span of matrices $\begin{pmatrix} \ell W \\ I_p \end{pmatrix}$ where W is a $1 \times q$ real row matrix such that $WW^* = t_1^2$ with $0 \leqq t_1^2 < \frac{\pi}{2}$. Here $t_1 = \tan a_1$. So again we arrive at the description of $\frac{1}{2}\Omega_u$ claimed in Proposition 13.2.4, and that completes the proof. □

13.2C Geometric interpretation

We give a geometric interpretation of Proposition 13.2.4.

Suppose first that $\mathbb{F}$ is $\mathbb{C}$ or $\mathbb{H}$. Since v and $-v$ have the same linear span, the antipode of x_0 in the circle $\sin(a)\ell e_b + \cos(a)e_{p+b}$ is given by $a = \pi/2$. Now $\frac{1}{2}\Omega_u \cap A_u(x_0)$ consists of all the $a_{\ell A}(x_0)$ for which every $|a_b| < \pi/4$, in other words, every $|\sin(a_b)| < |\cos(a_b)|$. Since

$$h\big(\sin(a)\ell e_b + \cos(a)e_{p+b},\ \sin(a)\ell e_b + \cos(a)e_{p+b}\big) = \sin^2(a) - \cos^2(a),$$

we have

$$\tfrac{1}{2}\Omega_u \cap A_u(x_0) = \{w \in A_u(x_0) \mid w \text{ is negative definite for } h\}.$$

All the structures here are $\breve{K}_0$-stable, and therefore

$$\tfrac{1}{2}\Omega_u = \{w \in G_u(x_0) \mid w \text{ is negative definite for } h\}.$$

Express this in terms of column span. The condition $|a_b| < \pi/4$, in other words $|\sin(a_b)| < |\cos(a_b)|$, ensures $\cos(a_b) \neq 0$. So

$$a_{\ell A}(x_0) = \operatorname{Span}\{\tan(a_1)\ell e_1 + e_{p+1}, \dots, \tan(a_p)\ell e_p + e_{2p}, e_{2p+1}, \dots, e_{p+q}\}.$$

In other words, $a_{\ell A}(x_0)$ is the column span of $\begin{pmatrix} \ell W \\ I_q \end{pmatrix}$, where

$$WW^* = \operatorname{diag}\{\tan^2(a_1), \dots, \tan^2(a_p)\},$$

and the condition $|a_b| < \pi/4$ for $\frac{1}{2}\Omega_u$ just says that each $|\tan^2(a_b)| < 1$, or equivalently that $I_p - WW^* \gg 0$. That completes the proof for the cases where $\mathbb{F}$ is $\mathbb{C}$ or $\mathbb{H}$.

Suppose that $\mathbb{F} = \mathbb{R}$. First consider the case where $p \geqq 2$. Then the antipode of x_0 in the torus $A_u(x_0)$ is cut out by

$$|a_b| = \pi \text{ for } 1 \leqq b \leqq p \text{ and } |a_b + a_c| = \pi \text{ for } 1 \leqq b < c \leqq p.$$

Up to K_0-conjugacy we may take each $a_b \geqq 0$. Then these antipode conditions simplify to $a_b + a_c = \pi$ for $1 \leqq b < c \leqq p$, and arguing as before we see that $a_{\ell A}(x_0) \in \frac{1}{2}\Omega_u$ if and only if $a_{\ell A}(x_0)$ is the column span of $\left(\begin{smallmatrix} \ell W \\ I_q \end{smallmatrix}\right)$, where $WW^* = \operatorname{diag}\{\tan^2(a_1), \dots, \tan^2(a_p)\}$, where $|a_i| + |a_j| < \pi/2$ for $1 \leqq i < j \leqq p$. The alternate formulation now is an easy exercise.

Now suppose that $p = 1$. Then the antipode of x_0 in the circle $A_u(x_0)$ is the point $|a_1| = \pi$, which is x_0 with the opposite orientation. Hence $a_{\ell A}(x_0) \in \frac{1}{2}\Omega_u$ if and only if $a_{\ell A}(x_0)$ is the column span of $\left(\begin{smallmatrix} \ell W \\ I_q \end{smallmatrix}\right)$, where $WW^* = \tan^2(a_1)$ with $|a_1| < \pi/2$. That just says $\tan^2(a_1) < \infty$, which is not very interesting. But in any case we now have the description of Proposition 13.2.4.

13.2D Matrix space interpretation

We now translate Proposition 13.2.4 into terms closer to those of the matrix space of Proposition 13.2.1. Define

$$\begin{aligned} \mathcal{W}(p,q;\mathbb{R}) &= \{x \in \widetilde{\mathcal{G}}(p,q;\mathbb{R}) \mid x \cap \operatorname{Span}\{e_1, \dots, e_p\} = 0\} \text{ and} \\ \mathcal{W}(p,q;\mathbb{F}) &= \{x \in \mathcal{G}(p,q;\mathbb{F}) \mid x \cap \operatorname{Span}\{e_1, \dots, e_p\} = 0\} \text{ if } \mathbb{F} \neq \mathbb{R}. \end{aligned} \tag{13.2.10}$$

Proposition 13.2.11. *Define $\mathcal{W}(p,q;\mathbb{F}) = \{x \in \mathcal{G}(p,q;\mathbb{F}) \mid x \cap \operatorname{Span}\{e_1, \dots, e_p\} = 0\}$ if $\mathbb{F} \neq \mathbb{R}$, and $\mathcal{W}(p,q;\mathbb{R}) = \{x \in \widetilde{\mathcal{G}}(p,q;\mathbb{R}) \mid x \cap \operatorname{Span}\{e_1, \dots, e_p\} = 0\}$. Then $\mathcal{W}(p,q;\mathbb{F})$ is a dense open subset of $\mathcal{G}(p,q;\mathbb{F})$, and*

$$\left(W \mapsto \textit{column span of } \left(\begin{smallmatrix} W \\ I_q \end{smallmatrix}\right)\right) \ (\textit{oriented if } \mathbb{F} = \mathbb{R})$$

identifies $\mathbb{F}^{p\times q}$ with $\mathcal{W}(p,q;\mathbb{F})$. The complexification $\mathbb{F}_{\mathbb{C}}^{p\times q} := \mathbb{F}^{p\times q} \otimes_{\mathbb{R}} \mathbb{C}$ is thus identified with a dense open subset $\mathcal{W}(p,q;\mathbb{F})_{\mathbb{C}}$ of $\Omega = G/K$ that contains both Ω_0 and $\frac{1}{2}\Omega_u$. Specifically, in this identification,

$$\Omega_0 = \{W \in \mathbb{F}^{p\times q} \mid I_p - WW^* \gg 0\}$$

and $\frac{1}{2}\Omega_u$ is given by Proposition 13.2.4:

$$\begin{aligned} &\textit{Case } \mathbb{F} \neq \mathbb{R}: \tfrac{1}{2}\Omega_u = \{\ell W \in \ell\,\mathbb{F}^{p\times q} \mid I_p - WW^* \gg 0\}, \\ &\textit{Case } \mathbb{F} = \mathbb{R},\ \ p \geqq 2: \tfrac{1}{2}\Omega_u \\ &\quad = \left\{\ell W \in \ell\,\mathbb{R}^{p\times q} \,\middle|\, I_{p(p-1)/2} - \tfrac{2}{\pi}{\bigwedge}^2\left(\arctan\left(\sqrt{WW^*}\right)\right) \gg 0\right\}, \\ &\textit{Case } \mathbb{F} = \mathbb{R},\ \ p = 1: \tfrac{1}{2}\Omega_u = \{\ell W \in \ell\,\mathbb{R}^{1\times q} \mid \sqrt{WW^*} < \infty\} = \ell\,\mathbb{R}^{1\times q}. \end{aligned}$$

The action of G on $\mathbb{F}_{\mathbb{C}}^{p\times q}$ is given, where defined (i.e., where $CW + D$ is invertible), by linear fractional transformations,

$$\left(\begin{smallmatrix} A & B \\ C & D \end{smallmatrix}\right) : W \mapsto (AW + B)(CW + D)^{-1}.$$

Proof. First, suppose that $\mathbb{F} \neq \mathbb{R}$. The complement $\mathcal{G}(p, q; \mathbb{F}) \setminus \mathcal{W}(p, q; \mathbb{F})$ consists of all elements $x \in \mathcal{G}(p, q; \mathbb{F})$ such that $e_1 \wedge \cdots \wedge e_p \wedge v_1 \wedge \ldots v_q = 0$ where $\{v_b\}$ is a basis of x. So that complement is a closed proper subvariety and $\mathcal{W}(p, q; \mathbb{F})$ is a dense open subset. The identification of $\mathbb{F}^{p\times q}$ with $\mathcal{W}(p, q; \mathbb{F})$, and Ω_0 inside it, is exactly as in the argument of Proposition 13.2.1. This identification is based on the use of the basis $\{e_a, e_b\}$, not on the basis $\{\ell e_a, e_b\}$. Thus, a q-dimensional $\mathbb{F}$-subspace of $\mathbb{F}_u^{p\times q}$ corresponds to a matrix $\left(\begin{smallmatrix} W \\ I_q \end{smallmatrix}\right)$ with $W \in \ell\,\mathbb{F}^{p\times q}$. In view of Proposition 13.2.4 this gives us the description $\{\ell W \in \ell\,\mathbb{F}^{p\times q} \mid I_p - WW^* \gg 0\}$ of $\frac{1}{2}\Omega_u$.

Let $g = \left(\begin{smallmatrix} A & B \\ C & D \end{smallmatrix}\right) \in G$. Its action on q-dimensional subspaces of $\mathbb{F}_{\mathbb{C}}^{p,q}$ is given on the column span of a matrix $\left(\begin{smallmatrix} W \\ I \end{smallmatrix}\right)$ by matrix multiplication, $g\left(\begin{smallmatrix} W \\ I \end{smallmatrix}\right) = \left(\begin{smallmatrix} AW+B \\ CW+D \end{smallmatrix}\right)$, so the corresponding action on $\mathbb{F}_{\mathbb{C}}^{p\times q}$ is $W \mapsto (AW + B)(CW + D)^{-1}$ whenever $CW + D$ is invertible.

If $\mathbb{F} = \mathbb{R}$ the argument is essentially the same, with $\mathcal{G}(p, q; \mathbb{F})$ replaced by $\widetilde{\mathcal{G}}(p, q; \mathbb{R})$, the column spans are taken to be oriented, and with the description of $\frac{1}{2}\Omega_u$ given by Proposition 13.2.4. □

Propositions 11.3.3 and 11.3.5 identify (in the nonholomorphic cases) the cycle space $\mathcal{M}_D$ with the universal domain $\mathcal{U}$, and, in this $SU(p, q; \mathbb{F})$ case, Proposition 13.2.11 gives the concrete geometric structure of $\mathcal{U}$.

Corollary 13.2.12. *Suppose that $\mathbb{F} = \mathbb{C}$ or $\mathbb{F} = \mathbb{H}$. Let $G_0 = SU(p, q; \mathbb{F})$, let $Z = G/Q$ be a complex flag manifold, and let $D = G_0(z_0)$ be an open orbit that is not of holomorphic type. Then the natural action of G_0 on the bounded symmetric domain $\mathcal{B}(p, q; \mathbb{F}) = \{W \in F^{p\times q} \mid I - WW^* \gg 0\}$ is by linear fractional transformations. The cycle space*

$$\mathcal{M}_D \cong \mathcal{B}(p, q; \mathbb{F}) \times \ell\mathcal{B}(p, q; \mathbb{F}) \cong \mathcal{B}(p, q; \mathbb{F}) \times \mathcal{B}(p, q; \mathbb{F})$$

where $\mathbb{C} = \mathbb{R} + \ell\mathbb{R}$ is used to complexify $\mathbb{F}$. In the $\mathcal{B}(p, q; \mathbb{F}) \times \ell\mathcal{B}(p, q; \mathbb{F})$ picture, the natural action of G_0 on $\mathcal{M}_D$ carries over to the holomorphic action by linear fractional transformations,

$$\left(\begin{smallmatrix} A & B \\ C & D \end{smallmatrix}\right) : W \mapsto (AW + B)(CW + D)^{-1},$$

where $W = W' + \ell W''$ with $W', W'' \in \mathcal{B}$.

Proof. At this point, everything is proved except the statements concerning the action of G_0 on the $\mathcal{B}(p, q; \mathbb{F}) \times \ell\mathcal{B}(p, q; \mathbb{F})$ picture of $\mathcal{M}_D$. For that, note that every element of $\mathcal{B}(p, q; \mathbb{F}) \times \ell\,\mathcal{B}(p, q; \mathbb{F})$ corresponds to a matrix

$$W = W' + \ell W'' \in \mathbb{F}_{\mathbb{C}}^{p\times q} \quad \text{with } W', W'' \in \mathcal{B}(p, q; \mathbb{F}).$$

The holomorphic extension of $\left(\begin{smallmatrix} A & B \\ C & D \end{smallmatrix}\right) : W \mapsto (AW + B)(CW + D)^{-1}$ from the domain $G_0.C_0 = \mathcal{B}(p, q; \mathbb{F})$ to $\mathcal{U} = \mathcal{B}(p, q; \mathbb{F}) \times \ell\mathcal{B}(p, q; \mathbb{F})$ is given by the same formula, as asserted. □

Corollary 13.2.13. *Suppose that* $\mathbb{F} = \mathbb{C}$ *or* $\mathbb{F} = \mathbb{H}$. *Then the group* $G_0 \times G_0$ *acts on* $\mathcal{M}_D$ *by* $\left(\left(\begin{smallmatrix} A' & B' \\ C' & D' \end{smallmatrix}\right), \left(\begin{smallmatrix} A'' & B'' \\ C'' & D'' \end{smallmatrix}\right)\right) : (W', \ell W'') \mapsto$

$$((A'W' + B')(C'W' + D')^{-1}, \ell(A''W'' + B'')(C''W'' + D'')^{-1}),$$

where $W', W'' \in \mathcal{B}(p, q; \mathbb{F})$.

In Corollary 13.2.12 we recover the complex structure of $\mathcal{M}_D$ by viewing the second $\mathcal{B}(p, q; \mathbb{F})$ factor of $\mathcal{M}_D$ as $\ell\,\mathcal{B}(p, q; \mathbb{F}) = \frac{1}{2}\Omega_u$, as in Proposition 13.2.11. In the hermitian nonholomorphic cases this is, of course, the same as the adapted complex structure.

Remark. Our considerations for the $SU(p, q; \mathbb{F})$ cycle spaces, $\mathbb{F} = \mathbb{C}$ or $\mathbb{F} = \mathbb{H}$, are valid for cycle spaces corresponding to groups G_0, where the corresponding symmetric space Ω_0 of noncompact type is naturally embedded in its compact dual Ω_u as $\frac{1}{2}\Omega_u$. An example, whose geometry is essentially the same as above, is that in which $G_0 = F_{4,B_4}$, the real form of F_4 with maximal compact subgroup Spin(9), where Ω_0 is the Cayley hyperbolic plane and Ω_u is the Cayley projective plane. ◇

13.3 Slice methods and trace transforms for $SU(2, 1)$ domains

Consider $V = \mathbb{C}^3$ with the hermitian form given by $\langle z, w\rangle = z_1\overline{w}_1 + z_2\overline{w}_2 - z_3\overline{w}_3$ in a "standard" basis $\{e_1, e_2, e_3\}$. Its group of isometries in $G = SL(3; \mathbb{C})$ is the real form $G_0 = SU(2, 1)$. The maximal compact subgroup K_0 in G_0 is the stabilizer of the decomposition $V = V^+ + V^-$, where $V^+ = \mathrm{Span}(e_1, e_2)$ and $V^- = \mathrm{Span}(e_3)$.

As is usual we refer to the elements $L \in \mathbb{P}(V)$ as lines and to the elements $P \in \mathbb{P}(V^*)$ as planes. A point in the full flag manifold Z is denoted $(L \subsetneqq P)$. The standard projections are $\pi : Z \to \mathbb{P}(V)$ and $\pi^* : Z \to \mathbb{P}(V^*)$.

Since G_0 preserves the hermitian form, it also preserves the decomposition $\mathbb{P}(V) = D_- \cup M_0 \cup D_+$ into sets of negative, isotropic and positive lines. Those are the G_0-orbits on $\mathbb{P}(V)$, given by $|z_1|^2 + |z_2|^2 < |z_3|^2$, $|z_1|^2 + |z_2|^2 = |z_3|^2$, and $|z_1|^2 + |z_2|^2 > |z_3|^2$, respectively.

Evidently $|z_1|^2 + |z_2|^2 < |z_3|^2$ requires $z_3 \neq 0$. Therefore we use inhomogeneous coordinates $u_i = z_i/z_3$ and express

$$D_- = \{(u_1, u_2) \in \mathbb{C}^2 \mid |u_1|^2 + |u_2|^2 < 1\}, \quad \text{where } u_i = z_i/z_3.$$

So we view D_- as the unit ball in $\mathbb{C}^2$, view $M \cong S^3$ as its topological boundary, and view D_+ as the complement of its closure in $\mathbb{P}(V)$.

The bounded domain D_- is holomorphically separable. Thus, a connected compact subvariety consists of a point. Hence, the base cycle is just a fixed point of $K_0 = S(U(2) \times U(1))$, in this case the line $V^- = [0 : 0 : 1] \in \mathbb{P}(V)$. Its orthocomplement is the plane $V^+ \in \mathbb{P}(V^*)$. The associated projective plane is a K_0-orbit and is therefore the base C_+ in D_+.

13.3A Comparison of slice methods

In [W12] the Korányi–Wolf boundary component theory was used to fiber K_0-equivariantly an open G_0-orbit D in a hermitian symmetric flag manifold $Z = G/KP_-$ over its base cycle C_0. The fiber S is a complex totally geodesic submanifold of D that is biholomorphic to a bounded symmetric domain. That fiber intersects every cycle $C \in \mathcal{M}_D$ in exactly one point. Such a "slice" S is optimal for purposes of the trace transform, in part because it is given geometrically and in part because its function theory is explicit in terms of certain subgroups of G_0. The method, however, is a bit technical and works only in a rather special setting, while the Schubert slice method works in general. Also, the slices are quite different: S has bounded holomorphic functions, while in general Schubert slices may not. Now we will compare the two methods for the orbit $D_+ \subset \mathbb{P}_2(\mathbb{C})$ of $G_0 = SU(2, 1)$.

For the Schubert slice we must choose a Borel subgroup $B \subset G$ that contains the component A_0N_0 of an Iwasawa decomposition $G_0 = K_0A_0N_0$. This is equivalent to B fixing a point in the boundary of D_-. That boundary is the sphere M_0, and we may assume that point is $L_1 := [1 : 0 : 1]$.

Every Borel subgroup of G has three orbits on $\mathbb{P}(V)$. Start with its fixed flag $(L \subsetneqq P)$. The orbits are the line $\{L\}$, the set of all lines in $P \setminus L$, and the lines in $\mathbb{P}(V) \setminus P$. In our case, $L = L_1 = [1 : 0 : 1]$ and $P = \mathrm{Span}(e_1 + e_3, e_2)$. Thus P is the projective line with equation $z_1 = z_3$, which is the complex tangent line to M_0 at L_1.

The base cycle C_+ in D_+ is just the plane $z_3 = 0$. The Schubert variety is P, and $P \cap C_+ = \{[0 : 1 : 0]\}$. Let L_2 denote this intersection point. Now the Schubert variety is the projective line determined by L_1 and L_2. Since the Schubert slices Σ are the components of the intersection of the Schubert variety at hand with the domain D, and since $P \cap D_+ = D \setminus L_1$, it is immediate that in this case there is only one Schubert slice, $\Sigma = P \setminus L_1$.

The slice of [W12] is determined without computation in this case, because there is a unique K_0-equivariant fibration of D_+ onto C_+. It is the restriction of the projection of $\mathbb{P}(V)$ from the base point L_0. Consequently, this slice is the intersection of D_+ with the projective line determined by L_0 and L_2 which is realized as the unit disk in the complex line.

13.3B Schubert slices in the flag manifold

The group $G_0 = SU(2, 1)$ has three open orbits in the flag manifold Z. We will describe their Schubert slices. For this it is convenient to regard these orbits as fibered over the corresponding orbits in $\mathbb{P}(V)$ and $\mathbb{P}(V^*)$.

The antiholomorphic map $\mathbb{P}(V) \to \mathbb{P}(V^*)$, given by $L \mapsto L^\perp$ on the level of subspaces of V and V^*, composes with complex conjugation to give a G_0-equivariant biholomorphic map $L \mapsto L^\perp$. That map sends orbits to orbits. There is no problem of complex structure, because, here, $\overline{G_0(z_0)} = G_0(\overline{z_0}) = G_0(z_0)$. Thus $D_-^\perp = \{P = L^\perp \mid L \in D_-\}$ can be regarded as the unit ball in $\mathbb{C}^2$,

$M_0^\perp = \{P = L^\perp \mid L \in M_0\}$ is its boundary, and $D_+^\perp = \{P = L^\perp \mid L \in D_+\}$ is the complement.

The G_0-orbits in Z are described in terms of their projections to $\mathbb{P}(V)$ and $\mathbb{P}(V^*)$. The three open orbits are $\widetilde{D}_- = \pi^{-1}(D_-)$, all flags $(L \subsetneq P)$ where the line $L \ll 0$ and the plane P has $\mathrm{sign}(P) = (1,1)$; $\widetilde{D}_+ = (\pi^*)^{-1}(D_-^\perp)$, where $L \gg 0$ and $\mathrm{sign}(P) = (2,0)$; and $\widetilde{D}_0 = \pi^{-1}(D_+) \cap (\pi^*)^{-1}(D_+^\perp)$, where $L \gg 0$ and $\mathrm{sign}(P) = (1,1)$. Here we only discuss $\widetilde{D}_-$ and $\widetilde{D}_0$.

The closed G_0-orbit $\widetilde{M}$ in Z consists of flags $(L \subsetneq P)$, where L is a null line and P is positive semidefinite with nullity 0. Let $L_1 = [1:0:1]$. Then $\pi^{-1}(L_1)$ meets the closed orbit in the single flag $(L_1 \subsetneq P_1)$, where $P_1 = \mathrm{Span}(e_1 + e_3, e_2)$. As in the previous section, our Iwasawa–Borel subgroup $B \subset G$ is the G-stabilizer of the flag $(L_1 \subsetneq P_1)$, which is the unique point in the closed G_0-orbit over L_1 in the manifold of full flags.

We will see in a moment that the cycles in all three open G_0-orbits in Z are one-dimensional. Therefore, from the point of view of Schubert slices, the two-dimensional B-Schubert varieties are the ones of relevance. Since $b_2(Z) = 2$, the open B-orbit in Z must have exactly two components on its boundary. One such is the preimage $\pi^{-1}(P_1)$ of the one-dimensional Schubert variety P_1 discussed above, and the other is the analogous π^* preimage.

Now we describe the cycles and the Schubert slices. The base cycle $\widetilde{C}_-$ in the domain $\widetilde{D}_- = \pi^{-1}(D_-)$ is just the π-fiber over $L_0 = [0:0:1]$. Let P_1^* be the one-dimensional B-Schubert cycle in $\mathbb{P}(V^*)$. It is a complex tangent line to the boundary of the bounded symmetric domain $D_-^\perp$ in $\mathbb{P}(V^*)$. The two-dimensional preimage $S^* = (\pi^*)^{-1}(P_1^*)$ is mapped by π onto $\mathbb{P}(V)$. The restriction $\pi|_{S^*}$ can be regarded as the blowup $S^* \to \mathbb{P}(V)$ of the B-fixed point L_1.

In any case one immediately sees that $\Sigma^* := S^* \cap \widetilde{D}_-$ is an exact slice for the fibration $\pi : \widetilde{D}_- \to D_-$, and that Σ^* is biholomorphic to the unit ball in $\mathbb{C}^2$.

We turn now to the more interesting orbit, $\widetilde{D}_0 = \pi^{-1}(D_+) \cap (\pi^*)^{-1}(D_+^\perp)$, where $L \gg 0$ and $\mathrm{sign}(P) = (1,1)$. Its boundary is the union of the boundaries of the other open orbits. These intersect and are tangent to each other along the closed G_0-orbit $\widetilde{M}$. Thus $\mathrm{bd}(\widetilde{D}_0)$ is singular along $\widetilde{M}$, but is otherwise smooth and pseudoconcave. Here we understand pseudoconcavity in the usual sense that the Levi form, restricted to the complex tangent space, has at least one negative eigenvalue.

Let ρ and ρ^* denote the obvious K_0-invariant strongly pseudoconcave exhaustion functions of D_+ and $D_+^\perp$. Here we understand "strongly pseudoconcave" to mean that the Levi form is negative definite on the complex tangent space at each point of the domain. Define $\widetilde{\rho} = \rho \circ \pi + \rho^* \circ \pi^*$. Note that it is a strongly pseudoconcave exhaustion function for $\widetilde{D}_0$.

The base cycle $\widetilde{C}_0$ in $\widetilde{D}_0$ is the intersection of the preimages of the base cycles in D_+ and $D_+^\perp$. In fact it is just the set of flags $(L \subsetneq P)$, where $P = V^+$ is fixed and L is an arbitrary line in P.

In the case of $\widetilde{D}_0$ the Schubert slices $\Sigma := \widetilde{D}_0 \cap S$ and $\Sigma^* = \widetilde{D}_0 \cap S^*$ have some intrinsic interest. Since the situation is symmetric, we only discuss Σ. Of course $\pi^*|_\Sigma : \Sigma \to D_+^\perp$ is biholomorphic. Thus, as an abstract manifold, Σ is the complement in $\mathbb{P}_2(\mathbb{C})$ of the union of a closed ball and a complex (projective) tangent line to one of its boundary points. In particular, it is not Stein. Let us underline this fact.

Remark. Schubert slices are not necessarily Stein manifolds. $\diamondsuit$

In the present case the boundary of Σ has three pieces: (1) the strongly pseudoconcave smooth boundary which, from the other point of view, is the strongly pseudoconvex boundary of the Schubert slice for the π^*-preimage of the hermitian symmetric space $D_-^\perp$ in $\mathbb{P}(V^*)$, (2) the fibers in Z over the B-fixed points in $\mathbb{P}(V)$, and (3) the fibers in Z over the B-fixed points in $\mathbb{P}(V^*)$. The fibers of (2) and (3) have different character: the one is just the fiber of the $\mathbb{P}_1(\mathbb{C})$-bundle fibration of S induced by π, and the other is blowdown by $\pi^*|_S$.

Here $S = \mathcal{O}\dot{\cup}Y$, where Y is the union of the fibers (2) and (3) mentioned above. Recall that the trace transform associates to any function $f \in \Gamma(S; \mathcal{O}(*Y))$ a function $\mathrm{Tr}(f)$ with poles along the incidence hypersurface $\mathcal{I}$.

Remark. Even in this simple case it would be very interesting to give a precise description of the subspace of $\mathcal{O}(\mathcal{M}_{\tilde{D}_0})$ generated by the images of the $\mathrm{Tr} : \Gamma(S; \mathcal{O}(*Y)) \to \mathcal{O}(\mathcal{M}_{\tilde{D}_0})$ as S ranges over all Iwasawa–Schubert varieties. Note that here $\mathcal{M}_{\tilde{D}_0} \cong D_- \times \overline{D_-}$. $\diamondsuit$

Part III

Analytic and Geometric Consequences

Overview

We now describe some of the analytic and geometric consequences of our results on cycle spaces.

First, using the methods and results of Part II we study a certain natural mapping from the q-cohomology of a homogeneous holomorphic vector bundle on a flag domain D to a space of sections of a canonically derived bundle on $\mathcal{M}_D$. That is our *double fibration transform*. Using structural results on Schubert slices, derived in Part II, we show that this double fibration transform is injective whenever the bundle over D is sufficiently negative.

Secondly, in contexts of representation theory and the theory of moduli of complex varieties we indicate how new viewpoints can be developed and results can be proved by using the principle of transferring problems from D to $\mathcal{M}_D$. For this one can apply the results of this monograph, in particular the availability of a concrete description of $\mathcal{M}_D$ through the identification $\mathcal{M}_D \cong \mathcal{U}$. We exemplify this by indicating connections to areas outside the basic theme of this monograph where cycle spaces of flag domains play a role.

Briefly, we explain certain aspects of representation theory to complex geometers and sketch complex geometric phenomena, in particular moduli theory, for specialists in representation theory. In this spirit, our main goal of this second "consequence" is to invite those who are working in one of these two areas to consider the interesting problems in the other.

The double fibration transform indicated above is a map $\mathcal{P}$ which arises by pulling back cohomology via μ^* to the universal family $\mathfrak{X}$ and then pushing it down to the level of sections by ν_*. It is closely related to the trace transform of Section 7.4.

Without going into detail, there are two points in the proof which should be underlined. The first is the fact that $\mathcal{M}_D$ is contractible. This follows directly from the identification $\mathcal{M}_D \cong \mathcal{U}$. It is perhaps of interest that we know of no other proof of this seemingly harmless fact.

The second point has to do with a method that should be of general use. We refer to this as the *Schubert fibration* $\mathcal{M}_D \to \Sigma$ defined by intersecting cycles with a Schubert slice Σ. It is a holomorphic map that is a real bundle and is equivariant with respect to the Iwasawa component A_0N_0 which defines Σ.

In addition to our work here on the double fibration transform, we indicate how cycle spaces of flag domains arise in moduli theory. This theory arises from the attempt to parameterize all integrable complex structures on a fixed compact orientable differentiable manifold. If one requires a Hausdorff parameter space which itself has complex structure and which has natural universality properties, then, at least as a first step, it is necessary to simultaneously parameterize additional structures. In our case these are called polarizations or markings and, under certain conditions which are discussed here, the resulting spaces of polarized or marked complex manifolds are embedded in flag domains by so-called *period maps*.

In the Griffiths theory these moduli spaces are embedded transversally to the cycles in the flag domain D. In other cases, such as that of K3 surfaces which is discussed here, the moduli space is the full flag domain.

The meaning of cycles in moduli theory is not well understood. For example, due to the results in Part II, we now know that the discrete subgroup Γ of G_0, which identifies complex varieties with the same complex structure but with different polarizations, acts properly and holomorphically on $\mathcal{M}_{\mathcal{D}}$. It would therefore seem appropriate to study, e.g., automorphic forms on the domains $\mathcal{U}$ in connection with moduli problems. Even doing this in the special case of the cycle space of the moduli space of marked K3 surfaces would be of interest.

Let us now give a more detailed outline of the structure of this part. Chapter 14 is devoted to the double fibration transform. The map $\mathcal{P}$ is introduced in a general context in Sections 14.1, 14.2 and 14.3. The main theorem on its injectivity, Theorem 14.3.8, requires the vanishing of a certain topological cohomology of the μ-fiber. In our cycle space context this is a direct consequence of the fact that the Schubert fibration is a topologically trivial bundle (see Theorem 14.5.2). A very brief sketch of the related representation theory is included in the last section. We do this in order to emphasize the possibility of constructing, analyzing, and perhaps even unitarizing, possibly-singular representations by the double fibration transform method.

In Chapter 15 we sketch the basics of the moduli theory in connection with variation of Hodge structure. The resulting period map has its image in a flag domain D. In this case $D = G_0/L_0$, where L_0 is compact, and the fibration $G_0/L_0 \to G_0/K_0$ has cycles as fibers. Here, the theory of Γ-automorphic forms can be discussed on D, because the action there is proper. However, even in the hermitian holomorphic case which arises most naturally in Griffiths' theory, it is appropriate to carry this out on $\mathcal{M}_D$.

The material in Chapter 16 shows how a certain flag domain, an open $SO(3, 19)$-orbit D in a 20-dimensional quadric Z, arises as the space of all marked K3-surfaces. After introductory remarks which indicate the importance of K3-surfaces within Kodaira–Enriques classification theory, we explain three classical methods for constructing families of them; see Section 16.2. Basic results on the period mapping, in particular, the fact that it realizes D as the space of marked K3-surfaces, are explained without proof. Finally in Section 16.4, using the methods of Part II, we study in detail the cycle space $\mathcal{M}_D$.

14

The Double Fibration Transform

We begin by reviewing some generalities on the double fibration transform. In the context of our spaces of q-dimensional cycles in open G_0-orbits D in flag manifolds, this is a mapping from q-cohomology with coefficients in a G_0-homogeneous holomorphic vector bundle $\mathbb{E}$ on D to the space of sections of a naturally derived bundle $\mathbb{E}'$ on the cycle space $\mathcal{M}_D$.

The advantage is that we are basically dealing with holomorphic vector-valued functions on $\mathcal{M}_D$ instead of the original cohomology classes. However, this double fibration transform is only of use if by applying it, we lose little or no information. In other words, the key point is to prove its injectivity.

The basic method for proving this injectivity has been known for some time, but, with the exception of certain hermitian cases, a topological ingredient was missing. The methods of Part II of this monograph now provide this ingredient, and we prove the appropriate injectivity result in the present chapter.

Our work here can be outlined as follows. In Sections 1 through 3 we recall the basics of the double fibration transform and prove the injectivity result (Theorem 14.3.8) which requires the topological ingredient mentioned above. We refer to this as the Buchdahl condition (14.2.2), which in our case requires the vanishing of certain Betti numbers of the fiber of the projection $\mu : \mathfrak{X} \to D$.

The double fibration transform amounts to pulling back cohomology classes to $\mathfrak{X}$ and pushing them down to $\mathcal{M}_D$ by the usual direct image procedure. One obtains sections in the direct image sheaf. In our setting, by analyzing the local G-action, we show that this is in fact a vector bundle (Theorem 14.4.4).

Sections 5 and 6 are devoted to proving the required vanishing of the Betti numbers of the μ-fiber. In fact, it is contractible. The method for proving this (see Theorem 14.5.2) involves the use of an A_0N_0-equivariant holomorphic map $\mathcal{M}_D \to \Sigma$ which is associated to any Schubert slice Σ. Due to their canonical nature, such fibrations may find other applications in the future.

Section 7 sketches some of the representation theory of real reductive Lie groups to which the double fibration transform has interesting potential applications.

14.1 Double fibration

Let D be a complex manifold (later it will be an open orbit of a real reductive group G_0 on a complex flag manifold $Z = G/Q$ of its complexification). We suppose that D fits into a **holomorphic double fibration**. This means that there are complex manifolds $\mathcal{M}$ and $\mathfrak{X}$ with maps

(14.1.1)
$$\begin{array}{ccc} & \mathfrak{X} & \\ {}^{\mu}\swarrow & & \searrow^{\nu} \\ D & & \mathcal{M} \end{array}$$

where μ is a holomorphic submersion and ν is a proper holomorphic map which is a locally trivial bundle. Given a holomorphic vector bundle $\mathbb{E} \to D$ with a certain extension property, we construct a holomorphic vector bundle $\mathbb{E}' \to \mathcal{M}$ and a transform

(14.1.2) $$\mathcal{P} : H^q(D; \mathcal{O}(\mathbb{E})) \to H^0(\mathcal{M}; \mathcal{O}(\mathbb{E}')),$$

under mild conditions on (14.1.1). This construction is fairly standard (see, for example, [BE], [PR1] and [M]), but we need several results specific to the case of flag domains, and those include the extension property to which we alluded above.

14.2 Pullback

The first step is to pull cohomology back from D to $\mathfrak{X}$. As usual, $\mathcal{O}_X \to X$ denotes the structure sheaf of a complex manifold X, and $\mathcal{O}(\mathbb{E}) \to X$ denotes the sheaf of germs of holomorphic sections of a holomorphic vector bundle $\mathbb{E} \to X$. Thus we have the sheaf $\mathcal{O}(\mathbb{E}) \to D$. Let $\mu^{-1}(\mathcal{O}(\mathbb{E})) \to \mathfrak{X}$ denote the inverse image sheaf. For every integer $r \geqq 0$ there is a natural map

(14.2.1) $$\mu^{(r)} : H^r(D; \mathcal{O}(\mathbb{E})) \to H^r(\mathfrak{X}; \mu^{-1}(\mathcal{O}(\mathbb{E})))$$

given on the Čech cocycle level by $\mu^{(r)}(c)(\sigma) = c(\mu(\sigma))$, where $c \in Z^r(D; \mathcal{O}(\mathbb{E}))$ and $\sigma = (w_0, \ldots, w_r)$ is a simplex. For every $q \geqq 0$ we consider the Buchdahl q-condition

(14.2.2)
$$\begin{array}{l} \text{the fiber } F \text{ of } \mu : \mathfrak{X} \to D \text{ is connected} \\ \text{and } H^r(F; \mathbb{C}) = 0 \text{ for } 1 \leqq r \leqq q-1. \end{array}$$

Proposition 14.2.3 (See [Bu]). *Fix $q \geqq 0$. If* (14.2.2) *holds, then* (14.2.1) *is an isomorphism for $r \leqq q-1$ and is injective for $r = q$. If the fibers of μ are cohomologically acyclic, then* (14.2.1) *is an isomorphism for all r.*

Let $\mu^*(\mathcal{O}(\mathbb{E})) := \mu^{-1}(\mathcal{O}(\mathbb{E}))\widehat{\otimes}_{\mu^{-1}(\mathcal{O}_D)}\mathcal{O}_{\mathfrak{X}} \to \mathfrak{X}$ denote the pullback sheaf. It is a coherent analytic sheaf of $\mathcal{O}_{\mathfrak{X}}$-modules. Here $\mu^*(\mathcal{O}(\mathbb{E})) = \mathcal{O}(\mu^*(\mathbb{E}))$ where $\mu^*(\mathbb{E})$ is the pullback bundle. On the sheaf level, $[\sigma] \mapsto [\sigma] \otimes 1$ defines a map $i : \mu^{-1}(\mathcal{O}(\mathbb{E})) \to \mu^*(\mathcal{O}(\mathbb{E}))$. For every $p \geqq 0$, that map specifies maps in cohomology, the coefficient morphisms

$$(14.2.4) \quad i_p : H^p(\mathfrak{X}; \mu^{-1}(\mathcal{O}(\mathbb{E}))) \to H^p(\mathfrak{X}; \mu^*(\mathcal{O}(\mathbb{E}))) = H^p(\mathfrak{X}; \mathcal{O}(\mu^*(\mathbb{E}))).$$

Our natural pullback maps are the compositions $j^{(p)} = i_p \cdot \mu^{(p)}$ of the maps (14.2.1) and (14.2.4):

$$(14.2.5) \qquad j^{(p)} : H^p(D; \mathcal{O}(\mathbb{E})) \to H^p(\mathfrak{X}; \mu^*(\mathcal{O}(\mathbb{E}))) \quad \text{for } p \geqq 0.$$

One can realize these sheaf cohomologies as Dolbeault cohomologies, and then the pullback maps (14.2.5) are given on the level of differential forms by the usual pullback $[\omega] \mapsto [\mu^*(\omega)]$.

14.3 Pushdown

In order to push the $H^q(\mathfrak{X}; \mu^*(\mathcal{O}(\mathbb{E})))$ down to $\mathcal{M}$, we assume that

$$(14.3.1) \qquad \nu : \mathfrak{X} \to \mathcal{M} \text{ is a proper map and } \mathcal{M} \text{ is a Stein manifold.}$$

The Leray direct image sheaves $\mathcal{R}^p(\mu^*(\mathcal{O}(\mathbb{E}))) \to \mathcal{M}$ are coherent [GR1]. As $\mathcal{M}$ is Stein,

$$(14.3.2) \qquad H^q(\mathcal{M}; \mathcal{R}^p(\mathcal{O}(\mathbb{E}))) = 0 \quad \text{for} \quad p \geqq 0 \text{ and } q > 0.$$

Thus the Leray spectral sequence collapses and gives

$$(14.3.3) \qquad H^p(\mathfrak{X}; \mu^*(\mathcal{O}(\mathbb{E}))) \cong H^0(\mathcal{M}; \mathcal{R}^p(\mu^*(\mathcal{O}(\mathbb{E})))).$$

Definition 14.3.4. The **double fibration transform** for the holomorphic double fibration (14.1.1) is the composition

$$(14.3.5) \qquad \mathcal{P} : H^p(D; \mathcal{O}(\mathbb{E})) \to H^0(\mathcal{M}; \mathcal{R}^p(\mu^*(\mathcal{O}(\mathbb{E}))))$$

of the maps (14.2.5) and (14.3.3).

In order for the double fibration transform (14.3.5) to be useful, two conditions should be satisfied. They are

$$(14.3.6) \quad \mathcal{P} : H^p(D; \mathcal{O}(\mathbb{E})) \to H^0(\mathcal{M}; \mathcal{R}^p(\mu^*(\mathcal{O}(\mathbb{E})))) \text{ should be injective,}$$

$$(14.3.7) \qquad \text{and there should be an explicit description of the image of } \mathcal{P}.$$

Assuming (14.3.1), injectivity of $\mathcal{P}$ is equivalent to injectivity of $j^{(p)}$ in (14.2.5). The most general way to approach this is the combination of vanishing and negativity in Theorem 14.3.8 below, based on the Buchdahl conditions (14.2.2).

The general (assuming (14.3.1) injectivity question uses a spectral sequence argument for the relative Dolbeault complex of the holomorphic fibration $\mu : \mathfrak{X} \to D$. See [WZ2] for the details. The end result is

Theorem 14.3.8. *Fix $q \geqq 0$. Suppose that the fiber F of $\mu : \mathfrak{X} \to D$ is connected and satisfies* (14.2.2). *Assume* (14.3.1) *that $\nu : \mathfrak{X} \to \mathcal{M}$ is proper and $\mathcal{M}$ is Stein, say with fiber C. Let $\Omega^r_\mu(\mathbb{E}) \to \mathfrak{X}$ denote the sheaf of relative $\mu^*\mathbb{E}$-valued holomorphic r-forms on $\mathfrak{X}$ with respect to $\mu : \mathfrak{X} \to D$. Suppose that $H^p(C; \Omega^r_\mu(\mathbb{E})|_C) = 0$ for $p < q$, and $r \geqq 1$. Then*

$$\mathcal{P} : H^q(D; \mathcal{O}(\mathbb{E})) \to H^0(\mathcal{M}; \mathcal{R}^q(\mu^*\mathcal{O}(\mathbb{E})))$$

is injective.

Remark 14.3.9. In the cases of interest to us, $\mathcal{P}$ has an explicit formula. We will see that the Leray derived sheaf is given by

(14.3.10)
$$\mathcal{R}^q(\mu^*(\mathcal{O}(\mathbb{E}))) = \mathcal{O}(\mathbb{E}'),$$

where $\mathbb{E}' \to \mathcal{M}$ has fiber $H^q(\nu^{-1}(C); \mathcal{O}(\mu^*(\mathbb{E})|_{\nu^{-1}(C)}))$ at C.

Let ω be an $\mathbb{E}$-valued $(0, q)$-form on D and $[\omega] \in H^q_{\bar\partial}(D, \mathbb{E})$ its Dolbeault class. Then $\mathcal{P}([\omega])$ is the section of $\mathbb{E}' \to \mathcal{M}$ whose value $\mathcal{P}([\omega])(C)$ at $C \in \mathcal{M}$ is $[\mu^*(\omega)|_{\nu^{-1}(C)}]$. In other words,

(14.3.11)
$$\mathcal{P}([\omega])(C) = [\mu^*(\omega)|_{\nu^{-1}(C)}] \in H^0_{\bar\partial}(\mathcal{M}; \mathbb{E}').$$

This is most conveniently interpreted by viewing $\mathcal{P}([\omega])(C)$ as the Dolbeault class of $\omega|_C$, and by viewing $C \mapsto [\omega|_C]$ as a holomorphic section of the holomorphic vector bundle $\mathbb{E}' \to \mathcal{M}$. ◇

14.4 Local G-structure of G_0-bundles

We first show that homogeneous G_0-bundles are locally homogeneous under the complex group G and use that to carry bundles over D to bundles over $\mathcal{M}_D$.

A G_0-homogeneous holomorphic vector bundle $\pi : \mathbb{E} \to D$ is automatically equipped with a local action of G on the total space $\mathbb{E}$ by holomorphic transformations. We now check local G-equivariance at the level of one-parameter subgroups. If $\xi \in \mathfrak{g}$, $\varepsilon > 0$, and U is a relatively compact subset of D such that $U_t := \exp(t\xi)(U) \subset D$ for $|t| < \varepsilon$, then the vector field $\widehat{\xi}_{\mathbb{E}}$ on $\mathbb{E}$ (generated by the local action of G) can be integrated with arbitrary initial values in the open set $\pi^{-1}(U)$, for all times t with $|t| < \varepsilon$. Since G_0 acts here as a group of holomorphic bundle transformations, the Identity Principle says that $g_t := \exp(t\xi)$ also acts by holomorphic bundle transformations on $\pi^{-1}(U)$. Thus the pullback bundle $g_t^*(\mathbb{E}|_{U_t}) \cong \mathbb{E}|_U$ for $|t| < \varepsilon$.

Proposition 14.4.1. *Every $C_1 \in \mathcal{M}_D$ has a neighborhood V in $\mathcal{M}_D$ such that $\mathbb{E}|_{C_1} \cong \mathbb{E}|_C$ for every $C \in V$.*

Proof. We have just seen that we can choose a neighborhood W of 0 in $\mathfrak{g}$ and a neighborhood U of C_1 in D such that if $\xi \in W$, then $\exp(\xi)(U) \subset D$ and $\exp(\xi)^* : \mathbb{E}|_{\exp(\xi)C_1} \cong \mathbb{E}|_{C_1}$. □

Corollary 14.4.2. *If $C_1, C_2 \in \mathcal{M}_D$ then $\mathbb{E}|_{C_1} \cong \mathbb{E}|_{C_2}$.*

Proof. This is immediate from Proposition 14.4.1 and the fact that $\mathcal{M}_D$ is arcwise connected. □

Example 14.4.3. We have made use of the fact that a G_0-homogeneous holomorphic vector bundle is locally a G-homogeneous vector bundle. The following shows that such a bundle is not in general the restriction of a globally defined G-homogeneous holomorphic vector bundle over $Z = G/Q$. Let $G_0 = SL(3;\mathbb{R})$, so that $G = SL(3;\mathbb{C})$, and let $Z = G/B$ where B is a Borel subgroup. The isotropy subgroup of G_0 in the open orbit D is a fundamental Cartan subgroup H_0. At the appropriate point of the open orbit, we have

$$\mathfrak{h}_0 = \text{all } \begin{pmatrix} a & \theta & 0 \\ -\theta & a & 0 \\ 0 & 0 & -2a \end{pmatrix} \text{ for } a, \theta \text{ real, and } H_0 = \text{all } \begin{pmatrix} e^a\cos\theta & \sin\theta & 0 \\ -\sin\theta & e^a\cos\theta & 0 \\ 0 & 0 & e^{-2a} \end{pmatrix}.$$

The one-dimensional representations $\chi_{c,n} : \begin{pmatrix} e^a\cos\theta & \sin\theta & 0 \\ -\sin\theta & e^a\cos\theta & 0 \\ 0 & 0 & e^{-2a} \end{pmatrix} \mapsto e^{in\theta}e^{ca}$, for $c \in \mathbb{C}$ and $n \in \mathbb{Z}$, define G_0-homogeneous holomorphic line bundles $\mathbb{L}_{c,n} \to D$. The bundle $\mathbb{L}_{c,n} \to D$ extends to a G-homogeneous holomorphic line bundle $\widetilde{\mathbb{L}}_{c,n} \to Z$ if and only if $\chi_{c,n}$ extends to a *well-defined* holomorphic character on the complex Cartan subgroup H. However, the extension is well defined if and only if $c \in 2\pi\mathbb{Z}$. ◇

From Example 14.4.3 it is easily seen that every G_0-homogeneous holomorphic vector bundle $\mathbb{E} \to D = G_0(z_0)$ extends to a G-homogeneous holomorphic vector bundle $\widetilde{\mathbb{E}} \to Z$ if and only if $G_0 \cap Q_{z_0}$ has compact center—which, of course, is automatic if G_0 has a compact Cartan subgroup, i.e., if rank $\mathfrak{k}$ = rank $\mathfrak{g}$. In [HW4] we incorrectly asserted that the extension always exists, as a consequence of [TW, Theorem 3.6]. However, the only use of the extension in [HW4] was to derive a proof of Corollary 14.4.2, for use as in Theorem 14.4.4, which we now prove by rather elementary considerations.

Theorem 14.4.4. *Let D be an open G_0-orbit on Z, let $\mathbb{E} \to D$ be a G_0-homogeneous holomorphic vector bundle, and let $q \geqq 0$. Then the Leray derived sheaf for ν is given by $\mathcal{R}^q(\mathcal{O}(\mu^*\mathbb{E})) = \mathcal{O}(\mathbb{E}')$, where $\mathbb{E}' \to \mathcal{M}_D$ is the G_0-homogeneous, holomorphic vector bundle with fiber $H^q(C; \mathcal{O}(\mathbb{E}|_C))$ over $C \in \mathcal{M}_D$ given by Corollary* 14.4.2.

Proof. The Leray construction gives sheaves $\mathcal{R}^q(\mathcal{O}(\mu^*\mathbb{E})) \to \mathcal{M}_D$ and Corollary 14.4.2 shows that, in our case, $\mathcal{R}^q(\mathcal{O}(\mu^*\mathbb{E})) = \mathcal{O}(\mathbb{E}')$, where $\mathbb{E}' \to \mathcal{M}_D$ is the bundle described above. □

14.5 The Schubert fibration

Let Σ be a Schubert slice defined by an Iwasawa–Borel subgroup B, and suppose that $\{z_0\} = \Sigma \cap C_0$ consists of our base point z_0. In the context of the double fibrations

(14.5.1)

$$\begin{array}{ccccc} & & \mathfrak{X}_D & & \\ & \swarrow^{\mu} & & \searrow^{\nu} & \\ D & & & & \mathcal{M}_D \end{array} \qquad \text{and} \qquad \begin{array}{ccccc} & & \mathfrak{X}_Z & & \\ & \swarrow^{\widetilde{\mu}} & & \searrow^{\widetilde{\nu}} & \\ Z & & & & \mathcal{M}_Z \end{array}$$

the projection ν carries the fiber $F = \mu^{-1}(z_0) = \{(z_0, C) : z_0 \in C\}$ biholomorphically onto the analytic subset $\{C \in \mathcal{M}_D : z_0 \in C\}$ of $\mathcal{M}_D$.

Now consider an arbitrary element $C \in \mathcal{M}_Z$ with $z_0 \in C$. By definition $C = g(C_0)$ for some $g \in G$. Since $z_0 \in C$, by adjusting g by an appropriate element of K_0 we may assume that $g \in Q = G_{z_0}$. Extend F to $\widetilde{F} = \widetilde{\mu}^{-1}(z_0)$. Now $\widetilde{F}$ is the Q-orbit of C_0 in $\mathcal{M}_Z$. Also, $\widetilde{F}$ is closed in $\mathcal{M}_Z$, for if a sequence $\{C_i\}$ in $\widetilde{F}$ converges to $C \in \mathcal{M}_Z$, then $z_0 \in C$, because $z_0 \in C_i$ for each i. Hence $C \in \widetilde{F}$.

In the double fibration (14.5.1), the Q-orbit of C_0 is identified with the fiber $\widetilde{F} = \widetilde{\mu}^{-1}(z_0)$. In particular, its open subset $F = \widetilde{F} \cap \mathcal{M}_D$ is a closed complex submanifold on $\mathcal{M}_D$.

Propositions 7.3.7 and 7.3.9 say that if Σ is a Schubert slice, then for every $C \in \mathcal{M}_D$ the intersection $C \cap \Sigma$ is transversal and consists exactly of one point. Thus we have an A_0N_0-equivariant map $\varphi : \mathcal{M}_D \to \Sigma$ defined by mapping C to its point of intersection with Σ. The fiber over $z_0 \in \Sigma$ is of course F.

Let $J_0 := (A_0N_0)_{z_0}$ be the A_0N_0-isotropy at the base point and note that

$$A_0N_0 \times_{J_0} F \to \mathcal{M}_D, \quad \text{defined by } [(a_0n_0, C)] \mapsto a_0n_0(C),$$

is well defined, smooth, and bijective. Thus $\varphi : \mathcal{M}_D \to \Sigma$ is naturally identified with the smooth A_0N_0-equivariant bundle

$$\pi_\Sigma : A_0N_0 \times_{J_0} F \to A_0N_0/J_0 = \Sigma.$$

In this sense, every Schubert slice defines a **Schubert fibration** of the cycle space $\mathcal{M}_D$.

Theorem 14.5.2. *Let B be an Iwasawa–Borel subgroup of G and Σ an associated Schubert slice for the open orbit D. Then the fibration $\pi_\Sigma : \mathcal{M}_D \to \Sigma$ is a holomorphic map onto a contractible base Σ and diffeomorphically realizes $\mathcal{M}_D$ as the product $\Sigma \times F$.*

Proof. Let $J := B_{z_0}$. The inclusions $A_0N_0 \hookrightarrow B$ and $F \hookrightarrow \widetilde{F}$ together define a map $A_0N_0 \times_{J_0} F \hookrightarrow B \times_J F$. That map realizes $A_0N_0 \times_{J_0} F \cong \mathcal{M}_D$ as an open subset of $B \times_J \widetilde{F}$. The latter is fibered over the open A_0N_0-orbit Σ in $\mathcal{O} = B(z_0)$ by the natural holomorphic projection $\pi : B \times_J \widetilde{F} \to B/J$. Since π_Σ is the restriction $\pi|_{\mathcal{M}_D}$, it follows that π_Σ is holomorphic as well.

The fact that Σ is a cell follows from the simple connectivity of the solvable group A_0N_0 and the fact that it is acting algebraically. □

Recall that $\Omega^r_\mu(\mathbb{E}) \to \mathfrak{X}_D$ denotes the sheaf of relative $\mu^*\mathbb{E}$-valued holomorphic r-forms on $\mathfrak{X}$ with respect to $\mu : \mathfrak{X}_D \to D$.

Corollary 14.5.3. *Suppose that* $\mathbb{E} \to D$ *is a holomorphic* G_0*-homogeneous vector bundle which is sufficiently negative so that* $H^p(C; \Omega^r_\mu(\mathbb{E})|_C) = 0$ *for* $p < q$, *and* $r \geqq 1$. *Then the double fibration transform*

$$\mathcal{P} : H^q(D; \mathcal{O}(\mathbb{E})) \to H^0(\mathcal{M}_D; \mathcal{O}(\mathbb{E}'))$$

is injective.

Proof. $\mathcal{M}_D$ is contractible by Proposition 11.3.5 and Theorem 11.3.7. Since Σ is likewise contractible and $\mathcal{M}_D$ is diffeomorphic to $\Sigma \times F$, it follows that F is cohomologically trivial. The Buchdahl conditions (14.2.2) follow. Proposition 14.2.3 now says that (14.2.1) is an isomorphism for all r. Composing with coefficient morphisms, the maps (14.2.4) also are isomorphisms. By Theorem 3.2.1 and Proposition 5.4.7, if $\mathcal{M}_D$ is not a point of $\mathcal{B}$ or $\overline{\mathcal{B}}$, then $\mathcal{M}_D = \mathcal{U}$. Thus we know that the conditions (14.3.1) are satisfied. The assertion now follows from Theorem 14.3.8. □

14.6 Contractibility of the fiber

With a bit more work one can see that the fiber F of $\mathcal{M}_D \to \Sigma$ is contractible, not just cohomologically trivial. We thank Peter Michor for showing us the following result for the C^∞–category, from which contractibility of F is immediate. His argument is based on the existence of complete Ehresmann connections for smooth fiber bundles.

Proposition 14.6.1. *Let* $p : M \to S$ *be a smooth fiber bundle with fiber* $F = p^{-1}(s_0)$. *If both* M *and* S *are contractible, then* F *is contractible.*

Proof. Since S is contractible and smooth, an approximation gives us a smooth contraction $h : [0, 1] \times S \to S$; here $h(0, s) = s$ and $h(1, s) = s_0$. Following [KMS, §9.9], the bundle $p : M \to S$ has a complete Ehresmann connection. Completeness means that every smooth curve in S has horizontal lifts to M. If $m \in M$, let $t \mapsto H(t, m)$ denote the horizontal lift of $t \mapsto h(t, p(m))$ such that $H(0, m) = m$. Note $H(1, m) \in F$. Fix a base point $m_0 \in M$ and a smooth contraction $I : [0, 1] \times M \to M$ of M to m_0; if $m \in M$, then $I(0, m) = m$ and $I(1, m) = m_0$. Denote $f_0 = H(1, m_0) \in F$. Define $J : [0, 1] \times F$ by $J(t, f) = H(t, I(t, f))$. Therefore $J(0, f) = f$ and $J(1, f) = f_0$. Thus J is a contraction of F to f_0, and consequently F is contractible. □

Now we can refine our requirements (14.3.6) and (14.3.7). Since Corollary 14.5.3 yields (14.3.6), this is a matter of refining (14.3.7). For this note that, since $\mathcal{M}_D$ is a contractible Stein manifold, the bundle $\mathbb{E}' \to \mathcal{M}_D$ is holomorphically trivial. Thus (14.3.7) is sharpened to the following two-part problem:

(14.6.2) find a canonical method of holomorphic trivialization for the $\mathbb{E}' \to \mathcal{M}_D$

and, in that canonical trivialization,

(14.6.3) find a canonical system of PDE to describe the image of $\mathcal{P}$.

Most of the material of this chapter is taken from [WZ2] and [HW4].

In the hermitian holomorphic case $\mathcal{M}_D$ is $\mathcal{B}$ or $\overline{\mathcal{B}}$ (Proposition 9.1.7), and one knew from [WZ2, Section 4] that F and $\mathcal{M}_D$ are contractible Stein manifolds. Thus the Buchdahl conditions were immediate.

In the hermitian nonholomorphic case, where $\mathcal{M}_D = \mathcal{B} \times \overline{\mathcal{B}}$ (see [HW3] or [WZ3], or see Proposition 9.1.8), there had only been partial information (see [WZ2, Theorem 6.6]) on the Buchdahl conditions, and that information had been based on contractibility of F in the cases where G_0 is a classical group of hermitian type. Specifically, in the cases where G_0 is a classical group of hermitian type, Wolf had verified by explicit calculation that F is obtained recursively starting with a bounded symmetric domain and building locally trivial bundles at each step (over the space of the previous step) whose fibers are bounded symmetric domains. Thus F was known to be an iterated fibration of bounded symmetric domains, in particular contractible, in those special cases.

Those hermitian cases were the only cases where the topology of F was known. In particular nothing was known in the non-hermitian cases. Now this matter is settled by Theorem 11.3.7.

One small remark. In some cases one knows that $H^q(D; \mathcal{O}(\mathbb{E}))$ is an irreducible representation space for a group under which all our constructions are equivariant, and one sees directly that $\mathcal{P}$ is an intertwining operator, thus zero or injective. In practice, however, we usually look for implications in the other directions.

14.7 Unitary representations of real reductive Lie groups

In this section we briefly indicate the role of the double fibration transform in the representation theory of real reductive Lie groups.

Harish-Chandra's analysis of the holomorphic discrete series can be viewed from the perspective of the double fibration transform as follows. Let G_0 be of hermitian type, $\mathcal{B} = G_0/K_0$. In this case, where $D = \mathcal{B} = \mathfrak{X}_D = \mathcal{M}_D$,

- the double fibration transform is the identity, (14.3.6) is settled,
- the canonical holomorphic trivializations (14.6.2) of the bundles over $D = \mathcal{M}_D$ are given by the universal factor of automorphy, and
- the system (14.6.3) of PDE defining the image of $\mathcal{P}$ consists of the $\overline{\partial}$ (here the Cauchy–Riemann) operator.

Let $\mathbb{E}_\lambda \to \mathcal{B}$ denote the homogeneous holomorphic hermitian vector bundle associated to the representation E_λ of K_0 of highest weight λ. By use of his system of strongly orthogonal noncompact positive roots, and the explicit holomorphic trivialization of $\mathbb{E}_\lambda \to \mathcal{B}$, Harish-Chandra proved (i) a holomorphic section of $\mathbb{E}_\lambda \to \mathcal{B}$ is $L^2(\mathcal{B})$ if and only if its K_0-isotypic components are $L^2(\mathcal{B})$, (ii) if some nonzero K_0-isotypic holomorphic section of $\mathbb{E}_\lambda \to \mathcal{B}$ is $L^2(\mathcal{B})$, then the constant section f_λ, value equal to the highest weight vector of E_λ, is $L^2(\mathcal{B})$, and (iii) f_λ is $L^2(\mathcal{B})$ if and only if $\langle \lambda + \rho, \beta \rangle < 0$, where ρ is half the sum of the positive roots and β is the maximal root.

Narasimhan and Okamoto [NO] extended the Harish-Chandra construction to "almost all" discrete series representations of a real group G_0 of hermitian type, again always working over $D = \mathcal{B} = \mathfrak{X}_D = \mathcal{M}_D$, where the double fibration transform is more or less invisible.

The double fibration transform first became visible, at least in degenerate form, in Schmid's holomorphic construction of the discrete series [S3], [S5]. There $Z = G/B$ for some Borel subgroup B and $D = G_0/T_0$ where T_0 is a compact Cartan subgroup, $T_0 \subset K_0 \subset G_0$. Only the "real form" $\phi : D \to G_0/K_0$ of the double fibration appears. The real symmetric space G_0/K_0 appears instead of the cycle $\mathcal{M}_D$ and correspondingly D appears instead of the incidence space $\mathfrak{X}_D$. Injectivity of this real double fibration transform $\mathcal{P}_{\mathbb{R}} : H^q(D; \mathbb{E}) \to H^0(G_0/K_0; \mathbb{E}')$ is given by Schmid's "Identity Theorem." This theorem says that, under appropriate restrictions, a Dolbeault class $[\omega] \in H^q(D; \mathbb{E})$ is zero if and only if the restriction $\omega|_C$ is cohomologous to zero for every fiber C of $\phi : D \to G_0/K_0$. This was extended a bit by Wolf [W3], for flag domains of the form $D \cong G_0/L_0$ with G_0 general reductive and L_0 compactly embedded in G_0.

The double fibration transform first appeared in modern form in the paper [WeW] of Wells and Wolf on Poincaré series and automorphic cohomology. The restriction there was that $D \cong G_0/L_0$ with L_0 compact, and a small extension of the Identity Theorem was used to, in effect, prove injectivity of the double fibration transform.

The Penrose transform applies to the case $D = SU(2,2)/S(U(1) \times U(1,2))$. There L_0 is noncompact, and perhaps that is the first case of noncompact L_0 to be studied carefully (see [BE]). Background work on interesting flag domains with noncompact isotropy includes parts of Berger's classification [Be] of semisimple symmetric spaces, Wolf's study [W0] of isotropic pseudo-riemannian manifolds, and [W2]. Important cases of construction of unitary representations using double fibration transforms on flag domains with noncompact isotropy were studied in Dunne–Zierau [DZ] and Patton–Rossi [PR2]. This area was first studied systematically in Wolf–Zierau [WZ2].

Finally, as noted in [W7], there are indications of a strong relation between the double fibration transforms of [WZ2] and the construction of unitary representations by indefinite harmonic theory of Rawnsley, Schmid, and Wolf [RSW].

15

Variation of Hodge Structure

In this chapter we indicate the theory of moduli spaces in the sense of Griffiths for Hodge manifolds X, in other words, what is called **variation of Hodge structure**. For a comprehensive treatment of the bases of this theory, see the original treatment in [Gr3] and [Gr4], a more up to date exposition in [V1] and [V2], or a summary of recent developments in [Gr6]. There also are expositions contained in [Gr5], [S3], [We1], and [We2]. We describe its tight connection with period matrix domains which in fact are very interesting flag domains. The first Hodge–Riemann bilinear relation specifies a complex flag manifold $Z = G/Q$ and the second Hodge–Riemann bilinear relation specifies an open G_0-orbit $D \subset Z$. The connection carries over to cycle spaces and double fibration transforms. Along the way we will sketch some relevant aspects automorphic cohomology theory.

There are two moduli space theories, the more traditional one based on the natural bigrading $H^r(X; \mathbb{C}) = \sum_{p+q=r} H^{p,q}(X; \mathbb{C})$ of the full Dolbeault cohomologies and the one we consider in this chapter based on the subspaces $H_0^r(X; \mathbb{C}) = \sum_{p+q=r} H_0^{p,q}(X; \mathbb{C})$ consisting of primitive cohomology classes in the sense of Lefschetz. Each has its own special features, each leads to an interesting automorphic cohomology theory, and each has an elegant formulation based on the appropriate period domain D and cycle space $\mathcal{M}_D$. In this chapter we concentrate on the moduli space theory based on primitive classes. In the next chapter we will study K3 surfaces and work out their moduli spaces from the more traditional viewpoint that uses the full Dolbeault cohomology.

Let X denote a compact Kähler manifold and let ω denote the Kähler form. A harmonic differential form ξ on X is called **primitive** if the interior product $\xi \lrcorner \omega = 0$, intuitively if ω is not an exterior product factor of ξ. A Dolbeault cohomology class on X is called **primitive** if it has a primitive harmonic representative. Alternatively, the Lefschetz operator $L : H^r(X; \mathbb{C}) \to H^{r+2}(X; \mathbb{C})$, given on the Dolbeault level by $L([\xi]) = [\omega \wedge \xi]$, has the property that L^{n-r} is injective if $0 \leqq r \leqq n$, where $n = \dim_{\mathbb{C}} X$. The primitive r-cohomology is given by $H_0^r(X; \mathbb{C}) = 0$ for $r > \dim_{\mathbb{C}} X$ and $H_0^r(X; \mathbb{C}) = \{[\xi] \in H^r(X; \mathbb{C}) \mid L^{n-r+1}[\xi] = 0\}$ for $r \leqq \dim_{\mathbb{C}} X$. Since ω has bidegree $(1, 1)$, we always have $H_0^r(X; \mathbb{C}) = \sum_{p+q=r} H_0^{p,q}(X; \mathbb{C})$. The connection between these two formulations of primitive cohomology is the **Lefschetz Theorem**

$$H^{p,q}(X;\mathbb{C}) = \bigoplus L^s H_0^{p-s,q-s}(X;\mathbb{C}),$$

where the sum is taken over those nonnegative s with $p+q-2s$ nonnegative.

Since ω is real, L sends $H^r(X;\mathbb{R})$ to $H^{r+2}(X;\mathbb{R})$. Thus we have the real primitive cohomology, defined as above by $\xi \lrcorner \omega = 0$ on the harmonic representative, given by $H_0^r(X;\mathbb{R}) = 0$ for $r > n$ and

$$H_0^r(X;\mathbb{R}) = \sum_{p+q=r} \left(H_0^{p,q}(X;\mathbb{C}) + H_0^{q,p}(X;\mathbb{C})\right) \cap H^r(X;\mathbb{R}).$$

The Lefschetz Theorem becomes

$$H_0^r(X;\mathbb{R}) = \bigoplus L^s \left(\left[(H_0^{p-s,q-s}(X;\mathbb{C}) + H_0^{q-s,p-s}(X;\mathbb{C})\right] \cap H^{r-2s}(X;\mathbb{R})\right),$$

where here the sum is taken over nonnegative p, q and s with $p+q=r$ and $r-2s$ nonnegative.

The decomposition $H_0^r(X;\mathbb{C}) = \sum_{p+q=r} H_0^{p,q}(X;\mathbb{C})$ by bidegree specifies the **Hodge filtration** $(F^0 \subset F^1 \subset \cdots \subset F^r)$ of $H_0^r(X;\mathbb{C})$, where $F^s = \sum_{i<s} H_0^{r-i,i}(X;\mathbb{C})$. This defines the complex flag $\mathcal{F}(X) = (F^0 \subset F^1 \subset \cdots \subset F^u)$, where $u = [\frac{r-1}{2}]$, the integer part of $(r-1)/2$. The primitive cohomology $H_0^r(X;\mathbb{C})$ carries a nondegenerate bilinear form b given (on Dolbeault representative differential forms) by

$$b(\xi,\eta) = (-1)^{r(r+1)/2} \int \omega^{n-r} \wedge \xi \wedge \eta.$$

Here ω is the Kähler form of X. Evidently $b(H_0^{p,q}(X;\mathbb{C}), H_0^{p',q'}(X;\mathbb{C})) = 0$ unless $p+p' = r = q+q'$. Define

$$c(\xi) = (\sqrt{-1})^{p-q}\xi \text{ for } \xi \in H_0^{q,p}(X;\mathbb{C}).$$

Then the **Hodge–Riemann bilinear relations** are the following conditions on the bilinear form b and an associated hermitian form w.

(15.1.1) b pairs $H_0^{p,q}(X;\mathbb{C})$ with its complex conjugate $H_0^{q,p}(X;\mathbb{C})$

and

(15.1.2) $w(\xi,\eta) := b(c(\xi),\overline{\eta})$ is positive definite on $H_0^r(X;\mathbb{C})$.

When $n = r = 1$, these are the classical period matrix conditions for Riemann surfaces.

The bilinear form b is traditionally denoted by Q, but we have reserved Q for a parabolic subgroup of a complex reductive group G. Also, the bilinear form is sometimes written without the $(-1)^{r(r+1)/2}$ factor, in which case that factor is inserted into the hermitian form w so that w is positive definite on $H_0^r(X;\mathbb{C})$ as in (15.1.2).

If $r = 2t$ even, then b is symmetric. For $i < t$ it is positive definite on the real form

$$(H_0^{r-i,i}(X;\mathbb{C}) \oplus H_0^{i,r-i}(X;\mathbb{C})) \cap H^r(X;\mathbb{R})$$

of $H_0^{r-i,i}(X;\mathbb{C}) \oplus H_0^{i,r-i}(X;\mathbb{C})$. It is negative definite on the real form $H_0^{t,t}(X;\mathbb{R})$ of $H_0^{t,t}(X;\mathbb{C})$. Thus the identity component of the isometry group of $(H_0^r(X;\mathbb{C}), b)$ is the complex special orthogonal group $G = SO(2h+k;\mathbb{C})$, where $k = \dim H_0^{t,t}(X;\mathbb{C})$ and $h = \sum_{i<t} h_0^{r-i,i}$ with $h_0^{p,q} = \dim H_0^{p,q}(X;\mathbb{C})$, and the identity component of the isometry group of $(H_0^r(X;\mathbb{R}), b)$ is its real form $G_0 = SO^0(2h,k)$. (Here note that the real special orthogonal group $SO(2h,k)$ has two components unless $hk = 0$.)

The dimension sequence $d_i = h_0^{r,0} + \cdots + h_0^{r-i,i}$ of the flag $\mathcal{F}(X)$ specifies the complex flag manifold

(15.1.3) $\quad Z = G/Q$: b-isotropic flags $(E^0 \subset E^1 \subset \cdots \subset E^{t-1})$ in $H_0^r(X;\mathbb{C})$.

The second Hodge–Riemann bilinear relation (15.1.2) shows that the isotropy subgroup L_0 of G_0 at $\mathcal{F}(X)$ is compact. More precisely, it gives us

(15.1.4) $$L_0 = (U(h_0) \times \cdots \times U(h_{t-1}) \times SO(k),$$

where $U(h_i)$ is the unitary group of $H_0^{r-i,i}(X;\mathbb{C})$ and $SO(k)$ is the orthogonal group of $H_0^{t,t}(X;\mathbb{R})$. The Hodge–Riemann bilinear relations (15.1.1) and (15.1.2) say that the flag $\mathcal{F}(X,\omega)$ belongs to open G_0-orbit

(15.1.5) $$D = \{\mathcal{E} \mid b \gg 0 \text{ on } (E^{t-1} \oplus \overline{E^{t-1}}) \cap H_0^r(X;\mathbb{R})\} \cong SO^0(2h,k)/L_0$$

in the flag manifold $Z = G/Q$ of (15.1.3).

If r is odd, say $r = 2t-1$, then b is antisymmetric. Hence $H_0^r(X;\mathbb{C})$ has even dimension $2m$ and the isometry group of $(H_0^r(X;\mathbb{C}), b)$ is the complex symplectic group $G = \mathrm{Sp}(m;\mathbb{C})$. The dimension sequence of the flag $\mathcal{F}(X)$ specifies the complex flag manifold $Z = G/Q$ consisting of all the flags $\mathcal{E} = (E^0 \subset E^1 \subset \cdots \subset E^t)$ in $H_0^r(X;\mathbb{C})$ with $b(E^t, E^t) = 0$. The isometry group of $(H_0^r(X;\mathbb{R}), b)$ is the real symplectic group $G_0 = \mathrm{Sp}(m;\mathbb{R})$. As above, G_0 has compact isotropy subgroup L_0 at $\mathcal{F}(X)$, necessarily of the form $U(h_0)\times\cdots\times U(h_t)$. The flag $\mathcal{F}(X)$ is an element of the open G_0-orbit $D = \{\mathcal{E} \mid b$ nondegenerate on each $(H_0^{r-i,i}(X;\mathbb{C}) + H_0^{i,r-i}(X;\mathbb{C}))\}$, which is realized as $\mathrm{Sp}(m;\mathbb{R})/(U(h_0)\times\cdots\times U(h_t))$, where $h_i = \dim H_0^{r-i,i}(X;\mathbb{C})$ as before.

Both for r even and for r odd, G_0 has compact isotropy subgroup L_0 on D, and as a result we have $L_0 \subset K_0$. Thus the holomorphic double fibration (14.5.1) simplifies quite a lot; for example, it is implemented by the projection $D = G_0/L_0 \to G_0/K_0 \subset \mathcal{M}_D$ onto a real form of $\mathcal{M}_D$.

The above discussion applies to every compact Kähler manifold. Now we are going to consider families of compact Kähler manifolds with fixed underlying real C^∞–manifold, but in which the complex structure and the Kähler form vary. Thus we will refine our notation and denote a compact Kähler manifold in the style $\underline{X} = (X, J, \omega)$, where X is the underlying C^∞ manifold, J is the almost complex structure, and ω is the Kähler form. Similarly, we will write $\mathcal{F}(\underline{X}) = \mathcal{F}(X, J, \omega)$ for the flag defined by the Hodge filtration. As $\underline{X}$ varies, however, we will need to ensure that these flags all live in the same flag manifold, and for this we introduce the notion of marked Kähler manifold.

Fix a finite-dimensional real vector space $V_{\mathbb{R}}$, a lattice $V_{\mathbb{Z}}$ in $V_{\mathbb{R}}$, an integral bilinear form $b_{\mathbb{Z}}$ on $V_{\mathbb{Z}}$, and an integer $r > 0$. We write $V_{\mathbb{C}}$ for the complexification of $V_{\mathbb{R}}$ and write $b_{\mathbb{R}}$ and $b_{\mathbb{C}}$ for the bilinear extensions of $b_{\mathbb{Z}}$ to $V_{\mathbb{R}}$ and $V_{\mathbb{C}}$. We suppose that these choices are made in such a way that there exists an isometry

$$(15.1.6) \qquad \varphi : \big(H_0^r(X; \mathbb{C}), b\big) \cong (V_{\mathbb{C}}, \ b_{\mathbb{C}})$$

for some compact Kähler manifold $\underline{X}$, where b is the intersection form discussed above. We refer to φ as a **mark** on $\underline{X}$, and refer to $(\underline{X}, \varphi)$ as a **marked** compact Kähler manifold. This terminology is standard for Riemann surfaces and K3 surfaces and is used (often implicitly) in most treatments of variation of Hodge structure.

Let $(\underline{X}, \varphi)$ be a marked compact Kähler manifold. For purposes of the discussion we suppose that $r = 2t$ even and retain the corresponding notation from the discussion above. But everything carries over with obvious modification for the cases where $r = 2t - 1$ odd. The isometry (15.1.6) carries the flag manifold $Z = G/Q$ of (15.1.3) to the flag manifold (which we also denote $Z = G/Q$) of all $b_{\mathbb{C}}$-isotropic flags $\mathcal{E} = (E^0 \subset E^1 \subset \cdots \subset E^{t-1})$ in $V_{\mathbb{C}}$ for the dimension sequence $d_i = h_0^{r,0} + \cdots + h_0^{r-i,i}$. It carries the open G_0-orbit D of (15.1.5) to the corresponding open G_0-orbit in the flag Z based on $(V_{\mathbb{C}}, b_{\mathbb{C}})$. That corresponding open G_0-orbit, which we also denote D, is specified by the condition that $b_{\mathbb{C}} \gg 0$ on $(E^{t-1} \oplus \overline{E^{t-1}}) \cap V_{\mathbb{R}}$.

Now consider a family $\{(\underline{X}_a, \varphi_a) \mid a \in A\}$ of marked compact Kähler manifolds, where the $\underline{X}_a$ all have the same Hodge numbers $h_0^{p,q} = \dim H_0^{p,q}(\underline{X}_a; \mathbb{C})$. Then we have a well-defined **period map**

$$(15.1.7) \qquad \varphi : A \to D \quad \text{defined by} \quad \varphi(a) = \varphi_a(\mathcal{F}(\underline{X}_a)).$$

Here we have not imposed any conditions on the way that $\underline{X}_a$ varies with $a \in A$, nor have we imposed any structure on the set A. Instead we have "related" the $\underline{X}_a$ by means of their markings.

Now we look at the geometric picture, where $\{(\underline{X}_a, \varphi_a) \mid a \in A\}$ is a deformation of some $\underline{X}_{a_0}$. For that we suppose that we have a locally trivial C^∞ fiber space $\pi : \mathcal{X} \to A$ whose base and total space are complex manifolds, whose fibers are compact complex submanifolds of the total space, and whose projection is a holomorphic map. Let X denote the underlying C^∞ manifold of the typical fiber, and denote the actual fibers, as complex manifolds, by (X, J_a) where $a \in A$. It is automatic (but certainly not trivial) that the Hodge numbers $h^{p,q} = \dim H^{p,q}((X, J_a); \mathbb{C})$ are locally constant in a.

We also suppose that $\mathcal{X}$ carries a 2-form ω whose restriction $\omega_a = \omega|_{X_a}$ is a Kähler form on (X, J_a), so $\underline{X}_a := (X, J_a, \omega_a)$ is a compact Kähler manifold.

Suppose that the parameter space A is contractible. Then restriction of cohomology classes gives isomorphisms $res_a : H^r(\mathcal{X}; \mathbb{C}) \cong H^r(X_a; \mathbb{C})$. Choose $a_0 \in A$ and define

$$\psi_a = res_{a_0} \circ res_a^{-1} : H^r(X_a; \mathbb{C}) \to H^r(X_{a_0}; \mathbb{C}).$$

By definition of the Kähler forms ω_a, ψ_a sends $[\omega_a]$ to $[\omega_{a_0}]$, and consequently it commutes with exterior multiplication by the Kähler forms. Thus it restricts to an isomorphism

$$\varphi_a : H_0^r(\underline{X}_a; \mathbb{C}) \cong H_0^r(\underline{X}_{a_0}; \mathbb{C})$$

of the primitive cohomologies.

As we indicated above, the Hodge numbers $h^{p,q} = \dim H^{p,q}(\underline{X}_a; \mathbb{C})$ are independent of $a \in A$; see [V1, §9.3.2]. Given that fact, $\varphi_a : H_0^{p,q}(\underline{X}_a; \mathbb{C}) \cong H_0^{p,q}(\underline{X}_{a_0}; \mathbb{C})$. Thus the Hodge numbers $h_0^{p,q} = \dim H_0^{p,q}(\underline{X}_a; \mathbb{C})$ are independent of $a \in A$.

We interpret the map $\varphi_a : H_0^r(\underline{X}_a; \mathbb{C}) \cong H_0^r(\underline{X}_{a_0}; \mathbb{C})$ as a mark on $\underline{X}_a$ where $V_\mathbb{R} := H_0^r(\underline{X}_{a_0}; \mathbb{R})$ and $b_\mathbb{R} := b_{a_0}|_{V_\mathbb{R}}$, $V_\mathbb{Z} := H_0^r(\underline{X}_{a_0}; \mathbb{Z})$ and $b_\mathbb{Z} := b_{a_0}|_{V_\mathbb{Z}}$, and $V_\mathbb{C} := H_0^r(\underline{X}_{a_0}; \mathbb{C})$ and $b_\mathbb{C} := b_{a_0}$. Then each $(\underline{X}_a, \varphi_a)$ is a marked compact Kähler manifold, and the period map (15.1.7) is a map of complex manifolds. A famous theorem of Griffiths [Gr3], [Gr4] says that it is holomorphic.

The integral homology $H_r(X; \mathbb{Z})$ maps naturally into $H_r(X; \mathbb{R})$ or $H_r(X; \mathbb{C})$ with kernel that is its torsion subgroup. Thus we view $H_r(X; \mathbb{Z})/(\text{torsion})$ as sitting inside $H_r(X; \mathbb{C})$. Similarly, the coefficient maps $\mathbb{Z} \to \mathbb{R}$ and $\mathbb{Z} \to \mathbb{C}$ define maps $H^r(X; \mathbb{Z}) \to H^r(X; \mathbb{R})$ and $H^r(X; \mathbb{Z}) \to H^r(X; \mathbb{C})$ whose kernels are the torsion subgroup of $H^r(X; \mathbb{Z})$. The classes in the images of those maps are called **integral**. By abuse of notation we write the lattice of integral elements as $H^r(X; \mathbb{Z})$.

Consider the case where the $\underline{X}_a = (X, J_a, \omega_a)$ are Hodge manifolds, in other words, where the $[\omega_a]$ are integral. Equivalent formulations: (X_a, J_a) is a projective algebraic variety in such a way that some positive integral multiple of $[\omega_a]$ is the pullback of the Chern class of the hyperplane bundle; $\underline{X}_a$ is a polarized Kähler manifold; $[\omega_a]$ is the Chern class of a positive line bundle on (X, J_a). In that case the Lefschetz operator $L_a : [\xi] \mapsto [\omega_a \wedge \xi]$ sends integral classes to integral classes. Thus one has a well-defined notion of primitive integral cohomology $H_0^r(\underline{X}_t; \mathbb{Z})$, and the period map (15.1.7), along with most of its ingredients, have interesting additional structure.

Choose a $\mathbb{Z}$-basis $\{\gamma_1, \dots, \gamma_v\}$ of $H_r(X; \mathbb{Z})/(\text{torsion})$. We view the γ_i as real r-cycles on X, suitable for integration of r-forms.

Fix a family $\{(\underline{X}_a, \varphi_a) \mid a \in A\}$ of marked compact Kähler manifolds. If $a \in A$ now the period map φ sends it to a flag $\varphi(a)$ in the vector space $V_\mathbb{C}$. Write $\varphi(a)$ as the $b_\mathbb{C}$-isotropic flag $(E_a^0 \subset E_a^1 \subset \cdots \subset E_a^{t-1})$ in $V_\mathbb{C}$ for the dimension sequence $d_i = h_0^{r,0} + \cdots + h_0^{r-i,i}$. Denote $u = d_{t-1}$. Use the markings φ_a to make a coherent choice of bases $\{\beta_a^1, \dots, \beta_a^u\}$ of the $H_0^r(\underline{X}_a; \mathbb{C})$, starting with a basis of $H_0^{r,0}(\underline{X}_a; \mathbb{C})$, then a basis of $H_0^{r-1,1}(\underline{X}_a; \mathbb{C})$, and continuing through the subspaces in the flag $\varphi(a)$. This defines a $u \times v$ period matrix

$$\Pi(a) = \Pi(\underline{X}_a, \varphi_a) := \begin{pmatrix} \int_{\gamma_1} \beta_a^1 & \cdots & \int_{\gamma_v} \beta_a^1 \\ \vdots & & \vdots \\ \int_{\gamma_1} \beta_a^u & \cdots & \int_{\gamma_v} \beta_a^u \end{pmatrix}, \tag{15.1.8}$$

which of course specifies $\varphi(a)$. Specifically, the period matrix map is a matrix formulation of the period map φ.

As in the case of period matrices of Riemann surfaces, one can change the basis $\{\gamma_i\}$ by any integral element of G_0 and change the basis $\{\beta_a^j\}$ by any element of G_0 that does not change $\varphi(a)$. Thus the moduli space for r-forms of marked compact Kähler manifolds (X, J, ω) with given Hodge numbers $h_0^{p,q} := \dim H_0^{p,q}(X; \mathbb{C})$, $p + q = r$, is contained in D and projects down into the arithmetic quotient as follows.

(15.1.9)

$$\begin{aligned}\Gamma\backslash D &= G_{\mathbb{Z}}\backslash G_0/L_0\\ &= SO(2h, k; \mathbb{Z})\backslash SO(2h, k)/(U(h_0) \times \cdots \times U(h_{t-1}) \times SO(k)) \text{ for } r \text{ even},\\ &= \mathrm{Sp}(m; \mathbb{Z})\backslash \mathrm{Sp}(m; \mathbb{R})/(U(h_0) \times \cdots \times U(h_t)) \text{ for } r \text{ odd},\end{aligned}$$

where $h_i = h_0^{r-i,i}$, and $\Gamma = G_{\mathbb{Z}}$ is defined by the lattice $V_{\mathbb{Z}}$ in $V_{\mathbb{R}}$.

In the deformation setting $\pi : \mathcal{X} \to A$, where the period map is holomorphic, the composition of the period map with the projection $D \to \Gamma\backslash D$ is a well-defined holomorphic map $A \to \Gamma\backslash D$.

Classically one constructs Γ-automorphic functions on D as quotients of Γ-invariant sections of holomorphic line bundles over D (automorphic forms of a given weight). Also classically D is a bounded symmetric domain $\mathrm{Sp}(g; \mathbb{R})/U(g)$ and one works in a fixed holomorphic trivialization of the line bundles over D where the Γ-invariance condition is expressed by a transformation law. In this way one constructs the function field of the moduli space $\Gamma\backslash D$.

The classical theory of automorphic functions must be modified in our context, because in general D has no nonconstant holomorphic functions [W2], and in general nontrivial homogeneous vector bundles over D have no nonzero holomorphic sections. Instead one considers sufficiently negative homogeneous holomorphic vector bundles $\mathbb{E} \to D$. Roughly speaking, those are the bundles whose L^2 cohomology, and whose sheaf cohomology, viewed as G_0-modules, have the same underlying Harish-Chandra module. Their cohomology occurs in degree $\dim C_0$, where $C_0 \cong K_0/L_0$. One looks for **automorphic cohomology**, meaning Γ-invariant classes in $H^q(D; \mathcal{O}(\mathbb{E}))$. That is a bit remote from the idea of a function field for $\Gamma\backslash D$, but the double fibration transform $\mathcal{P} : H^q(D; \mathcal{O}(\mathbb{E})) \to H^0(\mathcal{M}_D; \mathcal{O}(\mathbb{E}^\dagger))$ and the holomorphic trivialization of $\mathbb{E}^\dagger \to \mathcal{M}_D$ carry the automorphic cohomology space $H^q(D; \mathcal{O}(\mathbb{E}))^\Gamma$ to a space of holomorphic functions $\mathcal{M}_D \to H^q(C_0; \mathcal{O}(\mathbb{E}|_{C_0}))$ with a certain transformation law under Γ. In this sense $\Gamma\backslash\mathcal{M}_D$ can be a good replacement for $\Gamma\backslash D$ as a universal deformation space.

In much of the literature one considers only the situation where G_0 is of hermitian type and the bounded symmetric domain $\mathcal{B} = G_0/K_0$ is used instead of $\mathcal{M}_D$. (Of course they are the same if D is of hermitian holomorphic type.) When G_0 is not of hermitian type then again G_0/K_0 is used instead of $\mathcal{M}_D$, and it is considered somewhat of an obstacle that G_0/K_0 is not a complex manifold. Our use of $\Gamma\backslash\mathcal{M}_D$ addresses this point.

In connection with construction of automorphic cohomology, Wells showed by direct computation that $\mathcal{M}_D$ is a Stein manifold in one particular case ($r = 2$) [We1]. That result was extended in Wells–Wolf [WeW] to the more general situation of open G_0-orbits D of the form G_0/L_0 with L_0 compact, using a special case of the double fibration transform together with somewhat general methods of complex analysis (Andreotti–Grauert [AnG], Andreotti–Norguet [AN], Docquier–Grauert [DG]) associated to questions of holomorphic convexity and the Levi problem. The goal of [WeW] was construction of automorphic cohomology as convergent Poincaré ϑ-series $\vartheta_\Gamma(c) := \sum_{\gamma\in\Gamma} \gamma^*(c)$ where $c \in H^q(D; \mathcal{O}(\mathbb{E}))$ is a K_0-finite cohomology class. The relevant estimates were derived from semisimple representation theory, specifically from Hecht–Schmid [HSc] and Schmid [S4], and by passing between D and $\mathcal{M}_D$.

This theory of Poincaré ϑ-series and automorphic cohomology later was developed quite a bit. According to [W5], if Γ is any discrete subgroup of G_0, $\mathbb{E} \to D$ is sufficiently negative and $1 \leqq p \leqq \infty$, then every Γ-invariant $L^p(\Gamma\backslash D)$ class in $H^q(D; \mathcal{O}(\mathbb{E}))$ can be realized as a Poincaré series $\vartheta_\Gamma(c)$, where $c \in H^q(D; \mathcal{O}(\mathbb{E}))$ is $L^p(D)$. In particular this is close to the idea of catching all of the function field. The "sufficiently" part of the "sufficiently negative" condition on $\mathbb{E} \to D$ is relaxed in Wallach–Wolf [WaW] by construction of an appropriate reproducing kernel. Finite dimensionality of automorphic cohomology was proved by Williams [Wi1], [Wi2], [Wi3], using the index theory of Moscovici and Connes, for the case where $\Gamma\backslash D$ is compact. Despite this development, automorphic cohomology has not yet been effectively applied to variation of Hodge structure. We expect that the new information on the double fibration transform, presented in this monograph, will make a difference here.

16

Cycles in the K3 Period Domain

In this chapter we outline moduli space results for K3 surfaces which are marked with a basis for their integral homology. This moduli space, which is often called a period domain, is an open orbit D of $G_0 = SO^0(3, 19)$ in a 20-dimensional quadric. We compare it to the Griffiths domain discussed in Chapter 15. In particular, in the present case the analogous discrete group does not act properly. Thus it is appropriate to move to the level of the cycle space $\mathcal{M}_D$, which we compute through its identification with the universal domain $\mathcal{U}$.

Since we hope that this monograph will be of interest to colleagues and students coming from a wide range of backgrounds, we begin this chapter with a sketch of some basic information on complex surfaces in general and on K3 surfaces in particular. Then, in Section 2 we give three basic methods of construction for K3 surfaces. First, we explain the Kummer construction (see Proposition 16.2.1), which amounts to going to the quotient of a compact torus by the group of order two generated by $-\mathrm{id}$ and then blowing up the resulting 16 singularities in a canonical way. Then, by using the adjunction formula and the Lefschetz Theorem, we note that smooth surfaces of degree four in $\mathbb{P}_3$ are K3 surfaces (Proposition 16.2.2). Finally, we show that the 2 to 1 cover of $\mathbb{P}_2$ which is ramified over a smooth curve of degree six is also a K3 surface (Proposition 16.2.3).

In Section 3 the notion of a marking is discussed and then we explain the Torelli Theorem 16.3.3, which states that the space of all marked K3 surfaces can be naturally identified with a certain open orbit D of $G_0 = SO^0(3, 19)$. In this case, we have the standard hermitian form of signature $(3, 19)$ on $\mathbb{C}^{22}$. If b is the associated complex bilinear form, then the manifold $Z = G/Q$ is the 20-dimensional quadric defined by b in $\mathbb{P}_{21}$. The group G_0 acts on this quadric, and the K3 period domain, i.e., the space of all marked K3 surfaces, is the domain D of all positive lines in Z.

In Section 4 we discuss the relevant cycle spaces. Here the cycles themselves are quadric curves which arise as the (transversal) intersection of Z with a projective plane in $\mathbb{P}_{21}$.

As is explained in Section 4 (see Proposition 16.4.1) the cycles which are defined over the reals correspond to Calabi–Yau metrics on the K3 surface. Here "real" means that the above mentioned plane is defined over the reals, or equivalently, that cycle itself is a base cycle which is defined by a choice of the maximal compact subgroup K_0.

We close the chapter with considerations which revolve around the Schubert slices in D. In an interesting way this leads to related hermitian bounded domains of the group $SO(2, 18)$. Finally, using its identification with $\mathcal{U}$ and the matrix domain realization presented in Chapter 14, we give an explicit description of the cycle space $\mathcal{M}_D$. Of course it is our hope that this cycle space, and perhaps actions on it by discrete subgroups of G_0, will have interesting interpretations in moduli theory.

16.1 Position of K3 surfaces in the Kodaira classification

By **surface** we mean a connected compact complex manifold of complex dimension 2. Unlike the one-dimensional case many surfaces fail to be projective algebraic varieties. For example, in order that a torus $\Gamma\backslash\mathbb{C}^2$ have a nonconstant meromorphic function, the lattice Γ must be rather special.

Surfaces have been classified in a rough sense by Kodaira; see [BPV]. The analogous classification of projective algebraic surfaces came earlier, due at least in part to the availability of the algebraic case of the Riemann–Roch Theorem. That classification is generally credited to Enriques (see [Z]), but of course there were many other contributors.

One of the basic invariants of any compact complex manifold X is its Kodaira dimension κ_X, defined as follows.

If $\mathbb{L} \to X$ is a holomorphic line bundle, we have the graded ring $\mathcal{R}(\mathbb{L}) := \sum_{k \geqq 1} \Gamma(X; \mathcal{O}(\mathbb{L}^k))$ of sections of all the positive tensor powers of $\mathbb{L}$. If $\mathcal{R}(\mathbb{L}) = 0$, one defines the Kodaira dimension $\kappa(\mathbb{L})$ to be $-\infty$. If $\mathcal{R}(\mathbb{L}) \neq 0$, one considers its graded quotient field $\mathcal{Q}(\mathbb{L})$, the field generated by meromorphic functions on X that are quotients of sections of some positive power of $\mathbb{L}$. By definition, the Kodaira dimension $\kappa(\mathbb{L})$ is the transcendence degree of $\mathcal{Q}(\mathbb{L})$ over $\mathbb{C}$. A theorem of Thimm, Siegel and Remmert says that analytically dependent meromorphic functions on a compact complex manifold are algebraically dependent, and this implies $\kappa(\mathbb{L}) \leqq \dim_{\mathbb{C}}(X)$.

The Kodaira dimension $\kappa = \kappa_X$ of X is defined to be the Kodaira dimension $\kappa(\mathbb{K}_X)$ of the canonical line bundle $\mathbb{K}_X = \bigwedge^n \mathbb{T}^*_X$, $n = \dim_{\mathbb{C}}(X)$. If $n = 1$, the possibilities are $\kappa = -\infty, 0, 1$ corresponding to the possibilities $X = \mathbb{P}_1(\mathbb{C})$, $X = \Gamma\backslash\mathbb{C}$ torus, and genus $g(X) \geqq 2$, in other words whether X carries a Kähler metric of constant Gauss curvature $+1$, 0 or -1. Of course most Riemann surfaces fall into the last category.

The possibilities for Kodaira dimension of a surface are $\kappa = -\infty, 0, 1, 2$. More or less, as in the one-dimensional case, the class of surfaces with $\kappa = 2$ (surfaces of **general type**) is large and, in fact, is still an area of intense research activity.

The class of surfaces where $\kappa = -\infty$ contains the rational homogeneous surfaces and those that fiber over a one-dimensional base with fiber $\mathbb{P}_1(\mathbb{C})$. The algebraic surfaces in this class are rather well understood but there remain interesting questions on the complex manifold side. For example, are there surfaces in this class that have no one-dimensional subvarieties?

Surfaces in the class $\kappa = 1$ can be studied by their canonical fibration onto a one-dimensional base. The general fiber is a torus of complex dimension 1. Its complex structure depends on the point over which it is the fiber, and many interesting families of elliptic curves (with degeneration allowed) occur in this way.

The holomorphic tangent bundle of a complex torus $X = \Gamma\backslash\mathbb{C}^n$ is (holomorphically) trivial; in particular, $\mathbb{K}_X$ is trivial and thus $\kappa_X = 0$.

By definition, a K3 **surface** is a surface X with trivial canonical bundle and first Betti number $b_1(X) = 0$. Triviality of $\mathbb{K}_X$ implies $\kappa_X = 0$, as in the case of a complex torus. The class of surfaces with $\kappa = 0$ consists of the complex 2-tori, the K3 surfaces, the $(2:1)$ unramified quotients of K3 surfaces (these are called **Enriques surfaces**), and a few other special types of surfaces.

For a number of reasons, some of which will emerge in what follows, K3 surfaces play an extremely important role in complex geometry. For example, if X is a K3 surface, the condition that $\mathbb{K}_X$ be trivial says that X has a holomorphic symplectic form. The topological condition $b_1(X) = 0$ is rather strong here, because it turns out that X is simply connected. It is difficult to find examples of such manifolds in higher dimensions except for those built up from K3 surfaces. And in fact, looking at the surface picture for the first time, one might not even know where to look for K3 surfaces.

16.2 Three classes of examples

We look at three different constructions that produce K3 surfaces.

16.2A The Kummer construction

Consider a complex 1-torus $T_1 = \Gamma_1\backslash\mathbb{C}$. Let $\Lambda_1 \subset \mathrm{Aut}(T_1)$ be the group of order 2 generated by $z \mapsto -z$ on $\mathbb{C}$, and define $Z_1 = \Lambda_1\backslash T_1$. Then Z_1 is smooth, in fact biholomorphic to $\mathbb{P}_1(\mathbb{C})$, and its meromorphic function field $\mathcal{M}(Z_1) = \mathcal{M}(T_1)^{\Lambda_1} = \mathbb{C}(\wp)$ where $\wp$ is the Weierstrass $\wp$–function, and the quotient mapping $T_1 \to Z_1$ is the $(2:1)$ ramified cover $\wp : T_1 \to \mathbb{P}_1(\mathbb{C})$.

In higher dimensions such quotients tend to be singular. For example, let Λ be the subgroup of $GL(2;\mathbb{C})$ which is generated by $-I$. Then the ring $\mathbb{C}[z,w]^\Lambda$ of invariants is generated by the three functions $x_1 = z^2$, $x_2 = zw$ and $x_3 = w^2$. The only relation is the obvious one, $x_2^2 - x_1x_3 = 0$. Thus the quotient is the variety Z in $\mathbb{C}^3$ defined by this equation. If one removes the origin from Z, one obtains an unramified cover $\mathbb{C}^2 \setminus \{0\} \to Z \setminus \{0\}$, by $(z,w) \mapsto (z^2, zw, w^2)$. So the fundamental group $\pi_1(Z \setminus \{0\}) = \mathbb{Z}_2$. Thus the origin is a singularity in Z.

The space Z has a natural desingularization. For this, note that the natural map $\mathbb{C}^3 \setminus \{0\} \to \mathbb{P}_2(\mathbb{C})$ realizes $Z \setminus \{0\}$ as a principal $\mathbb{C}^*$ bundle over the smooth quadric $C = \{[x_1, x_2, x_3] \mid x_2^2 = x_1x_3\}$. Completing this to a holomorphic line bundle at the end that corresponds to the origin in Z, we obtain a complex manifold X with a natural holomorphic map $\pi : X \to Z$ that is biholomorphic except over the origin, and there $\pi^{-1}(0)$ is the copy of $C \cong \mathbb{P}_1(\mathbb{C})$ that is the zero-section of the line bundle.

The Kummer construction amounts to carrying out the above procedure on a torus $T = \Gamma \backslash \mathbb{C}^2$. Just as in the one-dimensional case let $\Lambda \subset \mathrm{Aut}_{\mathbb{C}}(T)$ be generated by the transformation $-I$ of $\mathbb{C}^2$. Every point of T has a local coordinate neighborhood in which Λ is given by $-I$.

The fixed point set of Λ on T consists of the 16 points that come from $\frac{1}{2}\Gamma(0)$, in other words from the $\pm\frac{1}{2}\gamma_i$, where $\{\gamma_1, \gamma_2, \gamma_3, \gamma_4\}$ is a generating set of the free abelian group Γ. The quotient $Z = \Lambda \backslash T$ has 16 singular points, the images of the fixed points of Λ on T. Since the desingularization process is a local procedure, we may blow up each of them as above, obtaining a complex manifold X with 16 special complex curves, and a natural holomorphic map $\pi : X \to Z$ that blows these curves down to the singular points in Z.

The manifold X is the **Kummer surface** associated to T, and we denote it by $\mathrm{Kum}(T)$. It is an interesting exercise to give a direct proof that $X = \mathrm{Kum}(T)$ is simply connected. One can show as follows that X has a nowhere vanishing holomorphic 2-form ω. In the standard linear coordinate (z, w) on $\mathbb{C}^2$, the form $dz \wedge dw$ is Λ-invariant, so it is well defined on T and thus also on the complement of the 16 special curves in X. One computes directly that it extends holomorphically to a nowhere vanishing form ω on X. Thus we have

Proposition 16.2.1. *The Kummer surface* $X = \mathrm{Kum}(T)$ *of a torus* $T = \Gamma \backslash \mathbb{C}^2$ *is a* $K3$ *surface.*

16.2B Quartic surfaces in $\mathbb{P}_3(\mathbb{C})$

Let $\mathbb{C}[z_0, z_1, \ldots, z_n]_{(d)}$ denote the set of homogeneous complex polynomials of degree d in $n+1$ variables. If $P_d \in \mathbb{C}[z_0, z_1, \ldots, z_n]_{(d)}$, then its zero set in $\mathbb{C}^{n+1} \setminus \{0\}$ is saturated by the fibration $\mathbb{C}^{n+1} \setminus \{0\} \to \mathbb{P}_n(\mathbb{C})$. The associated zero set $X = N(P_d) \subset \mathbb{P}_n(\mathbb{C})$ is a codimension 1 subvariety. It is of degree d in the sense that it meets every projective line in $\mathbb{P}_n(\mathbb{C})$ in d points, counting multiplicity.

Now let us assume that $X = N(P_d)$ is smooth. In particular this means that one can compute with its normal bundle to obtain a precise description of the canonical bundle, as follows. The embedding $X \hookrightarrow \mathbb{P}_n(\mathbb{C}) := Y$ gives an exact sequence

$$0 \to \mathbb{T}(X) \to \mathbb{T}(Y)|_X \to \mathbb{N}_X \to 0$$

of bundles over X, where $\mathbb{T}$ means tangent bundle and $\mathbb{N}$ means normal bundle. The sequence of dual bundles,

$$0 \to \mathbb{N}_X^* \to \mathbb{T}^*(Y)|_X \to \mathbb{T}^*X \to 0$$

is exact. The **adjunction formula** says that the tensor product of the top exterior powers of the two at the ends is equivalent to the top exterior power of the one in the middle. Therefore, $\mathbb{N}_X^* \otimes \bigwedge^{n-1} \mathbb{T}^*X \cong \bigwedge^n \mathbb{T}^*(Y)|_X$. In other words, canonical bundles satisfy $\mathbb{K}_X = \mathbb{K}_Y|_X \otimes [X]|_X$, where $[X] \to Y$ is the line bundle associated to the divisor given by X. As X is of degree d in $Y = \mathbb{P}_n(\mathbb{C})$, the line bundle $[X] \to Y$ is $\mathbb{H}^d$, where $\mathbb{H} \to Y$ is the hyperplane section bundle, the holomorphic line bundle whose sections are the linear homogeneous polynomials. Now

$$\mathbb{K}_X = \mathbb{K}_Y|_X \otimes \mathbb{H}^d|_X = \mathbb{H}^{-(n+1)}|_X \otimes \mathbb{H}^d|_X.$$

Thus, for $d = n + 1$ the canonical bundle is trivial. In particular, smooth surfaces of degree 4 in $\mathbb{P}_3(\mathbb{C})$ have trivial canonical bundle.

From a theorem of Lefschetz on hyperplane sections,[1] $\pi_1(X) = 1$, in particular $b_1(X) = 0$. Thus we have a large family of K3 surfaces.

Proposition 16.2.2. *A smooth quartic hypersurface in $\mathbb{P}_3(\mathbb{C})$ is a K3 surface.*

16.2C Galois coverings

We discuss the simplest means of constructing cyclic Galois coverings (mostly of interest when they are ramified) $\pi : X \to Y$ of a given compact connected complex manifold Y. The point is that K3 surfaces arise as Galois coverings in many different ways.

Given Y and a finite group Λ one looks for a connected complex manifold (or complex space) X with $\Lambda \subset \mathrm{Aut}_{\mathbb{C}}(X)$ and $Y = \Lambda\backslash X$. Here the complex structure on Y should be the quotient complex structure from X, in other words the structure sheaf $\mathcal{O}_Y$ should be given by germs of Λ-invariant of holomorphic functions on X. This is opposite to the viewpoint in the Kummer surface construction, because in that case one starts with the torus and goes down, while here we go up. If Λ is cyclic of order r, one can think of X as a sort of rth root of Y at the function space level.

To make this precise, let $\mathbb{L} \to Y$ be a holomorphic line bundle that has an rth root. In other words there is another holomorphic line bundle $\mathbb{L}_1 \to Y$ such that $\mathbb{L} = \mathbb{L}_1^r$. Then the natural map $\pi_1 : \mathbb{L}_1 \to \mathbb{L}$ is given on the fiber coordinate level by $z \mapsto z^r$. Evidently it is ramified along the zero section, and away from the zero section it is an $(r : 1)$ unramified covering.

Now let $s \in \Gamma(Y; \mathcal{O}(\mathbb{L}))$ such that $Z := \{y \in Y \mid s(y) = 0\}$ is a smooth submanifold of Y. To avoid trivial cases we suppose that (i) s is not identically zero and (ii) s is sometimes zero. In other words $\emptyset \neq Z \neq Y$. Then smoothness of Z means that $s(Y)$ is transversal to the zero section in the total space of $\mathbb{L}$. Define $X = \pi_1^{-1}(s(Y)) \subset \mathbb{L}_1$. It is clear that X is smooth at points not in the zero section, but it is smooth as well at zero section points because of the transversality. Also, X is connected because $Z \neq \emptyset$.

Thus restriction of π_1 gives us a Galois covering $\pi : X \to s(Y) \cong Y$, where the group of covering transformations is the cyclic group $\mathbb{Z}_r$. Now we look at a concrete example that is very interesting from the viewpoint of K3 surfaces.

[1] Here is the theorem in question. Let X be a compact complex manifold and $\mathbb{L} \to X$ a positive line bundle. Consider a holomorphic section s of $\mathbb{L} \to X$ such that $H := \{x \in X \mid s(x) = 0\}$ meets the zero-section transversally, and H and $X \cap H$ are smooth submanifolds of X. Then the relative homotopy $\pi_i(X, X \cap H) = 0$ for $i < \dim_{\mathbb{C}} X$.

Let C be a smooth sextic curve in $\mathbb{P}_2(\mathbb{C})$. So C is the zero set of a section $s \in \Gamma(\mathbb{P}_2(\mathbb{C}); \mathcal{O}(\mathbb{L}))$, where $\mathbb{L}$ is the sixth power $\mathbb{H}^6$ of the hyperplane bundle and where s satisfies the conditions described above. Then $\mathbb{H}^3 \to \mathbb{H}^6$ gives us a smooth submanifold $X \subset \mathbb{H}^3$ and a $(2:1)$ Galois covering $\pi : X \to s(\mathbb{P}_2(\mathbb{C})) \cong \mathbb{P}_2(\mathbb{C})$, as before with $\mathbb{L}_1 = \mathbb{H}^3$ and $\mathbb{L} = \mathbb{H}^6$. The covering group is $\mathbb{Z}_2$, and the covering is ramified over C. Now we show that X is a K3 surface.

Repeat the covering construction using $\mathbb{H} \to \mathbb{H}^3$ to obtain a smooth surface $X_1 \subset \mathbb{H}$ which is a $(3:1)$ ramified cover of X. The advantage is that X_1 is a compact submanifold of $\mathbb{H}$, and $\mathbb{H}$ can be compactified to $\mathbb{P}_3(\mathbb{C})$ by adding one point at ∞, or equivalently by adding a section at ∞ and blowing it down to a point. The surface X_1 can now be regarded as a surface in $\mathbb{P}_3(\mathbb{C})$.

The adjunction formula for $X_1 \subset \mathbb{P}_3(\mathbb{C})$ says $\mathbb{K}_{X_1} = \mathbb{H}^{d-4}_{\mathbb{P}_3(\mathbb{C})}|_{X_1}$, where X_1 is of degree d. We may take the homogeneous coordinate $[z_0 : z_1 : z_2 : z_3]$ on $\mathbb{P}_3(\mathbb{C})$ in such a way that the zero-section of $\mathbb{H}_{\mathbb{P}_3(\mathbb{C})} \to \mathbb{P}_3(\mathbb{C})$ is given by $z_0 = 0$ and the covering $\mathbb{H} \to \mathbb{H}^3$ is given as a quotient of the $\mathbb{Z}_3$ action

$$[z_0 : z_1 : z_2 : z_3] \to [z_0 : \zeta z_1 : \zeta z_2 : \zeta z_3].$$

We regard that zero section as a hyperplane U in $\mathbb{P}_3(\mathbb{C})$. Since the maps $\mathbb{H} \to \mathbb{H}^3$ and $\mathbb{H}^3 \to \mathbb{H}^6$ are biholomorphic on C, we take C in $\mathbb{P}_3(\mathbb{C})$. Then X_1 meets U transversally and $X_1 \cap U = C$. Also, as our ramified coverings do nothing to the zero-sections, C is a curve of degree 6 in U. Counting intersection points with lines, it is a curve of degree 6 in $\mathbb{P}_3(\mathbb{C})$. We have just checked that $d = 6$.

Since $d = 6$, the adjunction formula says $\mathbb{K}_{X_1} = \mathbb{H}^2_{\mathbb{P}_3(\mathbb{C})}|_{X_1}$. In other words $z_0^2|_{X_1}$ is a section $\widetilde{\gamma}$ of the canonical bundle $\mathbb{K}_{X_1}$. But it is invariant under the $\mathbb{Z}_3$ action of $\mathbb{H} \to \mathbb{H}^3$, and so it is the pullback of a section γ of the canonical bundle $\mathbb{K}_X$. We want to show that γ never vanishes which implies that the $\mathbb{K}_X$ is trivial.

Choose local coordinates on X and X_1 such that $X_1 \to X$ is given by $(z, w) \mapsto (z^3, w)$. In those coordinates $\widetilde{\gamma}$ has the form

$$f(z^3, w)d(z^3) \wedge dw = 3z^2 f(z^3, w)dz \wedge dw.$$

Since z_0^2 vanishes to order exactly 2 on C, the same holds for $\widetilde{\gamma}$, and thus $f(z, w)$ never vanishes on C. We have proved that γ never vanishes and, as desired, the $\mathbb{K}_X$ is trivial.

Consider the pullback $p^* : H^1(X; \mathbb{C}) \to H^1(X_1; \mathbb{C})$ on de Rham cohomology given by the covering $p : X_1 \to X$. If α is a closed 1-form on X and $p^*(\alpha) = d\widetilde{f}$ exact, we may average with respect to the $\mathbb{Z}_3$ action and assume $\widetilde{f}$ to be invariant. Then $\widetilde{f} = p^* f$ for a function $f : X \to \mathbb{C}$, and $\alpha = df$. Thus $p^* : H^1(X; \mathbb{C}) \to H^1(X_1; \mathbb{C})$ is injective. But $H^1(X_1; \mathbb{C}) = 0$ by the Lefschetz Theorem, because X_1 is a surface in $\mathbb{P}_3(\mathbb{C})$. Thus $b_1(X) = 0$, and therefore we have another large family of K3 surfaces.

Proposition 16.2.3. *A double cover of $\mathbb{P}_2(\mathbb{C})$, ramified along a smooth sextic curve, is a K3 surface.*

16.3 Parameterizing K3 surfaces

16.3A The moduli space of genus 1 Riemann surfaces

In order to indicate the procedure for describing the structure of the space of K3 surfaces, we first recall the situation for Riemann surfaces X of genus 1. This looks very similar to the construction of Chapter 15, but the analogous domain in the K3 case is very different. In the case of a genus 1 surface X, the space $\Omega(X)$ of holomorphic 1-forms has dimension 1. Since holomorphic 1-forms are closed, $\Omega(X)$ can be viewed as a complex line in the two-dimensional complex vector space $H^1(X;\mathbb{C})$.

Recall the cap product $(\cdot,\cdot)_X : H^1(X;\mathbb{Z}) \times H^1(X;\mathbb{Z}) \to \mathbb{Z}$. It is the dual to the intersection pairing on $H_1(X;\mathbb{Z})$, and if we extend coefficients (as in $H^1(X;\mathbb{C}) = H^1(X;\mathbb{Z}) \otimes_{\mathbb{Z}} \mathbb{C}$), it is given on the de Rham cohomology level as exterior product of closed 1-forms followed by integration over X. In any case it is antisymmetric. If we vary X, then $(\cdot,\cdot)_X$ changes to an equivalent $\mathbb{Z}$-valued antisymmetric bilinear form on $H^1(X;\mathbb{Z})$, but one must choose the equivalence.

In order to deal with equivalence of intersection pairing integral bilinear forms we will compare with the standard lattice $L = \mathbb{Z} \times \mathbb{Z}$ and its standard alternating pairing $(z, z') = z_1 z_2' - z_2 z_1'$. We define a **marking** φ of X to be an isomorphism

$$\varphi : (H^1(X;\mathbb{Z}), (\cdot,\cdot)_X) \to (L, (\cdot,\cdot)),$$

and denote the set of marked Riemann surfaces (X,φ) of genus 1 by $\widehat{\mathcal{M}}_1$.

Given $(X,\varphi) \in \widehat{\mathcal{M}}_1$, we identify $H^1(X;\mathbb{C})$ using the standard symplectic space $(\mathbb{C}^2, (\cdot,\cdot)_{\mathbb{C}})$ with the $\mathbb{C}$-bilinear extension of φ. Here $(\mathbb{C}^2, (\cdot,\cdot)_{\mathbb{C}})$ is the complexification of $(\mathbb{R}^2, (\cdot,\cdot)_{\mathbb{R}})$, and the $\mathbb{R}$-bilinear extension of φ identifies $H^1(X;\mathbb{R})$ with $(\mathbb{R}^2, (\cdot,\cdot)_{\mathbb{R}})$. Complex conjugation of $H^1(X;\mathbb{C})$ over $H^1(X;\mathbb{R})$ exchanges de Rham classes represented by holomorphic forms with those represented by antiholomorphic forms and gives $H^1(X;\mathbb{C}) = \Omega(X) \oplus \overline{\Omega(X)}$. The hermitian pairing

$$H^1(X;\mathbb{C}) \times H^1(X;\mathbb{C}) \to \mathbb{C} \text{ by } (\alpha,\beta) \mapsto \frac{1}{2\pi i}\int_X \alpha \wedge \overline{\beta}$$

carries over by φ to the standard hermitian pairing of signature $(1, 1)$ on $\mathbb{C}^2$.

Fix an orientation on the underlying differentiable manifold of X and restrict attention to those complex structures which induce that orientation. This gives us a subset of $\widehat{\mathcal{M}}_1$ which we denote as $\widehat{\mathcal{M}}_1^0$. Then $\widehat{\mathcal{M}}_1$ is the disjoint union of $\widehat{\mathcal{M}}_1^0$ and the corresponding subset using the conjugate complex structures, i.e., the opposite orientation, and this will correspond to the decomposition of $\widehat{\mathcal{M}}_1$ into its topological components. Given $(X,\varphi) \in \widehat{\mathcal{M}}_1^0$, we have the line $\varphi(\Omega(X))$ in $\mathbb{C}^2$, and it is positive with respect to the standard hermitian structure of signature $(1, 1)$ on $\mathbb{C}^2$.

Decompose the projective space of $\mathbb{C}^2$ as $\mathbb{P}_1(\mathbb{C}) = D_+ \cup D_0 \cup D_-$ according to whether the element of $\mathbb{P}_1(\mathbb{C})$, as a line in $\mathbb{C}^2$, is positive, null or negative with respect to the standard hermitian structure of signature $(1, 1)$.

Theorem 16.3.1. *The map $P : \widehat{\mathcal{M}}_1^0 \to D_+$, given by $P(X,\varphi) = \varphi(\Omega(X))$, is bijective.*

In fact this result is holomorphic in the sense that if one moves a surface of genus 1 in an appropriately holomorphic way, then the corresponding movement in D_+ is holomorphic. See the remarks after Theorem 16.3.3.

Results of the above type are extremely useful, but in this case one can prove more: *one can forget the markings*. More precisely, note that a given Riemann surface X of genus 1 occurs in D_+ with many different markings, corresponding to symplectic changes of basis in the lattice $L = \mathbb{Z} \times \mathbb{Z}$. Thus the space $\mathcal{M}_1$ of unmarked Riemann surfaces of genus 1, and its part $\mathcal{M}_1^0$ for the given orientation, can be identified with the respective quotients

$$\mathcal{M}_1 = \mathrm{Sp}(1; \mathbb{Z})\backslash\widehat{\mathcal{M}}_1 \text{ and } \mathcal{M}_1^0 = \mathrm{Sp}(1; \mathbb{Z})\backslash\widehat{\mathcal{M}}_1^0.$$

The group $\mathrm{Sp}(1; \mathbb{Z})$ acts properly on $\widehat{\mathcal{M}}_1$, so as a result these quotients are complex spaces which in general have rather simple singularities. In fact in this one-dimensional situation they are smooth. The Jacobi modular function identifies $\mathcal{M}_1^0$ with $\mathbb{C}$. Of course there is a great deal of important mathematics involving automorphic forms related to this picture.

16.3B The K3 period domain

Much of the above discussion can be carried out for K3 surfaces. However, the discrete group which identifies markings does not act properly, and therefore it is appropriate instead to move to the associated cycle space. Let us outline this situation.

Since the K3 spaces are simply connected the only interesting topological cohomology space at our disposal is $H^2(X; \mathbb{Z})$. It comes with a nondegenerate intersection pairing

$$(\cdot, \cdot)_X : H^2(X; \mathbb{Z}) \times H^2(X; \mathbb{Z}) \to \mathbb{Z}.$$

That pairing is symmetric and of signature (3, 19), and $(H^2(X; \mathbb{Z}), (\cdot, \cdot)_X)$ is isomorphic to the lattice

$$(L, (\cdot, \cdot)_L) = L_{1,-1} \oplus L_{1,-1} \oplus L_{1,-1} \oplus L_{\mathrm{rt}}(E_8) \oplus L_{\mathrm{rt}}(E_8) \tag{16.3.2}$$

where $L_{1,-1}$ is the standard $\mathbb{Z}^2$ with $((x, y), (x', y')) = xx' - yy'$ and $L_{\mathrm{rt}}(E_8)$ is the root lattice of E_8. The inner product on the root lattice E_8 is the negative multiple of the Killing form such that simple roots α have square norm $(\alpha, \alpha) = -2$. In abbreviated notation one often writes $L = 3H \oplus 2E_8$. Decompose the cohomology by bidegree,

$$H^2(X; \mathbb{C}) = H^{2,0}(X; \mathbb{C}) \oplus H^{1,1}(X; \mathbb{C}) \oplus H^{0,2}(X; \mathbb{C}).$$

Complex conjugation of $H^2(X; \mathbb{C})$ over $H^2(X; \mathbb{R})$ stabilizes $H^{1,1}(X; \mathbb{C})$ and exchanges $H^{2,0}(X; \mathbb{C})$ with $H^{0,2}(X; \mathbb{C})$. At this stage of the investigation one does not know that the K3 surfaces are Kähler. They are [Si], but the bidegree decomposition does not require Kähler in the case at hand.

The space $H^{2,0}(X; \mathbb{C})$ has dimension 1. Hence it is generated by any nonzero holomorphic 2-form γ. More or less as before, define $\widehat{\mathcal{M}}$ to be the space of marked K3 surfaces (X, φ), where here

$$\varphi : (H^2(X;\mathbb{Z}), (\cdot,\cdot)_X) \to (L, (\cdot,\cdot)_L)$$

is an isomorphism of lattices equipped with bilinear forms, here of signatures (3, 19). The scalar extension $\mathbb{C}^{22}$ has a symmetric bilinear form $(\cdot,\cdot)$ obtained from $(\cdot,\cdot)_L$. It thus has a hermitian form $\langle\alpha,\beta\rangle = (\alpha,\overline{\beta})$ of signature (3, 19), where the overline is complex conjugation of $\mathbb{C}^{22}$ over $\mathbb{R}^{22}$. Since γ is a 2-form, there is an implicit positivity, and the condition $\frac{1}{2\pi i}\int_X \gamma\wedge\overline{\gamma} > 0$ translates to the line $\varphi(H^2(X;\mathbb{C}))$ being positive definite in $(\mathbb{C}^{22}, \langle\cdot,\cdot\rangle)$.

We have $\gamma\wedge\gamma = 0$ from its bidegree, and that says

$$(\varphi(H^{2,0}(X;\mathbb{C})), \varphi(H^{2,0}(X;\mathbb{C}))) = 0,$$

in other words, $\varphi(H^{2,0}(X;\mathbb{C}))$ is an isotropic line in $(\mathbb{C}^{22}, (\cdot,\cdot))$. We translate this into the language of flag domains.

Let Z denote the quadric of $(\cdot,\cdot)$-isotropic lines in $\mathbb{C}^{22}$. Decompose $Z = D_+ \cup D_0 \cup D_-$ according to whether a line is positive, null or negative with respect to $\langle\cdot,\cdot\rangle$. The map

$$P : \widehat{\mathcal{M}} \to D_+ \text{ given by } P(X,\varphi) = \varphi(H^{2,0}(X;\mathbb{C}))$$

is called the **period map**, and D_+ is called the **period domain**,[2] for K3 surfaces. The major result here (see [BBD]) is the following.

Theorem 16.3.3 (Torelli Theorem). *The period map $P : \widehat{\mathcal{M}} \to D_+$ is bijective.*

This result has a holomorphic aspect, proved by Andreotti, as follows. Given a marked K3 surface (X_0, φ_0), there is a canonical local family $\pi : \mathcal{X} \to S$ of marked K3 surfaces, where S is open in $\mathbb{C}^{20}$, such that π has maximal rank, $\pi^{-1}(s) \in \widehat{\mathcal{M}}$ for every $s \in S$, and the map $S \to D_+$, given by $s \mapsto P(\pi^{-1}(s))$, is biholomorphic onto its image. See [BBD] for more information on this **Kuranishi family** of marked K3 surfaces as well as many other aspects of K3 surface moduli.

Now we come to the point where the analogy to genus 1 Riemann surfaces is no longer valid. The group that identifies pairs (X,φ_1) and (X,φ_2) of marked K3 surfaces is the integral orthogonal group $\Gamma = SO(L, (\cdot,\cdot)_L)$. Its action on D_+ is far from proper, so the quotient $\Gamma\backslash D_+$ is not a reasonable space, for example is not Hausdorff.

In some sense this is closely related to the fact that there are no nonconstant holomorphic functions on D_+, and that certain cohomology spaces are the appropriate places to realize group representations. Those representations carry over (by a double fibration transform) to representations on spaces of functions or vector bundle sections on the cycle space $\mathcal{M}_{D_+}$, and the action of Γ on that space is proper.

[2] Here is the reason for the word "period." Choose a basis $\{\xi_1, \ldots, \xi_{22}\}$ of the homology of the fixed underlying real manifold of X, which is dual to the standard basis of L. Then the coordinates of γ are computed as the integrals over the ξ_i, and in analogy with the one-dimensional case they are called "periods."

16.4 The cycle space $\mathcal{M}_{D_+}$

We have just seen that D_+ is naturally identified with the space of marked K3 surfaces. Here D_+ is the space of lines in the 20-dimensional nonsingular quadric $Z \subset \mathbb{P}_{21}(\mathbb{C})$ that are positive definite relative to a hermitian form $\langle \cdot, \cdot \rangle$ on $\mathbb{C}^{22}$. In fact Z is a homogeneous projective variety, i.e., a complex flag manifold, of the form G/Q, where $G = SO(22; \mathbb{C})$, and D_+ is an open orbit of the real form $G_0 = SO^0(3, 19)$ of G. We want to take a careful look at the cycle space $\mathcal{M}_{D_+}$ of D_+, and we start by describing the orbits of the various groups at hand.

16.4A The G_0- and K-orbits in Z

The group $G_0 = SO^0(3, 19)$ has four orbits in Z. The two open orbits are the spaces D_+ of positive lines in Z and the space D_- of negative lines in Z. They are separated by the real hypersurface D_0 of null lines. In turn, D_0 is the union of the closed orbit $Z_{\mathbb{R}}$ and its complement D_0' in D_0, and $Z_{\mathbb{R}}$ consists of the real points in Z.

The real group G_0 has maximal compact subgroup $K_0 = SO(3) \times SO(19)$. It has complexification $K = SO(3; \mathbb{C}) \times SO(19; \mathbb{C})$. The K-orbits dual to the G_0-orbits are given as follows. The base cycle in D_+ is $C_+ := K(z_+)$, where $z_+ = [1 : i : 0 : 0 : \cdots : 0]$. The base cycle in D_- is $C_- := K(z_-)$, where $z_- = [0 : 0 : 0 : 1 : i : 0 : \cdots : 0]$. The variety C_- is a quadric curve isomorphic to $\mathbb{P}_1(\mathbb{C})$, and C_+ is an 18-dimensional quadric. Those base cycles are the duals of the open G_0-orbits.

The dual of the closed G_0-orbit $Z_{\mathbb{R}}$ is the open K-orbit, which is the K-orbit of any point of $Z_{\mathbb{R}}$, say of $z_{\mathbb{R}} = [0 : 0 : 1 : 1 : 0 : \cdots : 0]$. It is affine algebraic and is naturally identified with the tangent bundle of $Z_{\mathbb{R}} = (S^2 \times S^{18})/\{\pm I\}$. Since the open K-orbit is affine, the remaining K-orbit is a hypersurface with two ends that close up to C_+ and C_-, and $z_0' = [0 : 1 : i : i : 1 : 0 : \cdots : 0]$ is a natural base point for this open K-orbit. Now the orbit is $K(z_0')$ and it is a $\mathbb{C}^*$ bundle over $C_+ \times C_-$, where each factor is blown down for partial compactification with the other.

16.4B Real cycles

The domain D_+ is our primary concern here. Its base cycle C_+ is easily described in terms of the embedding $Z \hookrightarrow \mathbb{P}_{21}(\mathbb{C})$. If V is the span of e_1, e_2 and e_3, then $C_+ = \mathbb{P}(V) \cap Z$. The group G_0 acts on $\mathbb{P}(V)$ as $SO(3, 19)$, and $\mathbb{P}(V)$ is defined over $\mathbb{R}$ in a manner invariant under G_0. Every cycle C in the component $\mathcal{C}_1(D_+)$ of C_+ has the same degree 2 as C_0, and thus spans a projective plane $\langle C \rangle \cong \mathbb{P}_2(\mathbb{C})$. One can check that C is a real point of $\mathcal{M}_{D_+}$ if and only if $\langle C \rangle$ is defined over $\mathbb{R}$.

Given a marked surface $(X, \varphi) \in D_+$, we consider the open cone $\underline{\kappa}$ of de Rham classes $[\omega] \in H^{1,1}(X; \mathbb{R})$ of Kähler forms ω on X. Recall $H^2(X; \mathbb{Z}) \cong 3H \oplus 2E_8$ from (16.3.2). The $2E_8$ summand contributes its 16 generators (corresponding to the simple roots of the two E_8 summands) to $H^{1,1}(X; \mathbb{Z})$. Those generators have square norm -2. In the case $X = \mathrm{Kum}(T)$ each of those 16 generators is given by a rational curve that results from desingularization of the torus quotient $\Lambda \backslash T$.

A Kähler class $[\omega]$ obviously has positive square norm in $H^{1,1}(X;\mathbb{R})$, because $\omega \wedge \omega$ is a volume form, and if $[\eta], [\zeta] \in H^2(X;\mathbb{Z})$, then their intersection pairing is $\int_X \eta \wedge \zeta$. Thus the Kähler classes determine a component in the cone $\mathcal{C}_+(X)$ of elements of positive square norm in $H^{1,1}(X;\mathbb{R})$, and the Kähler cone $\underline{\kappa}$ is an open subcone. In fact $\underline{\kappa}$ looks like a Weyl chamber cut out by the generators that come from simple roots of E_8. It is the set of all $[\alpha] \in \mathcal{C}_+(X)$ such that if $[\beta] \in H^{1,1}(X;\mathbb{R})$ is effective with $([\beta],[\beta]) = -2$, then $([\alpha],[\beta]) > 0$.

The isotropy subgroup of G_0 on D_+ at the point (X,φ) contains a copy of the $SO^0(1,19)$, and that $SO^0(1,19)$ acts faithfully on the tangent space $T_{(X,\varphi)}(D_+)$. This is particularly clear at the base point $z_0 \in C_+$. Let $\underline{\kappa}_1$ denote the elements of $\underline{\kappa}$ that correspond to Kähler metrics of volume 1: $\underline{\kappa}_1 := \{[\omega] \in \underline{\kappa} \mid ([\omega],[\omega]) = 1\}$. It can be shown that $T_{(X,\varphi)}(D_+)$ is canonically identifiable with $H^{1,1}(X;\mathbb{C})$, which has real form $H^{1,1}(X;\mathbb{R})$, and that $S0(1,19)$ stabilizes and acts transitively on $\underline{\kappa}_1$. Therefore we concentrate attention on the norm hypersurface $\underline{\kappa}_1$.[3]

The essential point now is the theorem of Yau [Y] that proved the Calabi Conjecture on Ricci-flat manifolds: Every Kähler class in $\underline{\kappa}$ contains a unique Kähler–Einstein metric. Here we identify the metric with its Kähler form ω and regard $\underline{\kappa}$ as the space of Kähler–Einstein metrics on X. Since the canonical bundle $\mathbb{K}_X$ is trivial, the Kähler–Einstein property says that the metric is Ricci flat.

Recall that $H^{2,0}(X;\mathbb{C})$ has dimension 1 so that it is generated by the de Rham class $[\gamma]$ of any nonzero holomorphic 2-form γ. Given a Kähler–Einstein class $[\omega] \in \underline{\kappa}_1$, we consider the subspace

$$V = V_{X,\varphi,[\omega]} := \mathrm{Span}_{\mathbb{C}}([\gamma],[\omega],\overline{[\gamma]})$$

of $L \otimes_{\mathbb{Z}} \mathbb{C} \cong \mathbb{C}^{22}$. Since $\omega = \overline{\omega}$, the space V is defined over $\mathbb{R}$. We now indicate how the cycle $C = D_+ \cap V$ is a 2-sphere and is the set of all marked K3 surfaces whose complex structure admits a Kähler–Einstein metric with the given Kähler form ω.

The Kähler–Einstein metric with Kähler form ω on the K3 surface X is Ricci-flat, and consequently it has holonomy $SU(2) = \mathrm{Sp}(1)$. The centralizer of the holonomy in the algebra of linear transformations of the real tangent space of X is a quaternion algebra. The unit sphere in the pure imaginary component of that quaternion algebra consists of almost complex structures that are invariant under parallel translation, thus integrable. In terms of the usual pure imaginary elements I, J and K of the quaternion algebra, $I^2 = J^2 = K^2 = -Id$, $IJ = K$, $KI = J$, and $JK = I$, these integrable almost complex structures are just the $aI + bJ + cK$ with a, b, c real and $a^2 + b^2 + c^2 = 1$. They form a 2-sphere. This sphere is exactly the cycle C. Moving the complex structure in this 2-sphere amounts to moving the K3 surface X in the cycle C.

Recall the Kähler cone $\underline{\kappa} = \underline{\kappa}(X,\varphi)$ of Kähler–Einstein structures on the underlying marked complex manifold (X,φ), and its real hypersurface $\underline{\kappa}_1$ of those that are normalized to have volume 1.

Because of our strong reliance on the existence of the Kähler–Einstein metric we refer to the cycles $C = D_+ \cap V$ constructed above, from $(X,\varphi) \in D_+$ and

[3] Note the similarity with Köcher's theory of tube domains and formally real Jordan algebras.

$[\omega] \in \underline{\kappa} = \underline{\kappa}(X, \varphi)$, as **Calabi–Yau quadrics**. We summarize our discussion as follows.

Proposition 16.4.1. *The set of Calabi–Yau quadrics can be identified with the set of real points in $\mathcal{M}_{D_+}$.*

The proof follows from the material sketched above and from the fact that the isotropy subgroup of G_0 at (X, φ) stabilizes and acts transitively on $\underline{\kappa}_1$. Note that this presents $\underline{\kappa}_1$ as the real hyperbolic 19-space $SO^0(1, 19)/SO(19)$. It should be underlined that the real points of $\mathcal{M}_{D_+}$ are just those cycles C in D_+ that span projective planes $\langle C \rangle$ in $\mathbb{P}_{21}(\mathbb{C})$ that are defined over $\mathbb{R}$. The base cycle C_+ is one such.

Remark. It would be desirable to have a differential geometric or complex analytic interpretation for cycles that are not real, and for what it means to move in an imaginary direction in the cycle space. $\diamond$

16.4C Schubert slices

Here we go explicitly through the Schubert slice construction for cycles in D_+. There are a number of simplifying factors in this case. For example, since the cycles are one-dimensional, the relevant Schubert varieties are hypersurfaces in Z. Since $b_2(Z) = 1$, these are just the complements of the open B-orbits.

As above, let Z be the 20-dimensional quadric $(z_0^2+z_1^2+z_2^2)-(z_3^2+\cdots+z_{21}^2) = 0$ in $\mathbb{P}_{21}(\mathbb{C})$, and decompose $Z = D_- \cup D_0 \cup D_+$ into the negative, null, and positive lines. Recall that the $D_\pm = G_0(z_\pm)$ are the open orbits, $z_+ = [1 : i : 0 : 0 : \cdots : 0]$ and $z_- = [0 : 0 : 0 : 1 : i : 0 : \cdots : 0]$. D_0 is the union of the closed orbit $Z_\mathbb{R} = G_0(z_\mathbb{R}), z_\mathbb{R} = [0 : 0 : 1 : 1 : 0 : \cdots : 0]$, with one other orbit, $D_0' = D_0 \setminus Z_\mathbb{R} = G_0(z_0'), z_0' = [0 : 1 : i : i : 1 : 0 : \cdots : 0]$.

Let B be an Iwasawa–Borel subgroup of G that fixes $z_\mathbb{R}$. Then the projective tangent space to Z at $z_\mathbb{R}$ is $PT = \{[z] \in \mathbb{P}_{21}(\mathbb{C}) \mid z_2 = z_3\}$. Of course $S := PT \cap Z$ is B-invariant. Note that S is the 1-codimensional B-Schubert variety in Z, and the components of $S \cap D_+$ are the Schubert slices. We are going to compute them and also the variety $Y := S \setminus \mathcal{O}$ that is used for the trace transform.

Let $\pi : \mathbb{P}_{21}(\mathbb{C}) \to \mathbb{P}_{20}(\mathbb{C})$ be the projection defined by the point $z_\mathbb{R}$. Geometrically this means that, given $p \in \mathbb{P}_{21}(\mathbb{C}) \setminus \{z_\mathbb{R}\}$, we identify all points on the projective line determined by p and $z_\mathbb{R}$. Of course this map is not defined at $z_\mathbb{R}$. In coordinates this map is given by $[z_0 : z_1 : \cdots : z_{21}] \mapsto [z_0 : z_1 : z_2 - z_3 : z_4 : \cdots : z_{21}]$. Let $[\xi_0 : \cdots : \xi_{20}]$ be homogeneous coordinates on the image space. The projective tangent space PT is mapped to the hyperplane $\xi_2 = 0$. We refer to that hyperplane as $\mathbb{P}_{19}(\mathbb{C})$ with homogeneous coordinates $[\eta_0 : \cdots : \eta_{19}]$.

The projection π maps the intersection $S = PT \cap Z$ to the quadric Z_1 in $\mathbb{P}_{19}(\mathbb{C})$ defined by $(\eta_0^2 + \eta_1^2) - (\eta_2^2 + \cdots + \eta_{19}^2) = 0$.

Now note that $\pi : S \to Z_1$ is equivariant with respect to the isotropy subgroup $SO^0(2, 18)$ of $G_0 = SO^0(3, 19)$ at $z_\mathbb{R}$. Also, of course, it is B-equivariant. Also note that if $p \in S$, then $p \in PT \cap D_+$ if and only if $\pi(p)$ is positive for the hermitian form

$(|\eta_0|^2 + |\eta_1|^2) - (|\eta_2|^2 + \cdots + |\eta_{19}|^2)$. The group $SO^0(2, 18)$ is the automorphism group of a bounded symmetric domain of tube type, and Z_1 is the associated hermitian symmetric flag manifold. Thus, if we regard π as a regular map from $\mathbb{P}_{21}(\mathbb{C}) \setminus \{z_{\mathbb{R}}\}$, we can describe the Schubert slices in D_+ as follows.

Proposition 16.4.2. *Let D'_+ and D''_+ denote the two $SO^0(2, 18)$-orbits in Z_1 that are bounded symmetric domains (of tube type). Then the B-Schubert slices in D_+ are their π-preimages, $\Sigma' = \pi^{-1}(D'_+)$ and $\Sigma'' = \pi^{-1}(D''_+)$.*

Finally, let us turn to the issue of describing Y. Since $\pi(S)$ is the full quadric Z_1, it follows that Y is the preimage (including $z_{\mathbb{R}}$) of the complement of the open B-orbit in Z_1. This is just the intersection $S_1 = PT_1 \cap Z_1$ of Z_1 with a projective tangent space. So Y is a cone over a cone in a straightforward way.

16.4D Matrix description of the K3 period domain cycle space

The K3 period domain is a flag domain of the group $G_0 = SO^0(3, 19)$, and thus belongs to the family of cycle spaces considered in Section 13.2. Its structure is given by Proposition 13.2.4 with $(p, q, \mathbb{F}) = (3, 19, \mathbb{R})$, as follows.

Corollary 16.4.3. *The cycle space $\mathcal{M}_{D_+} \cong \mathcal{B}_1 \times i\mathcal{B}_2$ where*

$\mathcal{B}_1$ is the real bounded symmetric domain $\{W_1 \in \mathbb{R}^{3\times 19} \mid I - W_1 W_1^* \gg 0\}$,

and

$$\mathcal{B}_2 = \{W_2 \in \mathbb{R}^{3\times 19} \mid W_2 W_2^* \text{ has eigenvalues } \tan^2(a_i), a_i \in \mathbb{R}, |a_i| + |a_j| < \pi/2 \text{ for } i \neq j\}.$$

The action of $G_0 = SO^0(3, 19)$ on $\mathcal{M}_{D_+}$ is given by linear fractional transformations,

$$\left(\begin{smallmatrix} A & B \\ C & D \end{smallmatrix}\right) : W \mapsto (AW + B)(CW + D)^{-1},$$

where $W = W_1 + iW_2$ with $W_1 \in \mathcal{B}$ and $W_2 \in \mathcal{B}'$.

Part IV

The Full Cycle Space

Overview

In Part II above it is shown that our group-theoretically defined cycle space $\mathcal{M}_D$ is a closed submanifold of the Barlet cycle space $\mathcal{C}(D)$. It was also shown that $\mathcal{M}_D$ is the associated G_0–bounded symmetric domain in the holomorphic hermitian case, and otherwise is naturally identified with the universal domain $\mathcal{U}$. This last result can be regarded as a negative result which says that very little of the flag domain geometry is reflected in this group-theoretically defined space of cycles.

Here we examine the position of $\mathcal{M}_D$ in the full cycle space $\mathcal{C}(D)$. The main result is the explicit determination of the module structure of the Zariski tangent space $T_{[C]}\mathcal{C}(Z)$ for all base cycles C in open G_0-orbits in arbitrary flag manifolds Z. Since $\mathcal{M}_D$ is the connected component containing $[C]$ of the intersection of the orbit $G([C])$ with $\mathcal{C}(D)$, it is of interest to understand the representation of K on the space V which is transversal to this orbit in the sense of a K-decomposition

$$T_{[C]}\mathcal{C}(Z) = T_{[C]}G([C]) \oplus V.$$

We show that $\mathcal{C}(Z)$ is smooth at $[C]$, and consequently, in the cases where it is nonzero, V really represents additional deformation parameters that are not due to the symmetry group G.

By explicit determination of the module structure we mean that the highest weights of the representation of K on V are given in explicit terms. In fact the representations which occur are very simple, but for a fixed $\mathfrak{g}_0$ several representations can occur. These depend on the cycle and the flag manifold at hand.

For certain real forms $\mathfrak{g}_0$ general methods show that in fact $\mathcal{M}_D$ is the component of $[C]$ in $\mathcal{C}(D)$, i.e., $V = 0$ (Theorem 18.4.13). However, the complementary list, given in Tables 18.5.1 and 18.5.2, is quite substantial, and for many real forms which occur in this list there are indeed cycles C and flag manifolds Z, where $V \neq 0$.

Occasionally G is dimension theoretically smaller than the full automorphism group $\mathrm{Aut}(Z)$, and therefore additional parameters appear for reasons of additional symmetry. However, this rarely happens, and when it does, we still have $\dim \mathcal{C}_{C_0}(Z) > \dim \mathrm{Aut}(Z)([C_0])$ in almost all cases. Thus there are indeed numerous series of real forms where the geometry of the flag domain is reflected in the full cycle space.

Let us now outline our method for carrying out the computations indicated above. Throughout, θ denotes the complex linear extension of the Cartan involution of $\mathfrak{g}_0$. It defines the Cartan decomposition $\mathfrak{g} = \mathfrak{k} + \mathfrak{s}$. A pair (C, Z) consisting of a closed K-orbit C in a G-flag manifold is given. The complex conjugation τ of $\mathfrak{g}$ over $\mathfrak{g}_0$ commutes with θ, and $\sigma := \tau\theta$ is complex conjugation of $\mathfrak{g}$ over a τ-stable θ-stable compact real form $\mathfrak{g}_u$.

In Chapter 17 we begin by choosing a base point z in C with isotropy algebra $\mathfrak{q}$ in $\mathfrak{g}$, such that $\mathfrak{q}$ contains a θ-invariant Borel subalgebra of $\mathfrak{g}$ which in turn contains a *reference* Borel subalgebra $\mathfrak{b}_{\mathfrak{k}}^{\mathrm{ref}}$ of $\mathfrak{k}$. Then $\mathfrak{b}$ contains a fundamental θ-stable τ-stable Cartan subalgebra $\mathfrak{h} \subset \mathfrak{q}$ of $\mathfrak{g}$, so $\mathfrak{t} := \mathfrak{h} \cap \mathfrak{k}$ is a Cartan subalgebra of $\mathfrak{k}$ contained in $\mathfrak{b}_{\mathfrak{k}}^{\mathrm{ref}}$.

The set $M(\mathfrak{t}, \mathfrak{s})$ of weights of $\mathfrak{t}$ on $\mathfrak{s}$ plays a key role in the considerations here. For this reason a description of the full representation of $\mathfrak{k}$ on $\mathfrak{s}$ is given in Section 17.4.

Our work here amounts to computing the representations of K on the cohomology spaces $H^0(C; \mathcal{O}(\mathbb{N}_Z(C)))$ and $H^1(C; \mathcal{O}(\mathbb{N}_Z(C)))$, where $\mathbb{N}_Z(C)$ is the normal bundle of the cycle C in Z. Direct computations show that this can be transferred to computing $H^1(C; \mathcal{O}(\mathbb{E}))$ and $H^2(C; \mathcal{O}(\mathbb{E}))$, where $\mathbb{E}$ is the homogeneous vector bundle on C defined by the isotropy representation of $Q_K := K \cap Q$ on $F := (\mathfrak{q}+\theta\mathfrak{q})\cap\mathfrak{s}$; see Proposition 17.5.1. Note that in the measurable case this simplifies, because in that case $\theta\mathfrak{q} = \mathfrak{q}$. Simplifications in the hermitian case are also noted in Chapter 17.

The most difficult aspects of our work in this part are contained in Chapter 18. After introducing the notation for the relevant root systems, we point out that the set of closed K-orbits in Z, i.e., the set of reference or base cycles in Z, is naturally parameterized by a double coset quotient of the Weyl group, $W^\theta = N_G(\mathfrak{t})/Z_G(\mathfrak{t})$. A convenient set of representatives W_1^θ of the equivalence classes in this quotient is used in an essential way in the final calculations in Chapters 19 and 20, where we transfer the results from the simplest choice of a cycle to results for the other cycles by the action of elements of W_1^θ.

The cohomology groups indicated above are computed by the algorithm of the Bott–Borel–Weil theorem. In the case at hand, where the bundle $\mathbb{E}$ is defined by a representation of the nonreductive isotropy group on F, it is necessary to first go to the filtration $\{F^j\}$ of F defined by the unipotent radical of Q_K, compute the cohomologies for the quotient bundles F^{j+1}/F^j as homogeneous bundles of the reductive part of Q_K, and then work back to the cohomology spaces of the original bundle $\mathbb{E}$.

This is a matter of computing Bott regular weights of index 1 and 2. A priori this might seem to be a difficult matter, but there are major simplifications and reductions which can be made. For example, for the quotients from the filtration, the String Lemma 18.4.4 shows that a weight μ of index 1 can only occur if it is the beginning of a string $\mu, \mu+\beta$, $\mu+2\beta$ in $M(\mathfrak{t}, \mathfrak{s})$, where β is a simple $\mathfrak{t}$-root on $\mathfrak{g}$ and $\mu+\beta$ is dominant. The conditions on weights of index 2 are even more restrictive.

One consequence of the String Lemma is that the list of real forms where the space V is possibly nonzero is now somewhat shorter; see Tables 18.5.1 and 18.5.2. Except for two cases where $\mathfrak{g}_0$ is complex and which are not difficult to handle, the algebra $\mathfrak{k}$ is at worst a direct sum of simple classical algebras. This is used in the concrete calculations of Chapters 19 and 20, where convenient matrix models are employed.

The next major reduction is given by the Cohomological Lemma 18.5.6 which precisely describes those representations that are allowed by the String Lemma and that are to be found in the cohomology of the bundle $\mathbb{E}$. With several exceptions, which must be handled case by case, for example where $\mathfrak{g}_0 = \mathfrak{so}(2p+1, 2q+1)$, this leads to the vanishing theorem which proves the smoothness of $\mathcal{C}(Z)$ at $[C]$ for any pair (C, Z) (Theorem 18.6.1).

In the last two chapters of this part we use the algorithm presented in Chapter 18 to compute the K-module V for every real form in Tables 18.5.1 and 18.5.2 and every pair (C, Z). In Section 20.5 we record the summary as Table 20.5.2. For every $\mathfrak{g}_0$ we give the representations on the transversal space V which occur for some pair (C, Z). In order to determine the pair, the reader must consult the summarizing theorems in the individual sections where the calculations are carried out.

17

Combinatorics of Normal Bundles of Base Cycles

In this chapter we translate the problem of giving the local description of the full cycle spaces to that of computing certain representations.

17.1 Characterization of compact K-orbits

We begin by giving a combinatorial criterion for an orbit $K(z) \subset Z$ to be compact, in terms of the corresponding isotropy subalgebra $\mathfrak{q}_z$. In view of the duality expressed by Proposition 4.3.3 this corresponds to the criterion of Theorem 4.2.2 for $G_0(z)$ to be open in Z. Recall that $\theta : G \to G$ and $\theta : \mathfrak{g} \to \mathfrak{g}$ denote the respective holomorphic extensions of the Cartan involutions $\theta : G_0 \to G_0$ and $\theta : \mathfrak{g}_0 \to \mathfrak{g}_0$, and that the product of any two of θ, τ, σ is equal to the third.

Proposition 17.1.1. *In the following* $\mathfrak{b}$ *is a Borel subalgebra of* $\mathfrak{g}$.

1. *If* $\mathfrak{b}$ *is* θ*-stable, then* $\mathfrak{b}_{\mathfrak{k}} := \mathfrak{b} \cap \mathfrak{k}$ *is a Borel subalgebra of* $\mathfrak{k}$.
2. *If* $\mathfrak{b}$ *contains a Borel subalgebra* $\mathfrak{b}_{\mathfrak{k}}$ *of* $\mathfrak{k}$*, then* $\mathfrak{b}$ *is* θ*-stable.*
3. *If* $z \in Z$*, then* $K(z)$ *is compact if and only if* $\mathfrak{q}_z$ *contains a* θ*-stable Borel subalgebra of* $\mathfrak{g}$.

Proof. For assertion 1, express $\mathfrak{b} = \mathfrak{h} + \sum_{\alpha(\xi)>0} \mathfrak{g}_{-\alpha}$, where $\mathfrak{h}$ is a θ-stable Cartan subalgebra of $\mathfrak{g}$ and ξ is a regular element of $\mathfrak{h}$. As $\theta(\mathfrak{b}) = \mathfrak{b}$ we can replace ξ by $\theta(\xi)$ and then by $\xi + \theta(\xi)$. Hence we may assume $\xi \in \mathfrak{k}$. In other words, $\mathfrak{t} := \mathfrak{h} \cap \mathfrak{k}$ is a Cartan subalgebra of $\mathfrak{k}$ and $\mathfrak{b}_{\mathfrak{k}} = \mathfrak{t} + \sum_{\nu(\xi)>0} \mathfrak{k}_{-\nu}$. Thus $\mathfrak{b}_{\mathfrak{k}}$ is a Borel subalgebra of $\mathfrak{k}$.

Assertion 2 is Lemma 4.3.2.

If $K(z)$ is closed in Z, then the proof of Proposition 4.3.3 constructs a θ-stable Borel subalgebra of $\mathfrak{g}$ that is contained in $\mathfrak{q}_z$. If $\mathfrak{q}_z$ contains a θ stable Borel subalgebra $\mathfrak{b} \subset \mathfrak{g}$, then $K(z)$ is closed in Z by Assertion 1. This completes the proof of Proposition 17.1.1. □

17.2 Base cycles and the arrangement of Borel subgroups

Given a flag manifold G/Q and a base cycle $C = K(z) \subset Z$ we wish to locally describe the cycle space $\mathcal{C}_{[C]}(Z)$ at the point $[C]$. In this and in the following sections we develop combinatorial tools which allow us to compute $T_{[C]}\mathcal{C}(Z)$ in an algorithmic way. Here we explain the first step toward transition of our geometric situation to the combinatorics of various root systems. Given an arbitrary pair (C, Z) (in other words, given a real form $\mathfrak{g}_0$ of a complex semisimple $\mathfrak{g}$, where we have selected a G-flag manifold Z and a base cycle $C = K(z)$) we attach to it certain Lie algebra data, $\mathfrak{b}_{\mathfrak{k}} \subset \mathfrak{b} \subset \mathfrak{q} \subset \mathfrak{g}$, which encode all geometric information of the pair (C, Z).

Let $\mathfrak{g}_0$ and a complex G-homogeneous flag manifold Z be given, let X be the corresponding G-homogeneous full flag manifold, and $\pi : X \to Z$ a fixed G-equivariant fibration. Proposition 17.1.1 gives us a useful description of closed K-orbits in a complex flag manifold Z : Let $C = K(z) \subset Z$ be such a closed orbit (i.e., a base cycle). Then $\pi^{-1}(C)$ contains a (possibly not unique) base cycle $C_X = K(x) \subset X$. Given any such C_X, we may assume that $\pi(x) = z$. The group data is now obtained as follows. The selection of x determines the corresponding isotropy group which is a Borel subgroup $B_{\mathfrak{k}} \subset K$. Further, the points $x \in X$ and $z \in Z$ give rise to the groups $B = B_x$ and $Q = Q_z$, which are a Borel and a parabolic subgroup of G. Hence, we have obtained the group data $B_{\mathfrak{k}} \subset B_x \subset Q_z \subset G$. Note that, according to Proposition 17.1.1, the Borel subgroup $B = B_x$ is θ-stable and $B_{\mathfrak{k}} = B \cap K$. In this situation we often refer to the Borel subgroup $B_{\mathfrak{k}}$ of K as "small," and the Borel subgroup B of G as "large." Note also that $Q_{\mathfrak{k}} := Q_z \cap K$ is a parabolic subgroup in K such that $C \cong K/Q_{\mathfrak{k}}$; see the following diagram:

(17.2.1)
$$\begin{array}{ccccc} K/B_K & \xrightarrow[\text{of } B=B_x]{\text{choice}} & K(x) = C_X \subset \pi^{-1}(C) \subset X = G/B_x \\ & & \downarrow \qquad \downarrow \qquad\qquad \downarrow \pi \\ K/Q_{\mathfrak{k}} \cong K(z) = C & & \hookrightarrow \qquad Z = G/Q_z. \end{array}$$

The subgroups $B_{\mathfrak{k}} \subset B \subset Q$ are not unique. However, any two Borel subgroups of K are conjugate, and the choice of such a (small) Borel subgroup amounts to a choice of a base point $x \in C_X$. Therefore, given $\mathfrak{g}_0$, or, equivalently, given $(\mathfrak{g}, \theta)$ (or $\mathfrak{g} = \mathfrak{k} + \mathfrak{s}$) we may fix once and for all a Borel subgroup $B_{\mathfrak{k}}^{\text{ref}}$ and consider only quadruples $B_{\mathfrak{k}}^{\text{ref}} \subset B_x \subset Q_z \subset G$. Finally, since all the above (sub)groups are connected, there is no lost in passing to their Lie algebras. Recapitulating, given $(\mathfrak{g}, \theta)$, to an arbitrary base cycle C in a G-flag manifold Z we associate the quadruple of Lie algebras

(17.2.2)
$$\mathfrak{b}_{\mathfrak{k}}^{\text{ref}} \subset \mathfrak{b} \subset \mathfrak{q} \subset \mathfrak{g}$$

we refer to any such quadruple of Lie algebras as **Lie algebra data associated to** (C, Z). It should be noted that once $\mathfrak{b}_{\mathfrak{k}}^{\text{ref}}$ is fixed, there may be several (but finitely many) large Borel subalgebras $\mathfrak{b}$ containing $\mathfrak{b}_{\mathfrak{k}}^{\text{ref}}$. Corollary 18.3.6 describes in terms of certain subgroups of the Weyl group $W(\mathfrak{g}, \mathfrak{h})$ the possibilities for the B which contain a given $B_{\mathfrak{k}}$. Once $\mathfrak{b}$ is selected, in every conjugacy classes of parabolic subalgebras in $\mathfrak{g}$ there is precisely one representative $\mathfrak{q}$ which contains $\mathfrak{b}$. In this situation, the given flag manifold Z determines $\mathfrak{q}$. Thus given Lie algebra data $\mathfrak{b}_{\mathfrak{k}}^{\text{ref}} \subset \mathfrak{b} \subset \mathfrak{q} \subset \mathfrak{g}$ uniquely determine a base cycle C in a flag manifold Z.

Geometrically summarized, a choice of Borel subgroup B containing $B_{\mathfrak{k}}$ amounts to a choice of closed K-orbit in G/B which lies over $K(z)$ by the projection $G/B \to G/Q$. Note that all such orbits are equivalent as abstract K-manifolds, but their positions in the cycle space $\mathcal{C}(X)$, or even in the homology of X, may be very different.

17.3 Normal bundles of base cycles

Consider a closed K-orbit (in other words a base cycle) $C = K(z)$ in Z. Then K acts naturally on the holomorphic normal bundle

$$\mathbb{N}_Z(C) := (\mathbb{T}Z|_C)/\mathbb{T}C,$$

and this gives $\mathbb{N}_Z(C)$ the structure of a K-homogeneous holomorphic vector bundle. In this section we study the structure of that bundle.

In the case of the full flag manifold $X = G/B$ every base cycle C is biholomorphically equivalent to the (small) full flag manifold $K/B_{\mathfrak{k}}$. The particular embedding $C \hookrightarrow X$ plays an essential role by the computation of the various cohomology groups associated with the given normal bundle. Let $\mathfrak{b} \supset \mathfrak{b}_{\mathfrak{k}} = \mathfrak{b}_{\mathfrak{k}}^{\text{ref}}$ be the corresponding Lie algebras as in Section 17.2 and $B \supset B_{\mathfrak{k}}$ the corresponding Lie groups. Recall $B = \theta(B)$, so that $\mathfrak{b} = \mathfrak{b}_{\mathfrak{k}} + \mathfrak{b}_{\mathfrak{s}}$. Therefore

$$\begin{aligned} \mathbb{N}_X(C) &= (\mathbb{T}X|_C)/\mathbb{T}C = (K \times_{B_{\mathfrak{k}}} \mathfrak{g}/\mathfrak{b})/(K \times_{B_{\mathfrak{k}}} \mathfrak{k}/\mathfrak{b}_{\mathfrak{k}}) \\ &= (K \times_{B_{\mathfrak{k}}} (\mathfrak{k}/\mathfrak{b}_{\mathfrak{k}} + \mathfrak{s}/\mathfrak{b}_{\mathfrak{s}}))/(K \times_{B_{\mathfrak{k}}} \mathfrak{k}/\mathfrak{b}_{\mathfrak{k}}) = K \times_{B_{\mathfrak{k}}} \mathfrak{s}/\mathfrak{b}_{\mathfrak{s}}. \end{aligned} \tag{17.3.1}$$

The $B_{\mathfrak{k}}$-module $\mathfrak{s}/\mathfrak{b}_{\mathfrak{s}}$ is the normal space $\mathbb{N}_X(C)_x$, and the module structure comes from the adjoint action of K on $\mathfrak{s}$.

The above description (17.3.1) of the normal bundle of C can be generalized to an arbitrary flag manifold $Z = G/Q$. In this general case there are two complications. First, for a given flag $Z = G/Q$ and real form G_0 of G, the base K-cycles may not be diffeomorphic or even of the same dimension. Second, although $\mathfrak{q}_z$ contains a θ-stable Borel subalgebra $\mathfrak{b}$, the parabolic $\mathfrak{q}_z$ need not be θ-stable.

To look at the latter point in more detail, consider the action θ as a graph automorphism of the Dynkin diagram derived from the positive root system $\Sigma^+ = \Sigma^+(\mathfrak{g}, \mathfrak{h})$ defined by $\mathfrak{b}$. Then $\theta(\Sigma^+) = \Sigma^+$, and we regard parabolic algebra $\mathfrak{q}_z$ as being $\mathfrak{q}_\Phi$ for some set Φ of simple roots of Σ^+.

Lemma 17.3.2. *The following conditions are equivalent:* (i) $\theta(\mathfrak{q}_z) = \mathfrak{q}_z$, (ii) $\theta(\Phi) = \Phi$, (iii) $\tau(\Phi) = -\Phi$, *and* (iv) *the open G_0-orbits on Z are measurable.*

Proof. Equivalence of (i) and (ii) is obvious. If $\alpha \in \mathfrak{h}_{\mathbb{R}}^*$, then $\tau(\alpha) = -\theta(\alpha)$. Thus (ii) and (iii) are equivalent. Equivalence of (iii) and (iv) follows from Theorem 4.5.1 and the observation that the condition (d) of Theorem 4.5.1 is equivalent to $\tau(\Phi) = -\Phi$. □

Theorem 4.5.1 and the classification of real simple Lie algebras now give us the following.

Proposition 17.3.3. *Let G_0 be a noncompact connected real simple Lie group. Then the only possibilities for $\mathfrak{g}_0$ for which there exists a parabolic subgroup $Q \subset G$ with $\theta(\mathfrak{q}_z) \neq \mathfrak{q}_z$ in the above setting are as follows:*

(i) $\mathfrak{g}_0 = \mathfrak{sl}(n;\mathbb{R})$ *(here* $\mathfrak{g} = \mathfrak{sl}(n;\mathbb{C})$ *and* $\mathfrak{k} = \mathfrak{so}(n;\mathbb{C})$, $n > 2$),
(ii) $\mathfrak{g}_0 = \mathfrak{sl}(n;\mathbb{H})$ $(= \mathfrak{su}^*(2n))$ *(here* $\mathfrak{g} = \mathfrak{sl}(2n;\mathbb{C})$ *and* $\mathfrak{k} = \mathfrak{sp}(n;\mathbb{C})$, $n > 1$),
(iii) $\mathfrak{g}_0 = \mathfrak{so}(2p+1, 2q+1)$ *(here* $\mathfrak{g} = \mathfrak{so}(2n+2;\mathbb{C})$ *and*

$$\mathfrak{k} = \mathfrak{so}(2p+1;\mathbb{C}) \oplus \mathfrak{so}(2q+1;\mathbb{C}),\ p+q > 0),$$

(iv) $\mathfrak{g}_0 = \mathfrak{e}_{6,\mathfrak{f}_4}$ $(= \mathfrak{e}_{6(-26)})$ *(here* $\mathfrak{g} = \mathfrak{e}_6(\mathbb{C})$ *and* $\mathfrak{k} = \mathfrak{f}_4(\mathbb{C})$),
(v) $\mathfrak{g}_0 = \mathfrak{e}_{6,\mathfrak{c}_4}$ $(= \mathfrak{e}_{6(6)})$ *(here* $\mathfrak{g} = \mathfrak{e}_6(\mathbb{C})$ *and* $\mathfrak{k} = \mathfrak{sp}(4;\mathbb{C})$), *and*
(vi) *G_0 is the underlying real group of a complex simple Lie group.*

If $\mathfrak{l}$ is a θ-stable subspace of $\mathfrak{g}$ we decompose $\mathfrak{l} = \mathfrak{l}_\mathfrak{k} + \mathfrak{l}_\mathfrak{s}$ under θ. In the setting of Section 17.1, where $K(z)$ is a closed K-orbit in Z, the normal bundle to the base cycle is described as follows.

Theorem 17.3.4. *Let $C = K(z)$ be a base cycle in Z. Then its holomorphic normal bundle is the K-homogeneous holomorphic vector bundle*

$$\mathbb{N}_Z(C) = K \times_{(Q_z \cap K)} (\mathfrak{s}/(\mathfrak{q}_z + \theta\mathfrak{q}_z)_\mathfrak{s}).$$

In the case where the open G_0-orbits on Z are measurable, this reduces to $\mathbb{N}_Z(C) = K \times_{(Q_z \cap K)} (\mathfrak{s}/\mathfrak{q}_\mathfrak{s})$.

Proof. In the measurable case, where the parabolic subalgebra $\mathfrak{q}_z = \mathfrak{q}$ is θ-stable, we have the decomposition $\mathfrak{q}_z = \mathfrak{q}_\mathfrak{k} + \mathfrak{q}_\mathfrak{s}$, and it is clear that $\mathbb{T}_z Z/\mathbb{T}_z C = (\mathfrak{g}/\mathfrak{q}_z)/(\mathfrak{k}/\mathfrak{q}_\mathfrak{k}) = \mathfrak{s}/\mathfrak{q}_\mathfrak{s}$ is the normal space. In that case, since $\mathfrak{q}_\mathfrak{s} = \theta\mathfrak{q}_\mathfrak{s}$, we have $\mathbb{N}_Z(C) = K \times_{Q_z \cap K} (\mathfrak{s}/(\mathfrak{q}_z + \theta\mathfrak{q}_z)_\mathfrak{s})$ as asserted.

In the general case, the computation of $\mathbb{T}_z Z/\mathbb{T}_z C$ is only a bit more involved and boils down to an exercise in linear algebra. Select a Cartan subalgebra $\mathfrak{t} \subset \mathfrak{k}$ which is contained in the isotropy algebra $\mathfrak{q} = \mathfrak{q}_z$ at z. Define $\tilde{\mathfrak{q}} := \mathfrak{q} \cap \theta\mathfrak{q}$ and select a complement $\mathfrak{f} \subset \mathfrak{q}$ of $\tilde{\mathfrak{q}}$ consisting of root spaces. Select a θ-stable complement $\mathfrak{r} \subset \mathfrak{g}$ of $\mathfrak{q}+\theta\mathfrak{q}$ (also consisting of root spaces). Note that $\mathfrak{r}$ can be identified with $\mathfrak{g}/(\mathfrak{q}+\theta\mathfrak{q})$ and $\mathfrak{r}_\mathfrak{s}$ with $\mathfrak{s}/(\mathfrak{q}+\theta\mathfrak{q})_\mathfrak{s}$. Further, note the following decompositions: $\mathfrak{q} = \tilde{\mathfrak{q}} \oplus \mathfrak{f}$, $\mathfrak{q}+\theta\mathfrak{q} = \tilde{\mathfrak{q}} \oplus \mathfrak{f} \oplus \theta\mathfrak{f}$ and $\mathfrak{g} = \tilde{\mathfrak{q}} \oplus \mathfrak{f} \oplus \theta\mathfrak{f} \oplus \mathfrak{r}$. Since $\mathfrak{f} \cap \mathfrak{k} = 0 = \mathfrak{f} \cap \mathfrak{s}$, we have $\mathfrak{f} \oplus \theta\mathfrak{f} = (\mathfrak{f}+\theta\mathfrak{f})_\mathfrak{k} \oplus \mathfrak{f}$. Now the typical fiber of $\mathbb{N}_Z(C)$ is

$$\begin{aligned}
\mathfrak{g}/(\mathfrak{k}+\mathfrak{q}) &= \Big((\mathfrak{q}+\theta\mathfrak{q}) + \mathfrak{r}\Big)\Big/\Big(\mathfrak{k} + \mathfrak{f} + (\mathfrak{q}\cap\theta\mathfrak{q})\Big) \\
&= \Big((\mathfrak{q}\cap\theta\mathfrak{q}) + (\mathfrak{f}+\theta\mathfrak{f}) + \mathfrak{r}\Big)\Big/\Big(\mathfrak{k} + \mathfrak{f} + (\mathfrak{q}\cap\theta\mathfrak{q})\Big) \\
&= (\mathfrak{f}+\theta\mathfrak{f}+\mathfrak{r})/(\mathfrak{k}+\mathfrak{f}) = ((\mathfrak{f}+\theta\mathfrak{f})_\mathfrak{k} + \mathfrak{f} + \mathfrak{r})/(\mathfrak{k}+\mathfrak{f}) = \mathfrak{r}_\mathfrak{s} = \mathfrak{s}/(\mathfrak{q}+\theta\mathfrak{q})_\mathfrak{s},
\end{aligned}$$

as asserted. □

Remark 17.3.5. The subalgebra $\widetilde{\mathfrak{q}} := \mathfrak{q} \cap \theta\mathfrak{q}$ (in the above proof) is parabolic, because $\mathfrak{q} = \mathfrak{q}_z$ contains a θ-stable Borel subalgebra $\mathfrak{b}_z$ by Proposition 17.1.1. Define $\widetilde{Z} := G/\widetilde{Q}$ and note that $\pi : \widetilde{Z} \to Z = G/Q$ is the minimal measurable fibration of [W9]. Define the points $z = 1Q \in Z$ and $\widetilde{z} = 1\widetilde{Q} \in \widetilde{Z}$. Note that $\widetilde{C} := K \cdot \widetilde{z}$ is a base cycle in $\widetilde{Z}$. By construction $\mathfrak{q} \cap \mathfrak{k} = \widetilde{\mathfrak{q}} \cap \mathfrak{k}$, giving an immediate proof of the fact [W9, Lemma 2.5] that the natural projection $\widetilde{Z} \to Z$ maps $\widetilde{C}$ biholomorphically onto C.

It will turn out that the components of C and $\widetilde{C}$ of their respective Barlet cycle spaces $\mathcal{C}(Z)$ and $\mathcal{C}(\widetilde{Z})$ satisfy $\dim \mathcal{C}_C(Z) \leqq \dim \mathcal{C}_{\widetilde{C}}(\widetilde{Z})$. ◇

The space $H^0(C; \mathcal{O}(\mathbb{N}_Z(C)))$ of global sections of the holomorphic normal bundle describes the Zariski tangent space of $\mathcal{C}(Z)$ at its point C, and the space $H^1(C; \mathcal{O}(\mathbb{N}_Z(C)))$ is the obstruction to the smoothness of $\mathcal{C}(Z)$ at C. The main goal of this part is to determine these cohomology groups for all complex flag manifolds $Z = G/Q$ and all pairs of associated symmetric subgroups (G_0, K), where G_0 is a noncompact real form of G, and all of the base cycles $C_j = K(z_j)$.

The bundles here are K-homogeneous vector bundles over a small flag manifolds $C = K/Q_K$, where Q_K denotes $Q \cap K$. In general let $\mathbb{E}(F) \to K/Q_K$ denote the homogeneous holomorphic vector bundle defined by a Q_K-module F. Thus, for example, Theorem 17.3.4 says that $\mathbb{N}_Z(C) = \mathbb{E}(\mathfrak{s}/(\mathfrak{q}_z + \theta(\mathfrak{q}_z))_{\mathfrak{s}})$.

If Q_K acts irreducibly on the holomorphic normal space $\mathfrak{s}/(\mathfrak{q}_z + \theta(\mathfrak{q}_z))_{\mathfrak{s}}$, then the Bott–Borel–Weil Theorem gives a complete description of the K-modules $H^0(C; \mathcal{O}(\mathbb{N}_Z(C)))$ and $H^1(C; \mathcal{O}(\mathbb{N}_Z(C)))$. However, in most cases the relevant Q_K-modules are not even completely reducible, and we must consider certain of their filtrations. The Bott–Borel–Weil Theorem does not apply directly in that case, and Griffiths' considerations of reducible bundles [Gr2] really require complete reducibility. We therefore need further tools in order to extract the information encoded in these filtrations and determine the various cohomology groups $H^j(C; \mathcal{O}(\mathbb{N}_Z(C))) = H^j(C; \mathcal{O}(\mathbb{E}((\mathfrak{q} + \theta\mathfrak{q})_{\mathfrak{s}})))$, at least for $j = 0$ and $j = 1$.

17.4 Module structure of the tangent space of a symmetric space

Here we recall some facts concerning the symmetric spaces $\Omega_0 = G_0/K_0$, their compact duals $\Omega_u = G_u/K_0$, and their complexifications $\Omega = G/K$. Our main point is to describe the isotropy (tangent space) representation of $\mathfrak{k}$ on the complexified tangent space $\mathfrak{s}$. This material is taken from [WG1, Proposition 2.11] and [WG1, Theorem 5.10]. Our description follows the treatment of [W6, Chapter 8]; there the tangent space representations appear in table (8.11.2) for rank $\mathfrak{g}$ = rank $\mathfrak{k}$ and in table (8.11.5) if rank $\mathfrak{g}$ > rank $\mathfrak{k}$.

The Cartan involution θ of $\mathfrak{g}_0$ preserves every simple ideal, and $(\mathfrak{g}_0, \theta)$ then breaks up as a direct sum of real simple Lie algebras with Cartan involution. This corresponds to the decomposition of $(\mathfrak{g}, \theta)$ as a direct sum of minimal θ-stable ideals. In more geometric terms, it corresponds exactly to the de Rham decomposition of Ω_0

(or, equivalently, Ω_u) as a riemannian product of irreducible riemannian manifolds. Thus we reduce our considerations to the irreducible case, i.e., to the case in which $\mathfrak{g}_0$ is simple.

Now assume that Ω_0 is irreducible, i.e., that $\mathfrak{g}_0$ is simple. If $\mathfrak{k}_0$ is semisimple, then Ω_0 does not have a G_0-invariant complex structure and the isotropy representation of $\mathfrak{k}$ on $\mathfrak{s}$ is irreducible. That is the non-hermitian case. If $\mathfrak{k}$ fails to be semisimple, then Ω_0 has a G_0-invariant complex structure and the isotropy representation of $\mathfrak{k}$ on $\mathfrak{s}$ is reducible. That is the hermitian case.

In the hermitian case we always have rank $\mathfrak{g}$ = rank $\mathfrak{k}$, the center of $\mathfrak{k}_0$ has dimension 1, and the decomposition of $\mathfrak{s}$ into irreducible $\mathfrak{k}$ modules is of the form $\mathfrak{s} = \mathfrak{s}_+ + \mathfrak{s}_-$ (See the first few paragraphs of Chapter 3.). Here $\mathfrak{s}_+$ represents the holomorphic tangent space of Ω_0 and $\mathfrak{s}_-$ represents the antiholomorphic tangent space. Further, $\mathfrak{k} + \mathfrak{s}_\pm$ are maximal parabolic subalgebras of $\mathfrak{g}$ with respective nilradicals $\mathfrak{s}_\pm$, and the elements of $\mathfrak{s}_\pm$ are Ad-nilpotent on all of $\mathfrak{g}$.

We first look at the tangent space representations of $\mathfrak{k}$ on $\mathfrak{s}$ in the cases rank $\mathfrak{g}$ = rank $\mathfrak{k}$. Let $\Psi = \{\psi_1, \dots, \psi_r\}$ denote the simple root system, in other words, the set of vertices of the Dynkin diagram $\Delta(\mathfrak{g}, \mathfrak{h})$; roots ψ_i and ψ_j are attached by $\frac{2\langle\psi_i,\psi_j\rangle}{\langle\psi_j,\psi_j\rangle}\frac{2\langle\psi_j,\psi_i\rangle}{\langle\psi_i,\psi_i\rangle}$ lines with an arrowhead pointing toward ψ_i in case $\langle\psi_i, \psi_i\rangle < \langle\psi_j, \psi_j\rangle$. Let ψ_0 denote the negative of the maximal root and express that maximal root as $\sum n_i\psi_i$. Form the **extended Dynkin diagram** $\widetilde{\Delta}(\mathfrak{g}, \mathfrak{h})$; its set of vertices is $\widetilde{\Psi} = \{\psi_0\} \cup \Psi = \{\psi_0, \psi_1, \dots, \psi_r\}$, and the rules for edges are the same as for $\Delta(\mathfrak{g}, \mathfrak{h})$. Now write 1 at the vertex for the negative ψ_0 of the maximal root and write n_i at the vertex for ψ_i, $1 \leqq i \leqq r$.

Recall the result of Borel and de Siebenthal [BoS] on the maximal subalgebras of maximal rank in $\mathfrak{g}_u$. First, they show that every maximal connected subgroup $L_u \subset G_u$, such that rank K_u = rank G_u, has a central element z such that L_u is the identity component of the centralizer of z in G_u. Then they look at the possibilities for z and L_u. They start with the **fundamental simplex**

$$\mathcal{D} = \{h \in \mathfrak{h}_{\mathbb{R}} \mid \psi_j(h) \geqq 0 \text{ for } 1 \leqq j \leqq r \text{ and } (-\psi_0)(h) \leqq 1\},$$

and note the result of Cartan which says that every element of G_u is conjugate to an element of $\exp(2\pi i\mathcal{D})$. The vertices of $\mathcal{D}$ are $\{v_0, \dots, v_r\}$, where $v_0 = 0$, $\psi_i(v_j) = 0$ for $i \neq j$, and $\psi_j(v_j) = \frac{1}{n_j}$ for $1 \leqq j \leqq r$. In other words, if $j > 0$, then v_j corresponds under the Killing form to $\frac{1}{n_j}\frac{2\xi_j}{\langle\psi_j,\psi_j\rangle}$, where ξ_j is the jth fundamental highest weight. Here is the theorem of Borel and de Siebenthal.

Theorem 17.4.1 ([BoS]). *Let the compact connected simple Lie group G_u, the fundamental simplex $\mathcal{D}$, and its vertices v_j be given as just above. Let $1 \leqq j \leqq r$.*

1. *Suppose $n_j = 1$. Let L_u be the centralizer of the circle group $T_u^1 = \{\exp(2\pi i t v_j) \mid t \in \mathbb{T}\}$. Then L_u is a maximal connected subgroup of maximal rank in G_u, $L_u = T_u^1 \cdot L_u'$ local direct product where L_u' is semisimple, and $\{\psi_1, \dots, \psi_{j-1}, \psi_{j+1}, \dots, \psi_r\}$ is a system of simple roots for $\mathfrak{l}_u'$. If $g \in T_u^1$ is not central in G_u, then L_u is the identity component of the centralizer of g.*

2. *Suppose that n_j is a prime $p > 1$. In this case $g_j := \exp(2\pi i v_j)$ has order p modulo the center of G_u, in other words, $\mathrm{Ad}(g_j)$ is an inner automorphism of order p. Let L_u be the identity component of the centralizer of g_j. Then L_u is a maximal connected subgroup of maximal rank in G_u, L_u is semisimple, and $\{\psi_0, \psi_1, \dots, \psi_{j-1}, \psi_{j+1}, \dots, \psi_r\}$ is a system of simple roots for $\mathfrak{l}_u$.*

3. *If L_u is a maximal connected subgroup of maximal rank in G_u, then L_u is conjugate to one of the groups described just above.*

The case $n_j = 1$ gives the hermitian symmetric spaces, and the case $n_j = 2$ gives the non-hermitian ones where rank $\mathfrak{g}$ = rank $\mathfrak{k}$. To see this explicitly we look at the possibilities for the extended Dynkin diagram and the coefficients n_j. Using the Bourbaki ordering on Ψ, they are given in Table 17.4.2 for the classical cases and in Table 17.4.3 for the exceptional cases.

This leads directly to the tangent space representations when rank $\mathfrak{g}$ = rank $\mathfrak{k}$. The result is more or less obvious in the hermitian case, and in the non-hermitian case it seems to have been written down for the first time in [WG1, Proposition 2.11] (or see [W6, Theorem 8.10.9]). The result is Theorem 17.4.4 below.

Table 17.4.2. Extended Dynkin Diagram $\widetilde{\Delta}(\mathfrak{g}, \mathfrak{h})$ with Coefficients

$A_r, r \geqq 1$	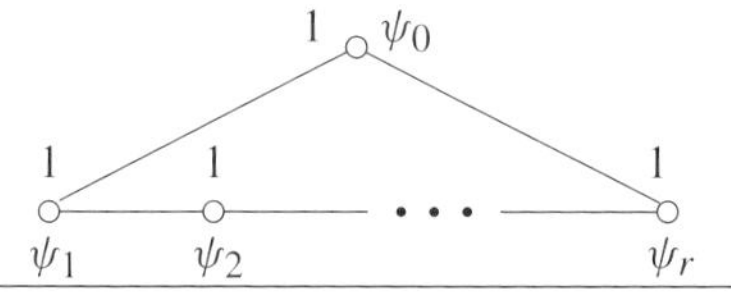
$B_r, r \geqq 2$	
$C_r, r \geqq 3$	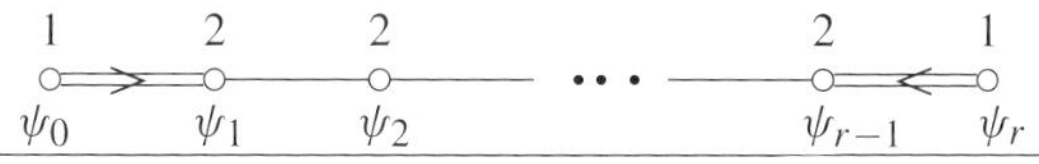
$D_r, r \geqq 4$	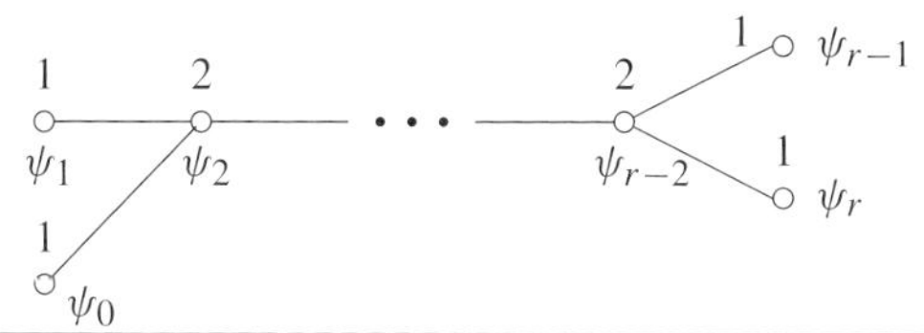

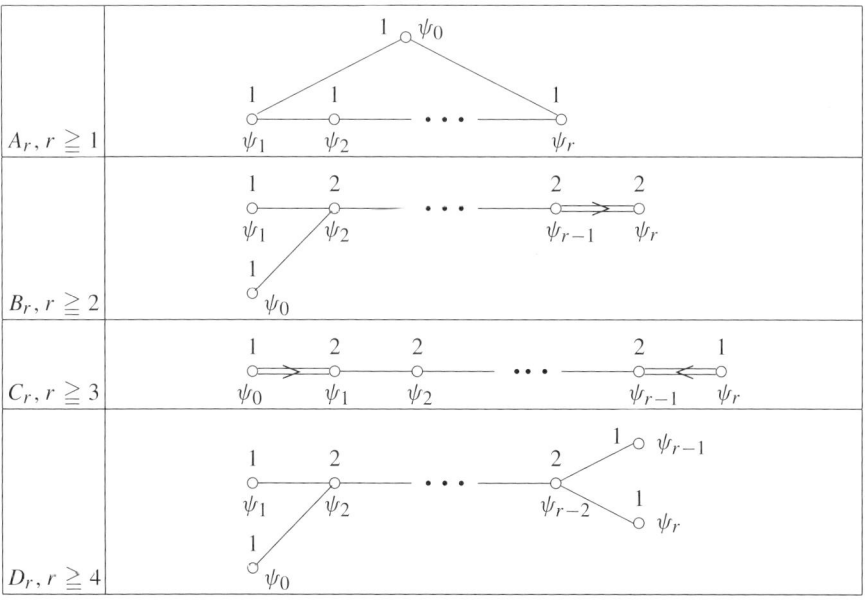

Table 17.4.3. Extended Dynkin Diagram $\widetilde{\Delta}(\mathfrak{g}, \mathfrak{h})$ with Coefficients

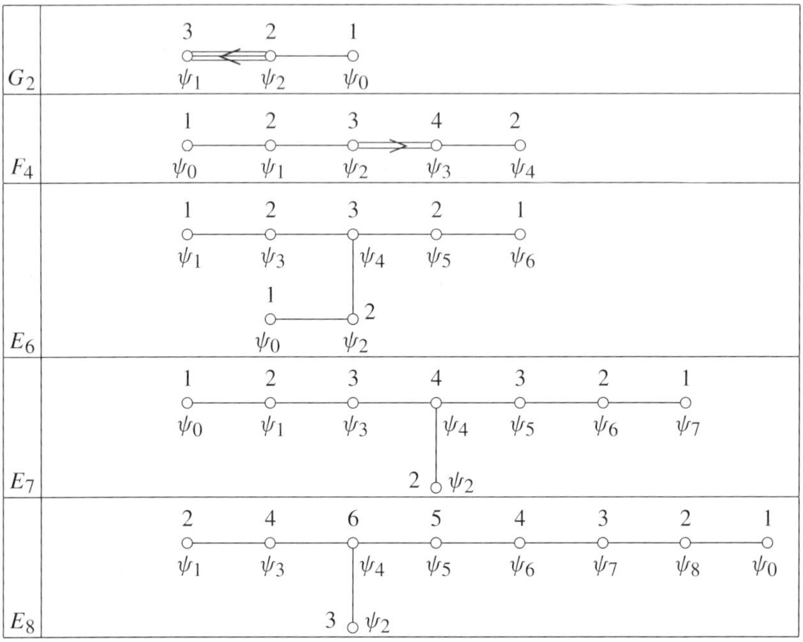

Theorem 17.4.4. *Suppose that* $\mathfrak{g}_0$ *is simple and* rank $\mathfrak{g} =$ rank $\mathfrak{k}$.

1. *Suppose first that* G_0 *is of hermitian type, so that* $\mathfrak{k} = \mathfrak{z}_{\mathfrak{k}} \oplus \mathfrak{k}'$, *where* $\mathfrak{k}' = [\mathfrak{k}, \mathfrak{k}]$ *is the semisimple component of* $\mathfrak{k}$, $\mathfrak{z}_{\mathfrak{k}}$ *is the center, and* $\dim \mathfrak{z}_{\mathfrak{k}} = 1$. *Denote* $\mathfrak{h}' = \mathfrak{h} \cap \mathfrak{k}'$. *Then* $\Delta(\mathfrak{k}', \mathfrak{h}')$ *is obtained from* $\Delta(\mathfrak{g}, \mathfrak{h})$ *by deleting a vertex* ψ_j *with* $n_j = 1$. *The tangent space representation is a direct sum* $\nu \oplus \nu^*$, *where* ν *is the representation on* $\mathfrak{s}_+$. *The representation* ν *is irreducible, its highest weight is the maximal root* $-\psi_0$, *and* $\nu = \chi \otimes \nu'$ *where* $\chi = -\psi_0|_{\mathfrak{z}_{\mathfrak{k}}}$ *and* ν' *represents* $\mathfrak{k}'$.

2. *In the non-hermitian case,* $\mathfrak{k}$ *is semisimple,* $\Delta(\mathfrak{k}, \mathfrak{h})$ *is obtained from* $\widetilde{\Delta}(\mathfrak{g}, \mathfrak{h})$ *by deleting a vertex* ψ_j, $(1 \leqq j \leqq r)$ *with* $n_j = 2$, *and the tangent space representation is the irreducible representation* $\nu_{-\psi_j}$ *of* $\mathfrak{k}$ *of highest weight* $-\psi_j$.

3. *Specifically,* $\Delta(\mathfrak{k}', \mathfrak{h}')$ *and the tangent space representation of* $\mathfrak{k}$ *are given by Tables* 17.4.5 *and* 17.4.6. (*Here recall that the* ξ_i *are the fundamental highest weights, and that we denote the representation of highest weight* $\sum n_i \xi_i$ *by writing* n_i *over the* i*th node of the Dynkin diagram whenever* $n_i \neq 0$. *Also recall that if* $\mathfrak{u}$ *is a semisimple* $Ad(\mathfrak{t})$*-module, then* $M(\mathfrak{t}, \mathfrak{u})$ *denotes the weights (joint eigenvalues) of* $\mathfrak{t}$ *on* $\mathfrak{u}$).

HERMITIAN CASE: $\nu = \chi \otimes \nu'$ *has highest weight* $-\psi_0$ *in the root numbering of* $\widetilde{\Delta}(\mathfrak{g}, \mathfrak{h})$. *In Table* 17.4.5 *the* β_i *are the reordering of those* ψ_j *to conform to the Bourbaki root order, and the* ξ_i *are the corresponding fundamental simple weights.*

NON-HERMITIAN CASE: *ν has highest weight $-\psi_j$, $n_j = 2$, in the root numbering of $\widetilde{\Delta}(\mathfrak{g}, \mathfrak{h})$. In Table 17.4.6 the β_i form the reordering of the ψ to conform to the Bourbaki root order, and the ξ_i are the corresponding fundamental simple weights.*

Table 17.4.5. Tangent Space Representations: Hermitian Cases

$\mathfrak{g}_0$	Representation of $\mathfrak{k}'$ on $\mathfrak{s}_+$	nonmax dominant weights in $M(\mathfrak{t}, \mathfrak{s}_+)$	χ in $M(\mathfrak{t}, \mathfrak{s}_+)$?
$\mathfrak{su}(1,q), q > 1$	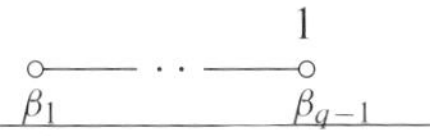		no
$\mathfrak{su}(p,q), 2 \leqq p \leqq q$	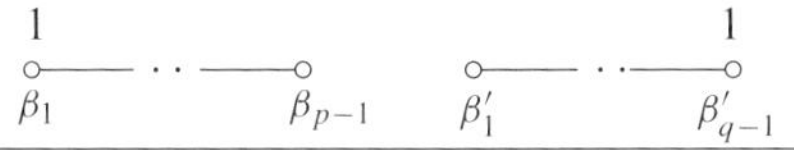		no
$\mathfrak{so}(2, 2q+1)$	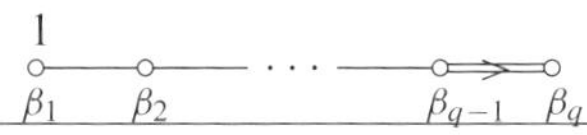	χ	yes
$\mathfrak{sp}(r; \mathbb{R})$	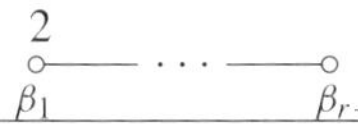	$\xi_2 + \chi$	no
$\mathfrak{so}(2, 2q), q \geqq 1$	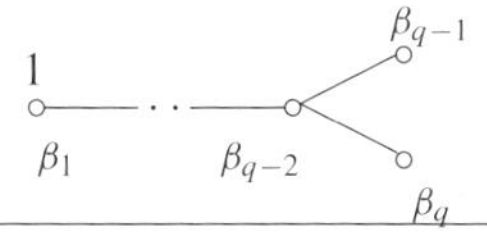		no
$\mathfrak{so}^*(2r), r \geqq 3$	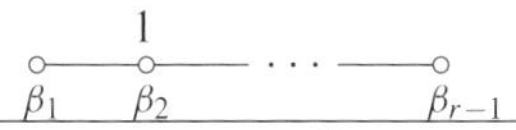		no
$\mathfrak{e}_{6,T_1D_5} \equiv \mathfrak{e}_{6(-14)}$	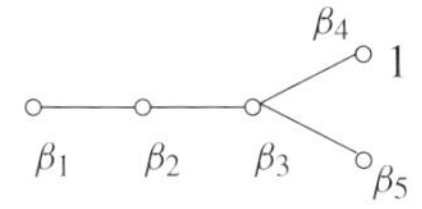		no
$\mathfrak{e}_{7,T_1E_6} \equiv \mathfrak{e}_{7(-25)}$	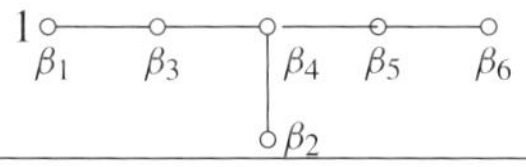		no

Proof. We take the description of the isotropy representation from [W6, Theorem 8.10.9]. In the hermitian case the tangent space representation has form $\nu \oplus \nu^*$. Here ν is irreducible with representation space $\mathfrak{s}_+$ that contains the maximal root space. The root order on $\Sigma(\mathfrak{k}, \mathfrak{h})$ is induced from $\Sigma^+(\mathfrak{g}, \mathfrak{h})$. Thus the maximal root is the highest weight of ν, as asserted.

In the non-hermitian case the tangent space representation ν is irreducible and $-\psi_i$ is one of its weights. If $0 < j \neq i$, then $\psi_j + (-\psi_i)$ is not a root, hence not a weight of ν. Furthermore, $\psi_0 + (-\psi_i)$ is not a root and hence it is not a weight of ν, because its negative would be of the form (maximal root) + (simple root). Thus $-\psi_i$ is the highest weight of a summand of the irreducible representation ν.

The additional information in Tables 17.4.5 and 17.4.6 follows by direct computation. □

Table 17.4.6. Tangent Space Representations: Equal Rank Non-Hermitian Cases

$\mathfrak{g}_0$	Representation of $\mathfrak{k}$ on $\mathfrak{s}$	nonmax dominant weights in $M(\mathfrak{t}, \mathfrak{s})$
$\mathfrak{so}(4, 2q+1)$	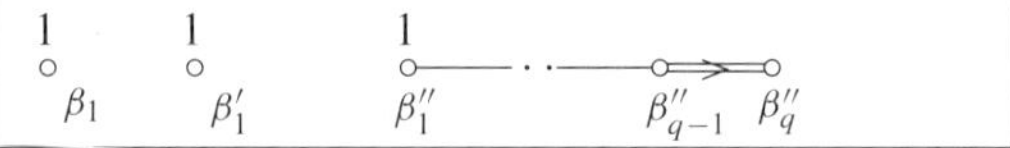	$\xi_1 + \xi_1'$ from $\mathfrak{so}(4) = A_1 A$
$\mathfrak{so}(2p, 3)$, $p > 2$	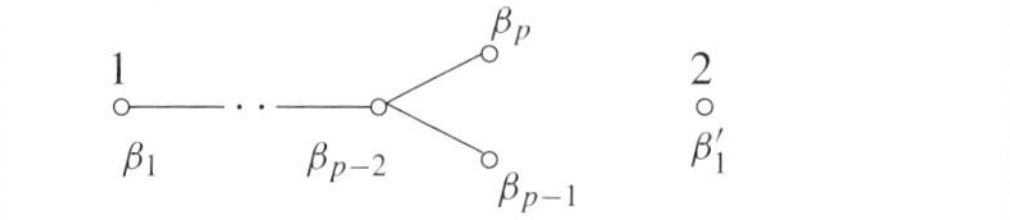	ξ_1 (vector rep) for $\mathfrak{so}(2p$
$\mathfrak{so}(2p, 2q+1)$ $(p>2,\ q\geqq 2)$	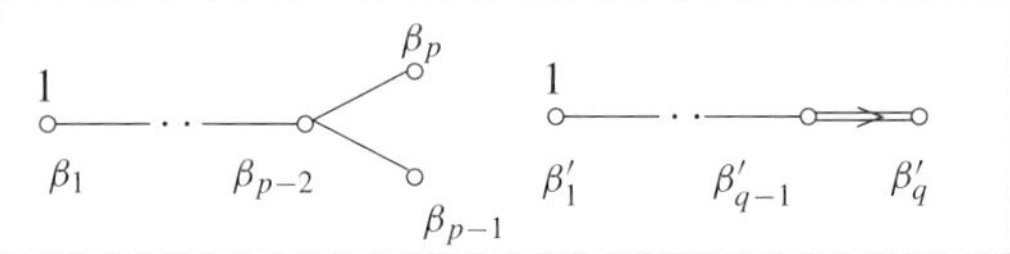	ξ_1 (vector rep) for $\mathfrak{so}(2p$
$\mathfrak{sp}(p, q)$ $(1\leqq p\leqq q)$	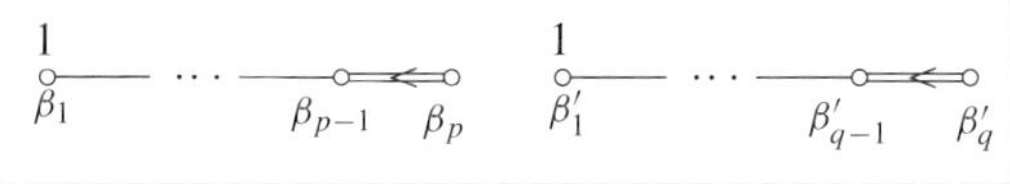	
$\mathfrak{so}(4, 2q)$, $q \geqq 2$	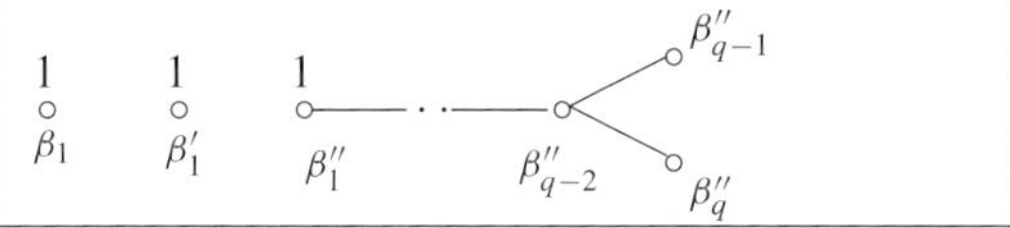	
$\mathfrak{so}(2p, 2q)$ $(2<p\leqq q)$	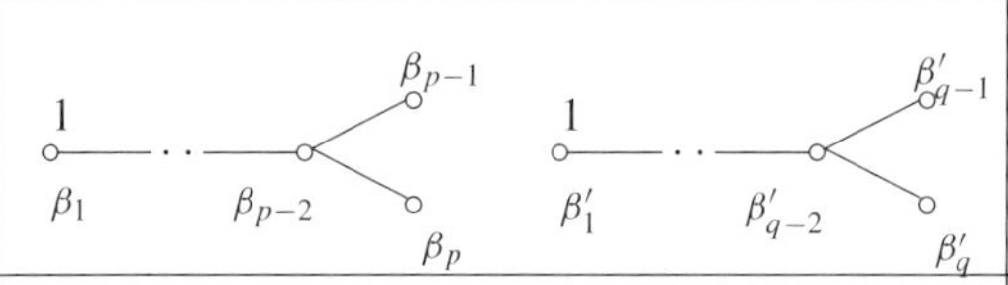	
$\mathfrak{g}_{2, A_1 A_1}$	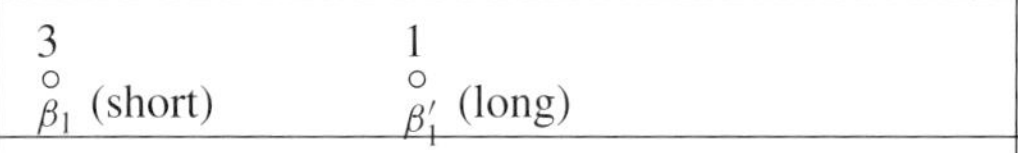	$\xi_1 + \xi_1'$
$\mathfrak{f}_{4, A_1 C_3} \equiv \mathfrak{f}_{4(4)}$	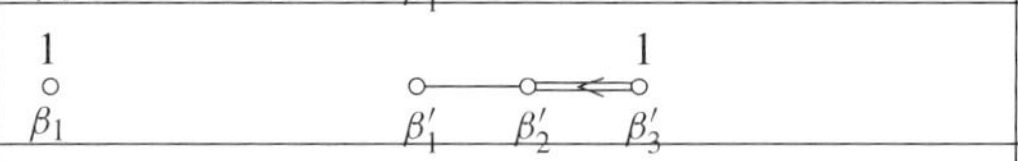	$\xi_1 + \xi_3'$
$\mathfrak{f}_{4, B_4}$	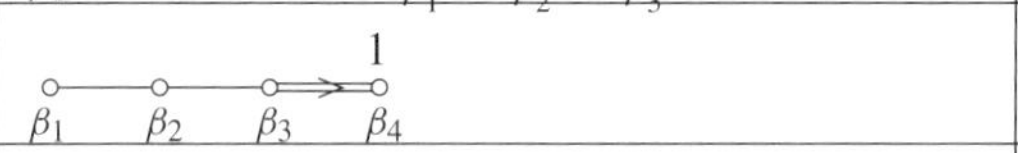	
$\mathfrak{e}_{6, A_1 A_5}$	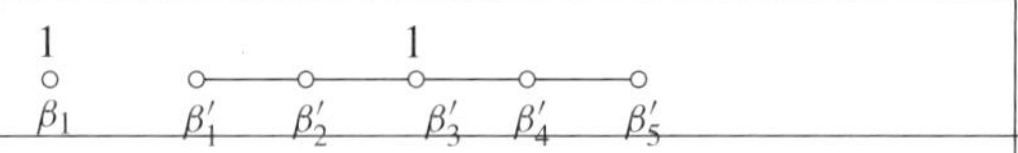	
$\mathfrak{e}_{7, A_1 D_6}$	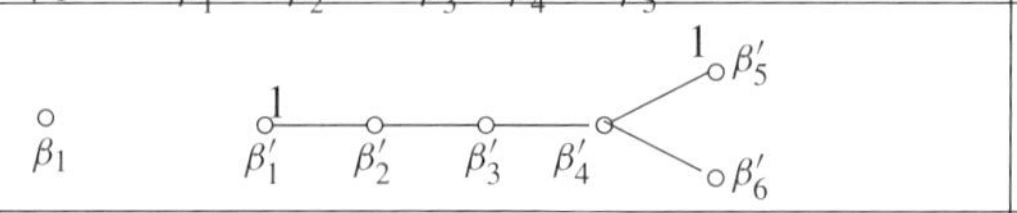	
$\mathfrak{e}_{7, A_7}$	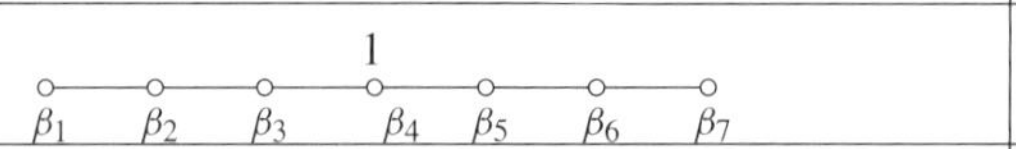	
$\mathfrak{e}_{8, A_1 E_7}$	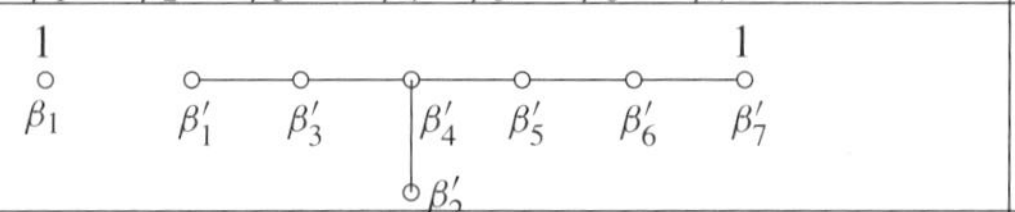	
$\mathfrak{e}_{8, D_8}$	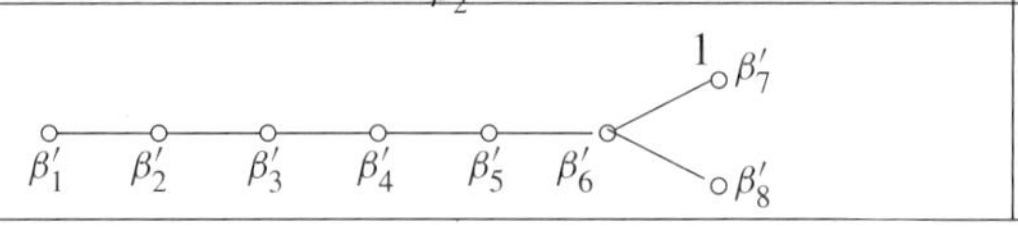	

Now we turn to the case rank $\mathfrak{g} >$ rank $\mathfrak{k}$.

Theorem 17.4.7 ([WG1, Theorem 5.10], or see [W6, pp. 291–292]). *Suppose that $\mathfrak{g}_0$ is simple and* rank $\mathfrak{g} >$ rank $\mathfrak{k}$. *Then $\mathfrak{k}$ is semisimple and acts irreducibly on $\mathfrak{s}$ as follows. If G_u/K_0 is the group manifold K_0, then the tangent space representation of $\mathfrak{k}$ on $\mathfrak{s}$ is just the adjoint representation of $\mathfrak{k}$; (those are listed in Table* 1.2.3. *If G_u/K_0 is not a group manifold, then the tangent space representation of $\mathfrak{k}$ on $\mathfrak{s}$ is given Table* 17.4.8 *below. There, the simple roots $\beta \in \Psi_{\mathfrak{k}}$ are restrictions of simple roots $\psi \in \Psi_{\mathfrak{g}}$ and are enumerated in the Bourbaki order as in Tables* 17.4.2 *and* 17.4.3 *for $\mathfrak{k}$, and the ξ are the fundamental simple weights of $\mathfrak{k}$ for the βs.*

Proof. We take the description of the isotropy representation from [W6, Section 8.11]. If $\mathfrak{g}$ is not simple, then G_u/K_0 is the group manifold K_0, with $\mathfrak{k}_0$ embedded diagonally in $\mathfrak{g}_u \cong \mathfrak{k}_0 \oplus \mathfrak{k}_0$ and $\theta(x, y) = (y, x)$. Then

$$\mathfrak{s}_u = \{(x, -x) \mid x \in \mathfrak{k}_0\},$$

and the tangent space representation is the adjoint representation of $\mathfrak{k}_0$. Those are listed in Table 1.2.3.

Now suppose that $\mathfrak{g}$ is simple. There are only a few possibilities. If $\mathfrak{g}_0 = \mathfrak{sl}(m; \mathbb{R})$, then the action of $\mathfrak{k}$ on $\mathfrak{s}$ is the representation of $\mathfrak{so}(m; \mathbb{C})$ on the space of $m \times m$ symmetric complex matrices of trace 0. If $\mathfrak{g}_0 = \mathfrak{sl}(m; \mathbb{H})$, then the action of $\mathfrak{k}$ on $\mathfrak{s}$ is the representation of $\mathfrak{sp}(m; \mathbb{C})$ on the space of $2m \times 2m$ antisymmetric complex matrices. If $\mathfrak{g}_0 = \mathfrak{so}(2p + 1, 2q + 1)$, it is the tensor product of the usual vector representations of $\mathfrak{so}(2p+1)$ and $\mathfrak{so}(2q+1)$. If $\mathfrak{g}_0 = \mathfrak{e}_{6,F_4}$, it is the unique irreducible representation of $\mathfrak{f}_4$ of degree 26, and if $\mathfrak{g}_0 = \mathfrak{e}_{6,C_4}$, it is the unique irreducible representation of $\mathfrak{c}_4$ of degree 42.

The additional information in Table 17.4.8 follows by direct computation. □

In Section 18.2B we will note a connection between classification of symmetric spaces, their tangent space representations (including most of the cases where rank $\mathfrak{k} <$ rank $\mathfrak{g}$), and a class of diagrams, due to Kač, that includes the extended Dynkin diagrams.

17.5 Shift of degree in the cohomology

From now on we assume that $\mathfrak{g}_0$ is a real simple Lie algebra. As noted above, the general semisimple case is reduced to this case. If $\mathfrak{g}_0$ is a direct sum of simple ideals, that sum decomposition gives direct product decompositions of G and Q, and everything splits:

- The flag manifold $Z = G/Q$ and the base cycles C are direct products.
- The normal bundle $\mathbb{N}_Z(C)$ of a base cycle C in Z is a direct sum of the normal bundles of the corresponding factors.
- The cohomology groups $H^k(C; \mathcal{O}(\mathbb{N}_Z(C)))$ are the direct sums of the corresponding cohomology groups for the simple factors.

Depending on the hermitian and non-hermitian cases, there are certain shifts in the degrees of cohomology.

Table 17.4.8. Tangent Space Representations When rank $\mathfrak{g}$ > rank $\mathfrak{k}$

$\mathfrak{g}_0$	Diagram of ν	maximal weight	nonmax dominant weights in $M(\mathfrak{t}, \mathfrak{s})$
$\mathfrak{sl}(3; \mathbb{R})$		$4\xi_1$	$2\xi_1,\ 0$
$\mathfrak{sl}(4; \mathbb{R})$	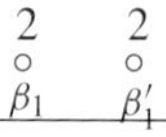	$2\xi_1 + 2\xi_1'$	$2\xi_1,\ 2\xi_1',\ 0$
$\mathfrak{sl}(5; \mathbb{R})$	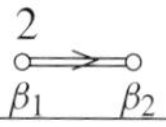	$2\xi_1$	$2\xi_2,\ \xi_1,\ 0$
$\mathfrak{sl}(6; \mathbb{R})$	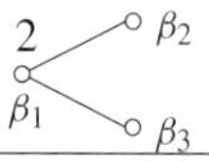	$2\xi_1$	$\xi_2 + \xi_3,\ 0$
$\mathfrak{sl}(2m; \mathbb{R}), m \geqq 4$	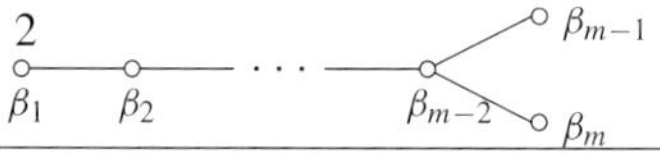	$2\xi_1$	$\xi_2,\ 0$
$\mathfrak{sl}(2m+1; \mathbb{R}), m \geqq 3$	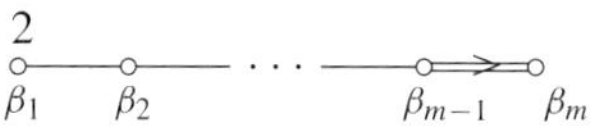	$2\xi_1$	$\xi_2,\ \xi_1,\ 0$
$\mathfrak{sl}(m; \mathbb{H}), m \geqq 3$	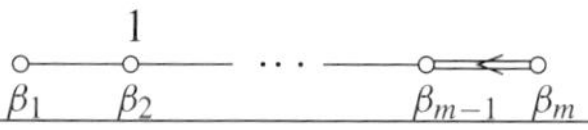	ξ_2	0
$\mathfrak{so}(1, 2q+1), q > 1$		ξ_1	0
$\mathfrak{so}(3, 2q+1), q > 1$	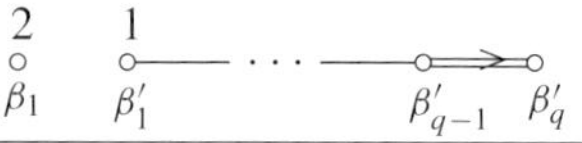	$2\xi_1 + \xi_1'$	$2\xi_1,\ \xi_1',\ 0$
$\mathfrak{so}(2p+1, 2q+1)$ $_{p,q>1}$	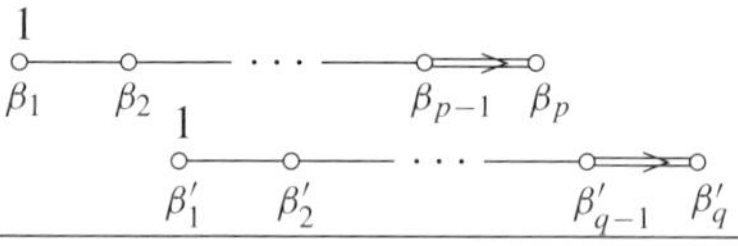	$\xi_1 + \xi_1'$	$\xi_1,\ \xi_1',\ 0$
$\mathfrak{e}_{6,F_4}$	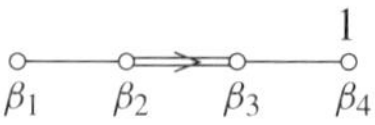	ξ_4	0
$\mathfrak{e}_{6,C_4}$	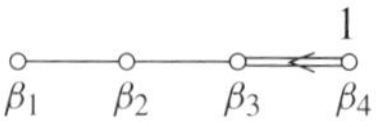	ξ_4	$\xi_2,\ 0$

The non-hermitian case

This is the case where $\mathfrak{k}$ acts irreducibly on $\mathfrak{s}$, equivalently where $\mathfrak{k}$ is semisimple. If rank $\mathfrak{k}$ < rank $\mathfrak{g}$, then we are in this case; in particular, all nonmeasurable examples belong to this class.

As before, K is the complex analytic subgroup of G with Lie algebra $\mathfrak{k}$, and $C = K(z) \subset Z$ is a base cycle such that the isotropy subgroup $Q_K := K \cap Q$ of K at z contains $B_{\mathfrak{k}}^{\text{ref}}$. We have the associated group data $B_{\mathfrak{k}} \subset B \subset Q = Q_z$ as in Section 17.2.

Proposition 17.5.1. *Suppose that G_0 is not of hermitian type. Consider a base cycle $K(z) = C \subset Z = G/Q$ and let $\mathfrak{b}_{\mathfrak{k}} \subset \mathfrak{b} \subset \mathfrak{q} \subset \mathfrak{g}$ be a choice of associated Lie algebra data. Then as K-modules*

$$H^0(C; \mathcal{O}(\mathbb{N}_Z(C))) = \mathfrak{s} \oplus H^1(C; \mathcal{O}(\mathbb{E}((\mathfrak{q} + \theta\mathfrak{q})_{\mathfrak{s}}))) \text{ and}$$
$$H^k(C; \mathcal{O}(\mathbb{N}_Z(C))) = H^{k+1}(C; \mathcal{O}(\mathbb{E}((\mathfrak{q} + \theta\mathfrak{q})_{\mathfrak{s}})))$$

for all $k > 0$.

Proof. The short exact sequence $(\mathfrak{q} + \theta\mathfrak{q})_{\mathfrak{s}} \hookrightarrow \mathfrak{s} \to \mathfrak{s}/(\mathfrak{q} + \theta\mathfrak{q})_{\mathfrak{s}}$ of Q_K-modules yields the exact sequence of K-homogeneous holomorphic vector bundles

$$0 \longrightarrow \mathbb{E}((\mathfrak{q} + \theta\mathfrak{q})_{\mathfrak{s}}) \longrightarrow \mathbb{E}(\mathfrak{s}) \longrightarrow \mathbb{E}(\mathfrak{s}/(\mathfrak{q} + \theta\mathfrak{q})_{\mathfrak{s}}) = \mathbb{N}_Z(C) \longrightarrow 0.$$

The associated long exact cohomology sequence begins with

(17.5.2)
$$H^0(C; \mathcal{O}(\mathbb{E}((\mathfrak{q} + \theta\mathfrak{q})_{\mathfrak{s}}))) \hookrightarrow H^0(C; \mathcal{O}(\mathbb{E}(\mathfrak{s}))) \longrightarrow H^0(C; \mathcal{O}(\mathbb{N}_Z(C))) \longrightarrow$$
$$H^1(C; \mathcal{O}(\mathbb{E}((\mathfrak{q} + \theta\mathfrak{q})_{\mathfrak{s}}))) \longrightarrow H^1(C; \mathcal{O}(\mathbb{E}(\mathfrak{s}))) \longrightarrow \dots$$

and then continues with

(17.5.3)
$$\dots\ H^k(C; \mathcal{O}(\mathbb{E}(\mathfrak{s}))) \to H^k(C; \mathcal{O}(\mathbb{N}_Z(C))) \to$$
$$H^{k+1}(C; \mathcal{O}(\mathbb{E}((\mathfrak{q} + \theta\mathfrak{q})_{\mathfrak{s}}))) \to H^{k+1}(C; \mathcal{O}(\mathbb{E}(\mathfrak{s}))) \dots$$

for $k \geqq 1$. Here $\mathbb{E}(\mathfrak{s}) \to C = K/Q_K$ is a trivial holomorphic vector bundle, because the representation $\mathfrak{q}_{\mathfrak{k}} \times \mathfrak{s} \to \mathfrak{s}$ extends to a representation of K. Thus every global holomorphic section of $\mathbb{E}(\mathfrak{s})$ is constant (with respect to the K-equivariant holomorphic trivialization) and $H^0(C; \mathcal{O}(\mathbb{E}(\mathfrak{s})))$ is just the irreducible $\mathfrak{k}$-module $\mathfrak{s}$. Here $\mathbb{E}(\mathfrak{s}) \to C$ is trivial and C is projective rational. Therefore $H^k(C; \mathcal{O}(\mathbb{E}(\mathfrak{s}))) = 0$ for $k > 1$.

Irreducibility of K on $\mathfrak{s}$ forces the inclusion

$$H^0(C; \mathcal{O}(\mathbb{E}((\mathfrak{q} + \theta\mathfrak{q})_{\mathfrak{s}}))) \hookrightarrow H^0(C; \mathcal{O}(\mathbb{E}(\mathfrak{s})))$$

of (17.5.2) to be zero or isomorphic. If it were an isomorphism, then $(\mathfrak{q} + \theta\mathfrak{q})_{\mathfrak{s}} = \mathfrak{s}$, and thus the typical fiber $\mathfrak{s}/(\mathfrak{q} + \theta\mathfrak{q})_{\mathfrak{s}}$ of $\mathbb{N}_Z(C) \to C$ would be zero. This forces

$C = Z$. Then the K_0-action (and thus the G_0-action) would be transitive on Z. We would therefore be in one of the two cases of Proposition 5.2.1. As discussed right after Proposition 5.2.1, we agreed to exclude those cases.

Thus $H^0(C; \mathcal{O}(\mathbb{E}((\mathfrak{q}+\theta\mathfrak{q})_{\mathfrak{s}}))) \hookrightarrow H^0(C; \mathcal{O}(\mathbb{E}(\mathfrak{s})))$ is zero. Now (17.5.2) gives us $0 \to \mathfrak{s} \to H^0(C; \mathcal{O}(\mathbb{N}_Z(C))) \to H^1(C; \mathcal{O}(\mathbb{E}((\mathfrak{q}+\theta\mathfrak{q})_{\mathfrak{s}}))) \to 0$, an exact sequence of K-modules, which splits because K is semisimple. This proves the assertion for 0-cohomology.

The second assertion follows from (17.5.3) because $H^k(C; \mathcal{O}(\mathbb{E}(\mathfrak{s}))) = 0$ for $k > 0$, by the above argument. □

The hermitian case

Here the reductive subalgebra $\mathfrak{k}$ has one-dimensional center $\mathfrak{z}_{\mathfrak{k}}$, and the action of $\mathfrak{k}$ on $\mathfrak{s}$ decomposes into the direct sum $\mathfrak{s} = \mathfrak{s}_- + \mathfrak{s}_+$ of two irreducible subrepresentations. The θ-stable τ-stable fundamental (relative to $\mathfrak{g}_0$) Cartan subalgebra $\mathfrak{h}$ of $\mathfrak{g}$ is contained in $\mathfrak{k}$ and we have the following decompositions:

- $\mathfrak{k} = \mathfrak{z}_{\mathfrak{k}} \oplus \mathfrak{k}^s$, where $\mathfrak{k}^s = [\mathfrak{k}, \mathfrak{k}]$ is semisimple.
- $\mathfrak{h} = \mathfrak{z}_{\mathfrak{k}} \oplus \mathfrak{h}^s$, where $\mathfrak{h}^s = \mathfrak{h} \cap \mathfrak{k}^s$ is a Cartan subalgebra in $\mathfrak{k}^s$.

Note that there is an element $\zeta \in \mathfrak{z}_{\mathfrak{k}} \cap \mathfrak{k}_0$ that induces the complex structure on $\mathfrak{s}_0$ and $\mathfrak{s}_\pm$ is the $(\pm i)$-eigenspace of $\mathrm{ad}(\zeta)$ on $\mathfrak{g}$. Also, since rank $\mathfrak{g}$ = rank $\mathfrak{k}$, Lemma 17.3.2 says that $\theta(\mathfrak{q}) = \mathfrak{q}$. Since $\mathfrak{h} \subset \mathfrak{q}$, it follows that

$$\mathfrak{q} = \theta(\mathfrak{q}) = \mathfrak{q}_{\mathfrak{k}} + \mathfrak{q}_{\mathfrak{s}_+} + \mathfrak{q}_{\mathfrak{s}_-}, \quad \text{where } \mathfrak{q}_{\mathfrak{s}_\pm} := \mathfrak{q} \cap \mathfrak{s}_\pm.$$

Proposition 17.5.4. *Suppose that G_0 is of hermitian type. Consider a base cycle $K(z) = C \subset Z = G/Q$ and let $\mathfrak{b}_{\mathfrak{k}} \subset \mathfrak{b} \subset \mathfrak{q} \subset \mathfrak{g}$ be a choice of associated Lie algebra data. Then we have K-module decompositions as follows.*

1. *If $k > 0$, then $H^k(C; \mathcal{O}(\mathbb{N}_Z(C))) = H^{k+1}(C; \mathcal{O}(\mathbb{E}(\mathfrak{q}_{\mathfrak{s}})))$.*
2. *If $\mathfrak{s}_- \not\subset \mathfrak{q}$ and $\mathfrak{s}_+ \not\subset \mathfrak{q}$, then*

$$H^0(C; \mathcal{O}(\mathbb{N}_Z(C))) = \mathfrak{s} \oplus H^1(C; \mathcal{O}(\mathbb{E}(\mathfrak{q}_{\mathfrak{s}_+}))) \oplus H^1(C; \mathcal{O}(\mathbb{E}(\mathfrak{q}_{\mathfrak{s}_-}))).$$

3. *If $\mathfrak{s}_+ \subset \mathfrak{q}$ but $\mathfrak{s}_- \not\subset \mathfrak{q}$, then $H^0(C; \mathcal{O}(\mathbb{N}_Z(C))) = \mathfrak{s}_- \oplus H^1(C; \mathcal{O}(\mathbb{E}(\mathfrak{q}_{\mathfrak{s}_-})))$, and*
4. *If $\mathfrak{s}_- \subset \mathfrak{q}$ but $\mathfrak{s}_+ \not\subset \mathfrak{q}$, then $H^0(C; \mathcal{O}(\mathbb{N}_Z(C))) = \mathfrak{s}_+ \oplus H^1(C; \mathcal{O}(\mathbb{E}(\mathfrak{q}_{\mathfrak{s}_+})))$.*

Proof. We first deal with the 0-cohomology. As in the proof of Proposition 17.5.1, the two short exact sequences $0 \to \mathfrak{q}_{\mathfrak{s}_\pm} \to \mathfrak{s}_\pm \to \mathfrak{s}_\pm/\mathfrak{q}_{\mathfrak{s}_\pm} \to 0$ induce long cohomology sequences. Since the bundles $\mathbb{E}(\mathfrak{s}_+) \to C$ and $\mathbb{E}(\mathfrak{s}_-) \to C$ are holomorphically trivial, those cohomology sequences fit together as follows:

$$\begin{array}{ccccccc} H^0(C; \mathcal{O}(\mathbb{E}(\mathfrak{q}_{\mathfrak{s}_+}))) & \stackrel{\iota^+}{\hookrightarrow} & H^0(C; \mathcal{O}(\mathbb{E}(\mathfrak{s}_+))) & & & H^1(C; \mathcal{O}(\mathbb{E}(\mathfrak{q}_{\mathfrak{s}_+}))) & \\ \oplus & & \oplus & \to H^0(C; \mathcal{O}(\mathbb{N}_Z(C))) \to & & \oplus & \to 0 \\ H^0(C; \mathcal{O}(\mathbb{E}(\mathfrak{q}_{\mathfrak{s}_-}))) & \stackrel{\iota^-}{\hookrightarrow} & H^0(C; \mathcal{O}(\mathbb{E}(\mathfrak{s}_-))) & & & H^1(C; \mathcal{O}(\mathbb{E}(\mathfrak{q}_{\mathfrak{s}_-}))) & \end{array}$$

The vector bundles $\mathbb{E}(\mathfrak{s}_\pm) \to C$ are holomorphically trivial. Thus the inclusion ι^+ (respectively, ι^-) is nonzero if and only if $\mathfrak{q}_{\mathfrak{s}_+} = \mathfrak{s}_+$ (respectively, $\mathfrak{q}_{\mathfrak{s}_-} = \mathfrak{s}_-$). Assertions 2, 3, and 4 now follow from the exactness of the cohomology sequences.

If $k > 0$, we have $H^k(C; \mathcal{O}(\mathbb{E}(\mathfrak{s}_\pm))) = 0$, and the first statement is immediate, as in the proof of Proposition 17.5.1. □

Remarks.

1. In the case where D is a bounded symmetric domain, $D \subset G/KS_- = G/Q = Z$ with C reduced to a point, the projection from $X = G/B$ to Z is especially interesting. We may assume in that case that $\mathfrak{b}_\mathfrak{k} \subset \mathfrak{b} \subset \mathfrak{k} + \mathfrak{s}_-$. Let $\widetilde{D} = G_0(1B)$, so that the projection fibers $\widetilde{D}$ over D and the fibers must be the maximal compact subvarieties $g\widetilde{C}$, where $g \in G_0$ and $\widetilde{C} = K(1B)$. Since D is Stein and contractible, and the structure group of the bundle $G/B \to G/Q$ is a complex Lie group acting holomorphically, Grauert's Oka principle implies that the fibration $\widetilde{D} \to D$ is holomorphically trivial. Hence $\widetilde{D} = \widetilde{C} \times D$ as complex manifolds. A further consequence is that in such a case the component $\mathcal{C}_{\widetilde{C}}(X)$ of the Barlet cycle space is biholomorphic to Z.
2. Consider a general flag manifold $Z' = G/Q'$ and an arbitrary base cycle $C' = K(1Q') \subset D' = G_0(Q')$. If $\mathfrak{s}_\pm \subset \mathfrak{q}'$, then the Borel subalgebra $\mathfrak{b}_\mathfrak{k} \oplus \mathfrak{s}_\pm \subset \mathfrak{q}'$. The geometric interpretation of the inclusions $\mathfrak{s}_\pm \subset \mathfrak{q}'$ is a bit complicated. One case is similar to the one described just above, i.e., the case where exists a G-equivariant fibration $\pi : Z' \to Z$ as above. If there is such a fibration, then the inclusion $\mathfrak{s}_\pm \subset \mathfrak{q}$. However, if Q' is not a Borel subgroup, there is a second possibility. This can be described as follows.
   Let Ψ be the simple root system for the positive system $\Sigma^+(\mathfrak{g}, \mathfrak{h})$ defined by the Borel subalgebra $\mathfrak{b}_\mathfrak{k} + \mathfrak{s}_+$. If Φ is the set of simple roots that defines $\mathfrak{q}'$ and the noncompact simple root belongs to Φ, then there is no G-equivariant fibration $Z' \to Z$. Only in this second case can $H^1(C; \mathbb{E}(\mathfrak{q}_\mathfrak{k}))$ be nonzero.

17.6 Equivariant filtrations

Let Q_K be a parabolic subgroup of K. In general, a finite-dimensional Q_K-module V_π is reducible but not completely reducible. We will encounter this phenomenon in most cases under consideration, in particular, where the module is the normal space $\mathbb{N}_Z(C)_z$ of a base K-cycle C in Z at some base point $z \in C$ where the module structure is given by the isotropy representation.

Let V be a Q_K-module. A Q_K-stable filtration

$$F^\bullet V : 0 = F^0 V \subsetneqq F^1 V \subsetneqq \cdots \subsetneqq F^\ell = V$$

is called **simple** or **irreducible** if the quotient modules $F^j V/F^{j-1} V$ are Q_K-irreducible for $1 \leqq j \leqq \ell$. Simple filtrations are often called **equivariant filtrations**, and it is an easy exercise in linear algebra to see (i) that they exist and (ii) that the unordered set $\{F^j V/F^{j-1} V\}$ of irreducible Q_K-modules is independent of choice of equivariant filtration.

We start with $T \subset B_K \subset Q_K = Q_K^{-n} Q_K^r$, where T is a τ-stable Cartan subgroup of K, B_K is a Borel subgroup of K, Q_K^{-n} is the unipotent radical of the parabolic subgroup Q_K of K, and Q_K^r is a reductive complement to Q_K^{-n} in Q_K. Since Q_K^{-n} is normal and unipotent in Q_K, it acts trivially in any irreducible Q_K-module, and thus a Q_K-module W is irreducible if and only if it is Q_K^r-irreducible.

Fix a Q_K-module V_π. There is a canonical filtration $F^\bullet V$, determined by the nilpotent subgroup Q_K^{-n}, defined recursively as follows:

$$(17.6.1) \qquad \begin{aligned} F^1 V &= \{v \in V \mid \pi(g)v = v \ \forall g \in Q_K^{-n}\} \\ F^{j+1} V &= \{v \in V \mid \pi(g)v - v \in F^j\}. \end{aligned}$$

This Q_K^{-n}-filtration is actually a Q_K-filtration. By construction, Q_K^{-n} acts trivially on the quotients F^j/F^{j-1}, and they are direct sums of irreducible Q_K-modules. Such filtrations will serve our purposes just as well as the simple equivariant filtrations.

18

Methods for Computing $H^1(C; \mathcal{O}(\mathbb{E}((\mathfrak{q} + \theta\mathfrak{q})_\mathfrak{s})))$

18.1 Guide to the computation

Recall our basic setup. Throughout $Z = G/Q$ is a complex flag manifold, $G_0 \subset G$ is a real form, $\mathfrak{g}_0 = \mathfrak{k}_0 + \mathfrak{s}_0$ is the decomposition of its Lie algebra under the Cartan involution θ, and $\mathfrak{g} = \mathfrak{k} + \mathfrak{s}$ is the complexified Cartan decomposition. Let N_z denote the normal space in Z of a closed orbit $C = K(z) \cong K/Q_K$. It is a Q_K-module, where Q_K acts by its isotropy representation on the tangent space of Z. More generally, if V_π is a Q_K-module arising in the context of normal bundles of closed K-orbits $C \subset Z$, then we have the corresponding K-homogeneous holomorphic vector bundle $\mathbb{E}(V_\pi) \to C$.

Every cohomology group $H^p(C; \mathcal{O}(\mathbb{E}(V_\pi)))$ is finite-dimensional because C is compact, and is a K-module because $\mathbb{E}(V_\pi) \to C$ is K-homogeneous. Our goal is to develop a method which enables us to determine this module structure in some algorithmic way. Any equivariant Q_K-filtration $F^\bullet V_\pi$ of V_π yields a filtration $\mathbb{E}(F^\bullet V_\pi)$ of $\mathbb{E}(V_\pi)$. Roughly speaking, our strategy is to deduce cohomology information for $\mathbb{E}(V_\pi) = \mathbb{N}_Z(C)$ from the Q_K-irreducible pieces of $\mathbb{E}(F^j V_\pi / F^{j-1} V_\pi)$. Of course we use the Bott–Borel–Weil Theorem to determine the cohomologies of the $\mathbb{E}(F^j V_\pi / F^{j-1} V_\pi)$.

Here are some of the central points of our procedure:

(A) Propositions 17.5.1 and 17.5.4 reduce the computation of $H^k(C; \mathcal{O}(\mathbb{N}_Z(C)))$ for $k = 0, 1$ to a computation of $H^k(C; \mathcal{O}(\mathbb{E}(V)))$ for $q = 1, 2$, where V is a certain Q_K-submodule of $\mathfrak{s}$. In order to understand such submodules in greater detail, we make strong use of the descriptions in of $\mathfrak{s}$ as a $\mathfrak{k}$-module Tables 17.4.5, 17.4.6, 1.2.3, and 17.4.8.

(B) Those descriptions lead to a determination of the $\mathfrak{t}$-weight system $M(\mathfrak{t}, \mathfrak{s})$ where $\mathfrak{t} := \mathfrak{h} \cap \mathfrak{k}$.

(C) In order to determine $H^k(C; \mathcal{O}(\mathbb{E}(V)))$, we first compute the cohomology spaces $H^k(C; \mathcal{O}(\mathbb{E}(F^j V / F^{j-1} V)))$. The Bott–Borel–Weil Theorem 1.6.8 says that these spaces vanish unless their weights μ, highest with respect to $\mathfrak{q}_\mathfrak{k} \cap \mathfrak{b}_\mathfrak{k}^+$, have the properties

(18.1.1)
regularity: $\mu + \rho_\mathfrak{k}$ is nonsingular, i.e., $\langle \mu + \rho_\mathfrak{k}, \alpha\rangle \neq 0$ for $\alpha \in \Sigma(\mathfrak{k}, \mathfrak{t})$,

where $\rho_\mathfrak{k}$ is half the sum of the positive roots of $\mathfrak{k}$, and then $\mu+\rho_\mathfrak{k}$ only contributes to cohomology of degree k when the

(18.1.2) **index:** $\mathrm{ind}(\mu + \rho_\mathfrak{k}) := \#\{\alpha \in \Sigma^+(\mathfrak{k}, \mathfrak{t}) \mid \langle \mu + \rho_\mathfrak{k}, \alpha\rangle < 0\}$

is k. We call such μ **Bott-regular of index** k. As noted in a footnote to the Bott–Borel–Weil Theorem, $\mathrm{ind}(\mu + \rho_\mathfrak{k})$ is the minimal length (as a product of simple root reflections) of the Weyl group element $w \in W(\mathfrak{k}, \mathfrak{t})$ such that $\langle w(\mu + \rho_\mathfrak{k}), \alpha\rangle > 0$ for all $\alpha \in \Sigma^+(\mathfrak{k}, \mathfrak{t})$.

(D) Only the regular highest weights of index 1 or 2 can contribute nontrivially to the modules $H^k(C; \mathcal{O}(\mathbb{E}(F^jV/F^{j-1}V)))$, $k = 1, 2$. The next major step is to determine all such weights. See Lemma 18.4.4 and Table 18.5.3.

(E) As explained in Section 17.2 given a pair (C, Z) consisting of a base cycle C in a flag manifold Z, we associate Lie algebras $\mathfrak{b}_\mathfrak{k}^{\mathrm{ref}} \subset \mathfrak{b} \subset \mathfrak{q} \subset \mathfrak{g}$. Given $\mu \in M(\mathfrak{t}, \mathfrak{s})$ as in (B) and a parabolic subalgebra $\mathfrak{q} \supset \mathfrak{b} \supset \mathfrak{b}_\mathfrak{k}$ of $\mathfrak{g}$, we must decide whether or not $\mu \in M(\mathfrak{t}, (\mathfrak{q}+\theta\mathfrak{q})_\mathfrak{s})$. If $\mu \in M(\mathfrak{t}, (\mathfrak{q}+\theta\mathfrak{q})_\mathfrak{s})$, we still have to decide whether μ is also a highest weight for the representation of Q_K^r on $(\mathfrak{q} + \theta\mathfrak{q})_\mathfrak{s}$. Putting this together gives a complete description of $H^k(C; \mathcal{O}(\mathbb{E}(F^jV/F^{j-1}V)))$ as $\mathfrak{k}$-module.

(F) Even when we know $H^k(C; \mathcal{O}(\mathbb{E}(F^jV/F^{j-1}V)))$ for all k, the nontrivial modules among them may or may not contribute to $H^k(C; \mathcal{O}(\mathbb{E}(V)))$. The Cohomological Lemma 18.5.6 is the main tool for determining whether the cohomology group $H^k(C; \mathcal{O}(\mathbb{E}(F^jV/F^{j-1}V)))$ contributes to $H^k(C; \mathcal{O}(\mathbb{E}(V)))$.

(G) Finally, having collected the necessary tools, we carry out the computation for all flag manifolds, real forms and base cycles. It turns out that for certain real forms the components of cycle spaces which contain the various base cycles C in the various flag manifolds Z are just the Zariski closures of the G-orbits $G(C)$ in that cycle space $\mathcal{C}(Z)$ (see Theorem 18.4.13).

However, there are many real forms for which there exist components in the cycle spaces of larger dimension than that of the corresponding G-orbit $G(C)$. We explicitly compute the module structure of $\mathbb{N}_Z(C)$ for all base cycles C and flag manifolds Z in Chapters 19 and 20.

18.2 Root systems and involutions

In Propositions 17.5.1 and 17.5.4 we saw that the cohomologies of the normal bundle $\mathbb{N}_Z(C)$ can be described as cohomology groups $H^\bullet(C; \mathcal{O}(\mathbb{E}(V)))$ for certain Q_K-submodules of the K-module $\mathfrak{s}$. Although the (tangent space) representation of $\mathfrak{k}$ on $\mathfrak{s}$ is given by Theorems 17.4.4 and 17.4.7, specifically in Tables 17.4.5, 17.4.6 and 17.4.8, we need a more explicit description of the $\mathfrak{t}$-weights on $\mathfrak{s}$ in order to pick out the ones that are highest weights for $\mathfrak{q}_\mathfrak{k}$ which lead to cohomology in degrees 0 and 1. We begin work on this in the present section.

18.2A Restricted root systems

Given $(\mathfrak{g}_0, \theta)$, we select a Cartan subalgebra $\mathfrak{t} \subset \mathfrak{k}$. The set of $\mathfrak{t}$-weights on $\mathfrak{s}$ is a subset of all $\mathfrak{t}$-weights $M(\mathfrak{t}, \mathfrak{g})$. A important step in our development of structural information on $M(\mathfrak{t}, \mathfrak{s})$ is the fact that $M(\mathfrak{t}, \mathfrak{g}) \setminus \{0\}$ is a root system.

We know from Lemma 4.2.1 that $\mathfrak{t}$ is regular in $\mathfrak{g}$, i.e., the centralizer $\mathfrak{h} := \mathfrak{z}_{\mathfrak{g}}(\mathfrak{t})$ is a Cartan subalgebra of $\mathfrak{g}$. Of course $\mathfrak{g}$ can be decomposed into joint eigenspaces with respect to any toral subalgebra. In particular, we have the joint eigenspace decomposition with respect to $\mathfrak{t}$:

$$\mathfrak{g} = \mathfrak{h} + \sum\nolimits_{\lambda \in M(\mathfrak{t}, \mathfrak{g}) \setminus \{0\}} \mathfrak{g}_\lambda .$$

Here, as before, $M(\mathfrak{t}, V_\pi)$ denotes the set of $\mathfrak{t}$-weights of a representation π of $\mathfrak{k}$ on V_π. If $\Sigma(\mathfrak{g}, \mathfrak{h})$ is the set of roots with respect to the centralizer $\mathfrak{h}$ of $\mathfrak{t}$, the elements of $M(\mathfrak{t}, \mathfrak{g})$ are the restrictions $\alpha|_{\mathfrak{t}}$ of $\alpha \in \Sigma(\mathfrak{g}, \mathfrak{h})$. In particular, the weights are real valued on $\mathfrak{t}_{\mathbb{R}} = i\mathfrak{t}_0 = \mathfrak{t} \cap \mathfrak{h}_{\mathbb{R}}$. Since $\mathfrak{t}$ is regular, every such restriction is nonzero. It can of course happen that two different roots in $\Sigma(\mathfrak{g}, \mathfrak{h})$ have the same restriction to $\mathfrak{t}$.

Recall the Definition 10.1.3 of an abstract root system. If $\mathfrak{a}_0 \subset \mathfrak{g}_0$ denotes a maximal split torus, then, as noted in Lemma 10.1.4, the **restricted root system** $\Sigma(\mathfrak{g}_0, \mathfrak{a}_0) := M(\mathfrak{a}_0, \mathfrak{g}_0) \setminus \{0\}$ is an abstract root system. The main remark here is that $\Sigma(\mathfrak{g}, \mathfrak{t}) := M(\mathfrak{t}, \mathfrak{g}) \setminus \{0\}$, considered as a subset of $\mathfrak{t}_{\mathbb{R}}^*$, also is a (possibly nonreduced) root system.

Since $M(\mathfrak{t}, \mathfrak{g}) \setminus \{0\}$ consists of all $\mathfrak{t}$-restrictions of roots in $\Sigma(\mathfrak{g}, \mathfrak{h})$, we still have $\mathfrak{h} \perp \mathfrak{g}_\lambda$ for all $\lambda \in M(\mathfrak{t}, \mathfrak{g}) \setminus \{0\}$ and $\mathfrak{g}_\lambda \perp \mathfrak{g}_\mu$ unless $\lambda + \mu = 0$, and $\mathfrak{g}_\lambda$ is paired nondegenerately with $\mathfrak{g}_{-\lambda}$ for all $\lambda \in M(\mathfrak{t}, \mathfrak{g})$. The complex conjugation $\sigma = \tau\theta$ of $\mathfrak{g}$ over $\mathfrak{g}_u$ defines a positive definite hermitian form $F(x, y) = -\langle x, \sigma(y)\rangle$ on $\mathfrak{g}$. Since $\sigma(\mathfrak{g}_\alpha) = \mathfrak{g}_{-\alpha}$ for every $\alpha \in \Sigma(\mathfrak{g}, \mathfrak{h})$, we have $\sigma(\mathfrak{g}_\lambda) = \mathfrak{g}_{-\lambda}$ for every $\lambda \in \Sigma(\mathfrak{g}, \mathfrak{t})$. Thus the decomposition $\mathfrak{g} = \mathfrak{h} + \sum_{\lambda \in \Sigma(\mathfrak{g}, \mathfrak{t}) \setminus \{0\}} \mathfrak{g}_\lambda$ is F-orthogonal. This discussion gives us

Lemma 18.2.1. *If $\lambda \in \Sigma(\mathfrak{g}, \mathfrak{t})$ and $0 \neq e \in \mathfrak{g}_\lambda$, then $\langle e, \sigma(e)\rangle < 0$, where we recall that $\sigma = \tau\theta$ is complex conjugation of $\mathfrak{g}$ over $\mathfrak{g}_u$.*

Further, for $\lambda \in \Sigma(\mathfrak{g}, \mathfrak{t})$ we have the normalized coroot $h_\lambda \in \mathfrak{t}_{\mathbb{R}}$ as in Section 10.1A. If As $\mathfrak{t}$ is fixed under θ, the root spaces $\mathfrak{g}_\lambda$, $\lambda \in M(\mathfrak{t}, \mathfrak{g})$, are θ-stable, and we decompose $\mathfrak{g}_\lambda = \mathfrak{k}_\lambda + \mathfrak{s}_\lambda$ under θ. If $0 \neq e \in \mathfrak{s}_\lambda$ then $\sigma(e) \in \mathfrak{s}$ because $\sigma\theta = \theta\sigma$, so $[e, \sigma(e)] \in \mathfrak{k}$. Similarly if $0 \neq e \in \mathfrak{k}_\lambda$ then $[e, \sigma(e)] \in \mathfrak{k}$. Since

$$\langle h, [e, \sigma(e)]\rangle = \lambda(h)\langle e, \sigma(e)\rangle = \langle h, h_\lambda\rangle\langle e, \sigma(e)\rangle = \langle h, \langle e, \sigma(e)\rangle h_\lambda\rangle$$

for every $h \in \mathfrak{t}$, we have the following.

Lemma 18.2.2. *If $0 \neq e \in \mathfrak{k}_\lambda$ or $0 \neq e \in \mathfrak{s}_\lambda$ then $0 \neq [e, \sigma(e)] \in \mathbb{R}{\cdot}h_\lambda$.*

Proposition 18.2.3. *$\Sigma(\mathfrak{g}, \mathfrak{t}) := M(\mathfrak{t}, \mathfrak{g}) \setminus \{0\}$ is a* (possibly nonreduced) *root system. If $\mathfrak{g}$ is simple, then $\Sigma(\mathfrak{g}, \mathfrak{t})$ is irreducible.*

Proof. We run through the root system axioms for $\Sigma(\mathfrak{g}, \mathfrak{t}) \subset \mathfrak{t}^*_\mathbb{R}$. The Killing form is positive definite on $\mathfrak{h}^*_\mathbb{R}$, hence also on its subspace $\mathfrak{t}^*_\mathbb{R}$. $\Sigma(\mathfrak{g}, \mathfrak{h})$ spans $\mathfrak{h}^*_\mathbb{R}$, so its $\mathfrak{t}$-restrictions span $\mathfrak{t}^*_\mathbb{R}$. The $\mathfrak{h}$-root reflections $s_\alpha : \beta \mapsto \beta - \frac{2\langle\alpha,\beta\rangle}{\langle\alpha,\alpha\rangle}\alpha$ preserve $\Sigma(\mathfrak{g}, \mathfrak{h})$ and the $\frac{2\langle\alpha,\beta\rangle}{\langle\alpha,\alpha\rangle}$ are integers. From the finite-dimensional representation theory of $\mathfrak{sl}(2; \mathbb{C})$, these two conditions are equivalent to the following condition: If $\alpha \in \Sigma(\mathfrak{g}, \mathfrak{h})$, then $\mathfrak{g}[\alpha] := [\mathfrak{g}_\alpha, \mathfrak{g}_{-\alpha}] + \mathfrak{g}_\alpha + \mathfrak{g}_{-\alpha}$ is isomorphic to $\mathfrak{sl}(2; \mathbb{C})$. Now let $\gamma \in \Sigma(\mathfrak{g}, \mathfrak{t})$ and let e_γ be a nonzero element of $\mathfrak{k}_\gamma$ or of $\mathfrak{s}_\gamma$. Lemma 18.2.1 says $\langle e_\gamma, \sigma(e_\gamma)\rangle < 0$. Denote $e_{-\gamma} = c\sigma(e_\gamma)$ where $c := -\frac{2}{\langle e_\gamma,\sigma(e_\gamma)\rangle\langle\gamma,\gamma\rangle}$. Then $\{[e_\gamma, e_{-\gamma}], e_\gamma, e_{-\gamma}\}$ is a standard generating set for a copy of $\mathfrak{sl}(2; \mathbb{C})$ in $\mathfrak{g}$. The adjoint action of this $\mathfrak{sl}(2; \mathbb{C})$ on $\mathfrak{g}$ gives the root system conditions (ii) and (iii) for Σ. □

From now on, we only write $\Sigma(\mathfrak{g}, \mathfrak{t})$ for $M(\mathfrak{t}, \mathfrak{g}) \setminus \{0\}$, in order to emphasize its root system properties. Our next step is to determine the set $M(\mathfrak{t}, \mathfrak{s})$ as a subset of $M(\mathfrak{t}, \mathfrak{g})$. This is implicit in [WG1] and [WG2], but it is formulated in a better way by Kač (see [Hel, Chapter X]), extending Theorem 17.4.1 of Borel and de Siebenthal.

Recall the action of θ on the simple root system Ψ discussed before Lemma 17.3.2. We view it as a graph automorphism of the Dynkin diagram $\Delta(\mathfrak{g}, \mathfrak{h})$. It is classical that $\mathrm{Aut}(\Delta(\mathfrak{g}, \mathfrak{h}))$ is canonically isomorphic to $\mathrm{Out}(\mathfrak{g}) := \mathrm{Aut}(\mathfrak{g})/\mathrm{Int}(\mathfrak{g})$, the group of outer automorphisms of $\mathfrak{g}$. When $\mathfrak{g}$ is simple, $\mathrm{Out}(\mathfrak{g})$ is trivial except for the cases listed in the following table.

Table 18.2.4. Outer Automorphisms of Order 2

Type of $\mathfrak{g}$	$\mathrm{Out}(\mathfrak{g})$	Action of θ on $\Delta(\mathfrak{g}, \mathfrak{h})$
$A_\ell, \ell > 1$	$\mathbb{Z}_2$	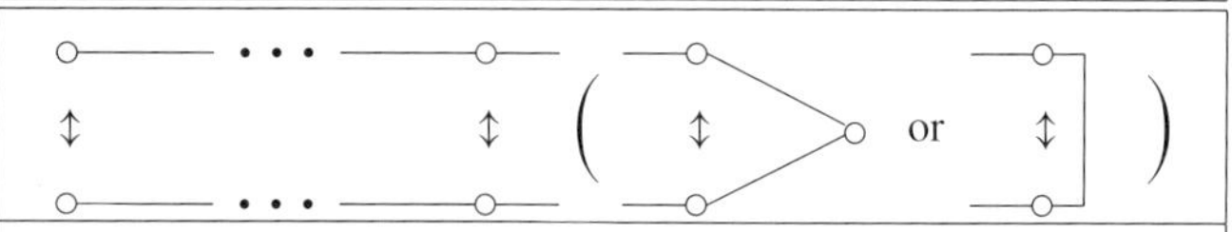
D_4	$\mathbb{D}_3$ dihedral	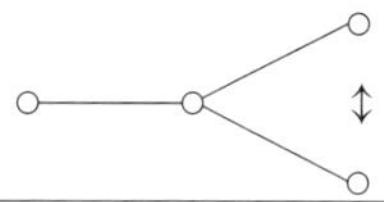
$D_\ell, \ell > 4$	$\mathbb{Z}_2$	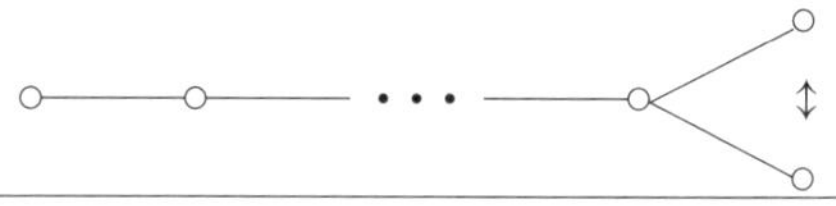
E_6	$\mathbb{Z}_2$	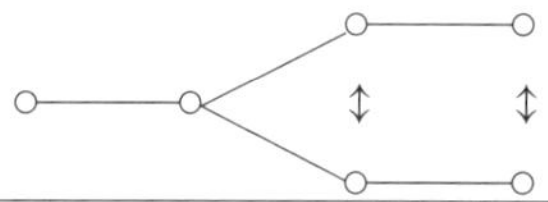

Compare [W6, p. 289]. Here note that any two elements of order 2 in $\mathrm{Out}(\mathfrak{g})$ are conjugate, so the action of θ does not depend on the choice of $\mathfrak{k}$. There is also one other graph automorphism, this time of order 3, and we mention it only for later reference.

Table 18.2.5. Triality Automorphism

Type of $\mathfrak{g}$	$\mathrm{Out}(\mathfrak{g})$	Action of triality on $\Delta(\mathfrak{g}, \mathfrak{h})$
D_4	$\mathbb{D}_3$ dihedral group	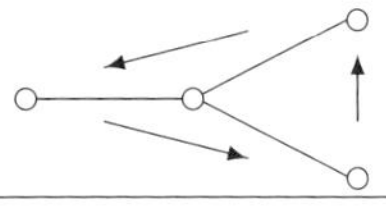

In all cases of Tables 18.2.4 and 18.2.5, the action on the extended Dynkin diagram $\widetilde{\Delta}(\mathfrak{g}, \mathfrak{h})$ is obtained from the action on $\Delta(\mathfrak{g}, \mathfrak{h})$ by fixing the vertex β_0 that is the negative of the highest root.

The root system $\Sigma(\mathfrak{g}, \mathfrak{t})$ is nonreduced if and only if it has a simple root β such that $2\beta \in \Sigma(\mathfrak{g}, \mathfrak{t})$. This can happen only when there exist $\psi', \psi'' \in \Sigma(\mathfrak{g}, \mathfrak{h})$ which restrict to β and some combination $\sum_{\psi \in \Psi} n_i \beta_i$ that restricts to 2β. But then $n_i = 0$ for $\psi' \neq \psi \neq \psi''$. Thus that combination must be $\psi' + \psi''$, and we must have $\psi' + \psi'' \in \Sigma(\mathfrak{g}, \mathfrak{h})$. That can happen only when ψ' and ψ'' are adjacent nodes in the Dynkin diagram $\Delta(\mathfrak{g}, \mathfrak{h})$ with $\theta(\psi') = \psi''$. A glance at Table 18.2.4 shows that this happens only when $\mathfrak{g}$ is of type A_ℓ with ℓ even.

If $\mathfrak{g}_0$ is simple but $\mathfrak{g}$ is not simple, then $\mathfrak{g}$ is the direct sum of two simple ideals, and every $\alpha \in \Sigma(\mathfrak{g}, \mathfrak{t})$ is the restriction of two roots, one from each simple ideal (extended by zero to the other simple ideal). Thus the resulting restricted root system is reduced, with each root of multiplicity 2.

We summarize this discussion as follows.

Proposition 18.2.6. *Let $\mathfrak{g}_0$ be simple. Then the root system $\Sigma(\mathfrak{g}, \mathfrak{t})$ is reduced except in the case where $(\mathfrak{g}_0, \mathfrak{k}_0) = (\mathfrak{sl}(2m+1, \mathbb{C}), \mathfrak{so}(2m+1))$. In that case, $\Sigma(\mathfrak{g}, \mathfrak{t})$ is of type BC_m, with root system*

$$\{\pm\varepsilon_i \pm \varepsilon_j \mid 1 \leqq i < j \leqq m\} \cup \{\pm\varepsilon_i \mid 1 \leqq i \leqq m\} \cup \{\pm 2\varepsilon_i \mid 1 \leqq i \leqq m\},$$

with one simple root system given by

$$B = \{\beta_1, \dots, \beta_m\}, \text{ where } \beta_i = \varepsilon_i - \varepsilon_{i+1} \text{ for } 1 \leqq i < m \text{ and } \beta_m = \varepsilon_m.$$

Fix a complex simple Lie algebra $\mathfrak{g}$, a Cartan subalgebra $\mathfrak{h}$, and a graph automorphism γ of the Dynkin diagram $\Delta(\mathfrak{g}, \mathfrak{h})$. Then γ comes from an automorphism (also denoted γ) of $\mathfrak{g}$ that preserves $\mathfrak{h}$, a positive root system $\Sigma^+(\mathfrak{g}, \mathfrak{h})$, and the associated simple root system $\Psi = \{\psi_1, \dots, \psi_\ell\}$. Thus γ fixes the maximal root $-\psi_0$, permutes $\widetilde{\Psi} = \Psi \cup \{\psi_0\}$, and defines aa graph automorphism of the extended Dynkin diagram.

Let $\mathfrak{t}$ denote the fixed point set of γ on $\mathfrak{h}$. Theorem 4.1.1 becomes

Lemma 18.2.7. *Let $\mathfrak{g}$ be semisimple and let γ be a semisimple (diagonalizable) automorphism of $\mathfrak{g}$. Let $\mathfrak{t}$ be a Cartan subalgebra of the fixed point set $\mathfrak{g}^{\gamma}$ of γ on $\mathfrak{g}$. Then $\mathfrak{t}$ is contained in a γ-invariant Cartan subalgebra of $\mathfrak{g}$. Further, $\mathfrak{t}$ contains regular elements of $\mathfrak{g}$, in other words, if $\alpha \in \Sigma(\mathfrak{g}, \mathfrak{h})$, then $\alpha|_{\mathfrak{t}} \neq 0$.*

The **restricted** Dynkin diagram $\Delta(\mathfrak{g}, \gamma)$ is obtained by restricting the ψ_j from $\mathfrak{h}$ to $\mathfrak{t}$. Let $B = \{\beta_1, \ldots, \beta_m\}$ denote the distinct restrictions with $\beta_j = \psi_j|_{\mathfrak{t}}$, and apply the usual Dynkin diagram rules to B. The diagrams for the root systems $\Sigma(\mathfrak{g}, \mathfrak{t})$ are special cases. The **restricted extended** Dynkin diagram $\widetilde{\Delta}(\mathfrak{g}, \gamma)$ is obtained by also restricting ψ_0, say to β_0. The coefficient $n_{\gamma,j}$ is the sum of the coefficients of the ψ_k that restrict to β_j, so $n_{\gamma,0} = 1$. The restricted extended Dynkin diagrams with coefficients are as follows.

CASE $\gamma = 1$. Then $\mathfrak{t} = \mathfrak{h}$, so $\Delta(\mathfrak{g}, \gamma) = \Delta(\mathfrak{g}, \mathfrak{h})$, $\beta_j = \psi_j$ and $\widetilde{\Delta}(\mathfrak{g}, \gamma) = \widetilde{\Delta}(\mathfrak{g}, \mathfrak{h})$ with the same coefficients. These are given by Tables 17.4.2 and 17.4.3.

CASE γ HAS ORDER 2. Then, from Tables 17.4.2, 17.4.3 and 18.2.4 we derive the diagrams of Table 18.2.8.

Table 18.2.8. Restricted Extended Diagram $\widetilde{\Delta}(\mathfrak{g}, \gamma)$ with Coefficients

<table>
<tr><th>Type of $\mathfrak{g}$</th><th>Restricted Extended Diagram</th></tr>
<tr><td>A_{2m-1}</td><td>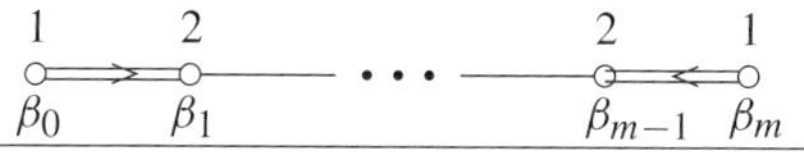</td></tr>
<tr><td>A_2</td><td>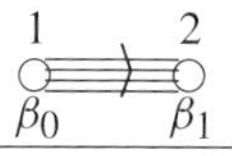</td></tr>
<tr><td>A_{2m}, $m > 1$</td><td>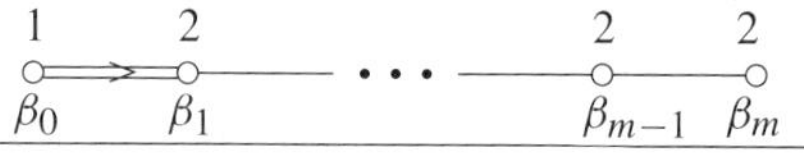</td></tr>
<tr><td>D_ℓ, $\ell \geqq 4$</td><td></td></tr>
<tr><td>E_6</td><td>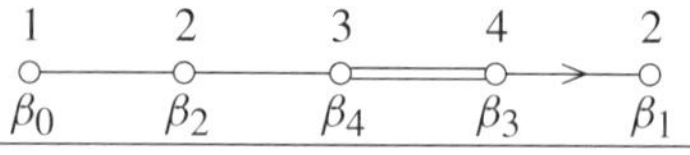</td></tr>
</table>

CASE γ HAS ORDER 3. The only instance of this is triality on D_4. This gives

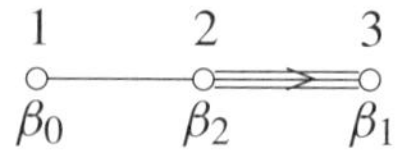

The second class of new diagrams consists of the affine diagrams. They appear implicitly in [dS], but we need the explicit details. These diagrams are the ones involved in Kač' extension of the Borel-de Siebenthal Theorem 17.4.1. In that extension, Kač defined them axiomatically and carried out the classification (see [Hel, Chapter X]). Here, we describe the result by comparison to the restricted extended Dynkin diagrams $\widetilde{\Delta}(\mathfrak{g}, \gamma)$. The **affine** Dynkin diagram $\Delta(\mathfrak{g}, \gamma, \text{aff})$ is constructed in the same way as $\widetilde{\Delta}(\mathfrak{g}, \gamma)$, except that instead of replacing β_j by $\beta_j = \psi_j|_\mathfrak{t}$, we replace ψ_j by μ_j which is defined as the sum over the orbit $\Gamma(\psi_j)$, where Γ is the group of graph automorphisms (necessarily of order 1, 2 or 3) generated by γ. However, if $2\beta_j \in \Sigma(\mathfrak{g}, \mathfrak{t})$, in other words if $\psi_j \neq \gamma(\psi_j)$ are adjacent in $\Delta(\mathfrak{g}, \mathfrak{h})$, then we use $2\mu_j$ instead of μ_j for the diagram. The affine Dynkin diagrams with coefficients, and a certain related algebra $\mathfrak{g}^\dagger$ that we will explain in Section 18.2B, are as follows.

Case $\gamma = 1$. If the graph symmetry γ is the identity, then $\Delta(\mathfrak{g}, \gamma) = \Delta(\mathfrak{g}, \mathfrak{h})$, and $\mu_j = \psi_j$ with the same coefficients, and the corresponding affine Dynkin diagrams are given by Tables 17.4.2 and 17.4.3.

Case γ has order 2. Suppose that the graph symmetry γ as order 2. Then, from Tables 17.4.2, 17.4.3 and 18.2.8 we have the affine Dynkin diagrams in Table 18.2.9.

Table 18.2.9. Affine Dynkin Diagrams $\Delta(\mathfrak{g}, \gamma, \text{aff})$

Type of $\mathfrak{g}$	Type of $\mathfrak{g}^\dagger$	Affine Dynkin Diagram with Coefficients
A_{2m-1}	D_{m+1}	
A_2	A_2	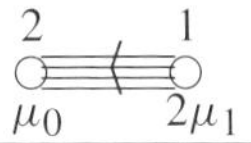
$A_{2m}, m > 1$	A_{2m}	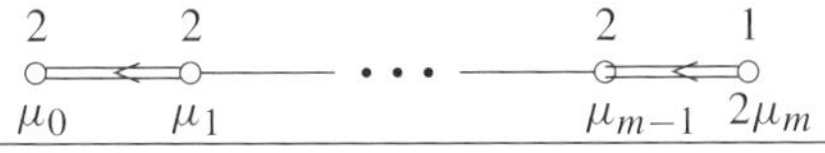
$D_\ell, \ell \geqq 4$	$A_{2\ell-3}$	
E_6	E_6	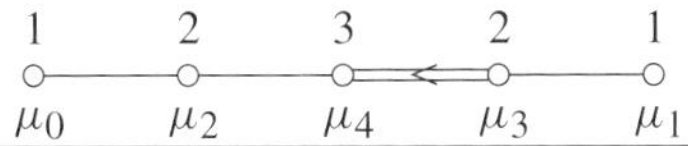

Case γ has order 3. The only case where the graph symmetry γ has order 3 is triality on D_4, which gives

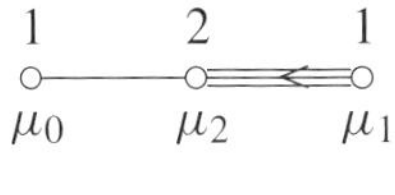

18.2B Digression: Affine Dynkin diagrams and classification of symmetric spaces

One can use Table 18.2.9 to organize the classification of symmetric spaces.

Fix a complex simple Lie algebra $\mathfrak{g}$ and an affine Dynkin diagram $\Delta(\mathfrak{g}, \gamma, \text{aff})$ with coefficients as in Table 18.2.9. We suppose that the vertices μ_j of $\Delta(\mathfrak{g}, \gamma, \text{aff})$ are numbered from 0 to m, with μ_0 coming from the negative of the highest weight of $\mathfrak{g}$. There is a single integral relation $\sum_{j=0}^{m} n(\mu_j)\mu_j = 0$, where the coefficients $n(\mu_j)$ are positive integers with no common factor $\neq 1$.

Removing a vertex μ_i with $n(\mu_i) = 1$, and ignoring the coefficients on the remaining vertices, one obtains the Dynkin diagram of a semisimple Lie algebra $\mathfrak{k}^{\dagger,i}$. If one already knows the classification of compact riemannian symmetric spaces, then one will observe that there is a unique (up to conjugacy) simple subalgebra $\mathfrak{g}^{\dagger} \subset \mathfrak{g}$ such that $\{\mathfrak{k}^{\dagger,i} \mid n(\mu_i) = 1\}$ consists exactly of the complexifications of Lie algebras $\mathfrak{k}_0^{\dagger}$ that correspond to compact symmetric spaces $G_u^{\dagger}/K_0^{\dagger}$ with rank $\mathfrak{k}^{\dagger} <$ rank $\mathfrak{g}^{\dagger}$. At first glance this cannot be used directly for the classification of symmetric spaces, because it requires adjustment in the cases where $\mathfrak{g}_0$ is simple but $\mathfrak{g}$ is not, and because it requires prior knowledge of the classification. But it does provide a uniform way of viewing the result.

However these problems are not very serious. First, one knows from the basic structure theory of symmetric spaces that the cases where $\mathfrak{g}_0$ is simple, but $\mathfrak{g}$ is not, correspond to the group manifolds of the compact connected simply connected simple Lie groups. Second, one can classify the admissible affine Dynkin diagrams and use them to construct a certain graded Lie algebra by a variation on the construction for affine Kač–Moody algebras, whose 0-level subalgebra is the appropriate algebra $\mathfrak{g}^{\dagger}$ (see [Hel, Section X.5]). It seems rather complicated, but it extends the Borel-de Siebenthal result to include outer automorphisms, as follows.

We assume that $\mathfrak{g}$ is simple. Let $c := \{c_0, \ldots, c_m\}$ be nonnegative integers with no common factor $\neq 1$. Then there is an automorphism θ_c of order k,

$$k = \text{Ord}(\theta_c) = \text{Ord}(\gamma)\sum\nolimits_0^m c_j n(\mu_j),$$

on $\mathfrak{g}^{\dagger}$, defined by its action on simple root vectors by the equation

$$\theta_c(e_{\mu_j}) = \exp(2\pi i c_j/k)e_{\mu_j}.$$

Conversely, every automorphism θ_c of finite order k arises in this way. The cases $k = 2$ are as follows.

$\text{Ord}(\gamma) = 1$. Then $\gamma = 1$, rank $\mathfrak{k}$ = rank $\mathfrak{g}$, and $\mathfrak{g} = \mathfrak{g}^{\dagger}$. In the non-hermitian case, $c_i = 0$ except for one index i, and there $n(\mu_i) = 2$. In the hermitian case there are indices $i \neq j$ with $c_i = 1 = c_j$ and $n(\mu_i) = 1 = n(\mu_j)$.

$\text{Ord}(\gamma) = 2$. Then rank $\mathfrak{k}^{\dagger} <$ rank $\mathfrak{g}^{\dagger}$. Here $\mathfrak{g} \neq \mathfrak{g}^{\dagger}$ in some cases, $\mathfrak{g} = \mathfrak{g}^{\dagger}$ in others, as indicated in Table 18.2.9. The coefficients c_j vanish except for one index i, and $c_i = 1 = n(\mu_i)$.

18.3 Various Weyl groups

In this section we review some basic properties of the various root systems that naturally arise in our context, and we define a subgroup of the (large) Weyl group $W = W(\mathfrak{g}, \mathfrak{h})$ which plays an important role for symmetric spaces G_0/K_0 when $\operatorname{rank} \mathfrak{k} < \operatorname{rank} \mathfrak{g}$,

Recall the decomposition $\mathfrak{g} = \mathfrak{k} + \mathfrak{s}$ as a sum of the eigenspaces of θ and the corresponding decomposition $\mathfrak{h} = \mathfrak{h}_\mathfrak{k} + \mathfrak{h}_\mathfrak{s}$ of a τ-stable θ-stable $\mathfrak{g}_0$-fundamental Cartan subalgebra of $\mathfrak{g}$. Here $\mathfrak{t} := \mathfrak{h}_\mathfrak{k}$ is a maximal toral subalgebra of $\mathfrak{k}$, and the Killing form is nondegenerate on $\mathfrak{k}$, $\mathfrak{s}$, $\mathfrak{t}$ and $\mathfrak{h}_\mathfrak{s}$. We will usually write $\mathfrak{t}$ rather than $\mathfrak{h}_\mathfrak{k}$.

The root systems which are in play here are the ordinary root system $\Sigma = \Sigma(\mathfrak{g}, \mathfrak{h})$, the restricted root system $\Sigma_\theta := \Sigma(\mathfrak{g}, \mathfrak{t})$ of Proposition 18.2.3, and the ordinary, root system $\Sigma_\mathfrak{k} := \Sigma(\mathfrak{k}, \mathfrak{t})$. They are related by

$$\text{(18.3.1)} \quad \operatorname{res} : \Sigma(\mathfrak{g}, \mathfrak{h}) = \Sigma \to \Sigma_\theta = \Sigma(\mathfrak{g}, \mathfrak{t}) \text{ and } \Sigma(\mathfrak{k}, \mathfrak{t}) = \Sigma_\mathfrak{k} \hookrightarrow \Sigma_\theta = \Sigma(\mathfrak{g}, \mathfrak{t}).$$

Since $\mathfrak{h}$ is $\mathfrak{g}_0$-fundamental in $\mathfrak{g}$, in other words, since $\mathfrak{t}$ contains regular elements of $\mathfrak{g}$, we have positive root systems $\Sigma^+ = \Sigma^+(\mathfrak{g}, \mathfrak{h})$, $\Sigma_\theta^+ = \Sigma^+(\mathfrak{g}, \mathfrak{t})$, and $\Sigma_\mathfrak{k}^+ = \Sigma^+(\mathfrak{k}, \mathfrak{t})$ consistent with the maps of (18.3.1):

$$\text{(18.3.2)} \quad \begin{aligned} &\operatorname{res} : \Sigma^+(\mathfrak{g}, \mathfrak{h}) = \Sigma^+ \to \Sigma_\theta^+ = \Sigma^+(\mathfrak{g}, \mathfrak{t}) \qquad \text{and} \\ &\Sigma^+(\mathfrak{k}, \mathfrak{t}) = \Sigma_\mathfrak{k}^+ \hookrightarrow \Sigma_\theta^+ = \Sigma^+(\mathfrak{g}, \mathfrak{t}). \end{aligned}$$

We write Ψ, Ψ_θ and $\Psi_\mathfrak{k}$ for the simple root systems of these consistent positive root systems.

Remarks.

1. Without the assumption that the τ-stable θ-stable Cartan subalgebra $\mathfrak{h} \subset \mathfrak{g}$ is $\mathfrak{g}_0$-fundamental, the restriction $\Sigma(\mathfrak{g}, \mathfrak{t})$ may not be a root system.
2. Independent of whether or not $\mathfrak{h}$ is $\mathfrak{g}_0$-fundamental, we have the decomposition $\Sigma(\mathfrak{g}, \mathfrak{h}) = \Sigma^{\mathrm{cpx}} \cup \Sigma^\theta \cup \Sigma^{-\theta}$, where $\Sigma^{\pm\theta} = \{\alpha \in \Sigma : \theta\alpha = \pm\alpha\}$ and $\Sigma^{\mathrm{cpx}} := \Sigma \setminus (\Sigma^\theta \cup \Sigma^{-\theta})$. The set Σ^θ should not be confused with $\Sigma_\theta = \Sigma(\mathfrak{g}, \mathfrak{t})$. In our situation, where $\mathfrak{t}$ is a Cartan subalgebra of $\mathfrak{k}$, we have $\Sigma^{-\theta} = \emptyset$. Also, $\Sigma^{\mathrm{cpx}} = \emptyset$ if and only if $\mathfrak{h}$ is contained either in $\mathfrak{k}$ or in $\mathfrak{s}$. Further, if $\mathfrak{h}$ is $\mathfrak{g}_0$-fundamental and $\alpha \in \Sigma^{\mathrm{cpx}}$, then $\alpha - \theta\alpha$ is not a root, because $(\alpha - \theta\alpha)|_\mathfrak{t} = 0$.
3. For arbitrary $\mathfrak{h}$ and $\alpha \in \Sigma^\theta$ we have $\mathfrak{g}_\alpha \subset \mathfrak{k}$ or $\mathfrak{g}_\alpha \subset \mathfrak{s}$. For that reason we write $\Sigma^\theta = \Sigma_\mathfrak{k}^\theta \cup \Sigma_\mathfrak{s}^\theta$.
4. For a $\mathfrak{g}_0$-fundamental $\mathfrak{h}$ the root spaces $\mathfrak{g}_\lambda$, $\lambda \in \Sigma(\mathfrak{g}, \mathfrak{t})$ are at most two-dimensional. Therefore, if $\alpha \in \Sigma(\mathfrak{g}, \mathfrak{h})$ and $\operatorname{res}\alpha := \alpha|_\mathfrak{t}$, then $\mathfrak{g}_{\operatorname{res}\alpha} = \mathfrak{g}_\alpha + \mathfrak{g}_{\theta\alpha}$. In particular, $\operatorname{res}\Sigma^{\mathrm{cpx}} \subset \Sigma(\mathfrak{k}, \mathfrak{t})$.

18.3A The group W^θ

We now discuss certain canonical subgroups of the Weyl group $W = W(\mathfrak{g}, \mathfrak{h})$. Recall that $W(\mathfrak{g}, \mathfrak{h})$ is the subgroup of $GL(\mathfrak{h})$ generated by the reflections s_α with respect to roots $\alpha \in \Sigma(\mathfrak{g}, \mathfrak{h})$. This (large) Weyl group can be identified with the quotient group $N_G(H)/H$, where H is the Cartan subgroup of G for $\mathfrak{h}$.

All the Cartan subgroups discussed below are stable under θ and τ, hence also under $\sigma = \theta\tau$. Consequently, those three involutions stabilize the normalizer $N_G(H)$ and thus act on $W(\mathfrak{g}, \mathfrak{h}) = N_G(H)/H$.

The Weyl group $W_{\mathfrak{k}} := W(\mathfrak{k}, \mathfrak{t})$ acts on the centralizer $\mathfrak{h}$ of $\mathfrak{t}$ and therefore can be considered as a subgroup of $W(\mathfrak{g}, \mathfrak{h})$. The subgroup which is most important for our purposes is the group W^θ, which in general lies between W and $W_{\mathfrak{k}}$.

Definition 18.3.3. The group $W^\theta = W^\theta(\mathfrak{g}, \mathfrak{h})$ is the subgroup of $W(\mathfrak{g}, \mathfrak{h})$ consisting of all elements that commute with θ.

This group can be described in a number of ways. For example,

$$\begin{aligned} W^\theta(\mathfrak{g}, \mathfrak{h}) &= \{w \in W(\mathfrak{g}, \mathfrak{h}) \subset GL(\mathfrak{h}) : w \circ \theta = \theta \circ w\} \\ &= \{w \in W(\mathfrak{g}, \mathfrak{h}) \subset GL(\mathfrak{h}) : w \circ \tau = \tau \circ w\} \\ &= \{n \in N_G(\mathfrak{h}) : \mathrm{Ad}(\theta(n))|_{\mathfrak{h}} = \mathrm{Ad}(n)|_{\mathfrak{h}}\}/H \\ &= \{n \in N_G(\mathfrak{h}) : \theta(n)^{-1}n \in H\}/H \\ &= N_G(\mathfrak{h}_0)/Z_G(\mathfrak{h}_0) = N_G(\mathfrak{t})/Z_G(\mathfrak{t}). \end{aligned}$$

We view $W(\mathfrak{k}, \mathfrak{t}) \subset W(\mathfrak{g}, \mathfrak{h})$ by means of

$$N_K(\mathfrak{h})/Z_K(\mathfrak{h}) = N_{K_0}(\mathfrak{h}_0)/Z_{K_0}(\mathfrak{h}_0) \cong W(\mathfrak{k}, \mathfrak{t}).$$

Then $W(\mathfrak{k}, \mathfrak{t}) \subset W^\theta(\mathfrak{g}, \mathfrak{h}) \subset W(\mathfrak{g}, \mathfrak{h})$.

If $\mathfrak{t}$ is already a Cartan subalgebra of $\mathfrak{g}$, in other words, if rank $\mathfrak{k}$ = rank $\mathfrak{g}$, then the action of θ on $W(\mathfrak{g}, \mathfrak{h})$ is trivial and $W^\theta(\mathfrak{g}, \mathfrak{h}) = W(\mathfrak{g}, \mathfrak{h})$. In general, however, $W^\theta(\mathfrak{g}, \mathfrak{h}) \subset GL(\mathfrak{h})$ need not be generated by reflections. Nevertheless the following proposition shows that this group, which plays an important role in our computation of the cohomology groups $H^j(C; \mathcal{O}(\mathbb{N}_Z(C)))$, is the Weyl group of a certain root system.

Proposition 18.3.4. *Let $\mathfrak{h}$ be a τ-stable θ-stable $\mathfrak{g}_0$-fundamental Cartan subalgebra of $\mathfrak{g}$.*

1. *The group $W^\theta(\mathfrak{g}, \mathfrak{h})$ stabilizes the decomposition*

$$\Sigma(\mathfrak{g}, \mathfrak{h}) = \Sigma^{\mathrm{cpx}} \cup \Sigma^\theta \cup \Sigma^{-\theta}.$$

   Furthermore, $W^\theta(\mathfrak{g}, \mathfrak{h})$ acts on the restricted root system $\Sigma(\mathfrak{g}, \mathfrak{t})$. However, in general it does not stabilize the decomposition $\Sigma^\theta = \Sigma^\theta_{\mathfrak{k}} \cup \Sigma^\theta_{\mathfrak{s}}$, and needs not to preserve the subsystem $\Sigma(\mathfrak{k}, \mathfrak{t}) \subset \Sigma(\mathfrak{g}, \mathfrak{t})$).
2. *The restriction $W^\theta(\mathfrak{g}, \mathfrak{h}) \to GL(\mathfrak{t})$ is a faithful representation of $W^\theta(\mathfrak{g}, \mathfrak{h})$. The restriction $W^\theta(\mathfrak{g}, \mathfrak{h}) \to GL(\mathfrak{h}_{\mathfrak{s}})$ is almost never faithful.*
3. *The action of $W^\theta(\mathfrak{g}, \mathfrak{h})$ on $\mathfrak{t}$ coincides with that of the Weyl group of $\Sigma(\mathfrak{g}, \mathfrak{t})$.*

Proof.

1. These statements are straightforward. Examples in which the decomposition $\Sigma(\mathfrak{g}, \mathfrak{t}) = \Sigma^\theta_\mathfrak{k} \cup \Sigma^\theta_\mathfrak{s}$ is not preserved by $W^\theta(\mathfrak{g}, \mathfrak{h})$ (and consequently $\Sigma(\mathfrak{k}, \mathfrak{t}) = \operatorname{res} \Sigma^\theta_\mathfrak{k} \cup \operatorname{res} \Sigma^{\mathrm{cpx}}$ is also not preserved) can be given as follows: If $\operatorname{rank} \mathfrak{g} = \operatorname{rank} \mathfrak{k}$, then $W^\theta = W$ and the decomposition $\Sigma^\theta_\mathfrak{k} \cup \Sigma^\theta_\mathfrak{s}$ is nothing but the decomposition of Σ into compact and noncompact roots. Clearly, W mixes $\Sigma^\theta_\mathfrak{k}$ and $\Sigma^\theta_\mathfrak{s}$. One example with $\operatorname{rank} \mathfrak{g} > \operatorname{rank} \mathfrak{k}$ is $\mathfrak{g} = \mathfrak{so}(2p + 2q + 2; \mathbb{C})$ and $\mathfrak{k} = \mathfrak{so}(2p + 1; \mathbb{C}) \times \mathfrak{so}(2q + 1; \mathbb{C})$. In the standard notation for roots of the Lie algebras of the orthogonal groups, as in Section 19.4, we have $\Sigma^\theta_\mathfrak{k} = \{\pm\varepsilon_j \pm \varepsilon_k\}_{1 \leqq j<k \leqq p} \cup \{\pm\varepsilon_j \pm \varepsilon_k\}_{p+1 \leqq j<k \leqq p+q}$. But $W^\theta(\mathfrak{g}, \mathfrak{h})$ contains $\mathfrak{S}_{p+q}$ which contains all permutations of $\varepsilon_1, \ldots, \varepsilon_{p+q}$.

2. Fix a θ-stable Borel $\mathfrak{b} \supset \mathfrak{h}$. Then $\mathfrak{b}$ decomposes under $\mathfrak{t}$ as a direct sum of root spaces, $\mathfrak{b} = \mathfrak{h} + \sum_{\lambda \in \Sigma^+(\mathfrak{g}, \mathfrak{t})} \mathfrak{g}_\lambda$ (some $\mathfrak{g}_\lambda$ may be two-dimensional). In this way, θ-stable Borel subalgebras determine θ-stable positive root systems $\Sigma^+(\mathfrak{g}, \mathfrak{h})$ and $\Sigma^+(\mathfrak{g}, \mathfrak{t}) := \operatorname{res}(\Sigma^+(\mathfrak{g}, \mathfrak{h}))$, and conversely.

Note that $W^\theta(\mathfrak{g}, \mathfrak{h})$ acts on the set of all θ-stable Borel subalgebras that contain a given Cartan subalgebra $\mathfrak{t}$ of $\mathfrak{k}$. In other words, if a Borel $\mathfrak{b}$ containing $\mathfrak{t}$ is θ-stable, then $w(\mathfrak{b}) = \mathfrak{b}(w(\Sigma_\mathfrak{b}))$ is θ-stable. This is a simple consequence of the root decomposition with respect to $\mathfrak{t}$. We write $w(\mathfrak{b})$ for $\operatorname{Ad}_{n_w}(\mathfrak{b})$, where $n_w \in N(T)$ is an arbitrary representative of w.

Let $w \in W^\theta \subset W$ such that $w|_\mathfrak{t} = \mathrm{id}$. Then w stabilizes the subset $\Sigma^+(\mathfrak{g}, \mathfrak{t})$, and consequently $w(\mathfrak{b}) = \mathfrak{b}$. Since $W(\mathfrak{g}, \mathfrak{h})$ acts simply transitively on the set of Borel subalgebras that contain $\mathfrak{h}$, we conclude that $w = \mathrm{id}$, and it follows that $W^\theta(\mathfrak{g}, \mathfrak{h}) \times \mathfrak{t} \to \mathfrak{t}$ is faithful.

3. Let $W(\mathfrak{g}, \mathfrak{t})$ denote the subgroup of $GL(\mathfrak{t})$ generated by all reflections s_λ for $\lambda \in \Sigma(\mathfrak{g}, \mathfrak{t})$. The first step is to show that $W(\mathfrak{g}, \mathfrak{t}) \subset W^\theta(\mathfrak{g}, \mathfrak{t}) = N_G(\mathfrak{t})/Z_G(\mathfrak{t})$. If $\lambda \in \Sigma(\mathfrak{g}, \mathfrak{t})$, define

$$\mathfrak{sl}_\lambda := \begin{cases} \mathfrak{s}_\lambda + \mathfrak{s}_{-\lambda} + \mathbb{C}h_\lambda & \text{if } \mathfrak{g}_\lambda \cap \mathfrak{s} \neq 0. \\ \mathfrak{k}_\lambda + \mathfrak{k}_{-\lambda} + \mathbb{C}h_\lambda & \text{otherwise.} \end{cases}$$

Select a σ-compatible $\mathfrak{sl}(2;\mathbb{C})$-triple $\{e_\lambda, e_{-\lambda}, h_\lambda\}$ from $\mathfrak{sl}_\lambda$: $e_{-\lambda} = -\sigma e_\lambda$ and $\theta(e_\lambda) = \pm e_\lambda$. Finally, define

$$n_\lambda := \exp i\tfrac{\pi}{2}(e_\lambda + e_{-\lambda}) \quad \text{or} \quad n_\lambda := \exp \tfrac{\pi}{2}(e_\lambda - e_{-\lambda}).$$

Then $\operatorname{Ad}(n_\lambda)$ stabilizes $\mathfrak{t}$ and acts on $\mathfrak{t}$ as a reflection in the hyperplane $\lambda^\perp := \{X \in \mathfrak{t} \mid \lambda(X) = 0\}$. This is a consequence of the following facts: (i) $\mathfrak{t} = \mathbb{C}h_\lambda + \lambda^\perp$, (ii) every element of $\exp \mathfrak{sl}_\lambda$ fixes $(\mathfrak{t}_\mathfrak{k})_{[\lambda]}$ pointwise, and (iii) $\operatorname{Ad}(n_\lambda)(h_\lambda) = -h_\lambda$. Only the last point requires proof. Compute

$$\begin{aligned} \operatorname{Ad}(n_\lambda)(h_\lambda) &= \sum \tfrac{(i\pi/2)^n}{n!} (\operatorname{ad}(e_\lambda + e_{-\lambda}))^n (h_\lambda) \\ &= \sum \tfrac{(i\pi/2)^{2k}}{2k!} (\operatorname{ad}(e_\lambda + e_{-\lambda}))^{2k} (h_\lambda) \\ &\quad + \sum \tfrac{(i\pi/2)^{2k+1}}{(2k+1)!} (\operatorname{ad}(e_\lambda + e_{-\lambda}))^{2k+1} (h_\lambda) \end{aligned}$$

$$= \sum \frac{(i\pi)^{2k}}{2k!} \cdot h_\lambda + \sum \frac{(i\pi)^{2k+1}}{(2k+1)!} \cdot (e_{-\lambda} - e_\lambda)$$
$$= \cos\pi \cdot h_\lambda + i\sin\pi \cdot (e_{-\lambda} - e_\lambda) = -h_\lambda.$$

This shows that $\mathrm{Ad}(n_\lambda) \in W^\theta(\mathfrak{g}, \mathfrak{h})$, and therefore $W(\mathfrak{g}, \mathfrak{t}) \subset W^\theta(\mathfrak{g}, \mathfrak{h})$.

The next step is to show that the above inclusion is an equality. For this observe that since $W^\theta(\mathfrak{g}, \mathfrak{h})$ permutes the set $\Sigma(\mathfrak{g}, \mathfrak{t})$ of restricted roots, it stabilizes the regular set $\mathfrak{t}_{\text{reg}} = \{\xi \in \mathfrak{t} \mid \lambda(\xi) \neq 0 \,\forall\, \lambda \in \Sigma(\mathfrak{g}, \mathfrak{t})\}$. Therefore it permutes the Weyl chambers, which are the connected components of $\mathfrak{t}_{\text{reg}} \cap \mathfrak{t}_{\mathbb{R}}$ in $\mathfrak{t}_{\mathbb{R}}$.

Now we show that $W^\theta(\mathfrak{g}, \mathfrak{h})$ acts simply transitively on the chambers in $\mathfrak{t}_{\mathbb{R}}$ of the small Weyl group. For this, assume to the contrary that $w \in W^\theta(\mathfrak{g}, \mathfrak{h})$ stabilizes such a Weyl chamber $C \subset \mathfrak{t}_{\mathbb{R}}$ (but perhaps not pointwise). Select $n_w \in N(H)$ with $\mathrm{Ad}(n_w) = w$ on $\mathfrak{t}$. Select a regular element $\xi \in C$ and let $\eta := \sum_{k=1}^m \frac{w^k \cdot \xi}{m}$, where m is the order of w in $W^\theta(\mathfrak{g}, \mathfrak{h})$. Then η is a regular element in C, C is convex, and $W^\theta(\mathfrak{g}, \mathfrak{h})$ acts linearly on $\mathfrak{t}$. By construction $w(\eta) = \eta$. Since η is regular, its centralizer $Z_G(\eta)$ coincides with the Cartan subgroup $H = \exp(\mathfrak{h})$. It follows that $n_w \in H$, i.e., $w = \mathrm{id}$. □

Remark 18.3.5. The construction of the simple three-dimensional algebra $\mathfrak{sl}_\lambda$ in the proof of Proposition 18.3.4 is similar to the corresponding construction of $\mathfrak{sl}_\alpha := [\mathfrak{g}_\alpha, \mathfrak{g}_{-\alpha}] + \mathfrak{g}_\alpha + \mathfrak{g}_{-\alpha}$ where $\alpha \in \Sigma(\mathfrak{g}, \mathfrak{h})$. In both cases γ determines the connected subgroup $SL_\gamma \subset G$. If G is simply connected and $\gamma \in \Sigma(\mathfrak{g}, \mathfrak{h})$, it is known that SL_γ is simply connected; but that fails when $\gamma \in \Sigma(\mathfrak{g}, \mathfrak{t})$ and $\dim \mathfrak{g}_\gamma = 2$. ◇

As an application of the above proposition, we note the following.

Corollary 18.3.6. *Let $\mathfrak{b}_\mathfrak{k}$ be a Borel subalgebra of $\mathfrak{k}$. Then the number of Borel subalgebras in $\mathfrak{g}$ that contain $\mathfrak{b}_\mathfrak{k}$ is exactly $|W^\theta(\mathfrak{g}, \mathfrak{h})|/|W(\mathfrak{k}, \mathfrak{t})|$. More geometrically, the number of K-cycles in G/B, or equivalently the number of open G_0-orbits in G/B, is equal to $|W^\theta(\mathfrak{g}, \mathfrak{h})|/|W(\mathfrak{k}, \mathfrak{t})|$.*

Proof. Since $W^\theta(\mathfrak{g}, \mathfrak{h}) = W(\mathfrak{g}, \mathfrak{t})$, it acts simply transitively on the set of Weyl chambers for $\Sigma(\mathfrak{g}, \mathfrak{t})$. Hence it likewise acts simply transitively on the set of all θ-stable Borel subalgebras of $\mathfrak{g}$ that contain $\mathfrak{h}$. □

Remark 18.3.7. As above let $\mathfrak{t}$ be a maximal toral subalgebra of $\mathfrak{k}$. The number of open G_0-orbits in G/B was also computed in [W2, Theorem 4.6] and [W2, Corollary 4.8] as a certain quotient $|N_G(\mathfrak{h}_0)/Z_G(\mathfrak{h}_0)|/|W(\mathfrak{k}, \mathfrak{t})|$. Of course $W^\theta(\mathfrak{g}, \mathfrak{h}) \cong N_G(\mathfrak{h}_0)/Z_G(\mathfrak{h}_0)$, but Corollary 18.3.6 realizes the number of Borel subalgebras that contain $\mathfrak{b}_\mathfrak{k}$ directly as a quotient of orders of Weyl groups.

Here we only consider the nontrivial situation $\Sigma_\theta \neq \Sigma$. The type of the root system Σ_θ (which, according to the above propositions determines W^θ) can be read off the affine Dynkin diagrams of Table 18.2.9. This is done as follows.

Consider the subdiagram, obtained by removing the root μ_0 from $\Delta(\mathfrak{g}, \gamma, \text{aff})$. This subdiagram is the ordinary Dynkin diagram of Σ_θ. Equivalently, Table 18.2.8 also provides us with such a Dynkin diagram. Note also that the Weyl group $W(\Sigma_\theta)$ is equal to $W(\Sigma_\theta^{\text{reduced}})$, where $\Sigma_\theta^{\text{reduced}}$ denotes a corresponding reduced root subsystem. ◇

For later use we define a section $s : W(\mathfrak{k}, \mathfrak{t})\backslash W^\theta(\mathfrak{g}, \mathfrak{h}) \to W^\theta(\mathfrak{g}, \mathfrak{h})$ as follows. Let $\mathfrak{b}_\mathfrak{k}$ be a fixed Borel subalgebra of $\mathfrak{k}$ and $\mathfrak{b} \supset \mathfrak{b}_\mathfrak{k}$ a fixed Borel subalgebra of $\mathfrak{g}$. Then $\mathfrak{b}$ is θ-stable. Given $w \in W^\theta(\mathfrak{g}, \mathfrak{h})$, $w(\mathfrak{b})$ is another θ-stable Borel subalgebra of $\mathfrak{g}$ that contains $\mathfrak{t}$. So $w(\mathfrak{b}) \cap \mathfrak{k}$ is a Borel subalgebra which may be different from $\mathfrak{b}_\mathfrak{k}$. Since $W(\mathfrak{k}, \mathfrak{t})$ acts simply transitively on the Borel subalgebras of $\mathfrak{k}$ that contain $\mathfrak{t}$, there is a unique element $w_\mathfrak{k} \in W(\mathfrak{k}, \mathfrak{t})$ such that $w_\mathfrak{k} w(\mathfrak{b}) \cap \mathfrak{k} = \mathfrak{b}_\mathfrak{k}$. We define

$$s : W(\mathfrak{k}, \mathfrak{t}) \backslash W^\theta(\mathfrak{g}, \mathfrak{h}) \to W^\theta(\mathfrak{g}, \mathfrak{h}) \quad \text{by } s(W(\mathfrak{k}, \mathfrak{t}){\cdot}w) := w_\mathfrak{k} w.$$

This defines a particular subset $W_1^\theta(\mathfrak{g}, \mathfrak{h}) \subset W^\theta(\mathfrak{g}, \mathfrak{h})$ by

$$W_1^\theta(\mathfrak{g}, \mathfrak{h}) = \text{ image of } s = \{w \in W^\theta(\mathfrak{g}, \mathfrak{h}) \mid w(\mathfrak{b} \cap \mathfrak{k}) = \mathfrak{b} \cap \mathfrak{k}\}. \tag{18.3.8}$$

18.4 Some distinguished weights

In Section 17.4 we described the highest weights of the irreducible $\mathfrak{k}$-modules that occur in $\mathfrak{s}$. The description was in terms of fundamental weights for $(\mathfrak{k}', \mathfrak{t}')$ where $\mathfrak{k}' = [\mathfrak{k}, \mathfrak{k}]$ is the semisimple component of $\mathfrak{k}$, $\mathfrak{t}$ is a τ-stable Cartan subalgebra of $\mathfrak{k}$, and $\mathfrak{t}'$ is the Cartan subalgebra $\mathfrak{t} \cap \mathfrak{k}'$ of $\mathfrak{k}'$. From these highest weights we know, at least in principle, the entire weight system $M(\mathfrak{t}', \mathfrak{s})$ of the representation of $\mathfrak{k}'$ on $\mathfrak{s}$.

One of our main goals is to make the last statement effective. For this we begin by recalling the notation. Let $\Lambda_{\mathfrak{k},wt}$ denote the lattice of all weights of $\mathfrak{k}$ relative to $\mathfrak{t}$, and $\Lambda_{\mathfrak{k},wt}^+$ be the set of dominant ones, i.e.,

$$\Lambda_{\mathfrak{k},wt}^+ = \{\lambda \in \Lambda_{\mathfrak{k},wt} \mid \langle \lambda, \alpha \rangle \geqq 0 \text{ for every } \alpha \in \Sigma^+(\mathfrak{k}, \mathfrak{t})\}.$$

Recall that the weights of representations π_λ of $\mathfrak{k}$ with highest weight λ form a saturated set:

$$\begin{aligned} M^+(\mathfrak{t}, \pi_\lambda) &:= M(\mathfrak{t}, \pi_\lambda) \cap \Lambda_{\mathfrak{k},wt}^+ = \{\mu \in \Lambda_{\mathfrak{k},wt}^+ \mid \mu \prec \lambda\} \text{ and} \\ M(\mathfrak{t}, \pi_\lambda) &= W(\mathfrak{k}, \mathfrak{t}) \cdot M^+(\mathfrak{t}, \pi_\lambda). \end{aligned} \tag{18.4.1}$$

Here $W(\mathfrak{k}, \mathfrak{t})$ is the Weyl group of $\mathfrak{k}$ relative to $\mathfrak{t}$ as before, and $\prec$ denotes the partial order on the weights defined by $\Sigma^+(\mathfrak{k}, \mathfrak{t})$: $\mu \prec \lambda$ if and only if

$$\lambda - \mu = \sum\nolimits_{\alpha \in \Sigma^+(\mathfrak{k}, \mathfrak{t})} n_\alpha \alpha,$$

where $n_\alpha \geqq 0$ for all α.

In practice, explicit determination of $M^+(\mathfrak{t}, \pi_\lambda)$ and $M(\mathfrak{t}, \pi_\lambda)$ from λ can be difficult, especially if $M^+(\mathfrak{t}, \pi_\lambda)$ contains many dominant weights. In our particular situation we have the advantage that (1) the weights in $M(\mathfrak{t}, \pi_\lambda)$ are elements of a root system and (2) there are at most three nonzero dominant weights.

To start we recall that the highest weight λ, which is the unique maximal element of $M(\mathfrak{t}, \pi_\lambda)$ relative to the partial order $\prec$ is also maximal with respect to the norm $||\cdot||$ given by the Killing form of $\mathfrak{g}$ (see, e.g., [Hu1, Section 13.4]).

Proposition 18.4.2. *If $\mu, \lambda \in \Lambda^+_{\mathfrak{k},wt}$, $\mu \neq \lambda$, and $\mu \prec \lambda$, then $||\mu|| < ||\lambda||$.*

Remarks.

- We apply the above statement as follows. In the Theorem 18.4.13, for a given complex symmetric pair $(\mathfrak{g}, \mathfrak{k})$, $\mathfrak{g} = \mathfrak{k} + \mathfrak{s}$, we need to decide whether or not all weights in $M(\mathfrak{t}, \mathfrak{s}) \setminus 0$ have the same length. If, according to the Tables 17.4.5, 17.4.6 and 17.4.8, there is a further dominant weight in the isotropy representation, we immediately deduce that there exist nonzero weights with at least two different lengths in $M(\mathfrak{t}, \mathfrak{s})$.
- Suppose that $\mathfrak{g}_0$ is not of hermitian type, so that the action of $\mathfrak{k}$ on $\mathfrak{s}$ is an irreducible representation π_λ. Then $M(\mathfrak{t}, \pi_\lambda)$ is a subset of $\Sigma(\mathfrak{g}, \mathfrak{t}) \cup \{0\}$, and $\Sigma(\mathfrak{g}, \mathfrak{t})$ is a possibly nonreduced root system. Note that $0 \in M(\mathfrak{t}, \pi_\lambda)$ if and only if rank $\mathfrak{k} <$ rank $\mathfrak{g}$. In view of Proposition 18.2.6 and Table 18.2.8, $M(\mathfrak{t}, (\mathfrak{q}+\theta\mathfrak{q})_\mathfrak{s})$ contains weights of at most 3 different lengths (including 0) if the root system is reduced and four different lengths in the one case where the root system is not reduced.

18.4A Bott's conditions

Given an irreducible $(K \cap Q)$-module V^μ of highest weight μ with respect to $B_K^+ \cap Q$, the Bott–Borel–Weil Theorem 1.6.8 gives precise conditions for the nonvanishing of cohomologies $H^\ell(C; \mathcal{O}(\mathbb{E}(V^\mu)))$. Using the positive root system $\Sigma^+(\mathfrak{k}, \mathfrak{t})$, we write $\Psi_\mathfrak{k}$ for the corresponding simple root system, and we define $\rho_\mathfrak{k} = \frac{1}{2}\sum_{\gamma \in \Sigma^+_\mathfrak{k}} \gamma = \xi_{\beta_1} + \cdots + \xi_{\beta_{r'}}$. Here, $\xi_{\beta_j} \in \mathfrak{t}^*$ denotes the fundamental weights, i.e., weights which are dual to the normalized coroots $h_{\beta_1}, \ldots, h_{\beta_{r'}}$. Bott's conditions for the nonvanishing of $H^\ell(C; \mathcal{O}(\mathbb{E}(V^\mu)))$ are

- $\mu + \rho_\mathfrak{k}$ is $\mathfrak{k}$-regular (18.1.1), i.e., $\langle \mu + \rho_\mathfrak{k}, \gamma \rangle \neq 0$ for all $\gamma \in \Sigma(\mathfrak{k}, \mathfrak{t})$, and
- the index (see 18.1.2) satisfies $\mathrm{ind}(\mu + \rho_\mathfrak{k}) = \ell$.

If $\varphi \in \mathfrak{t}^*_\mathbb{R}$ is $\mathfrak{k}$-regular there is a unique $w_\varphi \in W(\mathfrak{k}, \mathfrak{t})$ such that $w_\varphi(\varphi)$. is dominant. We denote that Weyl group translate by $[\varphi]$. The minimum number of simple root reflections in an expression $w_\varphi = s_{\beta_{j_1}} s_{\beta_{j_2}} \cdots s_{\beta_{j_\ell}}$ (the length of w_φ as a word in the simple root reflections) is equal to the index of φ.

We will need to understand the action of the Weyl group $W(\mathfrak{k}, \mathfrak{t})$ on the weight lattice $\Lambda_{\mathfrak{k},wt} \subset \mathfrak{t}^*_\mathbb{R}$. This amounts to a characterization of certain β-strings in weight systems $M(\mathfrak{t}, V_\varphi) \subset \Lambda_{\mathfrak{k},wt}$. The generating elements of $W(\mathfrak{k}, \mathfrak{t})$ are the reflections $s_\beta : v \mapsto v - \frac{2\langle v, \beta\rangle}{\langle \beta, \beta\rangle}\beta \cong v - \langle v|\beta\rangle\beta$.

18.4B Selected values of $\mu(h_\lambda)$

We now discuss the possible values of $\mu(h_\lambda) = \frac{2\langle \mu, \lambda\rangle}{\langle \lambda, \lambda\rangle}$, where λ and μ are weights of the representation $\mathfrak{k} \times \mathfrak{s} \to \mathfrak{s}$. It turns out that there are only a few possibilities. Since $M(\mathfrak{t}, (\mathfrak{q}+\theta\mathfrak{q})_\mathfrak{s}) \subset \Sigma(\mathfrak{g}, \mathfrak{t}) \cup \{0\}$ and $\Sigma(\mathfrak{g}, \mathfrak{t})$ is a (not necessarily reduced) root system, if $\lambda, \mu \in M(\mathfrak{t}, (\mathfrak{q}+\theta\mathfrak{q})_\mathfrak{s})$, then

$$\langle \mu \mid \lambda \rangle := \frac{2\langle \mu,\lambda\rangle}{\langle \lambda,\lambda\rangle} \in \{0,\ \pm 1,\ \pm 2,\ \pm 3,\ \pm 4\}.$$

We now single out the largest values.

1. Let $\frac{2\langle \mu,\lambda\rangle}{\langle \lambda,\lambda\rangle} = \pm 4$. Then $\mu = \pm 2\lambda$ and $\Sigma(\mathfrak{g}, \mathfrak{t})$ is not reduced. As observed in Proposition 18.2.6, this can happen if and only if the symmetric space G_0/K_0 is of the form $SL(2m+1; \mathbb{R})/SO(2m+1)$.
2. Let $\frac{2\langle \mu,\lambda\rangle}{\langle \lambda,\lambda\rangle} = \pm 3$. Then $\Sigma(\mathfrak{g}, \mathfrak{t})$ is reduced. From the classification of complex simple Lie algebras, it follows that $\Sigma(\mathfrak{g}, \mathfrak{t})$ is the root system of the exceptional simple Lie algebra $\mathfrak{g}_2$, where in fact there are μ and λ such that $||\mu||^2 = 3||\lambda||^2$. From Tables 17.4.2, 17.4.3 and 18.2.8, and of course the information in the case where $\mathfrak{g}_0$ is simple but $\mathfrak{g}$ is not, we see that either $\mathfrak{g}$ is of type $\mathfrak{g}_2$ with $K_0 = SO(4)$, or $\mathfrak{g} = \mathfrak{g}_2 \oplus \mathfrak{g}_2$ with $\mathfrak{g}_0$ embedded diagonally. Thus either G_u/K_0 is $\mathrm{G}_{2(-14)}/SO(4)$ or G_u/K_0 is the group manifold $\mathrm{G}_{2(-14)}$.

We will discuss these three examples separately in Section 19.3, Section 20.4, and Section 19.10. So we put them aside until then, and we suppose that $\Sigma(\mathfrak{g}, \mathfrak{t})$ is not isomorphic to either of the three examples described above. In other words, we assume for now that

(18.4.3) $\qquad \langle \mu \mid \lambda \rangle \in \{\pm 2,\ \pm 1,\ 0\}$ for any two elements $\lambda, \mu \in \Sigma(\mathfrak{g}, \mathfrak{t})$.

This assumption will significantly shorten some of the following arguments. As before, $\Sigma^+(\mathfrak{k}, \mathfrak{t})$ is our positive root system and $\Psi_{\mathfrak{k}}$ is its simple subsystem.

Lemma 18.4.4 (String Lemma). *Suppose that* $\Sigma(\mathfrak{g}, \mathfrak{t})$ *satisfies* (18.4.3), *in other words that* G_0/K_0 *is not isometric to* $SL(2m+1; \mathbb{R})/SO(2m+1)$, $\mathrm{G}_{2(2)}/SO(4)$, *nor* $\mathrm{G}_2(\mathbb{C})/\mathrm{G}_{2(-14)}$. *Let* $\mu \in M(\mathfrak{t}, \mathfrak{s})$. *Then*

1. *The weight* μ *is Bott-regular of index* 1, *if and only if there is a root* $\beta \in \Psi_{\mathfrak{k}}$ *such that* (i) $\frac{2\langle \mu,\beta\rangle}{\langle \beta,\beta\rangle} = -2$ *and* (ii) $\mu + \beta$ *is* $\Sigma^+(\mathfrak{k}, \mathfrak{t})$*-dominant. Equivalently, there exists* $\beta \in \Psi_{\mathfrak{k}}$ *such that* $\{\mu, \mu+\beta, \mu+2\beta\}$ *is a* β*-string in* $M(\mathfrak{t}, \mathfrak{s})$ *with* $\mu + \beta$ *dominant. Consequently,* $\mu + \beta$ *is not a highest weight of* $\mathfrak{k}$ *on* $\mathfrak{s}$, *and* $[\mu + \rho_{\mathfrak{k}}] - \rho_{\mathfrak{k}} = \mu + \beta$. *Two possibilities occur:*
   (a) *The dominant root* $\mu + \beta$ *is nonzero. Then* β *must be short.*
   (b) *The dominant root* $\mu+\beta$ *is zero. Then* μ *can be any element of* $-\Psi_{\mathfrak{k}} \cap M(\mathfrak{t}, \mathfrak{s})$.
2. *The weight* μ *is Bott-regular of index* 2 *if and only* $\mu = -\beta_1 - \beta_2$ *for two orthogonal simple roots* $\beta_1, \beta_2 \in \Psi_{\mathfrak{k}}$ *such that* $\beta_1 \pm \beta_2, \beta_1, \beta_2 \in M(\mathfrak{t}, \mathfrak{s})$. *Then* μ *is long and the* β_j *are short, and* $\lambda(\mu) = [\mu + \rho_{\mathfrak{k}}] - \rho_{\mathfrak{k}} = 0$.

In all cases the existence of μ *with nontrivial dominant weight* $\mu + \beta$ *implies that* $M(\mathfrak{t}, \mathfrak{s}) \setminus \{0\}$ *contains weights of* 2 *different lengths.*

Proof.

Assertion 1. The conditions that $\mathrm{ind}(\mu + \rho_{\mathfrak{k}}) = 1$ and $\mu + \rho_{\mathfrak{k}}$ is regular mean that there is precisely one simple root $\beta \in \Psi_{\mathfrak{k}}$ such that $s_\beta(\mu + \rho_{\mathfrak{k}})$ is regular dominant, i.e., $\frac{2\langle s_\beta(\mu \mid \rho_{\mathfrak{k}}),\beta'\rangle}{\langle \beta',\beta'\rangle} \geqq 1$ for all $\beta' \in \Psi_{\mathfrak{k}}$. We now check that this simple root β satisfies the conditions of Assertion 1.

The action of the reflection s_β is

$$s_\beta(\mu+\rho_{\mathfrak{k}}) = \mu+\rho_{\mathfrak{k}} - \frac{2\langle\mu+\rho_{\mathfrak{k}},\beta\rangle}{\langle\beta,\beta\rangle}\beta = \mu+\rho_{\mathfrak{k}} - \left(1+\frac{2\langle\mu,\beta\rangle}{\langle\beta,\beta\rangle}\right)\beta. \tag{18.4.5}$$

Since $\frac{2\langle s_\beta(\mu+\rho_{\mathfrak{k}}),\beta'\rangle}{\langle\beta',\beta'\rangle} \geqq 1$ for all $\beta' \in \Sigma^+(\mathfrak{k},\mathfrak{t})$, in particular for $\beta' = \beta$, now $\frac{2\langle\mu,\beta\rangle}{\langle\beta,\beta\rangle} \leqq -2$. Since $\frac{2\langle\mu,\beta\rangle}{\langle\beta,\beta\rangle} \geqq -2$ by (18.4.3), now $\frac{2\langle\mu,\beta\rangle}{\langle\beta,\beta\rangle} = -2$. Similarly, $\frac{2\langle\mu,\beta'\rangle}{\langle\beta',\beta'\rangle} \geqq \frac{2\langle\beta,\beta'\rangle}{\langle\beta',\beta'\rangle} \geqq 0$ for $\beta \neq \beta' \in \Psi_{\mathfrak{k}}$. In particular,

$$s_\beta(\mu+\rho_{\mathfrak{k}}) = \mu+\rho_{\mathfrak{k}}+\beta = [\mu+\rho_{\mathfrak{k}}],$$

and $\mu+\beta = [\mu+\rho_{\mathfrak{k}}]-\rho_{\mathfrak{k}}$ must be dominant. Since $\frac{2\langle\mu,\beta\rangle}{\langle\beta,\beta\rangle} = -2$, the β-string through μ has length 3. Note that $\mu+2\beta$ may or may not be dominant.

If $\mu+\beta$, the middle root in the string $\mu, \mu+\beta, \mu+2\beta$, is nonzero, then by elementary geometry of the two-dimensional root systems we conclude that β and the dominant weight $\beta+\mu$ are short. Since $\mu+2\beta$ is also a weight in $M(\mathfrak{t},\mathfrak{s})$, it follows that $\mu+\beta$ cannot be a highest weight. If $\mu+\beta = 0$, then $\mu = -\beta$. All elements $\beta' \in -\Psi_{\mathfrak{k}}$ have the property that $\operatorname{ind}(\beta'+\rho_{\mathfrak{k}}) = 1$ and $\beta'+\rho_{\mathfrak{k}}$ is regular. However, the subset $-\Psi_{\mathfrak{k}} \cap M(\mathfrak{t},\mathfrak{s})$ of $-\Psi_{\mathfrak{k}}$ may be proper or even empty.

If $\mu, \mu+\beta, \mu+2\beta$ is a β-string in $M(\mathfrak{t},\mathfrak{s})$, then by (18.4.3) it has maximal length, and therefore $\langle\mu \mid \beta\rangle = -2$. Applying the reflection s_β, (18.4.5) shows that $s_\beta(\mu+\rho_{\mathfrak{k}}) = \mu+\beta+\rho_{\mathfrak{k}}$. If $\mu+\beta$ is dominant and β simple, then $\mu+\rho_{\mathfrak{k}}$ is regular dominant and of index 1.

Assertion 2. We investigate the possibility that $\operatorname{ind}(\mu+\rho_{\mathfrak{k}}) = 2$ and $\mu+\rho_{\mathfrak{k}}$ is regular. By assumption, there exist 2 different roots $\beta_1, \beta_2 \in \Psi_{\mathfrak{k}}$ such that $s_{\beta_2}s_{\beta_1}(\mu+\rho_{\mathfrak{k}})$ is regular and dominant.

We check that $\frac{2\langle\mu,\beta_1\rangle}{\langle\beta_1,\beta_1\rangle} = \frac{2\langle\mu,\beta_2\rangle}{\langle\beta_2,\beta_2\rangle} = -2$ and $\langle\beta_1,\beta_2\rangle = 0$. Write $c_j := \frac{2\langle\mu,\beta_j\rangle}{\langle\beta_j,\beta_j\rangle}$ and, to simplify the notation, $b_{jk} := \frac{2\langle\beta_j,\beta_k\rangle}{\langle\beta_k,\beta_k\rangle}$. Then

$$\begin{aligned} s_{\beta_2}s_{\beta_1}(\mu+\rho_{\mathfrak{k}}) &= s_{\beta_2}(\mu+\rho_{\mathfrak{k}}-(c_1+1)\beta_1) \\ &= \mu+\rho_{\mathfrak{k}}-(c_1+1)\beta_1-(c_2+1-(c_1+1)b_{12})\beta_2. \end{aligned}$$

The conditions $\frac{2\langle s_{\beta_2}s_{\beta_1}(\mu+\rho_{\mathfrak{k}}),\beta_j\rangle}{\langle\beta_j,\beta_j\rangle} \geqq 1$ for $j = 1, 2$ yield the inequalities

$$\begin{aligned} -(c_1+1)-(c_2+1)b_{21}+(c_1+1)b_{12}b_{21} &\geqq 1 \\ \text{and} \qquad -(c_2+1)+(c_1+1)b_{12} &\geqq 1. \end{aligned} \tag{18.4.6}$$

Since $\Psi_{\mathfrak{k}}$ is a simple root subsystem of a reduced root system $\Sigma(\mathfrak{k},\mathfrak{t})$, and we have excluded the root system of type G_2, we have $b_{12}b_{21} \in \{0,1,2\}$. A direct check shows that any of the 3 choices $b_{12} = b_{21} = -1$, $b_{12} = -1, b_{21} = -2$ and $b_{12} = -2, b_{21} = -1$ contradicts our assumption (18.4.3).

When $b_{12} = b_{21} = 0$, (18.4.6) reduces to $-(c_1+1) \geqq 1$ and $-(c_2+1) \geqq 1$. Those inequalities together with (18.4.3) imply $\frac{2\langle\mu,\beta_1\rangle}{\langle\beta_1,\beta_1\rangle} = \frac{2\langle\mu,\beta_2\rangle}{\langle\beta_2,\beta_2\rangle} = -2$. Furthermore, $\langle\beta_1,\beta_2\rangle = 0$.

Note that $\{\mu, \mu+\beta_1, \mu+2\beta_1\}$ is a string in $M(\mathfrak{t}, \mathfrak{s})$ with $||\mu|| > ||\mu+\beta_1|| > 0$. (The case $\mu = -\beta_1$ can be excluded, for otherwise $\mu + \rho_{\mathfrak{k}}$ would be of index 1.) On the other hand, $\frac{2\langle\mu+\beta_1,\beta_2\rangle}{\langle\beta_2,\beta_2\rangle} = \frac{2\langle\mu,\beta_2\rangle}{\langle\beta_2,\beta_2\rangle} = -2$. Hence,

$$\{\mu+\beta_1, \mu+\beta_1+\beta_2, \mu+\beta_1+2\beta_2\}$$

is another string in $M(\mathfrak{t}, \mathfrak{s})$ with $||\mu+\beta_1|| > ||\mu+\beta_1+\beta_2||$. Since we assume (18.4.3) that $\Sigma(\mathfrak{g}, \mathfrak{t})$ contains roots of at most two different lengths, $\mu+\beta_1 \mid \beta_2 = 0$. As already mentioned, $\pm(\beta_1 \pm \beta_2), \pm\beta_j$ are all roots in $\Sigma(\mathfrak{g}, \mathfrak{t})$. Since β_1 and β_2 are orthogonal simple roots in $\Sigma(\mathfrak{k}, \mathfrak{t})$, $\pm(\beta_1+\beta_2)$ cannot be contained in $\Sigma(\mathfrak{k}, \mathfrak{t})$. Hence, $\pm(\beta_1 \pm \beta_2), \pm\beta_j \in M(\mathfrak{t}, \mathfrak{s})$, $\mu = -\beta_1 - \beta_2$ is long, and β_1 and β_2 must be short roots. □

For certain pairs $(\mathfrak{g}, \theta)$, the String Lemma is a key tool in the proof that for all G-homogeneous flag manifolds Z and all compact cycles $C = K(z) \subset Z$, the component of the cycle space $\mathcal{C}(Z)$ that contains $[C]$ is the topological (or algebraic) closure of the corresponding orbit $G([C])$ in the cycle space.

In order to prove such a result we need some preliminaries. As usual G_0 is a real form of G, θ is the corresponding Cartan involution, $Z = G/Q$ is a complex flag manifold, and $C \subset Z$ is a compact K-orbit. As explained in Section 17.2, this situation corresponds to a choice of a Borel subalgebra $\mathfrak{b}_{\mathfrak{k}} \subset \mathfrak{k}$, a Borel subalgebra $\mathfrak{b} \supset \mathfrak{b}_{\mathfrak{k}}$ of $\mathfrak{g}$, with $\mathfrak{b} \subset \mathfrak{q}$. We fix a θ-stable fundamental Cartan subalgebra $\mathfrak{h}_0$ of $\mathfrak{g}_0$, and thus the τ-stable Cartan subalgebra $\mathfrak{t} = \mathfrak{h} \cap \mathfrak{k}$ of $\mathfrak{k}$.

Following Propositions 17.5.1 and 17.5.4 we must compute the cohomologies $H^\ell(C; \mathcal{O}(\mathbb{E}((\mathfrak{q}+\theta\mathfrak{q})_{\mathfrak{s}})))$. Let $0 = F^0(\mathfrak{q}+\theta\mathfrak{q})_{\mathfrak{s}} \subset F^1(\mathfrak{q}+\theta\mathfrak{q})_{\mathfrak{s}} \subset \cdots$ be one of the filtrations as in Section 17.6. For simplicity, write

(18.4.7) $$\mathbb{F}^j := \mathbb{E}(F^j(\mathfrak{q}+\theta\mathfrak{q})_{\mathfrak{s}})$$

for the corresponding K-homogeneous holomorphic vector bundle over C. The short exact sequences

(18.4.8) $$0 \longrightarrow \mathbb{F}^j \longrightarrow \mathbb{F}^{j+1} \longrightarrow \mathbb{F}^{j+1}/\mathbb{F}^j \longrightarrow 0 \quad \text{for } j = 0, 1, 2\ldots$$

of K-homogeneous vector bundles induce the long exact cohomology sequences of K-modules

(18.4.9)
$$\begin{aligned}\to H^{\ell-1}(C; \mathcal{O}(\mathbb{F}^{j+1}/\mathbb{F}^j)) \xrightarrow{\chi} H^\ell(C; \mathcal{O}(\mathbb{F}^j)) \xrightarrow{\varphi} H^\ell(C; \mathcal{O}(\mathbb{F}^{j+1}))\\ \xrightarrow{\psi} H^\ell(C; \mathcal{O}(\mathbb{F}^{j+1}/\mathbb{F}^j)) \to H^{\ell+1}(C; \mathcal{O}(\mathbb{F}^j)) \to .\end{aligned}$$

Assume for some fixed ℓ (=1 or 2 in our applications) and for all $j \geqq 0$ that $H^\ell(C; \mathcal{O}(\mathbb{F}^{j+1}/F^j)) = 0$. This implies that all maps φ in the above sequences are surjective. Since $H^\ell(C; \mathcal{O}(\mathbb{F}^1)) = H^\ell(C; \mathcal{O}(\mathbb{F}^1/\mathbb{F}^0)) = 0$, we conclude that each $H^\ell(C; \mathcal{O}(\mathbb{E}((\mathfrak{q}+\theta\mathfrak{q})_{\mathfrak{s}}))) = 0$. We have therefore proved the following fact.

Lemma 18.4.10. *Suppose that $M(\mathfrak{t}, (\mathfrak{q}+\theta\mathfrak{q})_{\mathfrak{s}})$ does not contain a Bott-regular weight of index ℓ. Then $H^\ell(C; \mathcal{O}(\mathbb{E}((\mathfrak{q}+\theta\mathfrak{q})_{\mathfrak{s}}))) = 0$.*

The following is a direct consequence of the above remarks and the String Lemma.

Lemma 18.4.11. *If $\Sigma(\mathfrak{g},\mathfrak{t})$ satisfies the condition* (18.4.3), *then for each $j \geqq 0$, $H^2(C;\mathcal{O}(\mathbb{F}^j))$ is either* 0 *or is a sum of trivial one-dimensional K-modules.*

Now we look at the trivial K-submodules.

Lemma 18.4.12. *The K-module $H^0(C;\mathcal{O}(\mathbb{N}_Z(C)))$ does not contain any trivial K-submodules. In particular, as a submodule of $H^0(C;\mathcal{O}(\mathbb{N}_Z(C)))$, the cohomology space $H^1(C;\mathcal{O}(\mathbb{E}((\mathfrak{q}+\theta\mathfrak{q})_\mathfrak{s})))$ also has no trivial K-submodule.*

Proof. Every one-dimensional trivial K-submodule $L \subset H^0(C;\mathcal{O}(\mathbb{N}_Z(C)))$ corresponds to a K-fixed section $s : C \to K \times_{Q_K} (\mathfrak{s}/(\mathfrak{q}+\theta\mathfrak{q})_\mathfrak{s})$. A section s of this bundle corresponds to a $(K \cap Q)$-equivariant map $f_s : K \to \mathfrak{s}/(\mathfrak{q}+\theta\mathfrak{q})_\mathfrak{s}$, and s is K-fixed exactly when f_s is constant. For a K-fixed section s, the equivariance condition on f_s implies that $K \cap Q$ fixes the vector $f_s(K) \in \mathfrak{s}/(\mathfrak{q}+\theta\mathfrak{q})_\mathfrak{s}$.

The quotient $\mathfrak{s}/(\mathfrak{q}+\theta\mathfrak{q})_\mathfrak{s}$ decomposes into $\mathfrak{t}$-root spaces, and none of them is a $(K \cap Q)$-trivial because $\mathfrak{t} \subset \mathfrak{b}$ and $\mathfrak{s}/\mathfrak{h}_\mathfrak{s} \to \mathfrak{s}/(\mathfrak{q}+\theta\mathfrak{q})_\mathfrak{s}$ is a $\mathfrak{t}$-equivariant surjective map. Thus there are no $(K \cap Q)$-fixed vectors in $\mathfrak{s}/(\mathfrak{q}+\theta\mathfrak{q})_\mathfrak{s}$. In other words, the only K-fixed section $s : C \to K \times_{B_K} (\mathfrak{s}/(\mathfrak{q}+\theta\mathfrak{q})_\mathfrak{s})$ is the 0-section, and therefore $H^0(C;\mathcal{O}(\mathbb{N}_Z(C)))$ has no trivial K-submodule. □

Even if some of the cohomology groups in 18.4.9 are nonzero, but are only direct sums of trivial modules, we do not need to worry about cancellation and can conclude that the trivial submodules do not contribute to $H^1(C;\mathcal{O}(\mathbb{E}((\mathfrak{q}+\theta\mathfrak{q})_\mathfrak{s})))$.

Reviewing the discussion above, we come to our first result that gives a list of those real forms $\mathfrak{g}_0$ for which the component $\mathcal{C}_{[C]}(Z)$ of the cycle space coincides with the topological closure of the orbit $G([C])$ in $\mathcal{C}(Z)$. Given $\mathfrak{g}_0$, the result is independent of the choice of the flag manifold $Z = G/Q$ and closed K-orbit $C \subset Z$ under consideration.

The assumptions of the following theorem exclude the three cases of Section 18.4B that do not satisfy (18.4.3). We discuss them separately in Sections 19.3 and 20.4. Those are the cases where G_u/K_0 is $G_2/SO(4)$, $SU(2r+1)/SO(2r+1)$, or the group manifold G_2.

Theorem 18.4.13. *Suppose that $\mathfrak{g}_0$ is simple and that $\Sigma(\mathfrak{g},\mathfrak{t})$ satisfies* (18.4.3), *in other words that $\mathfrak{g}_0$ is not isomorphic to any of $\mathfrak{sl}(2m+1;\mathbb{R})$, $\mathfrak{g}_{2(2)}$, $\mathfrak{g}_2(\mathbb{C})$. Further suppose that the affine Dynkin diagram in Tables* 17.4.2 *and* 17.4.3 *for* rank $\mathfrak{k}$ = rank $\mathfrak{g}$, *or Table* 18.2.9 *for* rank $\mathfrak{k}$ < rank $\mathfrak{g}$, *satisfies one of the following two conditions.:*

1. *All roots in the underlying affine Dynkin diagram have the same length.*
2. *The roots in the underlying affine Dynkin diagram have two different lengths, but the (one or two) marked roots are short.*

Then for every flag manifold $Z = G/Q$ and every closed K-orbit $C \subset Z$, the Barlet cycle space $\mathcal{C}(Z)$ is smooth at $[C]$, and the component of $[C]$ in $\mathcal{C}(Z)$ coincides with the algebraic closure of the orbit $G([C])$.

Proof. If the nonzero weights in $M(\mathfrak{t}, \mathfrak{s})$ have equal length, there is no 3-string $\mu, \mu+\beta, \mu+2\beta$ with nonzero dominant $\mu+\beta$. Consequently, according to the String Lemma 18.4.4, there are no Bott-regular weights $\mu \in M(\mathfrak{t}, \mathfrak{s})$ of index 1 with $\lambda(\mu) = [\mu + \rho_{\mathfrak{k}}] - \rho_{\mathfrak{k}} \neq 0$. Hence, a glance at the exact sequence in (18.4.9) shows that for all j $H^1(C; \mathcal{O}(\mathbb{F}^j))$, $\mathbb{F}^j := \mathbb{E}(F^j(\mathfrak{q}+\theta\mathfrak{q})_{\mathfrak{s}})$, is 0 or isomorphic to a direct sum of trivial one-dimensional $\mathfrak{k}$-modules. Thus, together with Lemma 18.4.12 we conclude that $H^1(C; \mathcal{O}(\mathbb{E}((\mathfrak{q}+\theta\mathfrak{q})_{\mathfrak{s}}))) = 0$ for all C and $\mathfrak{q}$. In all these cases, Propositions 17.5.1 and 17.5.4 tell us that the Zariski tangent space $T_{[C]}\mathcal{C}(Z) = H^0(C; \mathcal{O}(\mathbb{E}((\mathfrak{q}+\theta\mathfrak{q})_{\mathfrak{s}})))$ is $\mathfrak{s}$ or one of $\mathfrak{s}_{\pm}$. Therefore, the dimensions of the Zariski tangent space of $\mathcal{C}(Z)$ at $[C]$ and the homogeneous space $G([C])$ are equal, so $\mathcal{C}(Z)$ is smooth at $[C]$. Since the G-action on $\mathcal{C}$ is algebraic, $G([C])$ is dense in the corresponding irreducible component of $\mathcal{C}(Z)$.

There is a nonzero weight φ in $M(\mathfrak{t}, \mathfrak{s})$ with $\|\varphi\| \neq \|\lambda^{\text{high}}\|$ (λ^{high} denotes the highest weight of $\mathfrak{k} \times \mathfrak{s} \to \mathfrak{s}$ or $\mathfrak{k} \times \mathfrak{s}_+ \to \mathfrak{s}_+$) if and only if there is a dominant weight $\widetilde{\varphi} \in M(\mathfrak{t}, \mathfrak{s})$ with $\|\widetilde{\varphi}\| = \|\varphi\| < \|\lambda^{\text{high}}\|$. This proves that the elements in $M(\mathfrak{t}, \mathfrak{s}) \setminus \{0\}$ have equal length if and only if there is no nonzero dominant weight $M(\mathfrak{t}, \mathfrak{s})$ (or $M(\mathfrak{t}, \mathfrak{s}_+)$, respectively), different from the highest weight ν. That information can be read directly from Tables 17.4.5, 17.4.6 and 17.4.8.

The data given by a marked affine Dynkin diagram is also sufficient to decide whether there exist dominant nonzero weights in $M(\mathfrak{t}, \mathfrak{s})$ which are shorter than the highest weight. This relies on the following basic facts: (1) There are as many different lengths of nonzero roots in $\Sigma(\mathfrak{g}, \mathfrak{t})$ as many different root lengths appear in the underlying affine Dynkin diagram of $\mathfrak{g}_0$. (2) The marked roots are the lowest weights of $\mathfrak{k} \times \mathfrak{s} \to \mathfrak{s}$ or $\mathfrak{k} \times \mathfrak{s}_+ \to \mathfrak{s}_+$, respectively. Now, if all roots in an affine Dynkin diagram have equal length or the marked roots are short (and consequently there are no shorter roots in $\Sigma(\mathfrak{g}, \mathfrak{t})$ than the marked ones) then all elements in $M(\mathfrak{t}, \mathfrak{s}) \setminus \{0\}$ have equal length. This completes the proof of the theorem. □

18.5 Computation of Bott-regular weights

The simple algebras $\mathfrak{g}_0$ not covered by Theorem 18.4.13 are listed in Table 18.5.1. In terms of combinatorics of roots they are those real forms for which the corresponding $\mathfrak{k}$-module $\mathfrak{s}$ contains nonzero $\mathfrak{t}$-weights of two different lengths. In terms of geometry of cycle spaces they are those real forms for which we may hope to find G-flag manifolds Z and cycles C such that the corresponding components $\mathcal{C}_{[C]}(Z)$ of the cycle space has dimension larger than that of the orbit $G{\cdot}[C]$. Of course, we need only to discuss the cases where $\mathfrak{g}_0$ is simple. In addition there are the three cases listed in Table 18.5.2, which will be considered separately.

Our next goal is to determination all Bott-regular weights of index 1 and 2 in the cases listed above. A priori, we do not know which real forms, for which there are nonzero weights in $M(\mathfrak{t}, \mathfrak{s})$ of at least two different lengths, admit a Bott-regular weights μ of index 1 and with $\lambda(\mu) \neq 0$ in $M(\mathfrak{t}, \mathfrak{s})$. Due to the first part of the String Lemma 18.4.4, the task of finding Bott-regular elements μ in $M(\mathfrak{t}, \mathfrak{s})$ of index 1, at least for all simple real forms $\mathfrak{g}_0 = \mathfrak{k}_0 + \mathfrak{s}_0$ which are different from $\mathfrak{sl}(2r+1, \mathbb{R})$, $\mathfrak{g}_{2(2)}$

Table 18.5.1. Real Forms with Elements in $M(\mathfrak{t}, \mathfrak{p}) \setminus 0$ of Two Different Lengths

$\mathfrak{g}$	$\mathfrak{g}_0$	$\mathfrak{k}_0$	
$\mathfrak{so}(2p+2q+1;\mathbb{C})$	$\mathfrak{so}(2p, 2q+1)$	$\mathfrak{so}(2p) \times \mathfrak{so}(2q+1)$	$p \geqq 1$
$\mathfrak{sp}(n;\mathbb{C})$	$\mathfrak{sp}(n;\mathbb{R})$	$\mathfrak{u}(n)$	
$\mathfrak{f}_4(\mathbb{C})$	$\mathfrak{f}_{4,C_1C_3} = \mathfrak{f}_{4(4)}$	$\mathfrak{sp}(1) \times \mathfrak{sp}(3)$	
$\mathfrak{sl}(2m;\mathbb{C})$	$\mathfrak{sl}(2m;\mathbb{R})$	$\mathfrak{so}(2m)$	
$\mathfrak{so}(2p+2q+2;\mathbb{C})$	$\mathfrak{so}(2p+1, 2q+1)$	$\mathfrak{so}(2p+1) \times \mathfrak{so}(2q+1)$	$p, q \geqq 1$
$\mathfrak{e}_6(\mathbb{C})$	$\mathfrak{e}_{6,C_4} = \mathfrak{e}_{6(6)}$	$\mathfrak{sp}(4)$	
$\mathfrak{so}(2m+1;\mathbb{C}) \oplus \mathfrak{so}(2m+1;\mathbb{C})$	$\mathfrak{so}(2m+1;\mathbb{C})$	$\mathfrak{so}(2m+1)$	$m \geqq 2$
$\mathfrak{sp}(m;\mathbb{C}) \oplus \mathfrak{sp}(m;\mathbb{C})$	$\mathfrak{sp}(m;\mathbb{C})$	$\mathfrak{sp}(m)$	$m \geqq 3$
$\mathfrak{f}_4(\mathbb{C}) \oplus \mathfrak{f}_4(\mathbb{C})$	$\mathfrak{f}_4(\mathbb{C})$	$\mathfrak{f}_{4(-52)}$	

Table 18.5.2. Remaining Series and Cases That Do Not Satisfy (18.4.3)

$\mathfrak{g}$	$\mathfrak{g}_0$	$\mathfrak{k}_0$
$\mathfrak{sl}_{2m+1}(\mathbb{C})$	$\mathfrak{sl}_{2m+1}(\mathbb{R})$	$\mathfrak{so}(2m+1)$
$\mathfrak{g}_2(\mathbb{C})$	$\mathfrak{g}_{2,A1A1} = \mathfrak{g}_{2(2)}$	$\mathfrak{so}(4)$
$\mathfrak{g}_2(\mathbb{C}) \oplus \mathfrak{g}_2(\mathbb{C})$	$\mathfrak{g}_2(\mathbb{C})$	$\mathfrak{g}_{2(-14)}$

and $\mathfrak{g}_2(\mathbb{C})$, is reduced to the determination of just when, for simple $\beta^\star \in \Psi_{\mathfrak{k}}$ and dominant $\lambda \in M(\mathfrak{t}, \mathfrak{s})$, both weights $\lambda \pm \beta^\star$ belong to $M(\mathfrak{t}, \mathfrak{s})$. Once such weights λ and $\beta^\star$ are detected, $\mu := \lambda - \beta$ is a Bott-regular of index 1 with $\lambda(\mu) = \lambda$. As a by-product of this computation, we have the following:

For all $\mathfrak{g}_0$ *in Tables* 18.5.1 *and* 18.5.2, *there exist*
Bott-regular weights μ *of index* 1 *with* $\lambda(\mu) \neq 0$.

From the general classification results we know all dominant weights in $M(\mathfrak{t}, \mathfrak{s})$. If $\Psi_{\mathfrak{k}}$ has roots of two lengths only the short simple roots are candidates for $\beta^\star$ as above. For example, in all cases under consideration (with the exception of those in Table 18.5.2 and the two series $(\mathfrak{sp}(2r;\mathbb{C}), \mathfrak{gl}(r;\mathbb{C}))$ and $(\mathfrak{sl}(2r;\mathbb{C}), \mathfrak{so}(2r;\mathbb{C}))$). In fact, in the remaining cases there are at most two such short roots, except in the case of $(\mathfrak{e}_6, \mathfrak{sp}(4))$, where there are three.

Although these general remarks restrict the set of candidates for $\beta^\star$we still need some more explicit knowledge of $M(\mathfrak{t}, \mathfrak{s})$ is order to decide when $\lambda \pm \beta^\star \in M(\mathfrak{t}, \mathfrak{s})$. The explicit determination of $M(\mathfrak{t}, \mathfrak{s})$ and $\mathfrak{g}_0$ as in Table 18.5.1 is significantly simplified by the following observations:

- All the Lie algebras $\mathfrak{k}$ that come into question here are Lie algebra direct sums $\mathfrak{k}^{(1)} \oplus \cdots \oplus \mathfrak{k}^{(\ell)}$ $(\oplus \mathfrak{z})$ of simple Lie algebras of *classical* type. Then the irreducible representations $\mathfrak{k} \times \mathfrak{s} \to \mathfrak{s}$ (or $\mathfrak{k} \times \mathfrak{s}^{\pm} \to \mathfrak{s}^{\pm}$ in the hermitian case) are tensor products of certain irreducible representations of the factors $\mathfrak{k}^{(j)}$. Suppose that

$\mathfrak{t}^{(1)} \oplus \cdots \oplus \mathfrak{t}^{(\ell)}$ $(\oplus\mathfrak{z})$ is the corresponding decomposition of the Cartan subalgebra $\mathfrak{t}$. The highest weights of the (tensor) representations $\mathfrak{k} \times \mathfrak{s} \to \mathfrak{s}$ are listed in Tables 17.4.5, 17.4.6 and 17.4.8 in the form $\lambda = \lambda^{(1)} + \cdots + \lambda^{(\ell)}$ $(\pm\chi)$. The highest weights $\lambda^{(j)}$ of the corresponding irreducible representations $\mathfrak{k}^{(j)} \times V^{(j)} \to V^{(j)}$ which are a priori only defined on $\mathfrak{t}^{(j)}$ are considered as trivially extended to the entire torus $\mathfrak{t}$.

- All the representations of the factors $\mathfrak{k}^{(j)}$ here are sufficiently "simple" to be understood with ad hoc methods: The "simplest" representation of a classical group L (of type A_ℓ, B_ℓ, C_ℓ, D_ℓ or a general linear group) is the **standard representation** $V^{\xi_1} \equiv V_L^{\mathrm{std}}$ where all weights $M(\mathfrak{t}_L, V_L^{\mathrm{std}})$ are explicitly known. Inspecting all relevant weights $\lambda^{(j)}$, it turns out that the representations $\mathfrak{k}^{(j)} \times V^{(j)} \to V^{(j)}$ either are fundamental (V^{ξ_k} and can be realized on appropriate subspaces of the wedge powers $\bigwedge^k V^{\xi_1}$ of the standard representation), or are appropriate subrepresentations of the symmetric square $S^2 V^{\xi_1}$. With that information the weights $M(\mathfrak{t}, \mathfrak{s})$ can be determined immediately. Consequently, it is straightforward to decide whether $\lambda \pm \beta^\star$ for a nonmaximal dominant $\lambda \in M(\mathfrak{t}, \mathfrak{s})$: This is the case if $\lambda \pm \beta^\star$ can be expressed as sums $\varphi_{(1)} + \cdots + \varphi_{(\ell)}(\pm\chi)$, where $\varphi_{(j)}$ denotes a weight of the corresponding irreducible representation of $\mathfrak{k}^{(j)}$ of the above type.
- We defer discussion of the series $(\mathfrak{sl}(2r+1;\mathbb{C}), \mathfrak{so}(2r+1;\mathbb{C}))$ to Section 19.3, the series $(\mathfrak{g}_2(\mathbb{C}), \mathfrak{so}(4;\mathbb{C}))$ to Section 20.4 and the series $(\mathfrak{g}_2(\mathbb{C}) \times (\mathfrak{g}_2(\mathbb{C}), (\mathfrak{g}_2(\mathbb{C}))$ to Section 19.10.

In summary, we list for all real simple Lie algebras of noncomplex type all β^*-strings of length 3 (as in the String Lemma 18.4.4) with dominant nonzero weight in the middle. They are given as columns in Table 18.5.3 on the next page. We also indicate the corresponding simple short root(s) $\beta^\star$ in $\Psi_\mathfrak{k}$.

Let $\beta_i^\star \in \Psi_\mathfrak{k}$ be the corresponding simple short roots in $\Psi_\mathfrak{k}$ such that $\mu_i, \mu_i + \beta_i^*, \mu_i + 2\beta_i^*$ as a 3-string as in the String Lemma (with an exception of $\mathfrak{g}_0 \in \{\mathfrak{g}_{2(2)}, \mathfrak{g}_2(\mathbb{C})\}$ where the string has length 4, see 20.4). Let us now review our situation. The main goal of this part of the monograph is to explicitly determine all flag G-manifolds Z and base cycles (closed K-orbits) $C \subset Z$ such that the space of holomorphic sections in the normal bundle $\mathbb{N}_Z(C)$, i.e., the Zariski tangent space $T_{[C]}\mathcal{C}(Z)$ of the cycle space $\mathcal{C}(Z)$ at the point $[C]$ is larger than the subspace coming from the G-orbit $G{\cdot}[C]$. In fact, it turns out that in all cases $\mathcal{C}(Z)$ is smooth at $[C]$ (see Theorem 18.6.1), and hence we can simply speak of a tangent space.

As explained in Section 17.2, given the geometric data (C, Z), we assign to it the Lie subalgebras $\mathfrak{b}_\mathfrak{k} \subset \mathfrak{b} \subset \mathfrak{q} \subset \mathfrak{g}$, where $\mathfrak{b}$ is a Borel subalgebra of $\mathfrak{g}$ containing a fixed Borel subalgebra $\mathfrak{b}_\mathfrak{k} \subset \mathfrak{k}$, and $\mathfrak{q}$ is a parabolic subalgebra $\mathfrak{q} \supset \mathfrak{b}$ which is the isotropy subalgebra at a base point $z \in C$. The tangent space $T_{[C]}\mathcal{C}(Z)$ is bigger than $T_{[C]}(G{\cdot}[C]) \cong \mathfrak{g}/\mathfrak{g}_{[C]}$ if and only if $H^1(C; \mathcal{O}(E((\mathfrak{q}+\theta\mathfrak{q})_\mathfrak{s}))) \neq 0$ (see Propositions 17.5.1 and 17.5.4).

Recall that $M(\mathfrak{t}, (\mathfrak{q} + \theta\mathfrak{q})_\mathfrak{s})$ denotes the set of $\mathfrak{t}$-weights that appear in $(\mathfrak{q} + \theta\mathfrak{q})_\mathfrak{s}$. By the Bott–Borel–Weil Theorem, a necessary (but not sufficient) condition for the nonvanishing of $H^1(C; \mathcal{O}(\mathbb{E}((\mathfrak{q} + \theta\mathfrak{q})_\mathfrak{s})))$, more precisely of $V^{\lambda(\mu)} \subset H^1(C; \mathcal{O}(\mathbb{E}((\mathfrak{q} + \theta\mathfrak{q})_\mathfrak{s})))$, where

Table 18.5.3. Dominant Weights and 3-Strings

Marked Affine Dynkin Diagrams of $\mathfrak{g}_0$	3-Strings $\mu + 2\beta^\star$, $\mu + \beta^\star$, μ
$\mathfrak{so}(2p, 2q+1)$ 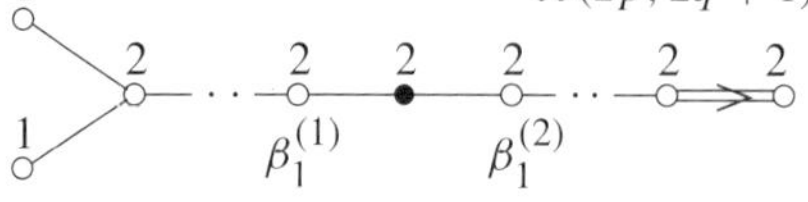	$\xi^{(1)} + \beta^{(2)}_{\mathrm{sh}}$ $(\neq \lambda^{\mathrm{high}})$ $\quad p > 2, q > 1$ $\xi^{(1)}$ $\xi^{(1)} - \beta^{(2)}_{\mathrm{sh}}$
$\mathfrak{so}(2p, 3)$ 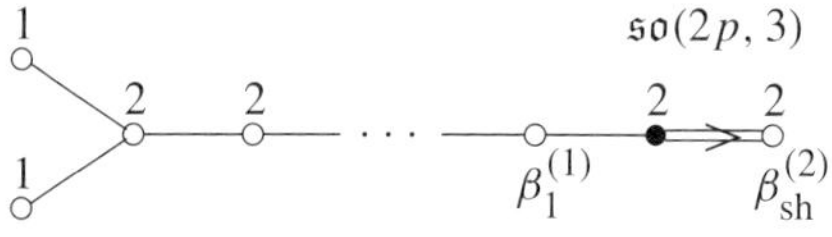	$\xi^{(1)} + \beta^{(2)}_{\mathrm{sh}} = \lambda^{\mathrm{high}}$ $\xi^{(1)}$ $\xi^{(1)} - \beta^{(2)}_{\mathrm{sh}}$
$\mathfrak{so}(4, 2q+1)$ 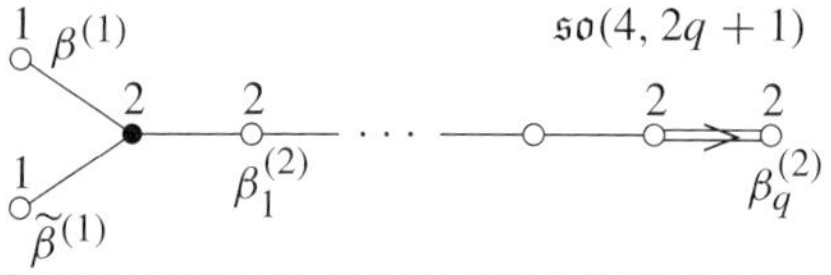	$\xi^{(1)} + \tilde{\xi}^{(1)} + \beta^{(2)}_{\mathrm{sh}}$ $(\neq \lambda^{\mathrm{high}})$ $\xi^{(1)} + \tilde{\xi}^{(1)}$ $\xi^{(1)} + \tilde{\xi}^{(1)} - \beta^{(2)}_{\mathrm{sh}}$
$\mathfrak{so}(2, 2q+1)$ 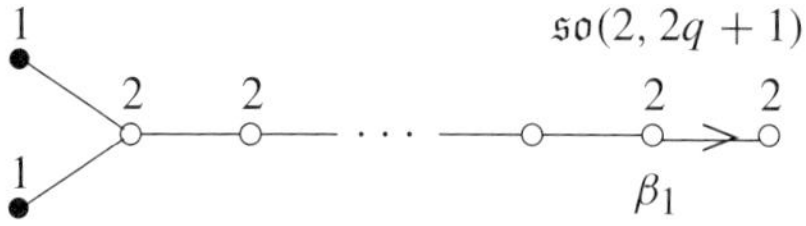	$\pm\chi + \beta_{\mathrm{sh}}$ $(\neq \lambda^{\mathrm{high}})$ $\pm\chi$ $\pm\chi - \beta_{\mathrm{sh}}$
$(r, \mathbb{R})$ 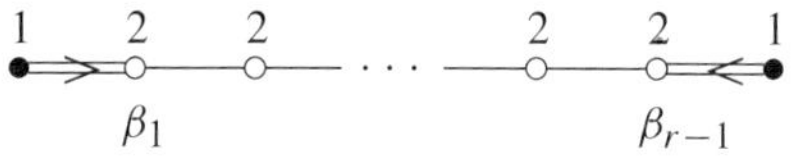	$\xi_\pm + \beta^\star_\pm \pm \chi$ $(= \lambda^{\mathrm{high}}_\pm)$ $\xi_\pm \pm \chi$ $\quad \xi_+ = \xi_2,\ \beta^\star_+ = \beta_1$ $\xi_\pm - \beta^\star_\pm \pm \chi$ $\quad \xi_- = \xi_{r-2},\ \beta^\star_- = \beta_{r-1}$
$\mathfrak{f}_{4(4)}$ 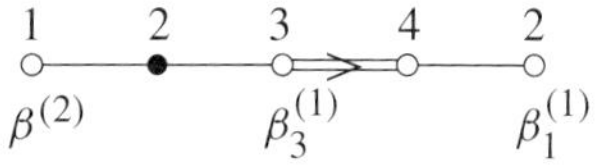	$\xi^{(1)}_1 + \xi^{(2)} + \beta^{(1)}_2$ $\quad (\neq \lambda^{\mathrm{high}})$ $\xi^{(1)}_1 + \xi^{(2)}$ $\xi^{(1)}_1 + \xi^{(2)} - \beta^{(1)}_2$
$\mathfrak{sl}(2r+1, \mathbb{R})$ 	$\xi_2 + \beta_1$ $(= \lambda^{\mathrm{high}} = 2\xi_1)$ ξ_2 $\xi_2 - \beta_1$
$\mathfrak{so}(3, 2q+1)$ 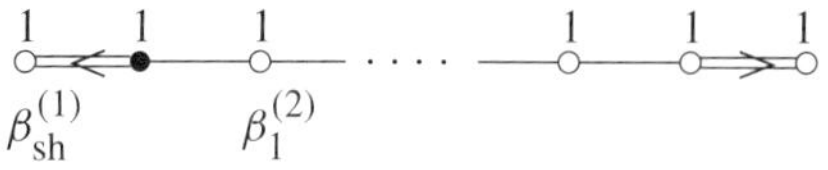	$\xi^{(2)}_1 + \beta^{(1)}$ $(= \lambda^{\mathrm{high}})$ $\quad 2\xi^{(1)} + \beta^{(2)}_{\mathrm{sh}}$ $\xi^{(2)}_1$ $\quad 2\xi^{(1)}$ $\xi^{(2)}_1 - \beta^{(1)}$ $\quad 2\xi^{(1)} - \beta^{(2)}_{\mathrm{sh}}$
$\mathfrak{so}(2p+1, 2q+1)$ 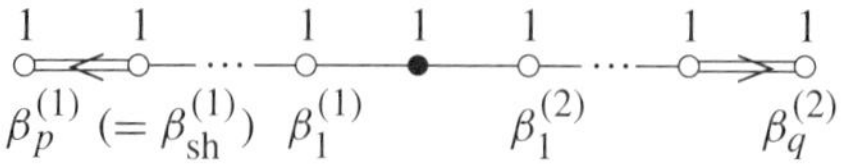	$\xi^{(2)}_1 + \beta^{(1)}_{\mathrm{sh}}$ $(= \lambda^{\mathrm{high}})$ $\quad \xi^{(1)}_1 + \beta^{(2)}_{\mathrm{sh}}$ $\xi^{(2)}_1$ $\quad \xi^{(1)}_1$ $\xi^{(2)}_1 - \beta^{(1)}_{\mathrm{sh}}$ $\quad \xi^{(1)}_1 - \beta^{(2)}_{\mathrm{sh}}$
$\mathfrak{e}_{6(6)}$ 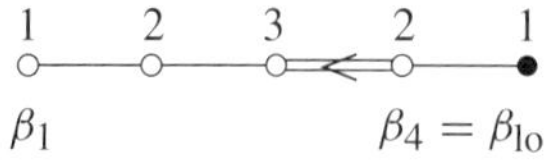	$\xi_2 + \beta_3$ $\quad (\neq \lambda^{\mathrm{high}} = \xi_4)$ ξ_2 $\xi_2 - \beta_3$

$$\lambda(\mu) := [\mu + \rho_{\mathfrak{k}}] - \rho_{\mathfrak{k}}, \tag{18.5.4}$$

is the existence of Bott-regular weights μ of index 1 in $M(\mathfrak{t}, (\mathfrak{q} + \theta\mathfrak{q})_{\mathfrak{s}}) \subset M(\mathfrak{t}, \mathfrak{s})$. We already know that $\lambda(\mu) \neq 0$, (see Lemma 18.4.12).

Table 18.5.3 specifies all such weights in $M(\mathfrak{t}, \mathfrak{s})$ for all symmetric pairs listed in Table 18.5.1 for which $\mathfrak{g}$ is simple. If $\mathfrak{g}$ is not simple, in other words if $\mathfrak{g}_0$ is the underlying real structure of a complex simple Lie algebra, then $\mathfrak{k}$ is isomorphic to that complex Lie algebra, and the representation $\mathfrak{k} \times \mathfrak{s} \to \mathfrak{s}$ is equivalent to the adjoint representation of $\mathfrak{k}$. In that case the set of dominant weights in $M(\mathfrak{t}, \mathfrak{s})$ is well known, and the corresponding Bott-regular elements are also easily found. We carry out the details of this case in Sections 19.6 through 19.10.

Now we recall the relevant combinatorial objects. In general we have three different root systems associated with $(\mathfrak{g}, \theta)$: $\Sigma_{\mathfrak{k}} := \Sigma(\mathfrak{k}, \mathfrak{t})$, $\Sigma_\theta := \Sigma(\mathfrak{g}, \mathfrak{t})$ and $\Sigma := \Sigma(\mathfrak{g}, \mathfrak{h})$. Given Borel subalgebras $\mathfrak{b}_{\mathfrak{k}} \subset \mathfrak{k}$ and $\mathfrak{b} \subset \mathfrak{g}$, they determine the positive root systems $\Sigma_{\mathfrak{k}}^+$, Σ_θ^+ and Σ^+ and in turn the simple root systems $\Psi_{\mathfrak{k}} =: \{\beta_1, \ldots, \beta_{r'}\} \subset \Sigma_{\mathfrak{k}}$, $\Psi_\theta =: \{\gamma_1, \ldots, \gamma_r\} \subset \Sigma_\theta$, and $\Psi = \{\psi_1, \ldots, \psi_m\} \subset \Sigma$. Also recall our convention for defining the roots determined by $\mathfrak{b}_{\mathfrak{k}}$ or $\mathfrak{b}$ as the *negative* roots. We write $\mathfrak{b}_{\mathfrak{k}}^+$ and $\mathfrak{b}^+$ for the opposite Borel subalgebras (with respect to the given Cartan subalgebras).

Further, let a parabolic subalgebra $\mathfrak{q} = \mathfrak{q}_z$ with $\mathfrak{q} \supset \mathfrak{b} \supset \mathfrak{b}_{\mathfrak{k}}$ be given. Define $\mathfrak{q}_{\mathfrak{k}} := \mathfrak{q} \cap \mathfrak{k}$ and $Q_K := Q \cap K$. This (small) parabolic subalgebra is determined by a subset $\Phi_{\mathfrak{k}} \subset \Psi_{\mathfrak{k}}$, in other words, $\mathfrak{q}_{\mathfrak{k}}^r = \mathfrak{t} + \sum_{\delta \in \langle \Phi_{\mathfrak{k}} \rangle_{\mathbb{Z}}} \mathfrak{k}^\delta$. We sometimes write $\mathfrak{q}_{\mathfrak{k}}$ as $(\mathfrak{q}_{\mathfrak{k}})_{\Phi_{\mathfrak{k}}}$ to underline this dependence.

Assume that a Bott regular weight μ in Table 18.5.3 belongs to $M(\mathfrak{t}, (\mathfrak{q} + \theta\mathfrak{q})_{\mathfrak{s}})$. The remainder of this section is devoted to establishing criteria which enable us to decide whether and when the representation space $V^{\lambda(\mu)}$ is contained in $H^1(K/Q_K, \mathcal{O}(\mathbb{E}((\mathfrak{q} + \theta\mathfrak{q})_{\mathfrak{s}})))$.

For the following lemma recall that given $\varphi \in \Sigma(\mathfrak{k}, \mathfrak{t})$, there are at most two roots in $\Sigma(\mathfrak{g}, \mathfrak{h})$, say $\varphi^\dagger$ and possibly $\varphi^\ddagger$, such that $\varphi^\dagger|_{\mathfrak{t}} = \varphi^\ddagger|_{\mathfrak{t}} = \varphi$.

Lemma 18.5.5. *Let μ be one of the Bott-regular weights in $M(\mathfrak{t}, \mathfrak{s})$ with dominant weight $\lambda(\mu) = [\mu + \rho_{\mathfrak{k}}] - \rho_{\mathfrak{k}} \neq 0$. Then*

1. *$\lambda(\mu) = \mu + n\cdot\beta^\star$ with $\mu(h_{\beta^\star}) < 0$ and $n > 0$ for precisely one short simple root $\beta^\star \in \Psi_{\mathfrak{k}}$,*
2. *$\mu(h_\beta) \geqq 0$ for all $\beta \in \Psi_{\mathfrak{k}} \setminus \{\beta^\star\}$, and in particular*
3. *if $\beta \in \Psi_{\mathfrak{k}} \setminus \{\beta^\star\}$, then $\mu + \beta \notin M(\mathfrak{t}, \mathfrak{s})$.*

Let $\mathfrak{q} \supset \mathfrak{b} \supset \mathfrak{b}_{\mathfrak{k}}$ be a parabolic subalgebra as above and $\mu \in M(\mathfrak{t}, (\mathfrak{q} + \theta\mathfrak{q})_{\mathfrak{s}})$. The weight μ is a highest weight with respect to $\mathfrak{b}_{\mathfrak{k}}^+ \cap \mathfrak{q}_{\mathfrak{k}}^r$ whenever the defining subset $\Phi_{\mathfrak{k}}$ of $\mathfrak{q}_{\mathfrak{k}}^r$ is contained in $\Psi_{\mathfrak{k}} \setminus \{\beta^\star\}$. Equivalently, $\langle\!\langle \Phi \rangle\!\rangle_{\mathbb{Z}}$ does not contain $\{(\beta^\star)^\dagger, (\beta^\star)^\ddagger\}$.

Proof. When $\mathfrak{g}_0$ satisfies the condition (18.4.3) the String Lemma and its proof show that there is precisely one $\beta^\star \in \Psi_{\mathfrak{k}}$ such that $\mu(h_{\beta^\star}) = -2$, $\mu = \lambda(\mu) - \beta^\star$ ($n = 1$ in the statement of the Lemma) and $\mu(h_\beta) \geq 0$ for all $\beta \in \Psi_{\mathfrak{k}} \setminus \{\beta^\star\}$; see (18.4.5) and the following paragraph in the proof of the String Lemma. Were $\mu + \beta$ an element in $M(\mathfrak{t}, \mathfrak{s})$, the previous inequality would imply $\mu - \beta \in M(\mathfrak{t}, \mathfrak{s})$, and by the geometry of the string $\mu-\beta, \mu, \mu+\beta$ we would have elements longer than μ. That would contradict the assumption (18.4.3). Hence, if $\beta^\star$ does not belong to $\Phi_\mathfrak{k}$, it follows that μ is highest with respect to $\mathfrak{b}_\mathfrak{k}^+ \cap \mathfrak{q}_\mathfrak{k}^r$. Finally, observe that $\mathfrak{g}^{\beta^{\star\dagger}} + \mathfrak{g}^{\beta^{\star\ddagger}} = \mathfrak{g}^{\beta^{\star\dagger}} + \theta\mathfrak{g}^{\beta^{\star\dagger}}$ is a θ-stable subspace, and consequently $\beta^\star \notin \Phi_\mathfrak{k}$ if and only if $\{\beta^{\star\dagger}, \beta^{\star\ddagger}\} \not\subset \langle\!\langle \Phi \rangle\!\rangle_\mathbb{Z} = \Sigma(\mathfrak{q}^r, \mathfrak{h})$.

The remaining cases are $\mathfrak{g}_0 = \mathfrak{sl}(2m+1;\mathbb{R})$, $\mathfrak{g}_0 = \mathfrak{g}_{2,SO(4)}$ and $\mathfrak{g}_0 = \mathfrak{g}_2(\mathbb{C})$. They are settled in Sections 19.3, 20.4, and 19.10. □

Let $F^j(\mathfrak{q}+\theta\mathfrak{q})_\mathfrak{s}$ be an equivariant filtration of the $\mathfrak{q}_\mathfrak{k}$-module $(\mathfrak{q}+\theta\mathfrak{q})_\mathfrak{s}$ as in Section 17.6, and for short write $\mathbb{F}^j$ for the corresponding homogeneous vector subbundles of $\mathbb{E} = \mathbb{E}((\mathfrak{q}+\theta\mathfrak{q})_\mathfrak{s})$. There is a $k \in \mathbb{N}$ such that $\mu \in M(\mathfrak{t}, F^k/F^{k+1})$. Since F^k/F^{k+1} is a direct sum of irreducible $\mathfrak{q}_\mathfrak{k}$-modules, we can apply the theorem of Bott. Consequently, $V^{\lambda(\mu)} \subset H^1(C; \mathcal{O}(\mathbb{F}^k/\mathbb{F}^{k+1}))$ if and only if the weight μ is highest with respect to the subset $\Phi_\mathfrak{k} = \Psi_\mathfrak{k} \cap \Sigma(\mathfrak{q}_\mathfrak{k}^r, \mathfrak{t})$. Note that $V^{\lambda(\mu)} \subset H^1(C; \mathcal{O}(\mathbb{F}^k/\mathbb{F}^{k+1}))$, but $V^{\lambda(\mu)}$ may be not contained in $H^1(C; \mathcal{O}(\mathbb{E}((\mathfrak{q}+\theta\mathfrak{q})_\mathfrak{s})))$. This is explained in greater detail in the following Cohomological Lemma.

We retain the above notation. In the assumptions of the following lemma we do not exclude the possibility of the existence of Bott-regular weights ν with $\lambda(\nu) = [\nu+\rho_\mathfrak{k}] - \rho_\mathfrak{k} = 0$.

Lemma 18.5.6 (Cohomological Lemma). *Let $\mu_i \in M(\mathfrak{t}, (\mathfrak{q}+\theta\mathfrak{q})_\mathfrak{s})$, $i = 1, 2$, be distinct Bott-regular weights of index 1 such that $\lambda(\mu_i) := [\mu_i + \rho_\mathfrak{k}] - \rho_\mathfrak{k} \neq 0$. Let $\beta_i^\star \in \Psi_\mathfrak{k}$ be the corresponding simple short roots in $\Psi_\mathfrak{k}$ such that $\mu_i, \mu_i+\beta_i^*, \mu_i+2\beta_i^*$ is a 3-string as in the String Lemma (with the obvious exception of $\mathfrak{g}_0 = \mathfrak{g}_{2(2)}$ or $\mathfrak{g}_2(\mathbb{C})$; see Section 20.4).*

1. *If $\beta_i^\star \in \Phi_\mathfrak{k}$, then $V^{\lambda(\mu_i)} \not\subset H^1(C; \mathcal{O}(\mathbb{E}((\mathfrak{q}+\theta\mathfrak{q})_\mathfrak{s})))$. In particular, if for all Bott-regular weights $\mu_i \in M(\mathfrak{t}, (\mathfrak{q}+\theta\mathfrak{q})_\mathfrak{s})$ with $\lambda(\mu_i) \neq 0$ every $\beta_i^\star$ also belongs to $\Phi_\mathfrak{k}$, then $H^1(C; \mathcal{O}(\mathbb{E}((\mathfrak{q}+\theta\mathfrak{q})_\mathfrak{s}))) = 0$.*
2. *If $\beta_i^\star \notin \Phi_\mathfrak{k}$ and $\mu_i + \beta_i^\star \notin M(\mathfrak{t}, (\mathfrak{q}+\theta\mathfrak{q})_\mathfrak{s})$, then*
$$V^{\lambda(\mu_i)} \subset H^1(C; \mathcal{O}(\mathbb{E}((\mathfrak{q}+\theta\mathfrak{q})_\mathfrak{s}))).$$
3. *If $\beta_i^\star \notin \Phi_\mathfrak{k}$, but $\mu_i + \beta_i^\star \in M(\mathfrak{t}, (\mathfrak{q}+\theta\mathfrak{q})_\mathfrak{s})$, then*
$$V^{\lambda(\mu_i)} \not\subset H^1(C; \mathcal{O}(\mathbb{E}((\mathfrak{q}+\theta\mathfrak{q})_\mathfrak{s}))).$$

Proof. Let $0 = F^0(\mathfrak{q}+\theta\mathfrak{q})_\mathfrak{s} \subset F^1(\mathfrak{q}+\theta\mathfrak{q})_\mathfrak{s} \subset \cdots \subset F^\ell(\mathfrak{q}+\theta\mathfrak{q})_\mathfrak{s} = (\mathfrak{q}+\theta\mathfrak{q})_\mathfrak{s}$ be an equivariant filtration of $(\mathfrak{q}+\theta\mathfrak{q})_\mathfrak{s}$. Given a Bott-regular weight of index 1, μ_j, there is precisely one k such that $\mu_j \in M(\mathfrak{t}, F^k(\mathfrak{q}+\theta\mathfrak{q})_\mathfrak{s}/F^{k-1}(\mathfrak{q}+\theta\mathfrak{q})_\mathfrak{s})$. Let $0 = \mathbb{F}^0 \subset \mathbb{F}^1 \subset \cdots$ denote the corresponding filtration of homogeneous K-bundle $\mathbb{E}((\mathfrak{q}+\theta\mathfrak{q})_\mathfrak{s})$ over $C \cong K/(K \cap Q)$.

For all k consider the long exact cohomology sequences of K-modules

$$\begin{aligned}
&\longrightarrow H^0(C; \mathcal{O}(\mathbb{F}^k/\mathbb{F}^{k-1})) \xrightarrow{\eta_1} H^1(C; \mathcal{O}(\mathbb{F}^{k-1})) \xrightarrow{\varphi} H^1(C; \mathcal{O}(\mathbb{F}^k)) \\
&\longrightarrow H^1(C; \mathcal{O}(\mathbb{F}^k/\mathbb{F}^{k-1})) \xrightarrow{\eta_2} H^2(C; \mathcal{O}(\mathbb{F}^{k-1})) \longrightarrow H^2(C; \mathcal{O}(\mathbb{F}^k)) \\
&\longrightarrow H^2(C; \mathcal{O}(\mathbb{F}^k/\mathbb{F}^{k-1})) \longrightarrow \ldots
\end{aligned} \tag{18.5.7}$$

All cohomology groups appearing in (18.5.7) are finite sums of irreducible K-modules and all maps are $\mathfrak{k}$-equivariant. We now analyze these maps in greater detail.

The String Lemma 18.4.4 and the Bott–Borel–Weil Theorem tell us that, for all k, $H^2(C; \mathcal{O}(\mathbb{F}^1))$ and $H^2(C; \mathcal{O}(\mathbb{F}^k/\mathbb{F}^{k-1}))$ are either zero or sums of one-dimensional trivial $\mathfrak{k}$-modules. Inductively, the using exactness of

$$H^2(C; \mathcal{O}(\mathbb{F}^{k-1})) \longrightarrow H^2(C; \mathcal{O}(\mathbb{F}^k)) \longrightarrow H^2(C; \mathcal{O}(\mathbb{F}^k/\mathbb{F}^{k-1})),$$

we conclude that $H^2(C; \mathcal{O}(\mathbb{F}^2))$, $H^2(C; \mathcal{O}(\mathbb{F}^3))$, etc., are zero or sums of one-dimensional trivial modules. Thus, for every k,

(18.5.8) $H^2(C; \mathcal{O}(\mathbb{F}^k))$ is 0 or a direct sum of one-dimensional trivial modules.

Assume now that for all Bott-regular weights μ_j of index 1 with $\lambda(\mu_j) \neq 0$, the corresponding simple roots $\beta_j^\star$ (as in the String Lemma) are in $\Phi_{\mathfrak{k}}$. As shown in the preceding lemma, in this case μ_j is not the highest weight of any $\mathfrak{q}_{\mathfrak{k}}^{\text{red}}$-module $F^k(\mathfrak{q}+\theta\mathfrak{q})_{\mathfrak{s}}/F^{k-1}(\mathfrak{q}+\theta\mathfrak{q})_{\mathfrak{s}}$, and therefore for all k the theorem of Bott implies that $V^{\lambda(\mu_j)} \not\subset H^1(K/Q_K; \mathcal{O}(\mathbb{F}^k/\mathbb{F}^{k-1}))$. For all k these cohomology groups are 0 or sums of trivial submodules. Thus an argument similar to that above shows that for all k (and in particular for $k = \ell$) $H^1(C; \mathcal{O}(\mathbb{F}^k))$ is also 0 or a direct sum of one-dimensional trivial modules. Employing Lemma 18.4.12 we see that all of the nonzero but trivial modules the $H^1(C; \mathcal{O}(\mathbb{F}^k))$ cancel. Therefore $H^1(C; \mathcal{O}(\mathbb{F}^\ell)) = 0$ and statement 1. is proved.

For statement 2 assume that $\mu_j \in M(\mathfrak{t}, F^k(\mathfrak{q}+\theta\mathfrak{q})_{\mathfrak{s}}/F^{k-1}(\mathfrak{q}+\theta\mathfrak{q})_{\mathfrak{s}})$ for some $k = n$ and $\beta_j^\star \notin \Phi_{\mathfrak{k}}$. In this case μ_j is highest with respect to $\mathfrak{q}_{\mathfrak{k}}^{\text{red}} \cap (\mathfrak{b}_{\mathfrak{k}})_+$. Again employing Bott's theorem, we conclude that

$$V^{\lambda(\mu_j)} \subset H^1(C; \mathcal{O}(\mathbb{F}^k/F^{k-1})).$$

By (18.5.8) the nontrivial submodule $V^{\lambda(\mu_j)}$ is contained in the kernel of η_2, and therefore $V^{\lambda(\mu_j)} \subset H^1(C; \mathcal{O}(\mathbb{F}^n))$. By assumption,

$$\lambda(\mu_j) \notin M(\mathfrak{t}, (\mathfrak{q}+\theta\mathfrak{q})_{\mathfrak{s}})$$

and consequently by the Borel–Weil Theorem [Ser], we have

$$V^{\lambda(\mu_j)} \not\subset H^0(C; \mathcal{O}(\mathbb{F}^k/\mathbb{F}^{k-1})) \quad \text{for all } k.$$

The exactness of the first row in (18.5.7) then guarantees that starting with $k = n+1$ the restrictions of the maps φ in (18.5.7) to $V^{\lambda(\mu_j)}$ must be injective for all $k = n+1, n+2, \ldots$. Hence, $V^{\lambda(\mu_j)} \subset H^1(C; \mathcal{O}(\mathbb{F}^\ell)) = H^1(C; \mathcal{O}(\mathbb{E}((\mathfrak{q}+\theta\mathfrak{q})_{\mathfrak{s}})))$ and the proof of (2) is complete.

For the last part of the lemma, we have to show that $V^{\lambda(\mu_j)} \not\subset H^1(C; \mathcal{O}(\mathbb{E}))$. Assume that the equivariant filtration is chosen as in (17.6.1). By construction of $F^\bullet(\mathfrak{q}+\theta\mathfrak{q})_{\mathfrak{s}}$, we have $\mu_j \in F^n/F^{n-1}$ and $\lambda(\mu_j) \in F^{n+1}/F^n$. In the present case the Bott–Borel–Weil Theorem implies that both $V^{\lambda(\mu)} \subset H^1(C; \mathcal{O}(\mathbb{F}^n/\mathbb{F}^{n-1}))$ and $V^{\lambda(\mu)} \subset H^0(C; \mathcal{O}(\mathbb{F}^{n+1}/\mathbb{F}^n))$. We would like to prove that

$$\varphi : H^1(C; \mathcal{O}(\mathbb{F}^n)) \to H^1(C; \mathcal{O}(\mathbb{F}^{n+1})) \text{ kills } V^{\lambda(\mu_j)},$$

but we cannot simply use exactness of the long cohomology sequence

$$H^0(C; \mathcal{O}(\mathbb{F}^n)) \hookrightarrow H^0(C; \mathcal{O}(\mathbb{F}^{n+1})) \to H^0(C; \mathcal{O}(\mathbb{F}^{n+1}/\mathbb{F}^n)) \\ \to H^1(C; \mathcal{O}(\mathbb{F}^n)) \xrightarrow{\varphi} H^1(C; \mathcal{O}(\mathbb{F}^{n+1})).$$

It might happen that $V^{\lambda(\mu_j)}$ is properly contained in $H^0(C; \mathcal{O}(\mathbb{F}^{n+1}))$, and in that case φ would be injective if restricted to $V^{\lambda(\mu_j)}$. Fortunately, however, we have the following fact.

Claim. $V^{\lambda(\mu_j)} \not\subset H^0(C; \mathcal{O}(\mathbb{F}^{n+1}))$, and therefore $V^{\lambda(\mu_j)} \not\subset H^1(C; \mathcal{O}(\mathbb{F}^{n+1}))$.

In order to prove this claim, we do not compute $H^0(C; \mathcal{O}(\mathbb{F}^{n+1}))$ directly but use the Bott–Borel–Weil Theorem. According to Theorem 1.6.11,

$$\mathrm{Hom}_G(V^{\lambda(\mu_j)}, H^0(C; \mathcal{O}(\mathbb{F}^{n+1}))) = \mathrm{Hom}_{\mathfrak{q}_{\mathfrak{k}}}(V^{\lambda(\mu_j)}, F^{n+1}(\mathfrak{q}+\theta\mathfrak{q})_{\mathfrak{s}}).$$

Therefore, it remains to show that $\mathrm{Hom}_{\mathfrak{q}_{\mathfrak{k}}}(V^{\lambda(\mu_j)}, F^{n+1}(\mathfrak{q}+\theta\mathfrak{q})_{\mathfrak{s}}) = 0$. To see this, let $\psi : V^{\lambda(\mu_j)} \to F^{n+1}(\mathfrak{q}+\theta\mathfrak{q})_{\mathfrak{s}}$ be a $\mathfrak{q}_{\mathfrak{k}}$-equivariant map. Since $\mathfrak{t} \subset \mathfrak{q}_{\mathfrak{k}}$, ψ maps weight spaces $V[\xi] \subset V^{\lambda(\mu_j)}$ to $\mathfrak{g}^{\xi} \cap \mathfrak{s} \subset F^{n+1}(\mathfrak{q}+\theta\mathfrak{q})_{\mathfrak{s}}$. Let $v^{\lambda} \in V^{\lambda(\mu_j)}$ be a nonzero highest vector. Since $(\mathfrak{b}_{\mathfrak{k}})_- \subset \mathfrak{q}_{\mathfrak{k}}$, the map ψ is uniquely determined by $\psi(v^{\lambda})$. Choose $X \in \mathfrak{g}_{-\beta^\star} \cap \mathfrak{k} = \mathfrak{k}_{-\beta^\star}$. By equivariance, $\psi(\pi(X)v^{\lambda}) = [X, \psi(v^{\lambda})]$. Since $\lambda(h_{\beta^\star}) = 0$ and λ is the highest weight, $\lambda - \beta^\star$ is not a weight of $V^{\lambda(\mu_j)}$, i.e., $\pi(X)v^{\lambda} = 0$. On the other hand, $\mathfrak{s}_{\lambda-\beta^\star} \subset F^{n+1}(\mathfrak{q}+\theta\mathfrak{q})_{\mathfrak{s}})$ and therefore $[\mathfrak{s}_{\lambda}, \mathfrak{k}_{-\beta^\star}] \neq 0$. Consequently, $\psi(v^{\lambda}) = 0$ and $\psi = 0$.

By the exactness of the long cohomology sequence, $V^{\lambda(\mu_j)} \not\subset H^1(C; \mathcal{O}(\mathbb{F}^k))$ for all $k \geqq n+1$, and the last statement of the lemma follows. □

18.6 Algorithm for computing the module structure of $T_{[C]}\mathcal{C}(Z)$

We retain the notation from the previous section: Let $\mathfrak{g}$ be a complex semisimple Lie algebra and $\mathfrak{g}_0 = \mathfrak{k}_0 + \mathfrak{s}_0$ is a real form that is a simple real Lie algebra. Let $(\mathfrak{g}, \theta)$ be the corresponding complex symmetric space. If $(\mathfrak{g}, \theta)$ does not belong to the lists in Tables 18.5.1 or 18.5.2, then we have already shown that $T_{[C]}\mathcal{C}(Z) = T_{[C]}(G([C]))$ for all G-flag manifolds Z and all base cycles $C \subset Z$; see Theorem 18.4.13. In particular, the cycle spaces $\mathcal{C}(Z)$ are smooth in that case. In fact this holds in complete generality.

Theorem 18.6.1. *For any flag manifold $Z = G/Q$ and any base cycle $C = K(z) \subset Z$, the cycle space $\mathcal{C}(Z)$ is smooth at $[C]$.*

Proof. It is necessary to discuss three cases. First, if $\dim T_{[C]}\mathcal{C} = \dim G{\cdot}[C]$, then $\mathcal{C}$ is smooth at $[C]$. This gives a proof of smoothness for all cycles and flags in the cases where the symmetric pair $(\mathfrak{g}, \mathfrak{k})$ is excluded by Theorem 18.4.13. It remains to deal with the symmetric pairs given in Tables 18.5.1 and 18.5.2. For spaces with rank $\mathfrak{g}$ = rank $\mathfrak{k}$ there are no Bott-regular weights of index 2 (see Chapter 19). The same is true if $\mathfrak{g}_0$ is of complex type. Thus it remains to handle the cases $\mathfrak{g}_0 = \mathfrak{sl}(\text{odd}, \mathbb{R}), \mathfrak{sl}(\text{even}, \mathbb{R}), \mathfrak{so}(2p+1, 2q+1)$ and the exceptional case $\mathfrak{e}_{6,C_4} = \mathfrak{e}_{6(6)}$. In Chapter 20 we prove the smoothness of $\mathcal{C}$ at $[C]$ case by case in the latter four cases by showing $H^1(C; \mathcal{O}(\mathbb{N}_Z(C))) = 0$ for all cycles and flags. □

If $(\mathfrak{g}, \theta)$ is listed either in Table 18.5.1 or in Table 18.5.2, flag manifolds Z and base cycles may (and in fact do, except for $\mathfrak{e}_{6(6)}$ and $\mathfrak{f}_{4(4)}$) exist such that $T_{[C]}\mathcal{C}(Z) \neq T_{[C]}(G([C]))$. Since the tangent space $T_{[C]}\mathcal{C}(Z)$ has the structure of a K-module coming from the isotropy representation of K on $T_{[C]}\mathcal{C}$, and since the group K is reductive, the tangent space decomposes as a direct sum of irreducible K-modules. At this stage we can already say even more: there is no trivial K-submodule in $T_{[C]}\mathcal{C}$; see Lemma 18.4.12. By Lemma 18.5.6, given a flag G-manifold Z and a cycle $C \subset Z$ we have

$$T_{[C]}\mathcal{C}(Z) = T_{[C]}(G{\cdot}[C]) + \sum_{\lambda(\mu_i)\in M(C,Z)} V^{\lambda(\mu_i)}. \qquad (18.6.2)$$

Here $V^{\lambda(\mu_i)}$ denotes the irreducible K-module with highest weight $\lambda(\mu_i)$, and $M(C, Z)$ is the set of these weights, where the μ_i are the Bott-regular weights of index 1 in $M(\mathfrak{t}, \mathfrak{s})$ which occur in the particular situation. These will be computed explicitly in the next chapter. The relation between μ_i and the highest weight is

$$\lambda(\mu_i) = [\mu_i + \rho_{\mathfrak{k}}] - \rho_{\mathfrak{k}} = \mu + n\beta_i^{\star} \neq 0, \ n \in \{1, 2\}$$

with $\beta_i^{\star}$ given as in Lemma 18.5.5; ($n = 2$ occurs only for $\mathfrak{g}_{2(2)}$ and $\mathfrak{g}_2(\mathbb{C})$). Our goal now is to give an explicit description of $M(C, Z)$ for every pair $C \subset Z$ and every complex symmetric pair $(\mathfrak{g}, \mathfrak{k})$ listed in the Tables 18.5.1 and 18.5.2.

Here we indicate an algorithm which enables us to determine the set $M(C, Z)$ (and therefore the tangent space $T_{[C]}\mathcal{C}(Z)$) for a given base cycle C and flag manifold Z. In the following chapter we explicitly carry out such computation for all pairs (C, Z) and for all complex symmetric pairs $(\mathfrak{g}, \mathfrak{k})$ given in Tables 18.5.1 and 18.5.2.

As in Section 17.2, let $\mathfrak{b}_{\mathfrak{k}}^{\text{ref}} \subset \mathfrak{b} \subset \mathfrak{q} \subset \mathfrak{g}$ be the Lie algebra data associated to (C, Z). We fix in each $\mathfrak{b}_{\mathfrak{k}}^{\text{ref}}$ a maximal toral subalgebra $\mathfrak{t}$. It determines the Cartan subalgebra $\mathfrak{h} = \mathfrak{z}_{\mathfrak{g}}(\mathfrak{t})$ of $\mathfrak{g}$, and $\mathfrak{t} = \mathfrak{h} \cap \mathfrak{k}$. Each pair $\mathfrak{b}_{\mathfrak{k}}^{\text{ref}} \subset \mathfrak{b}$ of Borel subalgebras (in $\mathfrak{k}$ and $\mathfrak{g}$, respectively) determines positive systems in the (in general three different) root systems $\Sigma_{\mathfrak{k}}(\mathfrak{t})$, $\Sigma_\theta(\mathfrak{t})$ and $\Sigma(\mathfrak{h})$ which in turn determine the root bases

$$\begin{aligned}
\Psi_{\mathfrak{k}} &= \{\beta_1, \ldots, \beta_{r'}\} \subset \Sigma_{\mathfrak{k}} = \Sigma(\mathfrak{k}.\mathfrak{t}),\\
\Psi_\theta &= \{\gamma_1, \ldots, \gamma_r\} \subset \Sigma_\theta = \Sigma(\mathfrak{g}, \mathfrak{t}),\\
\Psi &= \{\psi_1, \ldots, \psi_m\} \subset \Sigma = \Sigma(\mathfrak{g}, \mathfrak{h}).
\end{aligned}$$

While $\Psi_{\mathfrak{k}} = \Psi_{\mathfrak{k}}^{\text{ref}}$ remains fixed, every $\mathfrak{b} \supset \mathfrak{b}_{\mathfrak{k}}$ gives rise to a different root basis $\Psi_\theta = \Psi_\theta(\mathfrak{b})$.

Next, the parabolic subalgebra $\mathfrak{q} \subset \mathfrak{g}$ yields the (small) parabolic subalgebra $\mathfrak{q}_{\mathfrak{k}} = \mathfrak{q} \cap \mathfrak{k}$. As already explained in the paragraphs preceding Lemma 18.5.5, both parabolic subalgebras are uniquely described by the subsets

$$\Phi := \Phi(\mathfrak{q}) = \Sigma(\mathfrak{q},\mathfrak{h}) \cap \Psi \qquad \text{and} \qquad \Phi_{\mathfrak{k}} := \Phi_{\mathfrak{k}}(\mathfrak{q}_{\mathfrak{k}}) = \Sigma(\mathfrak{q}_{\mathfrak{k}},\mathfrak{t}) \cap \Psi_{\mathfrak{k}}.$$

The symmetric pairs $(\mathfrak{g},\mathfrak{k})$ under consideration are subdivided into two classes, depending on $\operatorname{rank}\mathfrak{g} = \operatorname{rank}\mathfrak{k}$ or $\operatorname{rank}\mathfrak{g} > \operatorname{rank}\mathfrak{k}$. If $\mathfrak{g}$ and $\mathfrak{k}$ are of equal rank, then clearly $\Sigma_\theta = \Sigma$ and $\Psi_\theta = \Psi$. Consequently, every parabolic subalgebra $\mathfrak{q} \supset \mathfrak{b}$ is θ-stable.

The more delicate case is when $\operatorname{rank}\mathfrak{g} > \operatorname{rank}\mathfrak{k}$ and therefore $\Sigma_\theta \neq \Sigma$. Note that the root system Σ_θ for $\mathfrak{g}_0$ of noncomplex type can be read off directly from the affine Dynkin diagrams as given in Table 18.2.9: Σ_θ is irreducible and the corresponding Dynkin diagram is obtained from the affine diagram in Table 18.2.9 by deleting the vertex labeled μ_0. Only for $\operatorname{rank}\mathfrak{g} > \operatorname{rank}\mathfrak{k}$ does the complex involution θ induce a nontrivial automorphism $\theta : \Psi \to \Psi$ of the simple root system of $(\mathfrak{g},\mathfrak{h})$. A parabolic subalgebra $\mathfrak{q} \supset \mathfrak{b} = \mathfrak{b}^-$ need not be θ-stable and the equation $\mathfrak{q} = \theta\mathfrak{q}$ holds if and only if $\theta\Phi = \Phi$. A θ-stable $\mathfrak{q}$ can also be described by the subset $\Phi_\theta = \Sigma(\mathfrak{q},\mathfrak{t}) \cap \Psi_\theta$. This notation is maintained in the following chapters.

18.6A Structure equations

We implement the Cohomological Lemma 18.5.6 in order to decide for given Lie algebra data $\mathfrak{b}_{\mathfrak{k}} \subset \mathfrak{b} \subset \mathfrak{q}$ and a Bott-regular weight $\mu_i \in M(\mathfrak{t},\mathfrak{s})$ of index 1, whether or not $V^{\lambda(\mu_i)}$ is a submodule of $H^1(C;\mathcal{O}(\mathbb{E}((\mathfrak{q}+\theta\mathfrak{q})_{\mathfrak{s}}))) \subset H^0(C;\mathcal{O}(\mathbb{N}_{Z/C}))$. In order to verify the conditions stated in the Cohomological Lemma it is sufficient have the following information:

If $\mathfrak{q} = \theta\mathfrak{q}$: the Bott-regular weights $\mu_i \in M(\mathfrak{t},\mathfrak{s})$, the corresponding dominant weights $\lambda(\mu_i)$ and the simple short roots $\beta_i^\star$ as linear combinations of the elements of Ψ_θ. More precisely, in the case $\mathfrak{q} = \theta\mathfrak{q}$ it is sufficient to have

$$\mu_i = \textstyle\sum_{\Psi_\theta} m^i_\gamma \gamma,\ \lambda(\mu_i) = \sum_{\Psi_\theta} \ell^i_\gamma \gamma = \mu_i + n\beta_i^\star,\ \beta_i^\star = \sum_{\Psi_\theta} b^i_\gamma \gamma. \tag{18.6.3}$$

In the case $\mathfrak{q} \neq \theta\mathfrak{q}$, it is sufficient to have the roots $\mu_i^\dagger, \mu_i^\ddagger, \lambda^\dagger, \lambda^\ddagger$, etc. as linear combinations of the elements of Ψ. Here, $\mu_i^\dagger, \mu_i^\ddagger \in \Sigma(\mathfrak{h})$ denote the roots with $\mu_i^\dagger|_{\mathfrak{t}} = \mu_i = \mu_i^\ddagger|_{\mathfrak{t}}$.

We refer to the above equations as the **structure equations** for μ, $\lambda(\mu)$ and $\beta^\star$ (or $\mu^\dagger, \mu^\ddagger, \lambda^\dagger$, etc.).

Since Ψ_θ is a root basis of the root system $\Sigma(\mathfrak{g},\mathfrak{t})$, Ψ is a root basis of $\Sigma(\mathfrak{g},\mathfrak{h})$, and $M(\mathfrak{t},\mathfrak{s}) \subset \Sigma(\mathfrak{g},\mathfrak{t}) \cup \{0\}$, in the above structure equations all coefficients are integers of equal sign.

We write $\operatorname{supp}_{\Psi_\theta}(\mu)$ for the positive support $\{\gamma \in \Psi_\theta \mid m_\gamma > 0\}$ and define supports $\operatorname{supp}_{\Psi_\theta}(\lambda(\mu))$ and $\operatorname{supp}_{\Psi_\theta}(\beta^\star)$ similarly. In the same way, we have some other supports $\operatorname{supp}_\Psi(\mu^\dagger) := \{\psi \in \Psi \mid m^\dagger_\psi > 0\}$, $\operatorname{supp}_\Psi(\lambda^\dagger) := \{\psi \mid \ell^\dagger_\psi > 0\}$, $\operatorname{supp}_\Psi(\lambda^\ddagger) := \{\psi \mid \ell^\ddagger_\psi > 0\}$, etc.

In the next subsection, assuming knowledge of the coefficients in (18.6.3), we explain how the structure equations give us sufficient information to apply the Cohomological Lemma.

18.6B Strategy of computation

Given a base cycle (C, Z) and data $\mathfrak{b}_{\mathfrak{k}} \subset \mathfrak{b} \subset \mathfrak{q}$, we have the corresponding bases of root systems, Ψ_θ and Ψ, and the subsets $\Phi_{\mathfrak{k}} \subset \Psi_{\mathfrak{k}}$, $\Phi \subset \Psi$, determined by the given parabolic subalgebra $\mathfrak{q}$ (and potentially $\Phi_\theta \subset \Psi_\theta$ if $\mathfrak{q}$ is θ-stable). In the following we show how the conditions of the Cohomological Lemma can be verified.

Let a Bott-regular element $\mu_i \in M(\mathfrak{t}, \mathfrak{s})$ and $\beta_i^\star \in \Psi_{\mathfrak{k}}$ (as in the String Lemma or Lemma 18.5.5) be given. If $\mathfrak{q}$ is θ-stable, then it is completely determined by the subset $\Phi_\theta \subset \Psi_\theta$. In such a case $\beta_i^\star \in \Phi_{\mathfrak{k}}$ if and only if $\mathrm{supp}_{\Psi_\theta}(\beta_i^\star) \subset \Phi_\theta$.

The coefficients in the structure equations (18.6.3) are either all nonnegative or all nonpositive. Suppose first that the coefficients for μ_i with respect to some choice of $\mathfrak{b}$ and $\leqq 0$. Since $\Sigma_\theta^- = \Sigma(\mathfrak{b}, \mathfrak{t}) \subset M(\mathfrak{t}, (\mathfrak{q}+\theta\mathfrak{q})_{\mathfrak{s}})$ it follows that $\mu_i \in M(\mathfrak{t}, (\mathfrak{q}+\theta\mathfrak{q})_{\mathfrak{s}})$ for all parabolics $\mathfrak{q} \supset \mathfrak{b}$.

Similarly, if the coefficients in the structure equation for μ_i (or $\lambda(\mu_i)$) are positive, then μ_i or $\lambda(\mu_i)$ belongs to $M(\mathfrak{t}, (\mathfrak{q} + \theta\mathfrak{q})_{\mathfrak{s}})$ if and only if $\mathrm{supp}_{\Psi_\theta}(\mu_i)$ (or $\mathrm{supp}_{\Psi_\theta}(\lambda(\mu_i))$, respectively) is a subset of Φ_θ.

The most delicate case is that in which $\theta\mathfrak{q} \neq \mathfrak{q}$, i.e., the open G_0-orbits in G/Q are not measurable in sense of Theorem 4.5.1. As before, $\mu_i \in M(\mathfrak{t}, (\mathfrak{q}+\theta\mathfrak{q})_{\mathfrak{s}})$ if and only if either $\mathrm{supp}_\Psi(\mu_i^\dagger)$ or $\mathrm{supp}_\Psi(\lambda_i^\dagger)$ belongs to Φ, the subset of Ψ which described the reductive part of $\mathfrak{q}$. We proceed similarly with the remaining key objects, i.e., with $\lambda(\mu_i)$ and $\beta_i^\star$

It remains only to determine the coefficients of the structure equations (18.6.3) in each situation $\mathfrak{b}_{\mathfrak{k}}^{\mathrm{ref}} \subset \mathfrak{b} \subset \mathfrak{q}$. In the cases when there are many Borel subalgebras $\mathfrak{b}$ containing the given $\mathfrak{b}_{\mathfrak{k}}^{\mathrm{ref}}$, we develop a method for computing the coefficients in (18.6.3) for all $\Psi_\theta(\mathfrak{b})$ once the coefficients are established for one particular $\Psi_{\theta,0} = \Psi_\theta(B_0)$. In that situation we use our knowledge of the particular shape of the subset $W_1^\theta \subset W^\theta$; see (18.3.8).

In the following chapters we discuss in complete detail each $(\mathfrak{g}, \theta)$ which is not excluded by Theorem 18.4.13.

19

Classification for Simple $\mathfrak{g}_0$ with rank $\mathfrak{k}$ < rank $\mathfrak{g}$

In this and the following chapter we carry out the detailed calculations of the tangent space of the cycle space at every base cycle, where $\mathfrak{g}_0$ belongs to one of the Tables 18.5.1 or 18.5.2. Recall if $\mathfrak{g}_0$ is not in one of these tables, then (Theorem 18.4.13) the corresponding component $\mathcal{C}_{[C]}(Z)$ of the cycle space coincides with the closure of the orbit $G{\cdot}[C]$.

In the present chapter we deal with those symmetric pairs for which the symmetric subalgebra $\mathfrak{k}$ has smaller rank than $\mathfrak{g}$. There are two distinct possibilities here. One is that the simple Lie algebra $\mathfrak{g}_0$ is absolutely simple, i.e., $\mathfrak{g}$ simple. The other is that $\mathfrak{g}_0$ is the underlying real Lie algebra structure of a complex simple Lie algebra $\mathfrak{l}$ and $\mathfrak{g} \cong \mathfrak{l} \times \mathfrak{l}$. In that case we say that $\mathfrak{g}_0$ is of **complex type**.

For convenience, we give a list of all complex symmetric pairs $(\mathfrak{g}, \mathfrak{k})$ and the corresponding real forms $\mathfrak{g}_0$ that satisfy rank $\mathfrak{k}$ < rank $\mathfrak{g}$ and are not excluded in Theorem 18.4.13.

$\mathfrak{g}_0$ absolutely simple, i.e., not of complex type:

- the series $\mathfrak{g}_0 = \mathfrak{sl}(2r; \mathbb{R})$, i.e., $(\mathfrak{g}, \mathfrak{k}) = (\mathfrak{sl}(2r; \mathbb{C}), \mathfrak{so}(2r; \mathbb{C}))$,
- the series $\mathfrak{g}_0 = \mathfrak{so}(2p+1, 2q+1)$, i.e.,
  $(\mathfrak{g}, \mathfrak{k}) = (\mathfrak{so}(2r; \mathbb{C}), \mathfrak{so}(2p+1; \mathbb{C}) \oplus \mathfrak{so}(2q+1; \mathbb{C}))$ with $p, q \geqq 1$ and $r = p + q + 1$,
- the case $\mathfrak{g}_0 = \mathfrak{e}_{6,C_4}$, where $(\mathfrak{g}, \mathfrak{k}) = (\mathfrak{e}_6(\mathbb{C}), \mathfrak{sp}(4; \mathbb{C}))$, and
- the series $\mathfrak{g}_0 = \mathfrak{sl}(2r+1; \mathbb{R})$, i.e.,
  $(\mathfrak{g}, \mathfrak{k}) = (\mathfrak{sl}(2r+1; \mathbb{C}), \mathfrak{so}(2r+1; \mathbb{C}))$.

$\mathfrak{g}_0$ is of complex type:

- the series $\mathfrak{g}_0 = \mathfrak{so}(2r+1, \mathbb{C})$,
- the series $\mathfrak{g}_0 = \mathfrak{sp}(r, \mathbb{C})$,
- the series $\mathfrak{g}_0 = \mathfrak{f}_4(\mathbb{C})$.

19.1 Strategy

Our strategy in the description of the module structure of $T_{[C]}\mathcal{C}$, i.e., the set $M(C, Z)$ of highest weights as in (18.6.2), is the following. Given a complex symmetric pair $(\mathfrak{g}, \mathfrak{k})$, a τ-stable Cartan subalgebra $\mathfrak{t} \subset \mathfrak{k}$ and the corresponding fundamental Cartan subalgebra $\mathfrak{h} = \mathfrak{z}_{\mathfrak{g}}(\mathfrak{t})$, we select

- a (small) Borel subalgebra $\mathfrak{b}_{\mathfrak{k}} = \mathfrak{b}_{\mathfrak{k}}^{\text{ref}} \subset \mathfrak{k}$ which contains $\mathfrak{t}$ and determines the simple roots $\Psi_{\mathfrak{k}} = \{\beta_1, \dots, \beta_r\}$, and
- a (large) Borel subalgebra $\mathfrak{b}^{\text{ref}} \subset \mathfrak{g}$ which contains $\mathfrak{b}_{\mathfrak{k}}$. It determines the bases $\Psi_\theta = \{\gamma_1, \dots, \gamma_r\}$ of $\Sigma(\mathfrak{g}, \mathfrak{t})$ and $\Psi = \{\psi_1, \dots, \psi_n\}$ of $\Sigma(\mathfrak{g}, \mathfrak{h})$. Note that the involution θ permutes the roots in Ψ.

First, we determine the coefficients in (18.6.3) with respect to this particular choice of $\mathfrak{b}_{\mathfrak{k}}^{\text{ref}} \subset \mathfrak{b}^{\text{ref}}$. Guided by the observation that the set $W_1^\theta \subset W^\theta$, as given in (18.3.8), parameterizes all Borel subalgebras $\mathfrak{b} = w{\cdot}\mathfrak{b}^{\text{ref}}$ which contain the given $\mathfrak{b}_{\mathfrak{k}}^{\text{ref}}$, we proceed as follows. The simple root systems defined by the Borel subalgebra $w{\cdot}\mathfrak{b}^{\text{ref}}$ are $\{w\gamma_1, \dots, w\gamma_r\}$ and $\{w\psi_1, \dots, w\psi_m\}$. The coefficients in the equations $\mu = \sum_{\Psi_\theta} m_\gamma \cdot w\gamma$, $\mu^\dagger = \sum_{\Psi_\theta} m_\psi^\dagger \cdot w\psi$ are the same as those in

$$(19.1.1) \qquad w^{-1}\mu = \textstyle\sum_{\Psi_\theta} m_\gamma \cdot \gamma \quad \text{or} \quad w^{-1}\mu^\dagger = \sum_{\Psi} m_\psi^\dagger \cdot \psi$$

The next steps in our computation are

- explicit determination of the reference bases $\{\gamma_1, \dots, \gamma_r\}$ and $\{\psi_1, \dots, \psi_m\}$,
- explicit determination of W_1^θ and its action on $\mathfrak{t}$, and, finally,
- description of $w^{-1}\mu$, $w^{-1}\lambda$ as linear combinations of elements of Ψ_θ^{ref} or Ψ^{ref}. This gives us the coefficients in the structure equations for $\mathfrak{b} = w{\cdot}\mathfrak{b}^{\text{ref}}$.

As w runs through W_1^θ, $w\mathfrak{b}$ runs through the set of all Borel subalgebras of $\mathfrak{g}$ that contain the fixed $\mathfrak{b}_{\mathfrak{k}}^{\text{std}}$. Equivalently, the Lie algebra data $\mathfrak{b}_{\mathfrak{k}} \subset w\mathfrak{b}^{\text{ref}} \subset \mathfrak{q}$ runs through the various data corresponding to all K-base cycles C in G-homogeneous flag manifolds Z (see Section 17.2).

In the following sections we give matrix realizations of the classical complex Lie algebras $\mathfrak{g}$ for which the Cartan subalgebras $\mathfrak{t}$ and $\mathfrak{h}$ consist of diagonal matrices.

Notation. *$\mathbb{M}_{p\times q}(A)$ denotes the algebra of $p \times q$ matrices with coefficients in a ring A, $\mathbb{B}_m^+ = \mathbb{B}_m^+(A)$ is the subalgebra of upper triangular matrices, and $\mathbb{B}_m^-$ denotes the algebra of the lower triangular matrices. Matrix elements are denoted as usual by A_{jk}, B_{jk},, etc. By $R_m \in \mathbb{M}_m(\mathbb{Z})$ we denote the symmetric matrix with 1 on the antidiagonal and 0 elsewhere, $I = I_m$ is the identity matrix and $I_{pq} := \begin{pmatrix} I_p & \\ & -I_q \end{pmatrix}$.*

19.2 The series for $\mathfrak{g}_0 = \mathfrak{sl}(2r; \mathbb{R})$

In this section we determine explicitly all coefficients in the structure equations 18.6.3 for the symmetric space $(\mathfrak{sl}(2r; \mathbb{C}), \mathfrak{so}(2r; \mathbb{C}))$. For this, we first select the reference

Borel subalgebras $\mathfrak{b}_{\mathfrak{k}}^{\text{ref}} \subset \mathfrak{so}(2r;\mathbb{C})$ and $\mathfrak{b}^{\text{ref}} \subset \mathfrak{sl}(2r;\mathbb{C})$ in an appropriate matrix realization of $(\mathfrak{g},\mathfrak{k})$.

Define $b(z,w) := z^t \cdot R_{2r} \cdot w$ which is a nondegenerate symmetric bilinear form on $\mathbb{C}^{2r}$. View $\mathfrak{k} = \mathfrak{so}(2r;\mathbb{C})$ as the Lie algebra of the isometry group of b. Here $\mathfrak{g} = \mathfrak{sl}(2r;\mathbb{C})$ is the Lie algebra of trace zero matrices in $\mathbb{M}_{2r}(\mathbb{C})$, and $\mathfrak{k}$ is the fixed point set of the involution $\theta : X \mapsto R_{2r}\cdot(-X^t)\cdot R_{2r}$ of $\mathfrak{g}$. To avoid confusion with matrix multiplication, we write $\theta(X)$ as ${}^{\theta}X$.

We choose Cartan subalgebra $\mathfrak{t}$ of $\mathfrak{k}$ to be $\mathfrak{h} \cap \mathfrak{k}$, where $\mathfrak{h}$ is the Cartan subalgebra of $\mathfrak{g}$ consisting of trace 0 diagonal matrices $\operatorname{diag}(\delta_1, \dots, \delta_{2r})$. Thus $\mathfrak{t} \subset \mathfrak{k}$ is given by

$$\left\{ \begin{pmatrix} \varepsilon_1 & & & & & \\ & \ddots & & & & \\ & & \varepsilon_r & & & \\ & & & -\varepsilon_r & & \\ & & & & \ddots & \\ & & & & & -\varepsilon_1 \end{pmatrix} \right\} \subset \left\{ \begin{pmatrix} A & B \\ C & {}^{\theta}A \end{pmatrix} \middle| \begin{matrix} B = {}^{\theta}B \\ C = {}^{\theta}C \end{matrix} \right\} \tag{19.2.1}$$

with $A, B, C \in \mathbb{M}_r(\mathbb{C})$. The reference Borel subalgebras are

$$\mathfrak{b}_{\mathfrak{k}}^{\text{ref}} = \mathfrak{so}(2r;\mathbb{C}) \cap \mathbb{B}_{2r}^{-} \quad \text{and} \quad \mathfrak{b}^{\text{ref}} = \mathfrak{sl}(2r;\mathbb{C}) \cap \mathbb{B}_{2r}^{-}.$$

In order to describe the various root systems, consider the diagonal entries $\varepsilon_1, \dots, \varepsilon_r$ of (19.2.1) as linear functionals on $\mathfrak{t}$. Evidently they are orthogonal with respect to the Killing form. Below, the various root systems and, with one exception, the corresponding root bases $\Psi_{\mathfrak{k}}$ and Ψ_θ are expressed in terms of the ε_j and ordered according to the Bourbaki order. The exception is $r = 3$, where $\mathfrak{k} = \mathfrak{so}(6,\mathbb{C}) \cong \mathfrak{sl}(4,\mathbb{C})$, but where we choose the ordering induced by the orthogonal series D_3 rather than by A_3.

The long roots in Σ_θ are $\pm 2\varepsilon_j$ and their root spaces are one-dimensional. The short roots in Σ_θ have two-dimensional root spaces. The simple system for $\Sigma(\mathfrak{g},\mathfrak{h})$ is determined by $\mathfrak{b}^{\text{ref}}$ and denoted by $\Psi = \{\psi_1, \dots, \psi_{2r-1}\}$ such that ψ_j are the consecutive simple roots in the Dynkin diagram o—o— ··· —o. In terms of the diagonal entries δ_j in $\mathfrak{h}$, considered as linear functionals on $\mathfrak{h}$ (subject to the single condition $\delta_1 + \cdots + \delta_{2r} = 0$), we have $\psi_k = \delta_k - \delta_{k+1}$, $1 \leqq k \leqq 2r-1$. The restrictions $\psi_j|_{\mathfrak{t}} = \gamma_j$ form the simple system for $\Sigma(\mathfrak{g},\mathfrak{t})$. This and other root systems and simple roots are

$$\Sigma_{\mathfrak{k}} = \{\pm(\varepsilon_j \pm \varepsilon_k)\} \quad \text{and} \quad \Psi_{\mathfrak{k}} = \left\{ \varepsilon_1 - \varepsilon_2, \varepsilon_2 - \varepsilon_3, \dots, \begin{matrix} \varepsilon_{r-1} - \varepsilon_r \\ \varepsilon_{r-1} + \varepsilon_r \end{matrix} \right\}$$
$$=: \left\{ \beta_1, \quad \beta_2, \quad \dots, \quad \begin{matrix} \beta_{r-1} \\ \beta_r \end{matrix} \right\},$$

$\Sigma_\theta \cup \{0\} = M(\mathfrak{t},\mathfrak{s}) = \{\pm(\varepsilon_j \pm \varepsilon_k), \pm 2\varepsilon_j, 0\}$, and

$\Psi_\theta = \{\varepsilon_1 - \varepsilon_2, \varepsilon_2 - \varepsilon_3, \dots, \varepsilon_{r-1} - \varepsilon_r,\ 2\varepsilon_r\} =: \{\gamma_1, \gamma_2, \dots, \gamma_r = \gamma_{\mathrm{lo}}\}$.

Before going further, let us discuss the low-dimensional cases. For $r = 2$ we have $\mathfrak{g}_0 = \mathfrak{sl}(4;\mathbb{R}) \cong \mathfrak{so}(3,3)$, which will be considered in Section 19.4. For $r = 1$ we have $\mathfrak{g}_0 = \mathfrak{sl}(2;\mathbb{R})$; that case is trivial from our point of view, because the only relevant flag manifold is the projective line $\mathbb{P}^1$, the cycles C are points, and we have $\mathcal{C}_{[C]}(\mathbb{P}^1) = \mathbb{P}^1 = SL(2;\mathbb{C})\cdot[C]$.

For the remainder of this section we assume that $r \geqq 3$. We use the above root bases. For a given root $\alpha \in \Sigma_\theta$ there are either one or two roots, $\alpha^\dagger$ and possibly $\alpha^\ddagger$ in Σ such that $\alpha^\dagger|_{\mathfrak{t}} = \alpha^\ddagger|_{\mathfrak{t}} = \alpha$. Looking at Table 18.5.3 we see that the only Bott-regular element μ and certain other weights which are relevant for the structure equations are the following:

$$
\begin{aligned}
\mu &= 2\xi_1 - 2\beta_1 = 2\varepsilon_2 \\
&= 2\gamma_2 + \cdots + 2\gamma_{r-1} + \gamma_{1o};\ \text{here } \mu^\dagger = \psi_2 + \cdots + \psi_{2r-2} \\
\beta^\star &= \beta_1 = \gamma_1;\ \text{here } \beta^{\star\dagger} = \psi_1 \text{ and } \beta^{\star\ddagger} = \psi_{2r-1} \\
\lambda &= \lambda(\mu) = \gamma_1 + 2\gamma_2 + \cdots + 2\gamma_{r-1} + \gamma_{1o}; \\
&\qquad \text{here } \lambda^\dagger = \psi_1 + \cdots + \psi_{2r-2} \text{ and } \lambda^\ddagger = \psi_2 + \cdots + \psi_{2r-1}.
\end{aligned}
\tag{19.2.2}
$$

For this particular series, $M(\mathfrak{t},\mathfrak{s})$ also contains Bott-regular weights of index 2, namely,

$$\mu_{\mathrm{II}} = -\beta_{r-1} - \beta_r = -2\varepsilon_{r-1}\ ; \quad \text{here} \quad \mu_{\mathrm{II}}^\dagger = -\psi_{r-1} - \psi_r - \psi_{r+1}.$$

The next step is to determine the various large Borel subalgebras which contain $\mathfrak{b}_{\mathfrak{k}}^{\mathrm{ref}}$. According to Corollary 18.3.6, this can be done in terms of certain Weyl groups.

Since $W_{\mathfrak{k}} \cong W(\mathfrak{t},\mathfrak{so}(2r)) \cong \mathfrak{S}_r \rtimes \mathbb{Z}_2^{r-1}$ and $W^\theta \cong W(\mathfrak{t},\mathfrak{sp}(r)) \cong \mathfrak{S}_r \rtimes \mathbb{Z}_2^r$, it follows that $|W^\theta/W_{\mathfrak{k}}| = 2$. Thus there are only two θ-stable Borel subalgebras containing the given $\mathfrak{b}_{\mathfrak{k}}^{\mathrm{ref}}$. Geometrically, this means that for $\mathfrak{g}_0 = \mathfrak{sl}(2r;\mathbb{R})$ there are exactly two base cycles in the full flag G/B and consequently at most two base cycles in a general flag manifold G/Q.

In order to find the second Borel subalgebra, we apply the reflection $s_{\gamma_{1o}}$. This is motivated by the observation that $\gamma_{1o} = \gamma_r \in \Psi_\theta \setminus \Psi_{\mathfrak{k}}$. The new simple systems which are determined by $\tilde{\mathfrak{b}} := s_{\gamma_{1o}}(\mathfrak{b}^{\mathrm{ref}})$ are

$$
\begin{aligned}
&\tilde{\gamma}_1 = \gamma_1,\ \ldots,\ \tilde{\gamma}_{r-2} = \gamma_{r-2},\ \tilde{\gamma}_{r-1} = \gamma_{r-1} + \gamma_{1o},\ \tilde{\gamma}_{1o} = -\gamma_{1o}; \\
&\tilde{\psi}_j = \psi_j \text{ for } j < r-1,\ \tilde{\psi}_{r-1} = \psi_{r-1} + \psi_r,\ \tilde{\psi}_r = -\psi_r,\ \tilde{\psi}_{r+1} = \psi_r + \psi_{r+1}.
\end{aligned}
$$

A simple check shows that the coefficients in the structure equations with respect to the root bases given by the second Borel subalgebra $\tilde{\mathfrak{b}}$ are identical with those in (19.2.2).

Since we know the coefficients of the structure equations for all pairs $C \subset Z$, we can use the Cohomological Lemma to prove the following.

Theorem 19.2.3. *Let $Z = G/Q$, where the parabolic subalgebra $\mathfrak{q} = \mathfrak{q}_\Phi$ is given by a subset $\Phi \subset \Psi$, and let $C \subset Z$ be any base cycle. Then in all cases $\mathcal{C}(Z)$ is smooth at $[C]$ and*

$$T_{[C]}\mathcal{C}(Z) = \begin{cases} T_{[C]}(G\cdot[C]) & \textit{if } \Phi \neq \{\psi_2, \psi_3, \ldots, \psi_{2r-2}\} \\ T_{[C]}(G\cdot[C]) \oplus V^{\xi_2}_{\mathfrak{so}(2r)} & \textit{if } \Phi = \{\psi_2, \psi_3, \ldots, \psi_{2r-2}\},\ r > 3, \\ T_{[C]}(G\cdot[C]) \oplus V^{\xi_2+\xi_3}_{\mathfrak{so}(6)} & \textit{if } \Phi = \{\psi_2, \psi_3, \ldots, \psi_{2r-2}\},\ r = 3, \end{cases}$$

where $V^{\xi_2}_{\mathfrak{so}(2r)}$ *(respectively,* $V^{\xi_2+\xi_3}_{\mathfrak{so}(6)}$*) indicates the representation of* $\mathfrak{k} = \mathfrak{so}(2r;\mathbb{C})$ *on* $\bigwedge^2 V^{\text{std}}_{\mathfrak{so}(2r)}$. *In particular, the set* $M(C, Z)$ *in* (18.6.2) *either is empty or coincides with* $\{\xi_2\}$ *(respectively,* $\{\xi_2 + \xi_3\}$*), where* ξ_j *is the* j*th fundamental highest weight of* $\mathfrak{so}(2r;\mathbb{C})$.

Proof. We have noted that the structure equations for any of at most two base cycles in a given flag manifold $Z = G/Q_\Phi$ are identical. Further, from the structure equations (19.2.2), $\mu \in M(\mathfrak{t}, (\mathfrak{q}_\Phi + \theta\mathfrak{q}_\Phi)_\mathfrak{s})$ if and only if $\Phi \supset \{\psi_2, \psi_3, \ldots, \psi_{2r-2}\}$. By the Cohomological Lemma 18.5.6, the submodule $V^{\xi_2}_{\mathfrak{so}(2r)}$ is contained in $H^1(C; \mathcal{O}(\mathbb{E}((\mathfrak{q}_\Phi + \theta\mathfrak{q}_\Phi)_\mathfrak{s}))) \subset T_{[C]}\mathcal{C}(Z)$ if and only if $\{\beta^{\star\dagger}, \beta^{\star\ddagger}\} \not\subset \Sigma(\mathfrak{q}^r_\Phi, \mathfrak{h})$ (Lemma 18.5.5) and neither of the roots $\lambda^\dagger, \lambda^\ddagger$ is contained in $\langle\!\langle \Phi \rangle\!\rangle_\mathbb{Z}$.

It remains to prove smoothness at all cycles C with $\dim T_{[C]}\mathcal{C}(Z) > \dim G\cdot[C]$. Note that $\mu_{\text{II}} \in M(\mathfrak{t}, \mathfrak{b}_\mathfrak{s}) \subset M(\mathfrak{t}, (\mathfrak{q}_\Phi + \theta\mathfrak{q}_\Phi)_\mathfrak{s})$. However, in the only relevant case where $\Phi = \{\psi_2, \psi_3, \ldots, \psi_{2r-2}\}$, both roots

$$\beta_{r-1} = \varepsilon_{r-1} - \varepsilon_r = \psi_{r-1}|_\mathfrak{t} = \psi_{r+1}|_\mathfrak{t} \quad \text{and}$$
$$\beta_r = \varepsilon_{r-1} + \varepsilon_r = (\psi_{r-1} + \psi_r)|_\mathfrak{t} = (\psi_r + \psi_{r+1})|_\mathfrak{t}$$

belong to $M(\mathfrak{t}, \mathfrak{q}^r_\mathfrak{k})$. Hence, μ_{II} is not highest with respect to $\mathfrak{q}_\mathfrak{k} \cap (\mathfrak{b}^{\text{ref}}_\mathfrak{k})^+$. Thus $H^2(C; \mathcal{O}(\mathbb{F}^{j+1}/\mathbb{F}^j)) = 0$ for all $j \geqq 0$ (compare Lemma 18.4.10) and consequently $H^2(C; \mathcal{O}(\mathbb{E}((\mathfrak{q}_\Phi + \theta\mathfrak{q}_\Phi)_\mathfrak{s}))) = H^1(C; \mathcal{O}(\mathbb{N}_{Z_\Phi}(C))) = 0$. □

19.3 The series for $\mathfrak{g}_0 = \mathfrak{sl}(2r+1;\mathbb{R})$

This series of real simple Lie groups was excluded from the considerations above, because the root system Σ_θ is not reduced and Condition 18.4.3 is not fulfilled. Thus we determine the dominant and Bott-regular weights by ad hoc methods. As in the preceding section, $b : \mathbb{C}^{2r+1} \times \mathbb{C}^{2r+1} \to \mathbb{C}$ is the nondegenerate symmetric bilinear form given by $b(z, w) := z^t \cdot R_{2r+1} \cdot w$ and $\theta : X \mapsto R_{2r+1}\cdot(-X^t)\cdot R_{2r+1} =: {}^\theta X$ is the Cartan involution of $\mathfrak{g}_0$ in terms of matrices. As before, we identify $\mathfrak{sl}(2r+1;\mathbb{C})$ with the space of trace zero matrices in $\mathbb{M}_{2r+1}(\mathbb{C})$. The matrix realization of $\mathfrak{t} \subset \mathfrak{k}$ is given by

$$\left\{ \begin{pmatrix} \varepsilon_1 & & & & & & \\ & \ddots & & & & & \\ & & \varepsilon_r & & & & \\ & & & 0 & & & \\ & & & & -\varepsilon_r & & \\ & & & & & \ddots & \\ & & & & & & -\varepsilon_1 \end{pmatrix} \right\} \subset \left\{ \begin{pmatrix} A & u & B \\ w & 0 & -u^t R_r \\ C & R_r w^t & {}^\theta A \end{pmatrix} \,\middle|\, \begin{matrix} B = {}^\theta B \\ C = {}^\theta C \end{matrix} \right\}. \tag{19.3.1}$$

Select $\mathfrak{b}_{\mathfrak{k}}^{\text{ref}} := \mathfrak{so}(2r+1;\mathbb{C})\cap\mathbb{B}_{2r+1}^-$ and $\mathfrak{b}^{\text{ref}} := \mathfrak{sl}(2r+1;\mathbb{C})\cap\mathbb{B}_{2r+1}^-$ as the reference Borel subalgebras. The root systems in question, and their simple subsystems $\Psi_{\mathfrak{k}}$ and Ψ_θ, can be expressed in terms of the ε_j as follows. For $r \geqq 2$ we have

$$\begin{aligned}
&\Sigma_{\mathfrak{k}} = \{\pm(\varepsilon_j \pm \varepsilon_k), \pm\varepsilon_j \mid j \neq k\} \text{ and}\\
&\Psi_{\mathfrak{k}} = \{\varepsilon_1 - \varepsilon_2, \varepsilon_2 - \varepsilon_3, \dots, \varepsilon_{r-1} - \varepsilon_r, \varepsilon_r\} =: \{\beta_1, \beta_2, \dots, \beta_{r-1}, \beta_{\text{sh}}\};\\
&\Sigma_\theta \cup \{0\} = M(\mathfrak{t},\mathfrak{s}) = \{\pm(\varepsilon_j \pm \varepsilon_k), \pm\varepsilon_j, \pm 2\varepsilon_j, 0 \mid j \neq k\} \text{ and}\\
&\Psi_\theta = \Psi_{\mathfrak{k}} =: \{\gamma_1, \dots, \gamma_{r-1}, \gamma_r = \gamma_{\text{sh}}\}.
\end{aligned}$$

The $\Psi_{\mathfrak{k}}$-dominant roots in Σ_θ can only be linear combinations of $\xi_1 = \varepsilon_1$ and $\xi_2 = \varepsilon_1 + \varepsilon_2$. Hence,

$$\Lambda^+_{\mathfrak{k},\text{wt}} \cap M(\mathfrak{t},\mathfrak{s}) = \left\{\begin{array}{lllll} 2\xi_1 &=& 2\varepsilon_1 &=& 2\sum_{\Psi_{\mathfrak{k}}}\beta_j\\ \xi_2 &=& \varepsilon_1+\varepsilon_2 &=& \beta_1 + 2\sum_{\Psi_{\mathfrak{k}}\setminus\beta_1}\beta_j\\ \xi_1 &=& 2\varepsilon_1 &=& \sum_{\Psi_{\mathfrak{k}}}\beta_j\\ 0 &&&& \end{array}\right\}.$$

If $r = 1$, then $\Sigma_{\mathfrak{k}} = \{\pm\varepsilon_1\} \supset \{\varepsilon_1\} = \{\beta\} = \Psi_{\mathfrak{k}}$ and $\Sigma_\theta = \{\pm\varepsilon_1, \pm 2\varepsilon_1\}$. The dominant weights in $\Lambda^+_{\mathfrak{k},\text{wt}} \cap M(\mathfrak{t},\mathfrak{s})$ are 2β, β and 0.

Finally, let $\psi_1, \dots, \psi_{2r}$ be the consecutive simple roots in $\Sigma(\mathfrak{g},\mathfrak{h})$ such that $\psi_j|_{\mathfrak{t}} = \beta_j = \gamma_j$ for $j = 1, \dots, r$.

Various Weyl groups. Concerning the various Weyl groups, note that $W^\theta = W_{\mathfrak{k}} = W(\mathfrak{t}, \mathfrak{so}(2r+1)) \cong \mathfrak{S}_r \ltimes \mathbb{Z}_2^r$. This follows from the fact that $\Sigma_{\mathfrak{k}}$ is a reduced version of Σ_θ. Therefore $|W^\theta/W_{\mathfrak{k}}| = 1$ and the standard Borel subalgebra $\mathfrak{b}$ chosen above is the only Borel that contains $\mathfrak{b}_{\mathfrak{k}}$. Hence, every flag manifold G/Q contains precisely one base cycle C (with respect to $\mathfrak{g}_0 = \mathfrak{sl}(2r+1,\mathbb{R})$).

Consider the trivial case $r = 1$. Since rank of $\mathfrak{k} = \mathfrak{so}(3,\mathbb{C})$ is 1 there are no weights of index 2. Both weights $-\beta$ and -2β are Bott-regular of index 1, while $\lambda(-\beta) = 0$ and $\lambda(-2\beta) = \beta = 2\xi \neq 0$.

The remainder of this section is devoted to the nontrivial case $r > 1$.

Bott-regular weights of index 1. Recall that $\langle\,,\,\rangle$ denotes the Killing form on $\mathfrak{g}$ and $\langle\varphi_1 \mid \varphi_2\rangle := \frac{2\langle\varphi_1,\varphi_2\rangle}{\langle\varphi_2,\varphi_2\rangle}$. Going through the first part of the proof of the String Lemma 18.4.4, one sees that it applies equally well to the situation where the numbers $\langle\varphi_1 \mid \varphi_2\rangle$, $\varphi_j \in M(\mathfrak{t},\mathfrak{s})$, take values in the set $\{0, \pm 1, \pm 2, \pm 4\}$. If $\beta \in \Phi_{\mathfrak{k}}$ is the simple root with a dominant $s_\beta(\mu + \rho_{\mathfrak{k}})$, then exactly as in the proof of the String Lemma, we obtain the inequality $\mu(h_\beta) \leqq -2$.

Now if $\mu(h_\beta) = -4$, then $\mu = -2\beta$, and a simple check shows that -2β cannot be Bott-regular of index 1 or 2 (recall that $r \geqq 2$). Hence, $\mu(h_\beta) = -2$ and we may proceed as in the proof of the String Lemma. Either $-\mu \in M(\mathfrak{t},\mathfrak{s}) \cap \Psi_{\mathfrak{k}} = \Psi_{\mathfrak{k}}$ and $\lambda(-\beta_j) = 0$ for all simple β_j, or μ is an edge of a 3-string $\mu, \mu+\beta^\star, \mu+2\beta^\star$ for some $\beta^\star \in \Psi_{\mathfrak{k}}$ such that $\mu + \beta^\star$ is nonzero dominant. In the latter case $\lambda(\mu) = \mu + \beta^\star$. Taking into consideration the above list of dominant weights in $M(\mathfrak{t},\mathfrak{s})$, a small calculation shows that there are two Bott-regular weights of index 1 such that $\lambda(\mu) \neq 0$. These are

$$
\begin{aligned}
\mu &= \xi_2 - \beta_1 = 2\varepsilon_2 = 2\beta_2 + \cdots + 2\beta_{r-1} + 2\beta_{sh} \\
&= (\psi_2 + \cdots + \psi_{2r-1})|_{\mathfrak{t}} \\
\lambda(\mu) &= \xi_2 = \varepsilon_1 + \varepsilon_2 = \beta_1 + 2\beta_2 + \cdots + 2\beta_{r-1} + 2\beta_{sh} \\
&= \textstyle\sum_1^{2r-1} \psi_j|_{\mathfrak{t}} \quad \text{or} \quad \sum_2^{2r} \psi_j|_{\mathfrak{t}} \\
\widetilde{\mu} &= \xi_1 - \beta_{sh} = \varepsilon_1 - \varepsilon_r = \beta_1 + \beta_2 + \cdots + \beta_{r-1} \\
&= \psi_1 + \cdots + \psi_{r-1}|_{\mathfrak{t}} \quad \text{or} \quad \psi_{r+2} + \cdots + \psi_{2r}|_{\mathfrak{t}} \\
\lambda(\widetilde{\mu}) &= \xi_1 = \varepsilon_1 = \beta_1 + \cdots + \beta_{r-1} + \beta_{sh} \\
&= \psi_1 + \cdots + \psi_r|_{\mathfrak{t}} \quad \text{or} \quad \psi_{r+1} + \cdots + \psi_{2r}|_{\mathfrak{t}}.
\end{aligned}
$$

Bott-regular weights of index 2. In order to find Bott-regular weights μ of index 2, note that the claim in the second part of the String Lemma also remains true for $(\mathfrak{g}, \mathfrak{k}) = (\mathfrak{sl}(2r+1;\mathbb{C}), \mathfrak{so}(2r+1;\mathbb{C}))$, i.e., in the case when $\langle \varphi_1 \mid \varphi_2 \rangle$ belong to $\{0, \pm 1, \pm 2, \pm 4\}$ for $\varphi_j \in M(\mathfrak{t}, \mathfrak{s})$. The inequalities (18.4.6) imply that the Bott-regular weights μ of index 2 necessarily satisfy

$$\mu(h_\beta) = \mu(h_{\beta'}) = -2 \tag{19.3.2}$$

for some pair β, β' of orthogonal simple roots. However, the argument in the proof of Lemma 18.4.4, which forces $\mu = -\beta - \beta'$ under the assumption (18.4.3), does not go through here, because the nonreduced root system $M(\mathfrak{t}, \mathfrak{s})$ contains roots of 3 different lengths. A correct argument, which excludes the possibility that $\mu \neq -\beta - \beta'$, goes as follows. Observe that $\mu + \beta + \beta' = s_\beta s_{\beta'}(\mu + \rho_{\mathfrak{k}}) - \rho_{\mathfrak{k}}$ is dominant. If it were not 0, it could only be the shortest nonzero dominant weight in $M(\mathfrak{t}, \mathfrak{s})$, in other words, $\mu + \beta + \beta' = \xi_1 = \varepsilon_1$. Since β and β' are orthogonal, a glance at Σ_θ shows that the only such combination of elements in Σ_θ (up to the order of β, β') is $\mu = \varepsilon_2 - \varepsilon_r = \sum_2^{r-1} \beta_j$, $\beta = \beta_1 = \varepsilon_1 - \varepsilon_2$ and $\beta' = \beta_r = \varepsilon_r$. In such a case, however, $\mu(h_{\beta_1}) = -1$ which violates (19.3.2).

The remaining possibility for μ to be Bott-regular of index 2 is then $\mu = -\beta - \beta'$. Since the simple roots β and β' must be orthogonal, $-\beta - \beta'$ would be a combination of at least three different ε_j, and therefore cannot be an element in Σ_θ. In conclusion, *in the case under consideration there are no Bott-regular weights μ of index 2 in $M(\mathfrak{t}, \mathfrak{s})$, i.e., $H^1(C; \mathcal{O}(\mathbb{N}_Z(C))) = 0$ for all flags Z and base cycles C.*

Summarizing, we have proved the following.

Theorem 19.3.3. *Suppose that $\mathfrak{g}_0 = \mathfrak{sl}(2r+1;\mathbb{R})$. Let $Z = G/Q$, where the parabolic subalgebra $\mathfrak{q} = \mathfrak{q}_\Phi$ with $\Phi \subset \Psi$, and let $C \subset Z$ be the (unique) base cycle. If $r > 1$, then*

$$
T_{[C]}\mathcal{C}(Z) = \begin{cases}
T_{[C]}(G\cdot[C]) \oplus V_{B_r}^{\xi_1} & \text{if } \{\psi_1, \ldots, \psi_{r-1}\} \subset \Phi \subset \Psi \setminus \{\psi_r, \psi_{r+1}\} \text{ or} \\
& \text{if } \{\psi_{r+2}, \ldots, \psi_{2r}\} \subset \Phi \subset \Psi \setminus \{\psi_r, \psi_{r+1}\} \\
T_{[C]}(G\cdot[C]) \oplus V_{B_r}^{\xi_2} & \text{if } \Phi = \{\psi_2, \psi_3, \ldots, \psi_{2r-1}\} \\
T_{[C]}(G\cdot[C]) & \text{otherwise.}
\end{cases}
$$

In particular, the set $M(C, Z)$ in the decomposition formula (18.6.2) is either empty or contains one element from $\{\xi_1, \xi_2\}$, where ξ_1, ξ_2 are the first and second fundamental weights of $\mathfrak{so}(2r+1;\mathbb{C})$.

If $r = 1$, *so* $\mathfrak{g}_0 = \mathfrak{sl}(3;\mathbb{R})$, *then* $T_{[C]}\mathcal{C}(Z) = T_{[C]}(G\cdot[C])$ *and* $\mathcal{C}_{[C]}(Z) \cong \mathbb{CP}_5$ *if* $Z = \mathbb{CP}_2$, *and* $T_{[C]}\mathcal{C}(Z) = T_{[C]}(G\cdot[C]) \oplus V^{\text{std}}_{\mathfrak{so}(3)}$ *if* $Z = SL(3;\mathbb{C})/B$.

19.4 The series for $\mathfrak{g}_0 = \mathfrak{so}(2p+1, 2q+1)$

Now we look at the cases corresponding to the series $\mathfrak{g}_0 = \mathfrak{so}(2p+1, 2q+1)$. Here $\mathfrak{g} = \mathfrak{so}(2(r+1);\mathbb{C})$, where $r = p+q$, and we assume $p \leqq q$. We only consider the case where $p, q \geqq 1$, disregarding the series $\mathfrak{so}(1, 2r+1)$ where the highest weight of $\mathfrak{k} \times \mathfrak{s} \to \mathfrak{s}$ is short. In the latter case Theorem 18.4.13 shows that there are no Bott-regular weights $\mu \in M(\mathfrak{t}, \mathfrak{s})$ with $\lambda(\mu) \neq 0$.

Again we choose appropriate matrix realizations. Define

$$\mathfrak{so}(2r+2;\mathbb{C}) = \{X \in \mathbb{M}_{2r+2}(\mathbb{C}) : R_{2r+2}\cdot(-X)^t\cdot R_{2r+2} = X\}.$$

Consider further the involutive automorphism of $\mathfrak{so}(2r+2;\mathbb{C})$, given by

$$\theta : X \mapsto \begin{pmatrix} I_{pq} & & \\ & R_2 & \\ & & -I_{qp} \end{pmatrix} \cdot X \cdot \begin{pmatrix} I_{pq} & & \\ & R_2 & \\ & & -I_{qp} \end{pmatrix} := {}^{\theta}X.$$

Then the fixed point set $\mathfrak{g}^{\theta}$ is $\mathfrak{k} = \mathfrak{so}(2p+1;\mathbb{C}) \times \mathfrak{so}(2q+1;\mathbb{C})$. It consists of all matrices

$$(19.4.1) \qquad \begin{pmatrix} A_p & 0 & u_p & 0 & B_p \\ 0 & A_q & u_q & B_q & 0 \\ v_p & v_q & \begin{matrix} 0 & 0 \\ 0 & 0 \end{matrix} & -u_q^t R & -u_p^t R \\ 0 & C_q & -Rv_q^t & {}^{\theta}A_q & 0 \\ C_p & 0 & -Rv_p^t & 0 & {}^{\theta}A_p \end{pmatrix},$$

where $A_\bullet, B_\bullet, C_\bullet \in \mathbb{M}_\bullet(\mathbb{C})$, $B = {}^{\theta}B$ and $C = {}^{\theta}C$, $u_\bullet \in \mathbb{M}_{\bullet\times 2}(\mathbb{C})$, $v_\bullet \in \mathbb{M}_{2\times\bullet}(\mathbb{C})$ with $u_{\bullet 1} = u_{\bullet 2}$ and $v_{1\bullet} = v_{2\bullet}$. The Cartan subalgebra $\mathfrak{h}$ of $\mathfrak{g}$ consists of the diagonal matrices in $\mathfrak{g}$,

$$(19.4.2) \qquad \mathfrak{h} = \{\mathrm{Diag}(\varepsilon_1, \ldots, \varepsilon_{r+1})\} := \left\{ \begin{pmatrix} \varepsilon_1 & & & & & \\ & \ddots & & & & \\ & & \varepsilon_{r+1} & & & \\ & & & -\varepsilon_{r+1} & & \\ & & & & \ddots & \\ & & & & & -\varepsilon_1 \end{pmatrix} \right\} \cong \mathbb{C}^{r+1}$$

and $\mathfrak{t}$ is the subalgebra given by $\varepsilon_{r+1} = 0$.

As before the ε_j, $1 \leqq j \leqq r+1$, denote the linear coordinate functions on $\mathfrak{h}$ or $\mathfrak{t}$. They are orthogonal with respect to the Killing form. The subalgebras $\mathfrak{b}^{\text{ref}}_{\mathfrak{k}} \subset \mathfrak{b}^{\text{ref}}$ consist of lower triangular matrices in the above matrix realizations of $\mathfrak{k}$ and $\mathfrak{g}$. The corresponding root systems and simple subsystems are given as follows.

$$
\begin{aligned}
\Sigma &= \{\pm\varepsilon_j \pm \varepsilon_k\}_{1 \leqq j < k \leqq r+1}, \\
\Psi &= \left\{\varepsilon_1 - \varepsilon_2, \ldots, \varepsilon_{r-1} - \varepsilon_r, \begin{matrix} \varepsilon_r - \varepsilon_{r+1} \\ \varepsilon_r + \varepsilon_{r+1} \end{matrix}\right\} =: \left\{\psi_1, \ldots, \psi_{r-1}, \begin{matrix} \psi_r \\ \psi_{r+1} \end{matrix}\right\}, \\
\Sigma_\theta &= \{\pm\varepsilon_j \pm \varepsilon_k\}_{1 \leqq j < k \leqq r} \cup \{\pm\varepsilon_j\}_{1 \leqq k \leqq r}, \\
\Psi_\theta &= \{\varepsilon_1 - \varepsilon_2, \ldots, \varepsilon_{r-1} - \varepsilon_r, \varepsilon_r\} = \{\gamma_1, \ldots, \gamma_{r-1}, \gamma_{\text{sh}}\}, \\
\Sigma_{\mathfrak{k}} &= \Sigma_{\mathfrak{k}}^{(1)} \cup \Sigma_{\mathfrak{k}}^{(2)} \quad \text{with} \\
\Sigma_{\mathfrak{k}}^{(1)} &= \{\pm\varepsilon_j \pm \varepsilon_k,\ i \pm \varepsilon_j\}_{1 \leqq j < k \leqq p} \quad \text{(respectively, } \{\pm\varepsilon_1\} \text{ for } p = 1), \\
\Sigma_{\mathfrak{k}}^{(2)} &= \{\pm\varepsilon_j \pm \varepsilon_k,\ \pm\varepsilon_j\}_{p+1 \leqq j < k \leqq p+q} \quad \text{(respectively, } \{\pm\varepsilon_r\} \text{ for } q = 1), \\
\Psi_{\mathfrak{k}}^{(1)} &= \{\varepsilon_1 - \varepsilon_2, \ldots, \varepsilon_{p-1} - \varepsilon_p,\ \varepsilon_p\} =: \{\beta_1^{(1)}, \ldots, \beta_{p-1}^{(1)}, \beta_{\text{sh}}^{(1)}\}, \\
\Psi_{\mathfrak{k}}^{(2)} &= \{\varepsilon_{p+1} - \varepsilon_{p+2}, \ldots, \varepsilon_{r-1} - \varepsilon_r,\ \varepsilon_r\}. =: \{\beta_1^{(2)}, \ldots, \beta_{q-1}^{(2)}, \beta_{\text{sh}}^{(2)}\}.
\end{aligned}
$$

Looking at Table 17.4.8, or arguing directly as explained in the few paragraphs preceding Table 18.5.3, we see that as a $\mathfrak{k}$-module, $\mathfrak{s} = V_{SO_{2p+1}}^{\text{std}} \otimes V_{SO_{2q+1}}^{\text{std}} = V^{\xi_1^{(1)}} \otimes V^{\xi_1^{(2)}}$ for $p, q > 1$ (or $= V^{2\xi^{(1)}} \otimes V^{\xi_1^{(2)}}$ for $p = 1 < q$, or $= V^{2\xi^{(1)}} \otimes V^{2\xi^{(2)}}$ for $p = q = 1$). The modules $V_{SO_{2p+1}}^{\text{std}}$ and $V_{SO_{2q+1}}^{\text{std}}$ decompose into the one-dimensional $\mathfrak{t}$-eigenspaces with weights $\{\pm\varepsilon_j, 0\}_{1 \leqq j \leqq p}$ and $\{\pm\varepsilon_j, 0\}_{p+1 \leqq j \leqq p+q}$, respectively. With the aid of this information the set $M(\mathfrak{t}, \mathfrak{s})$ can be immediately described. The structure equations with respect to the reference root basis Ψ for the dominant, Bott-regular and simple roots are given as follows (see also Table 18.5.3):

$$
\begin{aligned}
\mu &= \xi_1^{(1)} - \beta_{\text{sh}}^{(2)} = \varepsilon_1 - \varepsilon_r = \psi_1 + \cdots + \psi_{r-1}|_{\mathfrak{t}} \\
&\quad (= 2\xi^{(1)} - \beta_{\text{sh}}^{(2)} \quad \text{for } p = 1), \\
\lambda &= \xi_1^{(1)} = \varepsilon_1 = \psi_1 + \cdots + \psi_{r-1} + \psi_r|_{\mathfrak{t}} \\
&= \psi_1 + \cdots + \psi_{r-1} + \psi_{r+1}|_{\mathfrak{t}}, \\
\beta^\star &= \beta_{\text{sh}}^{(2)} = \varepsilon_r = \psi_r|_{\mathfrak{t}} = \psi_{r+1}|_{\mathfrak{t}}, \\
\widetilde{\mu} &= \xi_1^{(2)} - \beta_{\text{sh}}^{(1)} = \varepsilon_{p+1} - \varepsilon_p = -\psi_p|_{\mathfrak{t}} \quad (= 2\xi^{(2)} - \beta_{\text{sh}}^{(1)} \quad \text{for } q = 1), \\
\widetilde{\lambda} &= \xi_1^{(2)} = \varepsilon_{p+1} = \psi_{p+1} + \cdots + \psi_{r-1} + \psi_r|_{\mathfrak{t}} \\
&= \psi_{p+1} + \cdots + \psi_{r-1} + \psi_{r+1}|_{\mathfrak{t}}, \\
\widetilde{\beta}^\star &= \beta_{\text{sh}}^{(1)} = \varepsilon_p = \psi_p + \cdots + \psi_{r-1} + \psi_r|_{\mathfrak{t}} \\
&= \psi_p + \cdots + \psi_{r-1} + \psi_{r+1}|_{\mathfrak{t}}.
\end{aligned}
\tag{19.4.3}
$$

The only Bott-regular weight $\mu_{\text{II}} \in M(\mathfrak{t}, \mathfrak{s})$ of index 2 is, in view of the String Lemma,

$$
\mu_{\text{II}} = -\beta_{\text{sh}}^{(1)} - \beta_{\text{sh}}^{(2)} = \varepsilon_p - \varepsilon_r = -\psi_p - \cdots - \psi_r - \psi_{r+1}.
$$

There are quite a few Borel subalgebras $\mathfrak{b}$ of $\mathfrak{g}$ that contain the fixed $\mathfrak{b}_{\mathfrak{k}}^{\text{ref}}$. To see this we just compute the relevant Weyl groups. The Weyl group of Σ_θ is $W^\theta \cong$

$W(\mathfrak{so}(2r+1)) \cong \mathfrak{S}_r \ltimes \mathbb{Z}_2^r$. Identify $\mathfrak{t}$ with all r-tuples $(\varepsilon_1, \dots, \varepsilon_r)$. The action $W^\theta \times \mathfrak{t} \to \mathfrak{t}$ is given as follows. The permutation group component $\mathfrak{S}_r$ permutes the entries $\varepsilon_1, \dots, \varepsilon_r$, and $\mathbb{Z}_2^r$ acts on $(\varepsilon_1, \dots, \varepsilon_r)$ by sign changes.

For the other Weyl group, decompose $\mathfrak{k} = \mathfrak{k}_{(1)} \oplus \mathfrak{k}_{(2)}$. Then

$$W_{\mathfrak{k}} = W_{\mathfrak{k}(1)} \times W_{\mathfrak{k}(2)} = (\mathfrak{S}_p \ltimes \mathbb{Z}_2^p) \times (\mathfrak{S}_q \ltimes \mathbb{Z}_2^q),$$

and therefore $\frac{|W^\theta|}{|W_{\mathfrak{k}}|} = \frac{r!}{p!q!}$. Thus there are $\frac{r!}{p!q!}$ θ-stable Borel subalgebras $\mathfrak{b}$ of $\mathfrak{g}$ that contain the given $\mathfrak{b}_{\mathfrak{k}}$.

Each such $\mathfrak{b}$ gives rise to a base cycle $C = C_{\mathfrak{b}}$ in G/B. In order to compute $T_{[C]}\mathcal{C}$, we need to determine the coefficients in the structure equations with respect to all root bases given by the various $\mathfrak{b}$. As explained in the introduction to this chapter, we use the subset $W_1^\theta \subset W^\theta$ defined in (18.3.8).

Thus our next goal is to describe $W_1^\theta = \{w \in W^\theta \mid w \cdot \mathfrak{b}^{\mathrm{ref}} \supset \mathfrak{b}_{\mathfrak{k}}^{\mathrm{ref}}\}$. Before going into this, we should recall that the elements $w \in W_1^\theta$ are defined by the property that $w\mathfrak{b} \cap \mathfrak{k} = b_{\mathfrak{k}}^{\mathrm{ref}}$. It can happen, however, that $w \cdot \mathfrak{b}_{\mathfrak{k}}^{\mathrm{ref}} \neq \mathfrak{b}_{\mathfrak{k}}^{\mathrm{ref}}$.

Borel subalgebras $\mathfrak{b}_{\mathfrak{k}}^{\mathrm{ref}} = \mathfrak{b}_{(1)} \times \mathfrak{b}_{(2)}$, and $\mathfrak{b}$ determine the Weyl chambers

$$\mathcal{C}^+(\mathfrak{b}_{\mathfrak{k}}^{\mathrm{ref}}) = \{\xi \in \mathfrak{t}_{\mathbb{R}} \mid \beta(\xi) < 0 \quad \forall \beta \in \Sigma(\mathfrak{b}_{\mathfrak{k}}^{\mathrm{ref}}, \mathfrak{t})\}$$

and

$$\mathcal{C}^+(\mathfrak{b}) = \{\xi \in \mathfrak{t}_{\mathbb{R}} \mid \gamma(\xi) < 0 \quad \forall \gamma \in \Sigma(\mathfrak{b}, \mathfrak{t})\}.$$

Note that $\mathfrak{b} \supset \mathfrak{b}_{\mathfrak{k}}^{\mathrm{ref}}$ if and only if $\mathcal{C}^+(\mathfrak{b}) \subset \mathcal{C}^+(\mathfrak{b}_{\mathfrak{k}}^{\mathrm{ref}})$. Now $\mathfrak{t}_{\mathbb{R}}$ is identified with the real diagonal matrices as in (19.4.2). The two Weyl chambers consist of all $D := \mathrm{Diag}(\delta_1, \dots, \delta_r, 0)$ whose diagonal entries satisfy appropriate inequalities,

(19.4.4) $$\mathcal{C}^+(\mathfrak{b}_{\mathfrak{k}}^{\mathrm{ref}}) = \{D \mid \delta_1 > \cdots > \delta_p > 0 \quad \text{and} \quad \delta_{p+1} > \cdots > \delta_r > 0\}$$

and

(19.4.5) $$\mathcal{C}^+(\mathfrak{b}^{\mathrm{ref}}) = \{D \mid \delta_1 > \cdots > \delta_p > \delta_{p+1} > \cdots > \delta_{p+q} > 0\}.$$

We now search for those elements $w \in W^\theta$ such that $w \cdot \mathcal{C}^+(\mathfrak{b}^{\mathrm{ref}}) \subset \mathcal{C}^+(\mathfrak{b}_{\mathfrak{k}}^{\mathrm{ref}})$, in other words for diagonal matrices D whose entries satisfy (19.4.5) and such that the entries of $w \cdot D$ satisfy (19.4.4). We view $W^\theta \cong \mathfrak{S}_r \ltimes \mathbb{Z}_2^r$ as acting on the diagonal entries $\varepsilon_1, \dots, \varepsilon_r$ of $\mathrm{Diag}(\varepsilon_1, \dots, \varepsilon_r, 0)$. We claim that the elements from W_1^θ can be chosen from $\mathfrak{S}_r$. To see this, select elements $j_1 < j_2 < \cdots < j_p$ from $\{1, \dots, r\}$, let $\boldsymbol{j} := \{j_1, \dots, j_p\}$, and order the complementary set $\{1, \dots, r\} \setminus \boldsymbol{j}$ in increasing order, i.e., as $j_{p+1} < \cdots < j_{p+q}$. Define $w_{\boldsymbol{j}} \in \mathfrak{S}_r \subset W^\theta$ acting by permutation on the index set $\{1, \dots, r\}$ by

$$w_{\boldsymbol{j}}^{-1}(1) = j_1, \; w_{\boldsymbol{j}}^{-1}(2) = j_2, \; \dots, \; w_{\boldsymbol{j}}^{-1}(r) = j_r.$$

We claim that as $\boldsymbol{j} := \{j_1, \dots, j_p\}$ runs through all p-element subsets of $\{1, \dots, r\}$, the $w_{\boldsymbol{j}}$ exhaust W_1^θ. For this, note that the action of the $w_{\boldsymbol{j}}$ on the torus $\mathfrak{t} \cong \{\mathrm{Diag}(\varepsilon_1, \dots, \varepsilon_r, 0)\} \cong \{(\varepsilon_1, \dots, \varepsilon_r)\}$ is given by

(19.4.6) $$w_{\boldsymbol{j}}(\varepsilon_1, \dots, \varepsilon_r) = (\varepsilon_{w^{-1}(1)}, \dots, \varepsilon_{w^{-1}(r)}) = (\varepsilon_{j_1}, \dots, \varepsilon_{j_r}).$$

Given $\delta_1, \ldots, \delta_r$ which are subject to (19.4.5), for every $\boldsymbol{j}$ the components of $w_{\boldsymbol{j}}(\delta_1, \ldots, \delta_r)$ are subject to (19.4.4). Finally note that for any linear functional $\nu : \mathfrak{t} \to \mathbb{C}$, expressed as a linear combination of the ε_j, we have $w_{\boldsymbol{j}}^{-1}\nu(\varepsilon_1, \ldots, \varepsilon_r) = \nu(w_{\boldsymbol{j}}(\varepsilon_1, \ldots, \varepsilon_r)) = \nu(\varepsilon_1, \ldots, \varepsilon_r)$.

Having an explicit description of the elements in $W_1^\theta \subset W^\theta$, we apply them to $\mu, \lambda = \lambda(\mu), \beta^\star$ to compute $w_{\boldsymbol{j}}^{-1}\mu, w_{\boldsymbol{j}}^{-1}\lambda$. Given $\boldsymbol{j} \subset \{1, \ldots, r\}$ with $|\boldsymbol{j}| = p$, we obtain

(19.4.7)
$$
\begin{aligned}
w_{\boldsymbol{j}}^{-1}\mu &= \varepsilon_{j_1} - \varepsilon_{j_r} = \begin{cases} \psi_{j_1} + \psi_{j_1+1} + \cdots + \psi_{j_r-1}|_{\mathfrak{t}} & \text{if } j_1 < j_r, \\ -\psi_q|_{\mathfrak{t}} & \text{if } j_1 > j_r, \end{cases} \\
w_{\boldsymbol{j}}^{-1}\beta^\star &= \varepsilon_{j_r} = \psi_{j_r} + \psi_{j_r+1} + \cdots + \psi_{r-1} + \psi_r|_{\mathfrak{t}} \\
&= \psi_{j_r} + \psi_{j_r+1} + \cdots + \psi_{r-1} + \psi_{r+1}|_{\mathfrak{t}}, \\
w_{\boldsymbol{j}}^{-1}\lambda &= \varepsilon_{j_1} = \psi_{j_1} + \psi_{j_1+1} + \cdots + \psi_{r-1} + \psi_r|_{\mathfrak{t}} \\
&= \psi_{j_1} + \psi_{j_1+1} + \cdots + \psi_{r-1} + \psi_{r+1}|_{\mathfrak{t}}, \\
w_{\boldsymbol{j}}^{-1}\widetilde{\mu} &= \varepsilon_{j_{p+1}} - \varepsilon_{j_p} = \begin{cases} -\psi_p & \text{if } j_{p+1} > j_p, \\ \psi_{j_{p+1}} + \psi_{j_{p+1}+1} + \cdots + \psi_{j_p-1} & \text{if } j_{p+1} < j_p, \end{cases} \\
w_{\boldsymbol{j}}^{-1}\widetilde{\beta}^\star &= \varepsilon_{j_p} = \psi_{j_p} + \psi_{j_p+1} + \cdots + \psi_{r-1} + \psi_r|_{\mathfrak{t}} \\
&= \psi_{j_p} + \psi_{j_p+1} + \cdots + \psi_{r-1} + \psi_{r+1}|_{\mathfrak{t}}, \\
w_{\boldsymbol{j}}^{-1}\widetilde{\lambda} &= \varepsilon_{j_{p+1}} = \psi_{j_{p+1}} + \psi_{j_{p+1}+1} + \cdots + \psi_{r-1} + \psi_r|_{\mathfrak{t}} \\
&= \psi_{j_{p+1}} + \psi_{j_{p+1}+1} + \cdots + \psi_{r-1} + \psi_{r+1}|_{\mathfrak{t}}.
\end{aligned}
$$

Since we know the coefficients in the corresponding structure equations explicitly, the conditions of the Cohomological Lemma can be now checked for every $C \subset Z$.

Recall that given $\nu \in \Sigma(\mathfrak{g}, \mathfrak{t})$ we write $\nu^\dagger$ (and possibly also $\nu^\ddagger$) for the roots in $\Sigma(\mathfrak{g}, \mathfrak{h})$ such that $\nu^\dagger|_{\mathfrak{t}} = \nu = \nu^\ddagger|_{\mathfrak{t}}$. In the statement of the following theorem we use several times expressions of the form "$\mathrm{supp}(w^{-1}(\nu)) \subset \Phi$." They are abbreviations of the following relations:

- If $\nu \in \{\mu, \widetilde{\mu}, \lambda, \widetilde{\lambda}\}$, then "$\mathrm{supp}(w^{-1}(\nu)) \subset \Phi$" means that the support of at least one $(w^{-1}\nu)^\dagger$ or $(w^{-1}\nu)^\ddagger$ is a subset of Φ. According to our convention $\mathrm{supp}(\sum n_\psi \psi) = \{\psi \in \Psi \mid n_\psi > 0\}$.
- For $\nu = \beta^\star$, "$\mathrm{supp}(w^{-1}(\nu)) \subset \Phi$" means that $\mathrm{supp}(w^{-1}\beta^\star)^\dagger \cup \mathrm{supp}(w^{-1}\beta^\star)^\ddagger$ is contained in Φ. Similarly, for $\nu = \widetilde{\beta}^\star$, "$\mathrm{supp}(w^{-1}(\nu)) \subset \Phi$" means that $\mathrm{supp}(w^{-1}\widetilde{\beta}^\star)^\dagger \cup \mathrm{supp}(w^{-1}\widetilde{\beta}^\star)^\ddagger$ is contained in Φ.

In this sense, if $\nu \in \{\mu, \widetilde{\mu}, \lambda, \widetilde{\lambda}\}$, then "$\mathrm{supp}(w^{-1}(\nu)) \not\subset \Phi$" means that neither of the supports $\mathrm{supp}(w^{-1}\nu)^\dagger$, $\mathrm{supp}(w^{-1}\nu)^\ddagger$ is a subset of Φ. Similarly, "$\mathrm{supp}(w^{-1}(\beta^\star)) \not\subset \Phi$" means $\mathrm{supp}(w^{-1}(\beta^\star))^\dagger \cup \mathrm{supp}(w^{-1}(\beta^\star))^\ddagger \not\subset \Phi$.

Theorem 19.4.8. *In the series for $\mathfrak{g}_0 = \mathfrak{so}(2p+1, 2q+1)$, the cycle space $\mathcal{C}(Z)$ is smooth at $[C]$ for every base cycle C and every flag Z. The tangent space of the cycle space $\mathcal{C}(Z)$ at various $[C]$ is given as follows:*

1. *In the full flag manifold $Z = G/B$ there exist two base cycles $C_{(1)}, C_{(2)}$ with the following property. For every closed K-orbit C,*

$$T_{[C]}\mathcal{C}(X) = \begin{cases} T_{[C]}(G\cdot[C]) \oplus V^{\text{std}}_{SO(2p+1)} & \text{if } C = C_{(1)}, \\ T_{[C]}(G\cdot[C]) \oplus V^{\text{std}}_{SO(2q+1)} & \text{if } C = C_{(2)}, \\ T_{[C]}(G\cdot[C]) & \text{if } C \neq C_{(j)},\ j = 1, 2. \end{cases}$$

2. *Let* $\mathfrak{q} = \mathfrak{q}_\Phi$ *and* $Z = G/Q$*, where* $\Phi \subset \Psi$*. If* $\boldsymbol{j} = \{j_1, \dots, j_p\} \subset \{1, \dots, r\}$ *with* $j_1 < \cdots < j_p$*, and if and* $w_{\boldsymbol{j}} \in W^\theta_1$ *as in* (19.4.6)*, define the base cycle* $C(w_{\boldsymbol{j}}) := K\cdot[w_{\boldsymbol{j}}B^{\text{ref}}] \subset G/B^{\text{ref}}$*. Let* $\pi : G/B \to G/Q$ *be the canonical projection and set* $C := \pi(C(w_{\boldsymbol{j}}))$*. Then*

$$T_{[C]}\mathcal{C}(Z) = T_{[C]}(G\cdot[C]) \oplus \bigoplus_{\lambda \in \mathcal{M}(C,Z)} V^\lambda,$$

where $M(C,Z) \subset \{\lambda^{\text{std}}_{SO(2p+1)}, \lambda^{\text{std}}_{SO(2q+1)}\}$*, and* $M(C,Z)$ *is given as follows:*

(2a) $M(C,Z) = \emptyset$

(i) *if* $\text{supp}(w_{\boldsymbol{j}}^{-1}\mu) \not\subset \Phi$ *and* $\text{supp}(w_{\boldsymbol{j}}^{-1}\widetilde{\mu}) \not\subset \Phi$,

(ii) *or if* $\begin{cases} \textit{either } \text{supp}(w_{\boldsymbol{j}}^{-1}\beta^\star) \textit{ or } \text{supp}(w_{\boldsymbol{j}}^{-1}\lambda(\mu)) \subset \Phi \\ \textit{and} \\ \textit{either } \text{supp}(w_{\boldsymbol{j}}^{-1}(\widetilde{\beta}^\star)) \textit{ or } \text{supp}(w_{\boldsymbol{j}}^{-1}\lambda(\widetilde{\mu})) \subset \Phi \end{cases}$

(iii) *or if* $\begin{cases} \text{supp}(w_{\boldsymbol{j}}^{-1}\mu) \not\subset \Phi \\ \textit{and} \\ \textit{either } \text{supp}(w_{\boldsymbol{j}}^{-1}(\widetilde{\beta}^\star)) \textit{ or } \text{supp}(w_{\boldsymbol{j}}^{-1}\lambda(\widetilde{\mu})) \subset \Phi \end{cases}$

(iv) *or if* $\begin{cases} \text{supp}(w_{\boldsymbol{j}}^{-1}\beta^\star) \textit{ or } \text{supp}(w_{\boldsymbol{j}}^{-1}\lambda(\mu)) \subset \Phi \\ \textit{and} \\ \text{supp}(w_{\boldsymbol{j}}^{-1}\widetilde{\mu}) \not\subset \Phi. \end{cases}$

(2b) $M(C,Z) = \{\lambda^{\text{std}}_{SO(2p+1)}\}$ *if*

$$\begin{Bmatrix} \text{supp}(w_{\boldsymbol{j}}^{-1}\mu) \subset \Phi \textit{ and} \\ \text{supp}(w_{\boldsymbol{j}}^{-1}\beta^\star) \not\subset \Phi \textit{ and} \\ \text{supp}(w_{\boldsymbol{j}}^{-1}\lambda(\mu)) \not\subset \Phi \end{Bmatrix} \quad \textit{and} \quad \begin{Bmatrix} \text{supp}(w_{\boldsymbol{j}}^{-1}\widetilde{\mu}) \not\subset \Phi \textit{ or} \\ \text{supp}(w_{\boldsymbol{j}}^{-1}(\widetilde{\beta}^\star)) \subset \Phi \textit{ or} \\ \text{supp}(w_{\boldsymbol{j}}^{-1}\lambda(\widetilde{\mu})) \subset \Phi \end{Bmatrix}.$$

(2c) $M(C,Z) = \{\lambda^{\text{std}}_{SO(2q+1)}\}$ *if*

$$\begin{Bmatrix} \text{supp}(w_{\boldsymbol{j}}^{-1}\mu) \not\subset \Phi \textit{ or} \\ \text{supp}(w_{\boldsymbol{j}}^{-1}\beta^\star) \subset \Phi \textit{ or} \\ \text{supp}(w_{\boldsymbol{j}}^{-1}\lambda(\mu)) \subset \Phi \end{Bmatrix} \quad \textit{and} \quad \begin{Bmatrix} \text{supp}(w_{\boldsymbol{j}}^{-1}\widetilde{\mu}) \subset \Phi \textit{ and} \\ \text{supp}(w_{\boldsymbol{j}}^{-1}\widetilde{\beta}^\star) \not\subset \Phi \textit{ and} \\ \text{supp}(w_{\boldsymbol{j}}^{-1}\lambda(\widetilde{\mu})) \not\subset \Phi \end{Bmatrix}.$$

(2d) $M(C,Z) = \{\lambda^{\text{std}}_{SO(2p+1)}, \lambda^{\text{std}}_{SO(2q+1)}\}$ *if*

$$\begin{Bmatrix} \text{supp}(w_{\boldsymbol{j}}^{-1}\mu) \subset \Phi \textit{ and} \\ \text{supp}(w_{\boldsymbol{j}}^{-1}\beta^\star) \not\subset \Phi \textit{ and} \\ \text{supp}(w_{\boldsymbol{j}}^{-1}\lambda(\mu)) \not\subset \Phi \end{Bmatrix} \quad \textit{and} \quad \begin{Bmatrix} \text{supp}(w_{\boldsymbol{j}}^{-1}\widetilde{\mu}) \subset \Phi \textit{ and} \\ \text{supp}(w_{\boldsymbol{j}}^{-1}\widetilde{\beta}^\star)) \not\subset \Phi \textit{ and} \\ \text{supp}(w_{\boldsymbol{j}}^{-1}\lambda(\widetilde{\mu})) \not\subset \Phi. \end{Bmatrix}.$$

Proof. The decomposition of the Zariski tangent spaces as $\mathfrak{k}$-module given above is obtained as a direct application of Lemma 18.5.5 and the Cohomological Lemma together with the information given by the structure equations (19.4.7).

It remains only to prove the smoothness assertion. By Proposition 17.5.1 it suffices to show that $H^2(C; \mathcal{O}(\mathbb{E}((\mathfrak{q}+\theta\mathfrak{q})_{\mathfrak{s}}))) = 0$. Let $\mathfrak{t} \subset \mathfrak{b}_{\mathfrak{k}}^{\mathrm{ref}} \subset \mathfrak{b} \subset \mathfrak{q}$ be the Lie algebra data associated to $C \subset Z$, let $\mathfrak{q}_{\mathfrak{k}} = \mathfrak{q}_{\mathfrak{k}}^{r} \ltimes \mathfrak{q}_{\mathfrak{k}}^{-n}$ be the decomposition of $\mathfrak{q}_{\mathfrak{k}} = \mathfrak{q} \cap \mathfrak{k}$ into the reductive and nilpotent parts, and let

$$0 = F^0((\mathfrak{q}+\theta\mathfrak{q})_{\mathfrak{s}}) \subset F^1((\mathfrak{q}+\theta\mathfrak{q})_{\mathfrak{s}}) \subset \cdots =: F^0 \subset F^1 \subset \cdots$$

be the canonical filtration of $(\mathfrak{q}+\theta\mathfrak{q})_{\mathfrak{s}}$ as in (17.6.1). Since

$$\mu_{\mathrm{II}} = -\beta_{\mathrm{sh}}^{(1)} - \beta_{\mathrm{sh}}^{(2)} \in M(\mathfrak{t}, \mathfrak{s}_{\mathfrak{s}}) \subset M(\mathfrak{t}, (\mathfrak{q}+\theta\mathfrak{q})_{\mathfrak{s}}),$$

there exists precisely one k such that $\mu_{\mathrm{II}} \in M(\mathfrak{t}, F^k/F^{k-1})$. Now there is only one such Bott-regular weight of index 2. Hence the Bott–Borel–Weil Theorem 1.6.8 implies that $H^2(C; \mathcal{O}(\mathbb{F}^j/\mathbb{F}^{j-1}))$ vanishes for all j except possibly for $j = k$. Thus we have the exact $\mathfrak{k}$-equivariant sequences.

$$H^2(C; \mathcal{O}(\mathbb{F}^j)) \longrightarrow H^2(C; \mathcal{O}(\mathbb{F}^{j+1})) \longrightarrow 0 \quad \text{for } j \neq k-1, \quad \text{and}$$

$$H^1(C; \mathcal{O}(\mathbb{F}^{k+1})) \longrightarrow H^1(C; \mathcal{O}(\mathbb{F}^{k+1}/\mathbb{F}^k)) \xrightarrow{\nu} H^2(C; \mathcal{O}(\mathbb{F}^k)) \xrightarrow{\eta} H^2(C; \mathcal{O}(\mathbb{F}^{k+1})) \longrightarrow 0.$$

Consequently, if neither $+\beta_{\mathrm{sh}}^{(1)}$ nor $+\beta_{\mathrm{sh}}^{(2)}$ belongs to $M(\mathfrak{t}, (\mathfrak{q}+\theta\mathfrak{q})_{\mathfrak{s}})$, then $H^2(C; \mathcal{O}(\mathbb{F}^k)) = H^2(C; \mathcal{O}(\mathbb{F}^k/\mathbb{F}^{k-1})) = \mathbb{C}$. If one of these weights is in $M(\mathfrak{t}, \mathfrak{s})$, then μ_{II} is not $(\mathfrak{q}_{\mathfrak{s}} \cap \mathfrak{b}_{\mathfrak{k}}^{+})$-highest and $H^{(}C; \mathcal{O}(\mathbb{F}^k))$ vanishes. Thus in the latter case $H^2(C; \mathcal{O}(\mathbb{F}^k)) = 0$ for all j.

It remains to study the case $\dim H^2(C; \mathcal{O}(\mathbb{F}^k)) = 1$. Since $\mathfrak{b}$ is θ-stable, we have $-\beta_{\mathrm{sh}}^{(1)}, -\beta_{\mathrm{sh}}^{(2)} \in M(\mathfrak{t}, \mathfrak{b}_{\mathfrak{s}})$. Hence,$-\beta_{\mathrm{sh}}^{(1)}, -\beta_{\mathrm{sh}}^{(2)} \in M(\mathfrak{t}, F^{k+1}/F^k)$.

Inspecting the matrix realization (19.4.1), we see that for short roots $\alpha \in \Sigma(\mathfrak{k}, \mathfrak{t})$ we have $\dim \mathfrak{g}^{\alpha} = 2$, so $\alpha \in M(\mathfrak{t}, \mathfrak{s})$. Consequently, $-\Psi_{\mathfrak{k}} \cap M(\mathfrak{t}, \mathfrak{s}) = \{-\beta_{\mathrm{sh}}^{(1)}, -\beta_{\mathrm{sh}}^{(2)}\}$. In our situation neither of the roots $\beta_{\mathrm{sh}}^{(1)}, \beta_{\mathrm{sh}}^{(2)}$ is assumed to be in $M(\mathfrak{t}, (\mathfrak{q}+\theta\mathfrak{q})_{\mathfrak{s}})$, i.e., $-\beta_{\mathrm{sh}}^{(1)} - \beta_{\mathrm{sh}}^{(2)}$ is $(\mathfrak{q}_{\mathfrak{k}} \cap \mathfrak{b}_{\mathfrak{k}}^{+})$-highest. Again using the Bott–Borel–Weil Theorem, the isotypic component $H^1(C; \mathcal{O}(\mathbb{F}^{k+1}/\mathbb{F}^k))_0$ of the trivial $\mathfrak{k}$-representation is two-dimensional. By (18.4.11) we need only to discuss the trivial submodules.

Our next goal is to show that the map η in the above exact sequence is the zero map. This implies that $H^2(C; \mathcal{O}(\mathbb{F}^{k+1})) = 0$, so $\dim H^2(C; \mathcal{O}(\mathbb{F}^{j+1})) = 0$ for all $j \geqq k$. Since $\dim H^1(C; \mathcal{O}(\mathbb{F}^{k+1}/\mathbb{F}^k)) = 2$, the following statement completes the proof of smoothness of the cycle space at every $C \subset G/Q$.

Claim. The isotypic component $H^1(C; \mathcal{O}(\mathbb{F}^{k+1}((\mathfrak{q}+\theta\mathfrak{q})_{\mathfrak{s}})))_0$ is at most one-dimensional.

Proof of the Claim. According to Theorem 1.6.11, it is sufficient to compute the relative Lie algebra cohomology $H^1(\mathfrak{q}_\mathfrak{k}, \mathfrak{q}_\mathfrak{k}^r; F^{k+1}((\mathfrak{q}+\theta\mathfrak{q})_\mathfrak{s}))$. We show here that the cocycle space $Z^1(\mathfrak{q}_\mathfrak{k}, \mathfrak{q}_\mathfrak{k}^r; F^{k+1}((\mathfrak{q}+\theta\mathfrak{q})_\mathfrak{s}))$ is at most one-dimensional. By definition,

$$Z^1 \subset \mathrm{Hom}_{\mathfrak{q}_\mathfrak{k}^r}(\mathfrak{q}_\mathfrak{k}^{-n}, \quad F^{k+1}(\mathfrak{q}+\theta\mathfrak{q})_\mathfrak{s}) \subset \mathrm{Hom}_\mathfrak{t}(\mathfrak{q}_\mathfrak{k}^{-n}, \quad F^{k+1}(\mathfrak{q}+\theta\mathfrak{q})_\mathfrak{s}).$$

Since $\mathfrak{q}_\mathfrak{k}^{-n}$ and $F^{k+1}(\mathfrak{q}+\theta\mathfrak{q})_\mathfrak{s}$ are direct sums of one-dimensional $\mathfrak{t}$-eigenspaces, every element c in $\mathrm{Hom}_\mathfrak{t}(\mathfrak{q}_\mathfrak{k}^{-n}, F^{k+1}((\mathfrak{q}+\theta\mathfrak{q})_\mathfrak{s}))$ has the form $\sum c_\varphi I_\varphi$. Here I_φ is a fixed nontrivial element in the space of maps $\mathrm{Hom}(\mathfrak{k}_\varphi, \mathfrak{s}_\varphi)$ between root spaces for $\varphi \in \Sigma(\mathfrak{q}_\mathfrak{k}^{-n}, \mathfrak{t}) \cap M(\mathfrak{t}, F^{k+1})$ and $c = \sum c_\phi I_\phi$ is zero if restricted to the root spaces parameterized by $\Sigma(\mathfrak{q}_\mathfrak{k}^{-n}, \mathfrak{t}) \setminus M(\mathfrak{t}, F^{k+1})$. Recall that only short roots in $\Sigma_\mathfrak{k}$ have two-dimensional $\mathfrak{t}$-root spaces. These are the roots

$$\alpha_1^{(1)} = -\beta_{\mathrm{sh}}^{(1)}, \ \alpha_2^{(1)} = -\beta_{\mathrm{sh}}^{(1)} - \beta_{p-1}^{(1)}, \ldots, \alpha_p^{(1)} = -\beta_{\mathrm{sh}}^{(1)} - \beta_{p-1}^{(1)} - \cdots - \beta_1^{(1)},$$
$$\alpha_1^{(2)} = -\beta_{\mathrm{sh}}^{(2)}, \ \alpha_2^{(2)} = -\beta_{\mathrm{sh}}^{(2)} - \beta_{q-1}^{(2)}, \ \ldots, \alpha_q^{(2)} = -\beta_{\mathrm{sh}}^{(2)} - \beta_{q-1}^{(2)} - \cdots - \beta_1^{(2)}.$$

Let $\Phi_\mathfrak{k} = \Sigma(\mathfrak{q}_\mathfrak{k}, \mathfrak{t}) \cap \Psi_\mathfrak{k}$. By our assumption, $\{\beta_{\mathrm{sh}}^{(1)}, \beta_{\mathrm{sh}}^{(2)}\} \cap \Phi_\mathfrak{k} = \emptyset$. Consequently, all of the short roots $\alpha_\bullet^{(\bullet)}$ listed above belong to $\Sigma(\mathfrak{q}^{-n}, \mathfrak{t})$. In particular, $c = \sum c_j^\ell I_{\alpha_j^{(\ell)}} \in Z^1(\mathfrak{q}_\mathfrak{k}, \mathfrak{q}_\mathfrak{k}^r, F^{k+1}(\mathfrak{q}+\theta\mathfrak{q})_\mathfrak{s})$ for certain constants c_j^ℓ. The key point here is the observation that the cocycle condition

$$dc\,(X, Y) = [X, c(Y)] - [Y, c(X)] - c([X, Y]) = 0 \tag{19.4.9}$$

imposes sufficiently many conditions on the coefficients c_j^ℓ. To see this, observe that

$$[\mathfrak{k}_{\alpha_j^{(\ell)}}, \mathfrak{k}_{-\beta_{j'}^{(\ell)}}] = \mathfrak{k}_{\alpha_{j+1}^{(\ell)}} \quad \text{for} \quad j' = \begin{cases} p-j & \text{if } \ell = 1 \\ q-j & \text{if } \ell = 2 \end{cases}$$

and that $c(\mathfrak{k}_{-\beta_{j'}^{(\ell)}}) = 0$, because $\beta_{j'}^{(\bullet)}$ are long for all j'. Consequently Lemma 19.4.9 implies that $c_1^{(1)}$ determines $c_2^{(1)}, \ldots, c_p^{(1)}$ and $c_1^{(2)}$ determines $c_2^{(2)}, \ldots, c_q^{(2)}$. Finally, since $[\mathfrak{k}_{\alpha_1^{(1)}}, \mathfrak{s}_{\alpha_1^{(2)}}] \neq 0$ (we have assumed

$$\alpha_1^{(1)} + \alpha_1^{(2)} \in M(\mathfrak{t}, F^k) \subset M(\mathfrak{t}, F^{k+1})),$$

we see that the condition $\partial c = 0$ also imposes a relation between $c_1^{(1)}$ and $c_1^{(2)}$ by plugging in nontrivial $\xi \in \mathfrak{k}_{\alpha_1^{(1)}}$ and $\zeta \in \mathfrak{k}_{\alpha_1^{(2)}}$ in the cocycle condition. This proves that $\dim Z^1(\mathfrak{q}_\mathfrak{k}, \mathfrak{q}_\mathfrak{k}^r, F^{k+1}(\mathfrak{q}+\theta\mathfrak{q})_\mathfrak{s}) \leq 1$, and that proves the claim. □

19.5 The case $\mathfrak{g}_0 = \mathfrak{e}_{6,C_4}$

For this particular exceptional real form we determine the structure equations from the geometry and combinatorics of the relevant root systems.

Here we have $\Sigma(\mathfrak{g}, \mathfrak{h}) = \Sigma(\mathfrak{e}_6)$ and $\Sigma(\mathfrak{k}, \mathfrak{t}) = \Sigma(\mathfrak{sp}(4))$. In this case $\Sigma_\theta = \Sigma(\mathfrak{f}_4)$ because (1) Σ_θ has roots of 2 different lengths, (2) Σ_θ is reduced, and (3) Σ_θ contains $\Sigma(\mathfrak{k}, \mathfrak{t})$. Hence, Σ_θ can only be of type F_4 or C_4 (the root system B_4 does not have sufficiently many short roots). Counting dimensions, $\dim \mathfrak{k} = 36$, $\dim \mathfrak{s} = 42$, and using the fact that the dimension of a root space $\mathfrak{g}^\varphi$ for $\varphi \in \Sigma_\theta$ is at most two, it follows that C_4 is not possible.

Next we define $\mathfrak{b}^{\mathrm{ref}}$ from the simple root system

$$\Psi = \left\{ \begin{array}{ccccccccc} & & & & \psi_2 & & & & \\ & & & & | & & & & \\ \psi_1 & — & \psi_3 & — & \psi_4 & — & \psi_5 & — & \psi_6 \end{array} \right\}$$

and we set $\mathfrak{b}_{\mathfrak{k}}^{\mathrm{ref}} := \mathfrak{b}^{\mathrm{ref}} \cap \mathfrak{k}$. The only point is to express the corresponding simple system $\Psi_{\mathfrak{k}} = \{\beta_1 — \beta_2 — \beta_3 \Leftarrow \beta_4\}$ in terms of Ψ. These relations are given as follows.

$$\begin{aligned} \beta_1 &= \psi_2 + \psi_3 + \psi_4|_{\mathfrak{t}} = \psi_2 + \psi_4 + \psi_5|_{\mathfrak{t}}, \\ \beta_2 &= \psi_1|_{\mathfrak{t}} = \psi_6|_{\mathfrak{t}}, \\ \beta_3 &= \psi_3|_{\mathfrak{t}} = \psi_5|_{\mathfrak{t}}, \\ \beta_{\mathrm{lo}} &= \beta_4 = \psi_4. \end{aligned} \tag{19.5.1}$$

These can be deduced from Table 18.2.4 or from the affine Dynkin diagram for E_6 in Table 18.2.9.

Corollary 18.3.6 gives the number of Borel subalgebras $\mathfrak{b} \subset \mathfrak{g}$ that contain the fixed Borel subalgebra $\mathfrak{b}_{\mathfrak{k}}^{\mathrm{ref}}$. In our case it is

$$\left|\frac{W^\theta}{W_{\mathfrak{k}}}\right| = \frac{|W(F_4)|}{|W(C_4)|} = \frac{2^7 \cdot 3^2}{|\mathfrak{S}_4| \cdot |\mathbb{Z}_2^4|} = 3. \tag{19.5.2}$$

A simple check shows that the remaining two Borel subalgebras containing $\mathfrak{b}_{\mathfrak{k}}^{\mathrm{ref}}$ are $\mathfrak{b}' := s_{\psi_2}(\mathfrak{b}^{\mathrm{ref}})$ and $\mathfrak{b}'' := s_{\psi_2+\psi_4} \circ s_{\psi_2}(\mathfrak{b}^{\mathrm{ref}})$.

From Table 18.5.3 we see that $\mu = \xi_2 - \beta_3 = (\beta_1 + 2\beta_2 + 2\beta_3 + \beta_4) - \beta_3$ is the only Bott-regular weight of index 1, and the second part of the String Lemma implies that there are no Bott-regular weights of index 2. This can also checked directly by using the facts that (1) $\mathfrak{k} \times \mathfrak{s} \to \mathfrak{s}$ is the representation $V_{\mathrm{Sp}(4)}^{\xi_4}$, and (2) that representation can be described as the kernel of the contraction map

$$\iota_\omega : \bigwedge^4 V^{\mathrm{std}} \to \bigwedge^2 V^{\mathrm{std}},$$

where $V^{\mathrm{std}} = V^{\xi_1} = \sum_1^4 V_{\varepsilon_j} + \sum_1^4 V_{-\varepsilon_j}$ is the standard representation of $\mathfrak{k} = \mathfrak{isom}(V^{\mathrm{std}}, \omega) \cong \mathfrak{sp}(4, \mathbb{C})$, and V_t is the $\mathfrak{t}$-eigenspace with eigenvalue t. This gives all weights in $M(\mathfrak{t}, \mathfrak{s})$ in terms of linear combinations of $\varepsilon_1, \varepsilon_2, \varepsilon_3, \varepsilon_4$.

We have collected all of the information which is needed to compute the coefficients of the three systems of structure equations. In fact, we need only to write μ as a linear combination of elements in Ψ, Ψ', and Ψ''

$$
\begin{aligned}
\mu &= \psi_1 + \psi_2 + \psi_3 + 2\psi_4 + \psi_5 + \psi_6|_{\mathfrak{t}} \\
&= \psi_1' + \psi_2' + \psi_3' + 2\psi_4' + \psi_5' + \psi_6'|_{\mathfrak{t}} \\
&= \psi_1'' + \psi_2'' + \psi_3'' + \psi_4'' + \psi_5'' + \psi_6''|_{\mathfrak{t}}.
\end{aligned} \tag{19.5.3}
$$

Since $\operatorname{supp}_{\Psi^\bullet}(\mu^\dagger)$ coincides with $\Psi^\bullet$ in all three cases, μ never belongs to $M(\mathfrak{t}, (\mathfrak{q}+\theta\mathfrak{q})_{\mathfrak{s}})$ for any proper parabolic subalgebra $\mathfrak{q} \subset \mathfrak{g}$. The following theorem is the consequence of this fact, because we have previously shown that for all other real forms of $\mathfrak{e}_6(\mathbb{C})$ the relevant component of the cycle space is just the closure of the G-orbit of the base cycle (see Theorem 18.4.13).

Theorem 19.5.4. *Let G be the complex exceptional group E_6. Then for all flags $Z = G/Q$, all real forms $\mathfrak{g}_0$ of $\mathfrak{g}$, and all base cycles $C \subset Z$, we have $M(C, Z) = \emptyset$, in other words $T_{[C]}\mathcal{C}(Z) = T_{[C]}(G\cdot[C])$.*

19.6 Preliminaries for the cases where $\mathfrak{g}_0$ is of complex type

In the rest of this chapter we deal with the cases where the real form $\mathfrak{g}_0$ has a complex structure J such that $(\mathfrak{g}_0, J)$ is simple complex Lie algebra. We refer to such a $\mathfrak{g}_0$ as a real form of **complex type**. In this case the complexification $\mathfrak{g} = (\mathfrak{g}_0)^{\mathbb{C}}$ is the direct sum of two simple ideals. We briefly recall the details of this situation.

Let $\mathfrak{l}$ denote a simple complex Lie algebra of which $\mathfrak{g}_0$ is the underlying real structure. Its complexification has canonical decomposition $\mathfrak{l}^{\mathbb{C}} = \mathfrak{l}^{1,0} \oplus \mathfrak{l}^{0,1}$ into the $\pm i$ eigenspaces of $J^{\mathbb{C}}$. These are the two ideals in $\mathfrak{l}^{\mathbb{C}}$. The projection $\pi^{1,0} : \mathfrak{l} \to \mathfrak{l}^{1,0}$ is a $\mathbb{C}$-linear isomorphism, but there is no canonical $\mathbb{C}$-linear isomorphism between $\mathfrak{l}^{0,1}$ and $\mathfrak{l}$ unless we select an antiholomorphic involution in $\mathfrak{l}$.

In this case the Cartan involution $\theta : \mathfrak{l} \to \mathfrak{l}$ is J-antilinear, and we may identify $\mathfrak{l}^{\mathbb{C}}$ with $\mathfrak{l} \oplus \mathfrak{l}$ by the $\mathbb{C}$-linear isomorphism

$$
\mathfrak{l} \oplus \mathfrak{l} \to \mathfrak{l}^{\mathbb{C}}, \ (\xi, \zeta) \mapsto (\tfrac{1}{2}(\xi - iJ\xi), \tfrac{1}{2}(\theta\zeta + iJ\theta\zeta)).
$$

In this way $\mathfrak{l}$ is embedded as the antilinear diagonal in $\mathfrak{l}^{\mathbb{C}} = \mathfrak{l} \oplus \mathfrak{l}$ by $\xi \mapsto (\xi, \theta\xi)$, and the $\mathbb{C}$-linear extension $\theta : \mathfrak{l} \oplus \mathfrak{l} \to \mathfrak{l} \oplus \mathfrak{l}$ is given by reversing the components, $(\xi, \zeta) \to (\zeta, \xi)$. In particular, $\mathfrak{g} = \mathfrak{k} + \mathfrak{s}$, where $\mathfrak{k}$ is the diagonal in $\mathfrak{l} \oplus \mathfrak{l}$ and $\mathfrak{s} = \{(\xi, -\xi) : \xi \in \mathfrak{l}\}$.

Now select a Cartan subalgebra $\mathfrak{h}_{\mathfrak{l}} \subset \mathfrak{l}$ and consider the Cartan subalgebra $\mathfrak{h} := \mathfrak{h}_{\mathfrak{l}} \oplus \mathfrak{h}_{\mathfrak{l}}$ in $\mathfrak{l} \oplus \mathfrak{l}$. According to the above identifications the maximal torus $\mathfrak{t}$ in $\mathfrak{k}$ is the diagonally embedded $\mathfrak{h}_{\mathfrak{l}}$ in its centralizer in $\mathfrak{g}$, and the latter is the direct sum $\mathfrak{h}_{\mathfrak{l}} \oplus \mathfrak{h}_{\mathfrak{l}}$. Let $\mathrm{res} : \mathfrak{h}^* \oplus \mathfrak{h}^* \to \mathfrak{h}^*$ be the restriction map. Then $\Sigma(\mathfrak{k}, \mathfrak{t}) = \Sigma(\mathfrak{g}, \mathfrak{t}) = M(\mathfrak{t}, \mathfrak{s}) \setminus \{0\}$. Finally, select a Borel subalgebra $\mathfrak{b}_{\mathfrak{l}} \subset \mathfrak{l}$ which we identify with $\mathfrak{b}_{\mathfrak{k}}$ by the diagonal embedding in $\mathfrak{l} \oplus \mathfrak{l}$. It follows that the sum $\mathfrak{b} := \mathfrak{b}_{\mathfrak{l}} \oplus \mathfrak{b}_{\mathfrak{l}}$ is a θ-stable Borel subalgebra in $\mathfrak{g}$. In fact it is the only one which contains $\mathfrak{b}_{\mathfrak{k}}$. This is equivalent to the fact that there is only one closed K-orbit C in any flag manifold $G/Q = L/Q_1 \times L/Q_2$, when $\mathfrak{g}_0$ is a real form of complex type. Another coincidence, which considerably simplifies all computations, is that the isotropy representation $\mathfrak{k} \times \mathfrak{s} \to \mathfrak{s}$ coincides with the adjoint representation of $\mathfrak{k} = \mathfrak{l}$ on itself.

As already mentioned, every parabolic subalgebra $\mathfrak{q}$ that contains $\mathfrak{b} = \mathfrak{b}_{\mathfrak{l}} \oplus \mathfrak{b}_{\mathfrak{l}}$ is a direct sum $\mathfrak{q} = \mathfrak{q}_1 \oplus \mathfrak{q}_2$, where $\mathfrak{q}_j$ is a parabolic in $\mathfrak{l}$ that contains $\mathfrak{b}_{\mathfrak{l}}$. The algebra $\mathfrak{q}$ is θ-stable if and only if $\mathfrak{q}_1 = \mathfrak{q}_2$. In any case,

$$
\begin{aligned}
\mathfrak{q} + \theta\mathfrak{q} &= \sum\nolimits_{\alpha\in(\Phi^r_{Q_1}\cup\Phi^r_{Q_2})\cap\Sigma^+} (\mathfrak{l}_\alpha \oplus \mathfrak{l}_\alpha) \quad + \quad \mathfrak{b}_-,\\
\mathfrak{q} \cap \theta\mathfrak{q} &= \sum\nolimits_{\alpha\in(\Phi^r_{Q_1}\cap\Phi^r_{Q_2}\cap\Sigma^+)} (\mathfrak{l}_\alpha \oplus \mathfrak{l}_\alpha) \quad + \quad \mathfrak{b}_-,\\
\mathfrak{q} \cap \mathfrak{k} &= \sum\nolimits_{\alpha\in(\Phi^r_{Q_1}\cap\Phi^r_{Q_2}\cap\Sigma^+)} (\mathfrak{l}_\alpha \oplus \mathfrak{l}_\alpha)^\theta \quad + \quad \mathfrak{b}_{\mathfrak{k}},
\end{aligned}
\tag{19.6.1}
$$

where $\mathfrak{l}_\alpha$ are the root spaces in $\mathfrak{l}$ with respect to $\mathfrak{h}_{\mathfrak{l}}$.

Let Ψ denote the simple roots of $\mathfrak{l}$ and Φ_j the subsets of Ψ which generate $\Phi^r_{Q_j} \subset \Sigma$. The above decompositions allow us to easily verify the conditions of the Cohomological Lemma and the String Lemma.

As already mentioned, the String Lemma implies that the cycle space $\mathcal{C}_{[C]}(G/Q)$ coincides with the closure of the orbit $G{\cdot}[C]$ in all cases where the roots in the root system of $\mathfrak{l} \cong \mathfrak{g}_0$ are of equal lengths. Further, for all real forms of complex type the second part of the String Lemma tells us that there are no Bott-regular weights in $M(\mathfrak{t},\mathfrak{s})$ of index 2. Consequently,

The cycle spaces $\mathcal{C}_{[C]}(Z)$ are smooth for all real forms $\mathfrak{g}_0$ of complex type.

This completes the proof of Theorem 18.6.1. □

In the following we discuss the remaining three families of symmetric pairs in greater detail in order to detect all cases where the cycle space has greater dimension than the G-orbit of the base cycle.

19.7 The series for $\mathfrak{g}_0 = \mathfrak{so}(2r+1;\mathbb{C})$

Let $\beta_1,\ldots,\beta_r = \beta_{\mathrm{sh}}$ be the consecutive simple roots in the Dynkin diagram with respect to $\Sigma^+ = -\Sigma(\mathfrak{b}_{\mathfrak{l}},\mathfrak{h}_{\mathfrak{l}})$ as in Tables 17.4.2 and 17.4.3. As computed in Table 18.5.3, the only Bott-regular element of index 1 in $M(\mathfrak{t},\mathfrak{s})$ is $\mu = \beta_1 + \cdots + \beta_{r-1}$, $\beta^\star = \beta_r = \beta_{\mathrm{sh}}$ and $\lambda(\mu) = \beta_1 + \cdots + \beta_{r-1} + \beta_r = \xi_1$, where ξ_1 denotes the first fundamental weight. The representation with highest weight ξ_1 is the standard representation of the orthogonal group, $V^{\xi_1} = V^{\mathrm{std}}_{SO_{2r+1}}$.

As before let $\Phi_j \subset \Psi$, $j = 1,2$, denote the simple root systems that determine the reductive parts Q^r_j of the parabolic subgroup $Q = Q_1 \times Q_2$. Note that the case when one of the parabolics Q_j coincides with L is uninteresting, because in this case G_0, K and K_0 act transitively on Z and consequently the corresponding component in the cycle space is just a point. Recall that for every flag $Z = G/Q$ and real form of complex type there is the unique base cycle C. The following is therefore an immediate consequence of the Cohomological Lemma.

Theorem 19.7.1. *Let $Z = G/Q = L/Q_1 \times L/Q_2$ be the flag manifold and $\Phi_j \subset \Psi = \{\beta_1, \dots, \beta_{r-1}, \beta_{\text{sh}}\}$ the subsets defining Q_1 and Q_2. Then $T_{[C]}\mathcal{C}(Z) = T_{[C]}(G\cdot[C]) \oplus V^{\text{std}}_{SO_{2r+1}}$ if and only if either $\{\beta_1, \dots, \beta_{r-1}\} = \Psi_1$ and $Q_2 \neq L$ or $\{\beta_1, \dots, \beta_{r-1}\} = \Phi_2$ and $Q_1 \neq L$. In all other cases $T_{[C]}\mathcal{C}(Z) = T_{[C]}(G\cdot[C])$. In other words, the set $M(C, Z)$ as in (18.6.2) is either empty or consists of the single element ξ_1 which is the first fundamental weight of $\mathfrak{so}(2r+1; \mathbb{C})$.*

19.8 The series for $\mathfrak{g}_0 = \mathfrak{sp}(r; \mathbb{C})$

Let $\beta_1, \dots, \beta_r = \beta_{\text{lo}}$ be the consecutive simple roots in the Dynkin diagram with respect to $\Sigma^+ = -\Sigma(\mathfrak{b}_\mathfrak{l}, \mathfrak{h}_\mathfrak{l})$ in the standard notation as in Tables 17.4.2 and 17.4.3. Here $\mu = 2\beta_2 + \cdots + 2\beta_{r-1} + \beta_{\text{lo}}$, $\beta^\star = \beta_1$, and $\lambda(\mu) = \beta_1 + 2\beta_2 + \cdots + 2\beta_{r-1} + \beta_r = \xi_2$, where ξ_2 denotes the second fundamental weight and $V^{\xi_2} = (\bigwedge^2 V^{\text{std}}_{\text{Sp}_r})^{\text{irr}}$. As in the previous case, the theorem below directly follows from the Cohomological Lemma.

Theorem 19.8.1. *Let $Z = G/Q = L/Q_1 \times L/Q_2$ be the flag manifold and $\Phi_j \subset \Psi = \{\beta_1, \dots, \beta_{r-1}, \beta_{\text{lo}}\}$ the subsets defining Q_1 and Q_2. Then $T_{[C]}\mathcal{C}(Z) = T_{[C]}(G\cdot[C]) \oplus V^{\xi_2}$ if and only if either $\{\beta_2, \dots, \beta_r\} = \Phi_1$ and $Q_2 \neq L$ or $\{\beta_2, \dots, \beta_r\} = \Phi_2$ and $Q_1 \neq L$. In all other cases $T_{[C]}\mathcal{C}(Z) = T_{[C]}(G\cdot[C])$ In other words, the set $M(C, Z)$ is either empty or consists of the single element ξ_2 which is the second fundamental weight of $\mathfrak{sp}(r; \mathbb{C})$.*

19.9 The case $\mathfrak{g}_0 = \mathfrak{f}_4(\mathbb{C})$

As in the case we just considered, let $\beta_1, \dots, \beta_4$ be the simple roots of the exceptional Lie algebra $\mathfrak{f}_4$ as in Tables 17.4.2 and 17.4.3. The Bott-regular weight (see Table 18.5.3) of index one is $\mu = \beta_1 + 2\beta_2 + 2\beta_3 + 2\beta_4$. The simple root $\beta^\star$ as in the Cohomological Lemma is ψ_3. The dominant weight in the corresponding $\beta^\star$-string is $\alpha_1 + 2\alpha_2 + 3\alpha_3 + 2\alpha_4$. This is just the fourth fundamental weight ξ_4 of $\mathfrak{f}_4$. Note that the support of μ is the entire simple root system $\Psi_\mathfrak{l}$. It follows that if $M(\mathfrak{t}, (\mathfrak{q} + \theta\mathfrak{q})_\mathfrak{s})$ contains μ, then according to (19.6.1) $Q_1 = L$ or $Q_2 = L$. But in that case $\lambda(\mu) \in M(\mathfrak{t}, (\mathfrak{q} + \theta\mathfrak{q})_\mathfrak{s})$, and the Cohomological Lemma has the following consequence.

Theorem 19.9.1. *Let $Z = G/Q = L/Q_1 \times L/Q_2$ be an arbitrary $F_4 \times F_4$-flag manifold and let C be the unique base cycle. Then $T_{[C]}\mathcal{C}(Z) = T_{[C]}(G\cdot[C])$, and the component of the cycle space containing $[C]$ is the closure of the orbit $G\cdot[C]$, i.e., $M(C, Z) = \emptyset$.*

19.10 The case $\mathfrak{g}_0 = \mathfrak{g}_2(\mathbb{C})$

Since the roots of the exceptional Lie algebra $\mathfrak{g}_2$ do not satisfy the condition (18.4.3), we first have to directly determine the Bott-regular weights of indices one and two.

Since $M(\mathfrak{t}, \mathfrak{s})$ can be identified with the roots of the complex Lie algebra $\mathfrak{g}_2$ together with 0, we directly check the following (i.e., without using the String Lemma) for a given simple root system $\Psi_{\mathfrak{g}_2} = \{\beta_{\text{sh}}, \beta_{\text{lo}}\}$.

- The fundamental weights are $\xi_{\text{sh}} = \beta_{\text{lo}} + 2\beta_{\text{sh}}$ and $\xi_{\text{lo}} = 2\beta_{\text{lo}} + 3\beta_{\text{sh}}$.
- There are no weights of index 2.
- The only Bott-regular weight of index 1 is $\mu = \beta_{\text{lo}}$ and in this case $\lambda(\mu) = \mu + 2\beta_{\text{sh}} = \xi_{\text{sh}}$.

Even though the String Lemma in its present formulation cannot be applied to our situation, the Cohomological Lemma does indeed apply. Let Q_{lo} and Q_{sh} denote the two proper parabolic subgroups of G_2 with $\Phi = \{\beta_{\text{lo}}\}$ and $\Phi = \{\beta_{\text{sh}}\}$. In this notation we have the final result for real forms of complex type.

Theorem 19.10.1. *Let $Z = \mathrm{G}_2/Q_1 \times \mathrm{G}_2/Q_2$ be a $G_2 \times G_2$-flag manifold and $C \subset Z$ the (unique) base cycle. Then $T_{[C]}\mathcal{C}(Z) = T_{[C]}(G\cdot[C]) \oplus V^{\xi_{\text{sh}}}$ if and only if Q is one of the parabolic subgroups $Q_{\text{lo}} \times Q_2$ with $Q_2 \neq \mathrm{G}_2$ or $Q_1 \times Q_{\text{lo}}$ and $Q_1 \neq \mathrm{G}_2$. In all other cases the corresponding component of the cycle space in the closure of the orbit $G\cdot[C]$. In other words, $M(C, Z)$ is empty or consists of the first fundamental weight ξ_1 (compare Table 1.2.3) for $\mathfrak{g}_2(\mathbb{C})$.*

20

Classification for rank $\mathfrak{k}$ = rank $\mathfrak{g}$

In this chapter we investigate the series of real forms $\mathfrak{g}_0 = \mathfrak{k}_0 + \mathfrak{s}_0$ for which rank $\mathfrak{k}$ = rank $\mathfrak{g}$. This means that a Cartan subalgebra $\mathfrak{t}$ of $\mathfrak{k}$ is already a Cartan subalgebra of $\mathfrak{g}$. Consequently, all ad$(\mathfrak{t})$-eigenspaces $\mathfrak{g}_\varphi$, $\varphi \in \Sigma_\theta$, are one-dimensional and θ-stable. In particular, θ stabilizes every parabolic subalgebra $\mathfrak{q} \supset \mathfrak{t}$. Therefore, throughout this chapter we have

$$\mathfrak{t} = \mathfrak{h}, \quad \Sigma_\theta = \Sigma, \qquad \Psi_\theta = \Psi, \quad \text{and} \quad \mathfrak{q} = \mathfrak{q} + \theta\mathfrak{q}.$$

In this equal rank case, simple roots $\beta \in \Psi_{\mathfrak{k}} \subset \Sigma_{\mathfrak{k}}$ do not belong to $M(\mathfrak{t}, \mathfrak{s})$. According to the second part of the String Lemma 18.4.4, this implies that $M(\mathfrak{t}, \mathfrak{s})$ does not contain any Bott-regular weights of index 2. Therefore, an analysis of the second row in the exact sequence (18.5.7) shows that, for any given parabolic $\mathfrak{q} \supset \mathfrak{t}$ and filtration $F^0\mathfrak{q}_{\mathfrak{s}} \subset F^1\mathfrak{q}_{\mathfrak{s}} \subset \cdots$ the cohomology groups $H^2(C; \mathbb{F}^k)$ vanish for all k. Hence, due to Propositions 17.5.1 and 17.5.4, we have proved the following.

Proposition 20.0.1. *Let Z be any flag manifold and $C \subset Z$ a base cycle associated to a real form $\mathfrak{g}_0$ with the property that* rank $\mathfrak{g}$ = rank $\mathfrak{k}$. *Then $H^1(C; \mathbb{N}_Z(C)) = 0$ and consequently $\mathcal{C}(Z)$ is smooth at $[C]$.*

20.1 The series for $\mathfrak{g}_0 = \mathfrak{sp}(r; \mathbb{R})$

The only series of real forms of $\mathfrak{g} \cong \mathfrak{sp}(r; \mathbb{C})$ which admits closed K-orbits C in flag manifolds Z with cycle space component $\mathcal{C}_{[C]}(Z)$ larger than the closure of $G([C])$ are the hermitian real forms $\mathfrak{g}_0 \cong \mathfrak{sp}(r; \mathbb{R})$.

Let us fix the notation in this case. In the one-dimensional center $\mathfrak{z}_0$ of $\mathfrak{k}_0$ there exists an element J_0 which induces a complex structure on $\mathfrak{s}_0$ which can naturally be identified with the tangent space at $1 \cdot K_0$ of G_0/K_0. Then G_0/K_0 is holomorphically equivalent to a a bounded symmetric domain in $(\mathfrak{s}_0, \text{ad}(J_0))$.

Using $\mathfrak{k} = \mathfrak{k}' \oplus \mathfrak{z}_{\mathfrak{k}} = \mathfrak{k}' \oplus \mathbb{C} \cdot J_0$ we write $\chi : \mathfrak{t} \to \mathbb{C}$ for the linear functional given by $\chi(J_0) = i$ and $\chi(\mathfrak{t} \cap \mathfrak{k}') = 0$. The complexified space $\mathfrak{s} = (\mathfrak{s}_0, \text{ad}(J_0))^{\mathbb{C}}$ has the decomposition $\mathfrak{s} = \mathfrak{s}_+ + \mathfrak{s}_-$ where $\mathfrak{s}_\pm$ is the $(\pm i)$-eigenspace of $\mathrm{ad}(J_0)$. Since $\mathfrak{s}_\pm$ are direct sums of root spaces, we have a similar decomposition, $\mathfrak{q}_\mathfrak{s} = \mathfrak{q}_{\mathfrak{s}_+} + \mathfrak{q}_{\mathfrak{s}_-} := (\mathfrak{q} \cap \mathfrak{s}_+) + (\mathfrak{q} \cap \mathfrak{s}_-)$, for every parabolic algebra that contains $\mathfrak{t}$.

As a fine technical point, note that if $C = K(z) \subset Z = G/Q$ is a cycle, then the center Z_K of K acts trivially on it. However, Z_K (or $\mathfrak{z}_\mathfrak{k}$) acts nontrivially on the normal bundle $\mathbb{N}_Z(C) = \mathbb{E}(\mathfrak{q}_{\mathfrak{s}_+}) + \mathbb{E}(\mathfrak{q}_{\mathfrak{s}_-})$ of C. Precisely, the center $\mathfrak{z}_\mathfrak{k}$ acts by $\pm\chi$ on $\mathbb{E}(\mathfrak{q}_{\mathfrak{s}_\pm})$.

Let us give a matrix realization of $\mathfrak{t} \subset \mathfrak{k} \subset \mathfrak{g}$ and express the various root systems and weights in terms of the diagonal entries of $\mathfrak{t}$. To start with, define the skew-symmetric matrix

$$R^a := \begin{pmatrix} & & & & & -1 \\ & & & & 1 & \\ & & & -1 & & \\ & & 1 & & & \\ & \cdot^{\cdot^{\cdot}} & & & & \\ -1 & & & & & \\ 1 & & & & & \end{pmatrix} \in \mathbb{M}_{2r}$$

and the involutive automorphism $\eta : X \mapsto R^a(-X^t)R^a =: {}^\eta X$. The involution η defines the complex Lie algebra $\mathfrak{g} \cong \mathfrak{sp}(r, \mathbb{C})$ as a subalgebra of $\mathbb{M}_{2r}$: $\mathfrak{sp}(r, \mathbb{C}) = \{X \in \mathbb{M}_{2r}(\mathbb{C}) : X = {}^\eta X\}$ and $\mathfrak{g}_0 \cong \mathfrak{sp}(r, \mathbb{R}) = \mathfrak{sp}(r, \mathbb{C}) \cap \mathbb{M}_{2r}(\mathbb{R})$. As the complexification θ of an appropriate Cartan involution $\mathfrak{g}_0 \to \mathfrak{g}_0$, we select $\theta(X) = I_{r,r} \cdot X \cdot I_{r,r}$. In terms of matrices, the subalgebras $\mathfrak{k} \subset \mathfrak{g}$ are

$$\left\{ \begin{pmatrix} A & 0 \\ 0 & {}^\eta A \end{pmatrix} \middle| A \in \mathbb{M}_r(\mathbb{C}) \right\} \subset \left\{ \begin{pmatrix} A & S_+ \\ S_- & {}^\eta A \end{pmatrix} \middle| S_\pm = {}^\eta S_\pm \right\} \tag{20.1.1}$$

and $\mathfrak{k}$ has Cartan subalgebra $\mathfrak{t} = \mathfrak{h} \cong \mathbb{C}^r$, which consists of all matrices

$$\mathrm{Diag}(\varepsilon_1, \dots, \varepsilon_r) := \begin{pmatrix} \varepsilon_1 & & 0 & & \\ & \ddots & \vdots & & \\ & & \varepsilon_r \quad 0 & \cdots & 0 \\ 0 & \cdots & 0 \quad -\varepsilon_r & & \\ & & \vdots & \ddots & \\ & & 0 & & -\varepsilon_1 \end{pmatrix}.$$

Select $J_0 := \frac{i}{2} I_{r,r} \in \mathfrak{t}$. Then the matrices of ${}^{(}r; \mathbb{C})$ with $A = S_- = 0$, in (20.1.1), for the $+i$ eigenspace of $\mathrm{ad}(J_0)$. Select $\mathfrak{b}_\mathfrak{k}^{\mathrm{ref}} = \mathfrak{gl}(r; \mathbb{C}) \cap \mathbb{B}^-$ and $\mathfrak{b}^{\mathrm{ref}} = \mathfrak{sp}(r; \mathbb{C}) \cap \mathbb{B}^-$. The corresponding positive and simple root systems are

$$\begin{aligned}
&\Sigma_\mathfrak{k}^+ = \{\varepsilon_j - \varepsilon_k\}_{1 \leqq j < k \leqq r}, \\
&\Psi_\mathfrak{k} = \{\varepsilon_1 - \varepsilon_2, \dots, \varepsilon_{r-1} - \varepsilon_r\} =: \{\beta_1, \dots, \beta_{r-1}\}, \\
&\Sigma^+ = \{\varepsilon_j \pm \varepsilon_k\}_{1 \leqq j < k \leqq r} \cup \{2\varepsilon_j\}_{1 \leqq j \leqq r}, \\
&\Psi = \{\varepsilon_1 - \varepsilon_2, \dots, \varepsilon_{r-1} - \varepsilon_r, 2\varepsilon_r\} =: \{\psi_1, \dots, \psi_{r-1}, \psi_r = \psi_{\mathrm{lo}}\}, \\
&M(\mathfrak{t}, \mathfrak{s})^\pm = \{\pm(\varepsilon_j + \varepsilon_k)\}_{1 \leqq j < k \leqq r} \cup \{\pm 2\varepsilon_j\}_{1 \leqq j \leqq r}.
\end{aligned}$$

The Bott-regular weights of index 1 have the decomposition $\mu^{\pm} = \mu'^{\pm} \pm \chi$ with $\chi = \left(\frac{2}{r}\psi_1 + \frac{2\cdot 2}{r}\psi_2 + \frac{3\cdot 2}{r}\psi_3 + \cdots + \frac{2(r-1)}{r}\psi_{r-1} + \psi_{\mathrm{lo}}\right) = \frac{2}{r}(\varepsilon_1 + \cdots + \varepsilon_r)$. If $\mu'^{\pm} : \mathfrak{t} \to \mathbb{C}$ denote the semisimple parts which are trivially extended to $\mathfrak{z}_{\mathfrak{k}}$, then

$$\begin{aligned}\mu'^{+} &= 2\xi_1' - 2\beta_1 = \tfrac{2}{r}(-\beta_1 + (r-2)\beta_2 + (r-3)\beta_3 + \cdots + \beta_{r-1}),\\ \mu'^{-} &= 2\xi_{r-1}' - 2\beta_{r-1} = \tfrac{2}{r}(\beta_1 + 2\beta_2 + \cdots + (r-2)\beta_{r-2} - \beta_{r-1}).\end{aligned}$$

The following are the structure equations for the reference Borel subalgebras.

$$\begin{aligned}\mu^{+} &= 2\varepsilon_2 = 2\psi_2 + \cdots + 2\psi_{r-1} + \psi_{\mathrm{lo}},\\ \lambda(\mu^{+}) &= \varepsilon_1 + \varepsilon_2 = \psi_1 + 2\psi_2 + \cdots + 2\psi_{r-1} + \psi_{\mathrm{lo}},\\ \beta^{\star +} &= \varepsilon_1 - \varepsilon_2 = \psi_1,\\ \mu^{-} &= -2\varepsilon_{r-1} = -2\psi_{r-1} - \psi_{\mathrm{lo}},\\ \lambda(\mu^{-}) &= -\varepsilon_{r-1} - \varepsilon_r = -\psi_{r-1} - \psi_{\mathrm{lo}},\\ \beta^{\star -} &= \varepsilon_{r-1} - \varepsilon_r = \psi_{r-1}.\end{aligned} \tag{20.1.2}$$

The next step is to express the weights $\mu^{\pm}, \lambda^{\pm}, \beta^{\star\pm}$ in terms of simple roots coming from all the Borel subalgebras $\mathfrak{b}$ that contain $\mathfrak{b}_{\mathfrak{k}}^{\mathrm{ref}}$ and differ from our reference subalgebra $\mathfrak{b}^{\mathrm{ref}}$. Since $\mathfrak{k}$ has the same rank as $\mathfrak{g}$, it follows that $W^{\theta} = W = W(C_r) \cong \mathfrak{S}_r \ltimes \mathbb{Z}_2^r$ and $W_{\mathfrak{k}} \cong W(A_{r-1}) \cong \mathfrak{S}_r$. By Corollary 18.3.6 there are $|W|/|W_{\mathfrak{k}}| = 2^r$ Borel subalgebras $\widetilde{w}(\mathfrak{b})$ which contain a given Borel subalgebra $\mathfrak{b}_{\mathfrak{k}}$ of $\mathfrak{k}$.

As in Section 19.4, we explicitly determine the subset $W_1^{\theta} \subset W = \mathfrak{S}_r \ltimes \mathbb{Z}_2^r$ (see (18.3.8)). Identifying $\mathfrak{t}$ with $\mathbb{C}^r$, the subgroup $\mathfrak{S}_r$ acts on $\mathbb{C}^r$ by permutations and $\mathbb{Z}_2^r$ by change of signs. Let $\boldsymbol{j} := \{j_1, \ldots, j_k\}$ be an arbitrary subset of $\{1, \ldots, r\}$ and $\{j_{k+1}, \ldots, j_r\}$ its complement. We order these sets by $j_1 < j_2 < \cdots < j_k$ and $j_{k+1} < j_{k+2} < \cdots < j_r$. Here, the number $k = |\boldsymbol{j}|$ of elements in $\boldsymbol{j}$ is arbitrary; hence, in total there are 2^r such subsets. Given j we assign to it the sign change map $\mathbf{sc}_j : \mathfrak{t} \to \mathfrak{t}$, which acts on the diagonal entries $(\varepsilon_1, \ldots, \varepsilon_r) \mapsto (\mathrm{sc}_j(\varepsilon_1), \ldots, \mathrm{sc}_j(\varepsilon_r))$, where

$$\mathrm{sc}_j(\varepsilon_\ell) = \begin{cases} \varepsilon_\ell & \text{for } 1 \leqq \ell \leqq |\boldsymbol{j}|,\\ -\varepsilon_\ell & \text{for } |\boldsymbol{j}| + 1 \leqq \ell \leqq r \end{cases}$$

and the permutation map $p = p_j \in \mathfrak{S}_r$

$$\begin{aligned}&p(1) = j_1, \ldots, p(k) = j_k,\\ &p(k+1) = j_r,\ p(k+2) = j_{r-1},\ \ldots, p(r) = j_{k+1}\end{aligned}$$

which defines the Weyl group element $\mathbf{p}_j^{-1}(\varepsilon_1, \ldots, \varepsilon_r) := (\varepsilon_{p(1)}, \ldots, \varepsilon_{p(r)})$. Finally, define $w_j := \mathbf{sc}_j \circ \mathbf{p}_j^{-1} : \mathfrak{t} \to \mathfrak{t}$. Thus, in terms of the diagonal entries and $k = |\boldsymbol{j}|$ we have

$$w_j(\mathrm{Diag}(\varepsilon_1, \ldots, \varepsilon_r)) = \mathrm{Diag}(\varepsilon_{j_1}, \ldots, \varepsilon_{j_k}, -\varepsilon_{j_r}, -\varepsilon_{j_{r-1}}, \ldots, -\varepsilon_{j_{k+1}}). \tag{20.1.3}$$

To see that the 2^r elements w_j exhaust the set W_1^θ, we first write the inequalities defining the reference Weyl chambers ($D = \mathrm{Diag}(\delta_1, \ldots, \delta_r)$):

(20.1.4) $$\mathcal{C}^+(\mathfrak{b}_{\mathfrak{k}}^{\mathrm{ref}}) = \{D : \delta_1 > \cdots > \delta_r\},$$

(20.1.5) $$\mathcal{C}^+(\mathfrak{b}^{\mathrm{ref}}) = \{D : \delta_1 > \cdots > \delta_r > 0\}.$$

We see that for each $\boldsymbol{j}$ and $w_{\boldsymbol{j}}$ acting as in (20.1.3), the entries of elements of $w_{\boldsymbol{j}}(\mathcal{C}^+(\mathfrak{b}^{\mathrm{ref}}))$ satisfy condition (20.1.4), i.e., $w_{\boldsymbol{j}}(\mathcal{C}^+(\mathfrak{b}^{\mathrm{ref}})) \subset \mathcal{C}^+(\mathfrak{b}_{\mathfrak{k}}^{\mathrm{ref}})$. Hence W_1^θ consists precisely of all such $w_{\boldsymbol{j}}$.

The following equations explicitly give the coefficients of the structure equations for $\mathfrak{b} = w_{\boldsymbol{j}}(\mathfrak{b}^{\mathrm{ref}})$ $\boldsymbol{j} \subset \{1, \ldots, r\}$ arbitrary subsets and $k := |\boldsymbol{j}|$ (See Section 19.4 for an analogous case).

$$w_{\boldsymbol{j}}^{-1}\mu^+ = \begin{cases} 2\varepsilon_{j_2} = 2\psi_{j_2} + 2\psi_{j_2+1} + \cdots + 2\psi_{r-1} + \psi_{\mathrm{lo}} & \text{if } k \geqq 2, \\ -2\varepsilon_r = -\psi_{\mathrm{lo}} & \text{if } k = 1 \text{ and } j_1 \neq r, \\ -2\varepsilon_{r-1} = -(2\psi_{r-1} + \psi_{\mathrm{lo}}) & \text{if } k = 1 \text{ and } j_1 = r \\ -2\varepsilon_{r-1} = -(2\psi_{r-1} + \psi_{\mathrm{lo}}) & \text{if } k = 0, \end{cases}$$

$$w_{\boldsymbol{j}}^{-1}\lambda(\mu^+) = \begin{cases} \varepsilon_{j_1} + \varepsilon_{j_2} = \psi_{j_1} + \cdots + \psi_{j_2-1} + 2\psi_{j_2} + \\ \qquad + 2\psi_{j_2+1} + \cdots + 2\psi_{r-1} + \psi_{\mathrm{lo}} & \text{if } k \geqq 2, \\ \varepsilon_{j_1} - \varepsilon_r = \psi_{j_1} + \cdots + \psi_{r-1} & \text{if } k = 1 \text{ and } j_1 \neq r, \\ \varepsilon_r - \varepsilon_{r-1} = -\psi_{r-1} & \text{if } k = 1 \text{ and } j_1 = r, \\ -\varepsilon_r - \varepsilon_{r-1} = -\psi_{r-1} - \psi_{\mathrm{lo}} & \text{if } k = 0, \end{cases}$$

$$w_{\boldsymbol{j}}^{-1}\beta_+^\star = \begin{cases} \varepsilon_{j_1} - \varepsilon_{j_2} = \psi_{j_1} + \cdots + \psi_{j_2-1} & \text{if } k \geqq 2, \\ \varepsilon_{j_1} + \varepsilon_r = \psi_{j_1} + \cdots + \psi_{r-1} + \psi_{\mathrm{lo}} & \text{if } k = 1 \text{ and } j_1 \neq r, \\ \varepsilon_r + \varepsilon_{r-1} = \psi_{r-1} + \psi_{\mathrm{lo}} & \text{if } k = 1 \text{ and } j_1 = r, \\ -\varepsilon_r + \varepsilon_{r-1} = \psi_{r-1} & \text{if } k = 0, \end{cases}$$

$$w_{\boldsymbol{j}}^{-1}\mu^- = \begin{cases} 2\varepsilon_{j_{k+2}} = 2\psi_{j_{k+2}} + 2\psi_{j_{k+2}+1} + \cdots + 2\psi_{r-1} + \psi_{\mathrm{lo}} & \text{if } k < r-1, \\ -2\varepsilon_{r-1} = -2\psi_{r-1} - \psi_{\mathrm{lo}} & \text{if } k = r-1 \text{ and } j_r = r, \\ -2\varepsilon_r = -\psi_{\mathrm{lo}} & \text{if } k = r-1 \text{ and } j_r \neq r, \\ -2\varepsilon_{r-1} = -2\psi_{r-1} - \psi_{\mathrm{lo}} & \text{if } k = r, \end{cases}$$

$$w_{\boldsymbol{j}}^{-1}\lambda^- = \begin{cases} \varepsilon_{j_{k+1}} + \varepsilon_{j_{k+2}} = \psi_{j_{k+1}} + \cdots + \psi_{j_{k+2}-1} + \\ \qquad 2\psi_{j_{k+2}} + \cdots + 2\psi_{r-1} + \psi_{\mathrm{lo}} & \text{if } k < r-1, \\ -\varepsilon_{r-1} + \varepsilon_r = -\psi_{r-1} & \text{if } k = r-1 \text{ and } j_r = r, \\ \varepsilon_{j_r} - \varepsilon_r = \psi_{j_r} + \psi_{j_r+1} + \cdots + \psi_{r-1} & \text{if } k = r-1 \text{ and } j_r \neq r, \\ -\varepsilon_{r-1} - \varepsilon_r = -\psi_{r-1} - \psi_{\mathrm{lo}} & \text{if } k = r, \end{cases}$$

$$w_{\boldsymbol{j}}^{-1}\beta_-^* = \begin{cases} \varepsilon_{j_{k+1}} - \varepsilon_{j_{k+2}} = \psi_{j_{k+1}} + \cdots + \psi_{j_{k+2}-1} & \text{for } k < r-1, \\ \varepsilon_{r-1} + \varepsilon_r = \psi_{r-1} + \psi_{\mathrm{lo}} & \text{if } k = r-1 \text{ and } j_r = r, \\ \varepsilon_r + \varepsilon_{j_r} = \psi_{j_r} + \cdots + \psi_{r-1} + \psi_{\mathrm{lo}} & \text{if } k = r-1 \text{ and } j_r \neq r, \\ \varepsilon_{r-1} - \varepsilon_r = \psi_{r-1} & \text{if } k = r. \end{cases}$$

The above structure equations contain all necessary information in order to apply the Cohomological Lemma along the lines explained in Section 18.6B.

Theorem 20.1.6. *For the series $(\mathfrak{g}, \mathfrak{k}) = (\mathfrak{sp}(r; \mathbb{C}), \mathfrak{gl}(r; \mathbb{C}))$ the following are the tangent modules at the base cycles $[C]$ in the full cycle spaces.*

- *In the full flag $X = G/B^{\mathrm{ref}}$ there are exactly $r-1$ base cycles $C^j_{(1)} \subset Z$ and $r-1$ base cycles $C^k_{(2)} \subset Z$ such that for every closed K-orbit C*

$$T_{[C]}\mathcal{C}(X) = \begin{cases} T_{[C]}(G\cdot[C]) \oplus \bigwedge^2 V^{\mathrm{std}}_{GL(r;\mathbb{C})} & \text{if } C = C^j_{(1)} \text{ for some } j, \\ T_{[C]}(G\cdot[C]) \oplus \bigwedge^2 (V^{\mathrm{std}}_{GL(r;\mathbb{C})})^* & \text{if } C = C^k_{(2)} \text{ for some } k, \\ T_{[C]}(G\cdot[C]) & \text{otherwise.} \end{cases}$$

- *In the general case we have $\mathfrak{q} = \mathfrak{q}_\Phi$, with $\Phi \subset \Psi$, $Z = G/Q$, and the canonical projection $\pi : G/B^{\mathrm{ref}} \to G/Q$. For $\mathbf{j} = \{j_1, \ldots, j_k\} \subset \{1, \ldots, r\}$ with $j_1 < \cdots < j_k$ and $w_{\mathbf{j}} \in W^\theta_1$ as in (20.1.3), define the reference cycle $C(w_{\mathbf{j}}) := K\cdot[w_{\mathbf{j}} B^{\mathrm{ref}}] \subset G/B^{\mathrm{ref}}$ and set $C := \pi(C(w_{\mathbf{j}}))$. Then $T_{[C]}\mathcal{C}(Z) = T_{[C]}(G\cdot[C]) \oplus \bigoplus_{\lambda \in M(C,Z)} V^\lambda$ with*

$$M(C,Z) \subset \left\{\lambda^{\mathrm{high}}\left(\bigwedge^2 V^{\mathrm{std}}_{GL(r;\mathbb{C})}\right),\ \lambda^{\mathrm{high}}\left(\bigwedge^2 (V^{\mathrm{std}}_{GL(r;\mathbb{C})})^*\right)\right\}.$$

Explicit computation of $M(C,Z)$ reduces to a verification of the relations between the positive supports of $w_{\mathbf{j}}^{-1}(\mu^\pm)$, $w_{\mathbf{j}}^{-1}(\lambda(\mu^\pm))$, $w_{\mathbf{j}}^{-1}(\beta^\star_\pm)$ (which can be read off directly from the above structure equations) in reference to the set Φ, as in the Cohomological Lemma, exactly along the lines of Theorem 19.4.8.

20.2 The series for $\mathfrak{g}_0 = \mathfrak{so}(2p, 2q+1)$

This series can be handled by methods similar to those for the series in Section 19.4. We assume that $r = p + q \geqq 2$, because the case $r = 1$, where $\mathfrak{g} = \mathfrak{so}(3; \mathbb{C})$, is trivial. According to Theorem 18.4.13 we only need to consider $q \geqq 1$. Note that for $p = 1$ the corresponding series $\mathfrak{so}(2, 2r-1)$ is of hermitian type. For $p > 1$ the algebra $\mathfrak{so}(2p, 2q+1)$ is not of hermitian type.

For all p and q we have $\mathfrak{t} = \mathfrak{z}_{\mathfrak{g}}(\mathfrak{t}) = \mathfrak{h}$. To facilitate the explicit computations, we choose the matrix realizations of $\mathfrak{t} = \mathfrak{h}$ and $\mathfrak{g} = \mathfrak{so}(2r+1, \mathbb{C})$ and write $\mathrm{Diag}(\varepsilon_1, \ldots, \varepsilon_r)$ for a diagonal matrix in $\mathfrak{t}$. The complex involutive automorphism which defines $\mathfrak{k} \cong \mathfrak{so}(2p; \mathbb{C}) \times \mathfrak{so}(2q+1; \mathbb{C})$ is chosen as follows:

$$\theta(X) = \begin{pmatrix} I_p & & \\ & -I_{2q+1} & \\ & & I_p \end{pmatrix} \cdot X \cdot \begin{pmatrix} I_p & & \\ & -I_{2q+1} & \\ & & I_p \end{pmatrix} =: {}^\theta X.$$

Note that the diagonal Cartan subalgebra $\mathfrak{h}$ is pointwise fixed by the above θ. For convenience, we also describe the matrices in that realization of $\mathfrak{k}$. They are of the following form (compare to (19.4.1) for a similar realization of $\mathfrak{k}$).

$$\left\{ \begin{pmatrix} A_p & 0 & 0 & 0 & B_p \\ 0 & A_q & u_q & B_q & 0 \\ 0 & v_q & 0 & -u_q^t R & 0 \\ 0 & C_q & -Rv_q^t & {}^\theta A_q & 0 \\ C_p & 0 & 0 & 0 & {}^\theta A_p \end{pmatrix} \middle| \begin{array}{l} B_\bullet = {}^\theta B_\bullet \\ C_\bullet = {}^\theta C_\bullet \end{array} \right\}. \tag{20.2.1}$$

Here, $A_\bullet, B_\bullet, C_\bullet \in \mathbb{M}_\bullet(\mathbb{C})$, $u_q \in \mathbb{M}_{q\times 1}(\mathbb{C})$, and $v_q \in \mathbb{M}_{1\times q}(\mathbb{C})$. Finally, we select the standard Borel subalgebras $\mathfrak{b}_\mathfrak{k} \subset \mathfrak{k} \cong \mathfrak{so}(2p;\mathbb{C}) \times \mathfrak{so}(2q+1;\mathbb{C})$ and $\mathfrak{b} \subset \mathfrak{g} \cong \mathfrak{so}(2r+1;\mathbb{C})$, consisting of the lower-triangular matrices in $\mathfrak{k} \cong \mathfrak{so}(2p;\mathbb{C}) \times \mathfrak{so}(2q+1;\mathbb{C})$ and $\mathfrak{g} \cong \mathfrak{so}(2r+1;\mathbb{C})$, respectively. These subalgebras are stable under the involution θ. In the hermitian case $p = 1$, we select $J_0 := \mathrm{Diag}(i, 0, \ldots, 0) \in \mathfrak{z}_0'$, which determines the decomposition $\mathfrak{s} = \mathfrak{s}_+ + \mathfrak{s}_-$.

Next we give a (well-known) list of the various root systems and bases (with respect to the reference Borels). This is done in terms of the standard coordinate functions $\varepsilon_1, \ldots, \varepsilon_r$ on $\mathfrak{t}$.

$$\begin{aligned}
\Sigma &= \{\pm\varepsilon_j \pm \varepsilon_k\}_{1\leqq j<k\leqq r} \cup \{\pm\varepsilon_j\}_{1\leqq k\leqq r},\\
\Psi &= \{\varepsilon_1-\varepsilon_2, \ldots, \varepsilon_{r-1}-\varepsilon_r, \varepsilon_r\} =: \{\psi_1, \ldots, \psi_{r-1}, \psi_{\mathrm{sh}}\},\\
\Sigma_\mathfrak{k} &= \Sigma_\mathfrak{k}^{(1)} \cup \Sigma_\mathfrak{k}^{(2)} = \{\pm\varepsilon_j \pm \varepsilon_k\}_{1\leqq j<k\leqq p} \cup \{\pm\varepsilon_j \pm \varepsilon_k, \pm\varepsilon_j\}_{p+1\leqq j<k\leqq p+q},\\
\Psi_\mathfrak{k}^{(1)} &= \{\varepsilon_1-\varepsilon_2, \ldots, \varepsilon_{p-1}-\varepsilon_p, \varepsilon_{p-1}+\varepsilon_p\} =: \{\beta_1^{(1)}, \ldots, \beta_{p-1}^{(1)}, \beta_p^{(1)}\},\ p \geqq 3\\
&\qquad \{\varepsilon_1-\varepsilon_2, \varepsilon_1+\varepsilon_2\} =: \{\beta^{(1)}, \tilde\beta^{(1)}\},\ p = 2\\
&\qquad = \emptyset,\ p = 1,\\
\Psi_\mathfrak{k}^{(2)} &= \{\varepsilon_{p+1}-\varepsilon_{p+2}, \ldots, \varepsilon_{r-1}-\varepsilon_r, \varepsilon_r\} =: \{\beta_1^{(2)}, \ldots, \beta_{q-1}^{(2)}, \beta_{\mathrm{sh}}^{(2)}\}.
\end{aligned}$$

The Bott-regular weights of index 1 are listed in Table 18.5.3, rows 1 through 4. All weights in $M(\mathfrak{t},\mathfrak{s})$ can easily be determined using the fact that $\mathfrak{k} = \mathfrak{k}_{(1)} \times \mathfrak{k}_{(2)} \cong \mathfrak{so}(2p;\mathbb{C})\times\mathfrak{so}(2q+1;\mathbb{C})$ and that $\mathfrak{s}$ is isomorphic to $V^{\mathrm{std}}_{\mathfrak{so}(2p)} \otimes V^{\mathrm{std}}_{\mathfrak{so}(2q+1)}$ as a $\mathfrak{k}$-module. In particular, for $p = 1$, $V^{\mathrm{std}}_{so(2)} = W_+ \oplus W_-$ is a direct sum of two one-dimensional modules in compliance with the fact that $\mathfrak{so}(2, 2r-1)$ is hermitian. If $p > 1$, it follows that $V^{\mathrm{std}}_{\mathfrak{so}(2p)}$ is irreducible. The dominant and Bott-regular weights of index 1 (since rank $\mathfrak{k}$ = rank $\mathfrak{g}$, there are none of index 2) are the following.

In the non-hermitian cases we have

$$\begin{aligned}
\mu &= \xi_1^{(1)} - \beta_{\mathrm{sh}}^{(2)} = \varepsilon_1 - \varepsilon_r = \psi_1 + \cdots + \psi_{r-1} \text{ for } p \geqq 3\\
&= \xi^{(1)} + \tilde\xi^{(1)} - \beta_{\mathrm{sh}}^{(2)} \text{ for } p = 2,\\
\lambda(\mu) &= \xi_1^{(1)} = \varepsilon_1 = \psi_1 + \cdots + \psi_{r-1} + \psi_{\mathrm{sh}}\\
&= \xi^{(1)} + \tilde\xi^{(1)} \text{ for } p = 2,\\
\beta^\star &= \varepsilon_r = \psi_{\mathrm{sh}}.
\end{aligned}$$

In the hermitian case (the case $p = 1$), $\chi = \varepsilon_1$ is a root in $\Sigma(\mathfrak{g}, \mathfrak{h})$ and we have

$$\begin{aligned}
\mu_\pm &= \begin{matrix} \chi - \beta^{(2)}_{\mathrm{sh}} \\ -\chi - \beta^{(2)}_{\mathrm{sh}} \end{matrix} = \begin{matrix} \varepsilon_1 - \varepsilon_r \\ -\varepsilon_1 - \varepsilon_r \end{matrix} = \begin{matrix} \psi_1 + \cdots + \psi_{r-1} \\ -\psi_1 - \cdots - \psi_{r-1} - 2\psi_{\mathrm{sh}}, \end{matrix} \\
\lambda(\mu_\pm) &= \pm\varepsilon_1 = \pm(\psi_1 + \cdots + \psi_{r-1} + \psi_{\mathrm{sh}}), \\
\beta^\star &= \varepsilon_r = \psi_{\mathrm{sh}}.
\end{aligned}$$

Note that in this case $\lambda'_\pm = \lambda(\mu_\pm)|_{\mathfrak{t}'} = 0$, and the corresponding irreducible representation $V^{\lambda_\pm}$ of $\mathfrak{k} = \mathfrak{z}_\mathfrak{k} \oplus \mathfrak{k}'$ annihilates $\mathfrak{k}'$ and is one-dimensional. This is the only class of hermitian real forms (and corresponding complex symmetric pairs) for which the elements $\pm\chi$, a priori defined on $\mathfrak{z}_\mathfrak{k}$ and trivially extended to $\mathfrak{t}$, are roots in Σ and admit a 3-string $\pm\chi - \beta^\star, \pm\chi, \pm\chi + \beta^\star$. A geometric consequence of this fact is that there may be nontrivial sections in $H^0(C; \mathcal{O}(\mathbb{N}_Z(C)))$ which are fixed by the semisimple factor K' of K.

Lemma 18.4.12 says that $H^0(C; \mathcal{O}(\mathbb{N}_Z(C)))$ has no trivial K-submodules. Also, K' acts transitively on C and its center Z_K acts trivially, but Z_K acts nontrivially on $\mathbb{N}_Z(C)$.

The next step is to determine explicitly the structure equations (18.6.3) for all choices of Borel subalgebras $\mathfrak{b} \subset \mathfrak{g}$ such that $\mathfrak{b} \cap \mathfrak{k} = \mathfrak{b}^{\mathrm{ref}}_\mathfrak{k}$. The number of different Borel subalgebras with such property is given by Corollary 18.3.6, i.e.,

$$\frac{|W|}{|W_\mathfrak{k}|} = \frac{|\mathfrak{S}_r \ltimes \mathbb{Z}_2^r|}{|\mathfrak{S}_p \ltimes \mathbb{Z}_2^{p-1}| \cdot |\mathfrak{S}_q \ltimes \mathbb{Z}_2^q|} = \frac{r! \cdot 2^r}{p! 2^{p-1} \cdot q! 2^q} = 2\frac{r!}{p!q!}.$$

In order to determine all such Borel subalgebras, we follow our usual strategy and explicitly determine the subset $W_1^\theta \subset W \cong \mathfrak{S}_r \ltimes \mathbb{Z}_2^r$. We use an explicit description of the various Weyl chambers, associated with the given Borel subalgebras.

In both the hermitian and the non-hermitian cases, the Weyl chambers in $\mathfrak{h}_\mathbb{R} = \mathfrak{t}_\mathbb{R}$ for the reference Borel subalgebras consist of all $D := \mathrm{Diag}(\delta_1, \ldots, \delta_r)$ subject to the conditions

$$C^+(\mathfrak{b}^{\mathrm{ref}}_\mathfrak{k}) = \left\{ D \,\middle|\, \begin{matrix} \delta_1 > \cdots > \delta_p, \quad \delta_{p-1} + \delta_p > 0 \\ \text{and} \quad \delta_{p+1} > \cdots > \delta_r > 0 \end{matrix} \right\}, \quad p > 1, \tag{20.2.2}$$

$$C^+(\mathfrak{b}^{\mathrm{ref}}_\mathfrak{k}) = \mathfrak{z}_\mathbb{R} \times C^+(\mathfrak{b}^{\mathrm{ref}}_{\mathfrak{k}'}) = \{D \mid \varepsilon_2 > \cdots > \varepsilon_r > 0\}, \quad p = 1, \tag{20.2.3}$$

and

$$C^+(\mathfrak{b}^{\mathrm{ref}}) = \{D \mid \delta_1 > \cdots > \delta_p > \delta_{p+1} > \cdots > \delta_{p+q} > 0\}. \tag{20.2.4}$$

In the hermitian case, $\mathfrak{k} = \mathfrak{k}' \oplus \mathfrak{z}_\mathfrak{k}$ and $\mathfrak{t} = \mathfrak{t}' \oplus \mathfrak{z}_\mathfrak{k}$, we have $\mathfrak{t}' = \{D \mid \varepsilon_1 = 0\}$.

Next we explicitly construct elements in the cross-section $W_1^\theta \subset W$ of (18.3.8). Just as in the Section 19.4, select elements $j_1 < j_2 < \cdots < j_p$ from $\{1, \ldots, r\}$, write $\boldsymbol{j} := \{j_1, \ldots, j_p\}$ (there are $r!/(p!q!)$ such subsets), and order the complementary set $\{1, \ldots, r\} \setminus \boldsymbol{j}$ in increasing order, i.e., as $j_{p+1} < \cdots < j_{p+q}$. Define $w_{\boldsymbol{j}} \in \mathfrak{S}_r \subset W^\theta$, acting by permutation on the index set $\{1, \ldots, r\}$ by

$$w_j^{-1}(1) = j_1,\ w_j^{-1}(2) = j_2,\ \ldots,\ w_j^{-1}(r) = j_r.$$

Counting w_js, the above construction gives only half of W_1^θ. In order to obtain the other half, consider the following element in $W = \mathfrak{S}_r \ltimes \mathbb{Z}_2^r$ which acts by change of sign

$$\tilde{s}_p(\varepsilon_1, \ldots, \varepsilon_p, \varepsilon_{p+1}, \ldots, \varepsilon_r) := (\varepsilon_1, \cdots - \varepsilon_p, \varepsilon_{p+1}, \ldots, \varepsilon_r).$$

The action of w_j and $\tilde{s}_p \circ w_j$ on $D := \mathrm{Diag}(\varepsilon_1, \ldots, \varepsilon_r) \in \mathfrak{h}$ is then

$$\begin{aligned} w_j(D) &= \mathrm{Diag}(\varepsilon_{j_1}, \ldots, \varepsilon_{j_{p-1}}, \varepsilon_{j_p}, \varepsilon_{j_{p+1}}, \ldots, \varepsilon_{j_r}), \\ \tilde{s}_p w_j(D) &= \mathrm{Diag}(\varepsilon_{j_1}, \ldots, \varepsilon_{j_{p-1}}, -\varepsilon_{j_p}, \varepsilon_{j_{p+1}}, \ldots, \varepsilon_{j_r}). \end{aligned} \tag{20.2.5}$$

By inspecting the inequalities of (20.2.2) and (20.2.3), we have all the possibilities for $w_j\mathfrak{b}^{\mathrm{ref}}$, $\tilde{s}_p w_j \mathfrak{b}^{\mathrm{ref}}$ containing $\mathfrak{b}_{\mathfrak{k}}^{\mathrm{ref}}$.

We are now in the position to compute all of the coefficients in the structure equations for μ, $\lambda(\mu)$ and $\beta^\star$. The non-hermitian and hermitian cases are dealt with separately. For $2 \leqq p \leqq r-1$ we have

$$\begin{aligned} w_j^{-1}(\mu) = (\tilde{s}_p w_j)^{-1}(\mu) &= \varepsilon_{j_1} - \varepsilon_{j_r} \\ &= \begin{cases} \psi_{j_1} + \psi_{j_1+1} + \cdots + \psi_{j_r-1} & \text{if } j_1 < q+1, \\ -\psi_q & \text{if } j_1 = q+1, \end{cases} \\ w_j^{-1}(\lambda(\mu)) = (\tilde{s}_p w_j)^{-1}(\lambda(\mu)) &= \varepsilon_{j_1} = \psi_{j_1} + \psi_{j_1+1} + \cdots + \psi_{\mathrm{sh}}, \\ w_j^{-1}(\beta^\star) = (\tilde{s}_p w_j)^{-1}(\beta^\star) &= \varepsilon_{j_r} = \psi_{j_r} + \psi_{j_r+1} + \cdots + \psi_{\mathrm{sh}}. \end{aligned}$$

Theorem 20.2.6 (Non-hermitian case).

- *The full flag manifold $X = G/B^{\mathrm{ref}}$ has base cycles*

$$C_1 := C(w_j) = K(w_j B^{\mathrm{ref}}) \quad \textit{and} C_2 := C(\tilde{s}_p w_j) = K(\tilde{s}_p w_j B^{\mathrm{ref}})$$

with $j = \{q+1, q+2, \ldots, r\}$, w_j and $\tilde{s}_{pi} w_j$ as in (20.2.5)*, such that for every closed K-orbit $C \subset X$,*

$$T_{[C]}\mathcal{C}(X) = \begin{cases} T_{[C]}(G{\cdot}[C]) \oplus V_{SO(2p)}^{\mathrm{std}} & \textit{if } C = C_1 \textit{ or } C = C_2, \\ T_{[C]}(G{\cdot}[C]) & \textit{otherwise}. \end{cases}$$

- *In the general case $Z = G/Q_\Phi$, with the projection $\pi : G/B^{\mathrm{ref}} \to G/Q_\Phi$ and a base cycle $C = \pi(K(w_j{\cdot}B^{\mathrm{ref}}))$ or $C = \pi(K(\tilde{s}_p w_j{\cdot}B^{\mathrm{ref}}))$, the set $M(C, Z)$ in the decomposition* (18.6.2) *is empty or equal to $\{\lambda_{SO(2p)}^{\mathrm{std}}\}$. As explained in Theorem* 19.4.8 *the explicit formula for $M(C, Z)$ is obtained by comparing the various (positive) supports of $w_j^{-1}\mu$, $w_j^{-1}\lambda$ etc (which can be read directly from the preceding structure equations) and Φ, and using the Cohomological Lemma, along the lines as in Theorem* 19.4.8.

For $p = 1$ the structure equations are

$$w_j^{-1}(\mu_+) = \begin{cases} \varepsilon_{j_1} - \varepsilon_r = \psi_{j_1} + \cdots \psi_{r-1} & \text{if } j_1 \neq r, \\ \varepsilon_r - \varepsilon_{r-1} = -\psi_{r-1} & \text{if } j_1 = r, \end{cases}$$

$$w_j^{-1}(\mu_-) = \begin{cases} -\varepsilon_{j_1} - \varepsilon_r = -(\psi_{j_1} + \cdots + \psi_{r-1} + 2\psi_{\mathrm{sh}}) & \text{if } j_1 \neq r, \\ -\varepsilon_r - \varepsilon_{r-1} = -\psi_{r-1} - 2\psi_{\mathrm{sh}} & \text{if } j_1 = r, \end{cases}$$

$$(\tilde{s}_1 w_j)^{-1}(\mu_+) = \begin{cases} -\varepsilon_{j_1} - \varepsilon_r = -(\psi_{j_1} + \cdots \psi_{r-1} + 2\psi_{\mathrm{sh}}) & \text{if } j_1 \neq r, \\ -\varepsilon_r - \varepsilon_{r-1} = -\psi_{r-1} - 2\psi_{\mathrm{sh}} & \text{if } j_1 = r, \end{cases}$$

$$(\tilde{s}_1 w_j)^{-1}(\mu_-) = \begin{cases} \varepsilon_{j_1} - \varepsilon_r = \psi_{j_1} + \cdots + \psi_{r-1} & \text{if } j_1 \neq r, \\ \varepsilon_r - \varepsilon_{r-1} = -\psi_{r-1} & \text{if } j_1 = r, \end{cases}$$

$$w_j^{-1}(\lambda_\pm) = \begin{cases} \pm\varepsilon_{j_1} = \pm(\psi_{j_1} + \cdots + \psi_{r-1} + \psi_{\mathrm{sh}}) & \text{if } j_1 \neq r, \\ \pm\varepsilon_r = \pm\psi_{\mathrm{sh}} & \text{if } j_1 = r, \end{cases}$$

$$(\tilde{s}_1 w_j)^{-1}(\lambda_\pm) = \begin{cases} \mp\varepsilon_{j_1} = \mp(\psi_{j_1} + \cdots + \psi_{r-1} + \psi_{\mathrm{sh}}) & \text{if } j_1 \neq r, \\ \mp\varepsilon_r = \mp\psi_{\mathrm{sh}} & \text{if } j_1 = r, \end{cases}$$

$$w_j^{-1}(\beta^\star) = \varepsilon_{j_r} = \psi_{j_r} + \cdots + \psi_{\mathrm{sh}} = (\tilde{s}_1 w_j)^{-1}(\beta^\star).$$

Theorem 20.2.7 (Hermitian case).

- *There are two base cycles in the full flag manifold $X = G/B^{\mathrm{ref}}$, namely $C_+ := K(w_{\{r\}}B^{\mathrm{ref}})$ and $C_- := K(\tilde{s}_1 w_{\{r\}}B^{\mathrm{ref}})$ ($w_{\{r\}}$ and $\tilde{s}_1 w_{\{r\}}$ as in* (20.2.5)) *such that $T_{[C_\pm]}\mathcal{C}(X) = T_{[C_\pm]}(G{\cdot}[C_\pm]) \oplus V^{\pm\chi}$. Here, $V^{\pm\chi}$ are the $\mathfrak{k}$-modules for the one-dimensional representations $\pm\chi$. For all remaining base cycles $C \neq C_\pm$ we have $T_{[C]}\mathcal{C}(Z) = T_{[C]}(G{\cdot}[C])$.*
- *In the case of a general flag manifold $Z = G/Q_\Phi$ the tangent space $T_{[C]}\mathcal{C}(Z)$ either agrees with that of the G-orbit of $[C]$ or $T_{[C]}\mathcal{C}(Z) = T_{[C]}(G{\cdot}[C]) \oplus V^{\pm\chi}$ as C and Z varies. An explicit determination of the set $M(C, Z) \subset \{+\chi, -\chi\}$ for every C and Z is obtained by comparing the various (positive) supports of $w_j^{-1}\mu_\pm$, $w_j^{-1}\lambda_\pm$ etc (which can be read directly from the preceding structure equations) and Φ, exactly along the lines as in Theorem* 19.4.8.

20.3 The case $\mathfrak{g}_0 = \mathfrak{f}_{4,C_3A_1}$

Inspection of the real forms $\mathfrak{g}_0 = \mathfrak{k}_0 + \mathfrak{s}_0$ of $\mathfrak{f}_4(\mathbb{C})$ and the corresponding highest weights of the representation $\mathfrak{k} \times \mathfrak{s} \to \mathfrak{s}$ shows that the only real form which admits in $M(\mathfrak{t}, \mathfrak{s})$ weights of two different length is $\mathfrak{g}_0 = \mathfrak{f}_{4(4)} = \mathfrak{f}_{4,C_3A_1}$. This can also be deduced by looking at the abstract affine Dynkin diagrams $\Delta(\mathfrak{g}, \nu, \mathrm{aff})$ which here is the extended Dynkin diagram of $\mathfrak{f}_4(\mathbb{C})$. The marked affine Dynkin diagram for $(\mathfrak{g}, \mathfrak{k})$ for $\mathfrak{g}_0 = \mathfrak{f}_{4(4)}$ is given in Table 18.5.3, row 6.

We use some combinatorial and geometric properties of the Dynkin diagrams to determine the structure equations. We start by selecting $\mathfrak{t} = \mathfrak{h} \subset \mathfrak{b}^{\mathrm{ref}} \subset \mathfrak{f}_4(\mathbb{C})$

and let $\Psi = \{\psi_1 — \psi_2 \Longrightarrow \psi_3 — \psi_4\}$ be the corresponding simple roots. That (large) Borel subalgebra determines $\mathfrak{b}_{\mathfrak{k}}^{\mathrm{ref}} := \mathfrak{b}^{\mathrm{ref}} \cap \mathfrak{k}$ which in turns gives the basis $\Psi_{\mathfrak{k}} = \{\beta_1^{(1)} — \beta_2 \Longleftarrow \beta_3\} \cup \{\beta^{(2)}\}$ of $-\Phi(\mathfrak{b}_{\mathfrak{k}}^{\mathrm{ref}}, \mathfrak{t})$. In this case the subalgebra $\mathfrak{k} \cong \mathfrak{sp}(3;\mathbb{C}) \oplus \mathfrak{sp}(1;\mathbb{C})$ is of classical type. From Table 17.4.6 we read off that

$$\mathfrak{s} = V_{\mathrm{Sp}(3)}^{\xi_3^{(1)}} \otimes V_{SL(2;\mathbb{C})}^{\xi_1^{(2)}} \quad \text{as } \mathfrak{k}\text{-module},$$

and that $\xi_1^{(1)} + \xi_1^{(2)}$ is the nonmaximal dominant weight. So $\mu = \xi_1^{(1)} + \xi_1^{(2)} - \beta_2^{(1)}$ is the only Bott-regular weight of index 1. The elements of $\Psi_{\mathfrak{k}} = \Psi_{\mathfrak{k}}^{(1)} \cup \Psi_{\mathfrak{k}}^{(2)}$ are expressed in terms of Ψ as follows. This can either be immediately read off from the marked affine Dynkin diagram in Table 18.5.3, row 6, or found in the tables in [Kna, Appendix C.4].

$$\begin{aligned}
&\beta_1^{(1)} = \psi_4, \quad \beta_2^{(1)} = \psi_3, \quad \beta_3^{(1)} = \beta_{\mathrm{lo}}^{(1)} = \psi_2,\\
&\beta^{(2)} = 2\psi_1 + 3\psi_2 + 4\psi_3 + 2\psi_4,\\
&\mu = (\beta_1^{(1)} + \beta_2^{(1)} + \tfrac{1}{2}\beta_3^{(1)} + \tfrac{1}{2}\beta^{(2)}) - \beta_2^{(1)} =\\
&\quad = \psi_1 + 2\psi_2 + 2\psi_3 + 2\psi_4.
\end{aligned} \tag{20.3.1}$$

Next, we look for the structure equations with respect to nonreference Borel subalgebras $\mathfrak{b} \supset \mathfrak{b}_{\mathfrak{k}}^{\mathrm{ref}}$. By Corollary 18.3.6, we have

$$\frac{|W|}{|W_{\mathfrak{k}}|} = \frac{4!\cdot 2\cdot 3\cdot 4\cdot 2}{|\mathfrak{S}_3 \ltimes \mathbb{Z}_2^3 \times \mathbb{Z}_2|} = 12.$$

In order to find the remaining 11 Borel subalgebras, or equivalently 11 simple root systems $\Psi^{\mathrm{ref}} = \Psi_{[1]}, \dots, \Psi_{[12]}$ such that $\Sigma_{[d]}^{+} \supset \Psi_{\mathfrak{k}}^{\mathrm{ref}}$, we successively apply reflections s_α on Σ^{ref}. We use the elementary fact that if α is simple with respect to $\Sigma_{[d]}^{+}$, then $s_\alpha(\Sigma_{[d]}^{+}) = \Sigma_{[d]}^{+} \setminus \{\alpha\} \cup \{-\alpha\}$. Hence, if $\Sigma_{[d]}^{-} \supset \Phi(\mathfrak{b}_{\mathfrak{k}}^{\mathrm{ref}})$, in order to decide if $s_\alpha(\Sigma_{[d]}^{-}) \supset \Sigma(\mathfrak{b}^{\mathrm{ref}})$, we only need to check if $\alpha \notin \Sigma_{\mathfrak{k}}$.

We now compute μ, $\lambda(\mu)$ and $\beta^\star = \psi_3$ with respect to the simple system $\Psi_{[d]}$ for $d = 2, \dots, 12$. Each of the following bases $\Psi_{[k]}$ arises from some $\Psi_{[j]} \subset \Sigma_{[j]}^{+}$ by applying a simple reflection s_α with $\alpha \in \Psi_{[j]}$. We express each root basis $\Psi_{[d]} = \{\psi_1^{[d]}, \dots, \psi_4^{[d]}\}$ in terms of Ψ^{ref} and write $\nu = (n_1, n_2, n_3, n_4)$ for $\nu = \sum n_k\psi_k$, where $\psi_1, \dots, \psi_4$ are the reference simple roots.

$\psi_1^{[2]} = -(1000)$	$\psi_2^{[2]} = (1100)$	$\psi_3^{[2]} = (0010)$	$\psi_4^{[2]} = (0001)$
$\mu = (1222)^{[2]}$	$\lambda = (1232)^{[2]}$	$\beta^\star = (0010)^{[2]}$	
$\psi_1^{[3]} = (0100)$	$\psi_2^{[3]} = -(1100)$	$\psi_3^{[3]} = (1110)$	$\psi_4^{[3]} = (0001)$
$\mu = (1122)^{[3]}$	$\lambda = (1232)^{[3]}$	$\beta^\star = (0110)^{[3]}$	
$\psi_1^{[4]} = (0100)$	$\psi_2^{[4]} = (1120)$	$\psi_3^{[4]} = -(1110)$	$\psi_4^{[4]} = (1111)$
$\mu = (1122)^{[4]}$	$\lambda = (1232)^{[4]}$	$\beta^\star = (0110)^{[4]}$	
$\psi_1^{[5]} = (0100)$	$\psi_2^{[5]} = (1120)$	$\psi_3^{[5]} = (0001)$	$\psi_4^{[5]} = -(1111)$
$\mu = (1231)^{[5]}$	$\lambda = (1120)^{[5]}$	$\beta^\star = (0111)^{[5]}$	
$\psi_1^{[6]} = (1220)$	$\psi_2^{[6]} = -(1120)$	$\psi_3^{[6]} = (1121)$	$\psi_4^{[6]} = -(1111)$
$\mu = (1231)^{[6]}$	$\lambda = (1220)^{[6]}$	$\beta^\star = (0011)^{[6]}$	
$\psi_1^{[7]} = -(1220)$	$\psi_2^{[7]} = (0100)$	$\psi_3^{[7]} = (1121)$	$\psi_4^{[7]} = -(1111)$
$\mu = (1231)^{[7]}$	$\lambda = (1220)^{[7]}$	$\beta^\star = (0011)^{[7]}$	
$\psi_1^{[8]} = -(1220)$	$\psi_2^{[8]} = (0100)$	$\psi_3^{[8]} = (0010)$	$\psi_4^{[8]} = (1111)$
$\mu = (1232)^{[8]}$	$\lambda = (1222)^{[8]}$	$\beta^\star = (0010)^{[8]}$	
$\psi_1^{[9]} = (1220)$	$\psi_2^{[9]} = -(1120)$	$\psi_3^{[9]} = (0010)$	$\psi_4^{[9]} = (1111)$
$\mu = (1232)^{[9]}$	$\lambda = (1222)^{[9]}$	$\beta^\star = (0010)^{[9]}$	
$\psi_1^{[10]} = (1220)$	$\psi_2^{[10]} = (1122)$	$\psi_3^{[10]} = -(1121)$	$\psi_4^{[10]} = (0010)$
$\mu = (1221)^{[10]}$	$\lambda = (1220)^{[10]}$	$\beta^\star = (0001)^{[10]}$	
$\psi_1^{[11]} = -(1220)$	$\psi_2^{[11]} = (2342)$	$\psi_3^{[11]} = -(1121)$	$\psi_4^{[11]} = (0010)$
$\mu = (1221)^{[11]}$	$\lambda = (1220)^{[11]}$	$\beta^\star = (0001)^{[11]}$	
$\psi_1^{[12]} = (2342)$	$\psi_2^{[12]} = -(1122)$	$\psi_3^{[12]} = (0001)$	$\psi_4^{[12]} = (0010)$
$\mu = (1121)^{[12]}$	$\lambda = (1120)^{[12]}$	$\beta^\star = (0001)^{[12]}$	

From the above structure equations we see that support $\mathrm{supp}_{\Psi_{[d]}}(\mu)$ agrees with $\Psi_{[d]}$ for all $d = 1, \ldots, 12$. Hence, we have the following

Theorem 20.3.2. *If G is the exceptional group with Lie algebra $\mathfrak{g} = F_4$, then for all G-flags Z, all real forms $\mathfrak{g}_0 \subset \mathfrak{g}$ and all corresponding base cycles C we have $T_{[C]}\mathcal{C}(Z) = T_{[C]}(G\cdot[C])$.*

20.4 The case $\mathfrak{g}_0 = \mathfrak{g}_{2,A_1A_1}$

The only one real form $\mathfrak{g}_0$ of $G_2(\mathbb{C})$ which is of noncompact type is the normal real form $\mathfrak{g}_{2(2)}$. The corresponding complex symmetric pair is $(\mathfrak{g}_2(\mathbb{C}), \mathfrak{so}(4;\mathbb{C}))$. In the case under consideration the ranks of $\mathfrak{g}$ and $\mathfrak{k}$ are equal 2 and therefore we can easily visualize the root systems Σ and $\Sigma_{\mathfrak{k}}$. The marked affine Dynkin diagram for $(\mathfrak{g}, \mathfrak{k})$ is 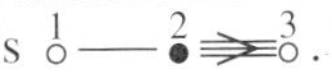. This tells us that $3\xi_{\mathrm{sh}} + \xi_{\mathrm{lo}} = 3\xi^{(1)} + \xi^{(2)}$ is the highest weight for the representation of $\mathfrak{k} \cong A_1 \oplus A_1$ of $\mathfrak{s}$. In order to determine the structure equation, select the reference simple root system $\Psi = \{\psi_{\text{sh}}, \psi_{\text{lo}}\}$ of $\Phi(\mathfrak{g}, \mathfrak{t})$ which gives the reference Borel subalgebra $\mathfrak{b}^{\text{ref}}$. Set $\mathfrak{b}_{\mathfrak{k}} := \mathfrak{b} \cap \mathfrak{k}$. The corresponding simple root system $\Psi_{\mathfrak{k}} = \{\beta^{(1)}, \beta^{(2)}\}$ of the root system of $\mathfrak{k}$ is related to $\psi_{\text{sh}}, \psi_{\text{lo}}$ as follows: $\beta^{(1)} = \psi_{\text{sh}}$ and $\beta^{(2)} = 3\psi_{\text{sh}} + 2\psi_{\text{lo}}$.

According to Corollary 18.3.6, given $\mathfrak{b}_{\mathfrak{k}}^{\text{ref}} \supset \mathfrak{t}$, the number of different Borel subalgebras $\mathfrak{b} \supset \mathfrak{t}$, which contain $\mathfrak{b}_{\mathfrak{k}}$ is $|W|/|W_{\mathfrak{k}}| = |W(\text{G}_2)|/|W(\mathfrak{so}(4;\mathbb{C}))| = 12/4 = 3$. Below, we depict the three positive systems. They are given by $\mathfrak{b}^{\text{ref}} = \mathfrak{b}_{[1]}$, $\mathfrak{b}_{[2]}$ and $\mathfrak{b}_{[3]}$, each of which containing $\mathfrak{b}_{\mathfrak{k}}^{\text{ref}}$. The light-gray cone is the Weyl chamber of $\mathfrak{b}_{\mathfrak{k}}^{\text{ref}}$ and the dark-gray cones correspond to the three Weyl chambers which are determined by the three Borels.

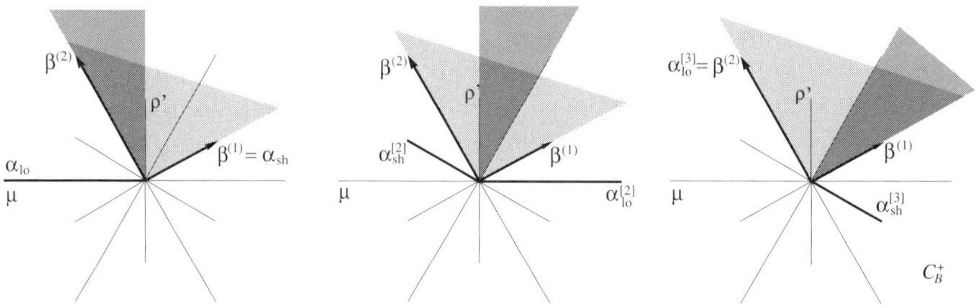

The present situation is not covered by the String Lemma. However, the information needed for the Cohomological Lemma (see Section 18.6B) can be directly derived from the above information. A simple check shows that $\mu := \psi_{\text{lo}} = -\frac{3}{2}\beta^{(1)} - \frac{1}{2}\beta^{(2)}$ is the only Bott-regular weight of index one such that

$$\lambda(\mu) = [\mu + \rho_{\mathfrak{k}}] - \rho_{\mathfrak{k}} = \rho_{\mathfrak{k}} = \xi_1^{(1)} + \xi_1^{(2)} = \tfrac{1}{2}\beta^{(1)} + \tfrac{1}{2}\beta^{(2)} \neq 0.$$

(Of course, $-\beta^{(1)}$ and $-\beta^{(2)}$ are also Bott-regular weights of index one, but in these cases $[-\beta^{(j)} + \rho_{\mathfrak{k}}] - \rho_{\mathfrak{k}} = 0$; by Lemma 18.4.12, trivial representations do not contribute to $H^0(C; \mathcal{O}(\mathbb{N}_Z(C))))$. Note that $\lambda(\mu)$ is the highest weight of the tensor representation $V^{\text{std}}_{SL(2;\mathbb{C})} \otimes V^{\text{std}}_{SL(2;\mathbb{C})}$, i.e., the highest weight of $V^{\text{std}}_{SO(4;\mathbb{C})}$.

There are 3 nonisomorphic flag manifolds which are homogeneous under the complex exceptional Lie group G_2, namely, the full flag manifold $X = G/B$, $Z_{\text{sh}} := G/Q_\Phi$, with $\Phi = \{\psi_{\text{sh}}\}$, and $Z_{\text{lo}} := G/Q_\Phi$ with $\Phi = \{\psi_{\text{lo}}\}$.

Theorem 20.4.1. *The infinitesimal cycle spaces for the case $\mathfrak{g}_0 = \mathfrak{g}_{2,A_1A_1}$ are listed below. The set of highest weights which may occur for the isotropy representation transversal to $G{\cdot}[C]$ consists of the single weight $\lambda(V^{\text{std}}_{SO(4)}) = \xi_1^{(1)} + \xi_1^{(2)}$.*

- *There are three closed K-orbits in the full flag manifold X. For two, the corresponding components $\mathcal{C}_{[C_j]}(X)$ of the cycle spaces are bigger than the closures of $G{\cdot}[C_j]$ and at the infinitesimal level $T_{[C_j]}\mathcal{C}(X) = T_{[C_j]}(G{\cdot}[C_j]) \oplus V^{\text{std}}_{SO(4)}$. For the third, $T_{[C]}\mathcal{C}(X) = T_{[C]}(G{\cdot}[C])$.*

- *There are two closed K-orbits in Z_{sh}. The corresponding components of the cycle space $\mathcal{C}(Z_{\text{sh}})$ are $T_{[C_1]}\mathcal{C}(Z_{\text{sh}}) = T_{[C_1]}(G\cdot[C_1])$ and $T_{[C_2]}\mathcal{C}(Z_{\text{sh}}) = T_{[C_2]}(G\cdot[C_2]) \oplus V^{\text{std}}_{SO(4)}$.*
- *In the flag Z_{lo} there are two closed K-orbits and each has $T_{[C]}\mathcal{C}(Z_{\text{lo}}) = T_{[C]}(G\cdot[C]) \oplus V^{\text{std}}_{SO(4)}$.*

Remark. The Lie algebra $\mathfrak{aut}(Z_{\text{lo}})$ of the automorphism group of $Z_{\text{lo}} = \mathrm{G}_2/Q_{\text{lo}}$ is $\mathfrak{so}(7;\mathbb{C})$, which is considerably larger than $\mathfrak{g}_2(\mathbb{C})$. On the other hand, $\mathfrak{aut}(Z_{\text{sh}}) = \mathfrak{g}_2(\mathbb{C})$. ◇

20.5 Final table

In this section we summarize the computation of $T_{[C]}\mathcal{C}(Z)$ for all base cycles C and flag manifolds Z. We have observed quite early on (see Theorem 18.4.13) that for all real forms not contained in Tables 18.5.1 and 18.5.2, we have

(20.5.1) $\quad T_{[C]}\mathcal{C}(Z) = T_{[C]}(G\cdot[C])$ for all corresponding base cycles and flags.

In the last two chapters we showed that there are still some real forms among those in these tables for which (20.5.1) nevertheless holds. In the following table, we list all those real forms for which there exist cycles and flags such that $T_{[C]}\mathcal{C}(Z) = T_{[C]}(G\cdot[C]) \oplus \bigoplus_{\lambda\in M(C,Z)} V^{\lambda}$ with $M(C,Z) \neq \emptyset$. For a given real form $\mathfrak{g}_0$ we list all possible nonempty sets $M(C,Z)$ which occur. These are the sets of highest weights that occur for the component of the K-isotropy representation at $[C]$ which is transversal to the orbit $G\cdot[C]$. For the precise detail on the cycles C and an flag manifolds Z where these occur, we refer to the corresponding theorems in Chapters 19 and 20.

We write $\xi_1, \xi_2, \ldots$ for the fundamental weights of a simple Lie algebras of types $A_r, \ldots, G_2$ using the Bourbaki order. In the case when $\mathfrak{k} = \mathfrak{k}' \oplus \mathfrak{z}_{\mathfrak{k}}$ is not semisimple (i.e., $\mathfrak{g}_0$ is of hermitian type) $\chi : \mathfrak{t} \to \mathbb{C}$ denotes the linear functional determined by $\chi(J_0) = i$ and $\chi(\mathfrak{t}') = 0$, where J_0 denotes the element in $\mathfrak{k}_0$ which defines the complex structure on the real tangent space $\mathfrak{s}_0$ at a base point of the associated bounded symmetric domain G_0/K_0.

Table 20.5.2. Transversal Isotropy Representation on the Cycle Space

Real form $\mathfrak{g}_0$	Type of $\mathfrak{k}$	Occurring nonempty sets $M(C, Z)$
$\mathfrak{sl}(3;\mathbb{R})$	A_1	$\{2\xi_1\}$
$\mathfrak{sl}(2r+1,\mathbb{R})$ $r \geqq 2$	B_r	$\{\xi_1\}$, $\{\xi_2\}$
$\mathfrak{sl}(4;\mathbb{R})$	$A_1^{(1)} \times A_1^{(2)}$	$\{2\xi^{(1)}\}$, $\{2\xi^{(2)}\}$, $\{2\xi^{(1)}, 2\xi^{(2)}\}$
$\mathfrak{sl}(6;\mathbb{R})$	A_3	$\{\xi_1+\xi_3\}$
$\mathfrak{sl}(2r;\mathbb{R})$ $r \geqq 4$	D_r	$\{\xi_2\}$
$\mathfrak{so}(3, 2q+1)$ $q \geqq 2$	$A_1^{(1)} \times B_q^{(2)}$	$\{2\xi_1^{(1)}\}$, $\{\xi_1^{(2)}\}$, $\{2\xi_1^{(1)}, \xi_1^{(2)}\}$
$\mathfrak{so}(2p+1, 2q+1)$ $p, q \geqq 2$	$B_p^{(1)} \times B_q^{(2)}$	$\{\xi_1^{(1)}\}$, $\{\xi_1^{(2)}\}$, $\{\xi_1^{(1)}, \xi_1^{(2)}\}$
$\mathfrak{so}(2, 2q+1)$ $q \geqq 1$	$B_q \times \mathbb{C}$	$\{\chi\}$, $\{-\chi\}$
$\mathfrak{so}(4, 2q+1)$ $q \geqq 1$	$A_1^{(1)} \times A_1^{(2)} \times B_q$	$\{\xi_1^{(1)} + \xi_1^{(2)}\}$
$\mathfrak{so}(6, 2q+1)$ $q \geqq 1$	$A_3^{(1)} \times B_q^{(2)}$	$\{\xi_2^{(1)}\}$
$\mathfrak{so}(2p, 2q+1)$ $p \geqq 4, q \geqq 1$	$D_p^{(1)} \times B_q^{(2)}$	$\{\xi_1^{(1)}\}$
$\mathfrak{sp}(r;\mathbb{R})$ $r \geqq 3$	$A_{r-1} \times \mathbb{C}$	$\{\xi_2' + \chi\}$, $\{\xi_{r-2}' - \chi\}$ $\{\xi_2' + \chi, \xi_{r-2}' - \chi\}$
$\mathfrak{g}_{2(2)}$	$A_1^{(1)} \times A_1^{(2)}$	$\{\xi_1^{(1)} + \xi_1^{(2)}\}$
$\mathfrak{so}(2r+1;\mathbb{C})$ $r \geqq 2$	B_r	$\{\xi_1\}$
$\mathfrak{sp}(r;\mathbb{C})$, $r \geqq 3$	C_r	$\{\xi_2\}$
$\mathfrak{g}_2(\mathbb{C})$	G_2	$\{\xi_1\}$

References

[AkG] D. N. Akhiezer and S. Gindikin, On the Stein extensions of real symmetric spaces, *Math. Ann.*, **286** (1990), 1–12.

[AnG] A. Andreotti and H. Grauert, Théorèmes de finitude pour la cohomologie des espaces complexes, *Bull. Soc. Math. France*, **90** (1962), 193–259.

[AN] A. Andreotti and F. Norguet, Problème de Levi et convexité holomorphe pour les classes de cohomologie, *Ann Scuola Norm. Sup. Pisa*, **20** (1966), 197–241.

[AB] M. Atiyah and R. Bott, The Yang-Mills equations over Riemann surfaces, *Philos. Trans. Roy. Soc. London Ser.* A, **308** (1982), 523–615.

[Ara] S. Araki, On root systems and an infinitesimal classification of irreducible symmetric spaces, *J. Math. Osaka City Univ.*, **13** (1962), 197–241.

[Ba] L. Barchini, Stein extensions of real symmetric spaces and the geometry of the flag manifold, *Math. Ann.*, **326** (2003), 331–346.

[B1] D. Barlet, Espace analytique réduit des cycles analytiques complexes compacts d'un espace analytique complexe de dimension finie, in *Fonctions de plusieurs variables complexes* II (*Séminaire François Norguet*, 1974–1975), Lecture Notes in Mathematics 482, Springer-Verlag, Berlin, New York, Heidelberg, 1975, 1–158.

[B2] D. Barlet, Convexité de l'espace des cycles, *Bull. Soc. Math. France*, **106** (1978), 373–397.

[BCR] J. Bochnak, M. Coste, and M.-F. Roy, *Real Algebraic Geometry*, Ergebnisse der Mathematik und ihrer Grenzgebiete (3) 36, Springer-Verlag, Berlin, 1998.

[BKa] D. Barlet and M. Kaddar, Incidence divisor, *Internat. J. Math.*, **14** (2003), 339–359.

[BK] D. Barlet and V. Koziarz, Fonctions holomorphes sur l'espace des cycles: La méthode d'intersection, *Math. Res. Lett.*, **7** (2000), 537–550.

[BM1] D. Barlet and J. Magnusson, Intégration de classes de coholomologie méromorphes et diviseurs d'incidence, *Ann. Sci. École Norm. Sup.*, **31** (1998), 811–841.

[BM2] D. Barlet and J. Magnusson, Transfer de l'amplitude du fibré normal au diviseur d'incidence, *J. Reine Angew. Math.*, **513** (1999), 71–95.

[BM3] D. Barlet and J. Magnusson, Integration of meromorphic cohomology classes and applications, *Asian J. Math.*, **8** (2004), 173–214.

[BE] R. J. Baston and M. G. Eastwood, *The Penrose Transform: Its Interaction with Representation Theory*, Clarendon Press, Oxford, UK, 1989.

[BER] M. Salah Baouendi, P. Ebenfelt, and L. P. Rothschild, *Real Submanifolds in Complex Space and Their Mappings*, Princeton Mathematical Series 47, Princeton University Press, Princeton, NJ, 1999.

[BBD] A. Beauville, J.-P. Bourguignon, and M. Demazure, *Géométrie des Surfaces K3: Modules et Périodes* (*Séminaire Palaiseau* 1981–1982), Astérisque 185, Société Mathématique de France, Paris, 1985.

[Be] M. Berger, Les espaces symétriques noncompacts, *Ann. Sci. École Norm. Sup.* (3), **74** (1957), 85–177.

[Besse] A. L. Besse, *Einstein Manifolds*, Ergebnisse der Mathematik und ihre Grenzgebiete (3) 10, Springer-Verlag, Berlin, 1987.

[Bir] D. Birkes, Orbits of linear algebraic groups, *Ann. Math.*, **93** (1971), 459–475.

[Bo1] A. Borel, Groupes linéaires algébriques, *Ann. Math.*, **64** (1956), 20–82.

[Bo2] A. Borel, *Linear Algebraic Groups*, 2nd enlarged ed., Graduate Texts in Mathematics 126, Springer-Verlag, New York, 1991.

[BoHC] A. Borel and Harish-Chandra, Arithmetic subgroups of alagebraic groups, *Ann. Math.*, **75** (1962), 485–535.

[BoH1] A. Borel and F. Hirzebruch, Characteristic classes and homogeneous spaces I, *Amer. J. Math.*, **30** (1958), 458–538.

[BoH2] A. Borel and F. Hirzebruch, Characteristic classes and homogeneous spaces II, *Amer. J. Math.*, **31** (1958), 315–382.

[BoS] A. Borel and J. de Siebenthal, Les sous-groupes fermés de rang maximum des groupes de Lie clos, *Comm. Math. Helv.*, **23** (1949), 200–221.

[BoT] A. Borel and J. Tits, Groupes réductifs, *Publ. Math. IHES*, **27** (1965), 55–150.

[Bot1] R. Bott, An application of the Morse theory to the topology of Lie groups, *Bull. Soc. Math. France*, **84** (1956), 251–281.

[Bot2] R. Bott, Homogeneous vector bundles, *Ann. Math.* (2), bf 66 (1957), 203–248.

[Bou] N. Bourbaki, *Éléments de mathématique: Groupes et algèbres de Lie*, Hermann, Paris, 1975, Chapitres 7 et 8.

[Bre] G. E. Bredon, *Introduction to Compact Transformation Groups*, Pure and Applied Mathematics 46, Academic Press, New York, 1972.

[Br] R. Bremigan, Quotients for algebraic group actions over non-algebraically closed fields, *J. Reine Angew. Math.*, **453** (1994), 21–47.

[BF] R. Bremigan and G. Fels, CR geometry and Olshanskii domains, to appear.

[BL] R. Bremigan and J. Lorch, Orbit duality for flag manifolds, *Manuscripta Math.*, **109** (2002), 233–261.

[Bru] F. Bruhat, Sur les représentations induites des groupes de Lie, *Bull. Soc. Math. France*, **84** (1956), 97–205.

[Bu] N. Buchdahl, On the relative DeRham sequence, *Proc. Amer. Math. Soc.*, **87** (1983), 363–366.

[Bur] D. Burns, Curvature of Monge–Ampère foliations and parabolic manifolds, *Ann. Math.* (2), **115** (1982), 349–373.

[BHH] D. Burns, S. Halverscheid, and R. Hind, The geometry of Grauert tubes and complexification of symmetric spaces, *Duke J. Math.*, **118** (2003), 465–491.

[BPV] W. Barth, C. Peters, and A. VandeVen, *Compact Complex Surfaces*, Ergebnisse der Mathematik und ihrer Grenzgebiete (3) 4, Springer-Verlag, Berlin, 1984.

[Car] H. Cartan, Quotient d'un espace analytique par un groupe d'automorphismes, in *Algebraic Geometry and Topology: A Symposium in Honor of S. Lefschetz*, Princeton University Press, Princeton, NJ, 1957, 90–102.

[C] R. J. Crittenden, Minimum and conjugate points in symmetric spaces, *Canadian J. Math.*, **14** (1962), 320–328.

[dCP] C. De Concini and C. Procesi, Complete symmetric varieties, in *Invariant Theory* (*Montecatini*, 1982), Lecture Notes in Mathematics 996, Springer-Verlag, Berlin, 1983, 1–44.

[dS] J. de Siebenthal, Sur les groupes de Lie compacts non connexes, *Com.. Math. Helv.*, **31** (1956), 41–89.

[DG] F. Docquier and H. Grauert, Levisches Problem und Rungescher Satz für Teilgebiete Steinscher Mannifaltigkeiten, *Math. Ann.*, **140** (1960), 94–123.

[D] J. Dufresnoy, Théorie nouvelle des familles complex normales: Applications à l'étude des fonctions algébroïdes, *Ann. Sci. École Norm. Sup.*, **61** (1944), 1–44.

[DZ] E. G. Dunne and R. Zierau, Twistor theory for indefinite Kähler symmetric spaces, *Contemp. Math.*, **154** (1993), 117–132.

[Fe1] G. Fels, A remark on homogeneous locally symmetric spaces, *Transform. Groups*, **2**-3 (1997), 269–277.

[Fe2] G. Fels, *On Complex Analytic Cycle Spaces of Flag Domains*, Habilitationsschrift, Universität Tübingen, Tübingen, Germany, 2004.

[FH] G. Fels and A. T. Huckleberry, Characterization of cycle domains via Kobayashi hyperbolicity, *Bull. Soc. Math. France*, **133** (2005), 121–144.

[Fr] A. Frölicher, Zur Differentialgeometrie der komplexen Strukturen, *Math. Ann.*, **129** (1955) 50–95.

[GF] H. Grauert and K. Fritzsche, *From Holomorphic Functions to Complex Manifolds*, Springer-Verlag, Heidelberg, 2001.

[GM] S. Gindikin and T. Matsuki, Stein extensions of riemannian symmetric spaces and dualities of orbits on flag manifolds, *Transform. Groups*, **8** (2003), 333–376.

[Gra] H. Grauert, On Levi's problem and the imbedding of real-analytic manifolds, *Ann. Math.*, **68** (1958), 460–472.

[GR1] H. Grauert and R. Remmert, *Coherent Analytic Sheaves*, Grundlehren der mathematischen Wissenschaften, Springer-Verlag, New York, 1984.

[GR2] H. Grauert and R. Remmert, *Theory of Stein Spaces*, reprint of 1st ed., Classics in Mathematics, Springer-Verlag, New York, 1979, 2004.

[Gr1] P. A. Griffiths, Some geometric and analytic properties of homogeneous complex manifolds I: Sheaves and cohomology, *Acta Math.*, **110** (1963), 115–155.

[Gr2] P. A. Griffiths, Some geometric and analytic properties of homogeneous complex manifolds II: Deformation and bundle theory, *Acta Math.*, **110** (1963), 157–208.

[Gr3] P. A. Griffiths, Periods of integrals on algebraic manifolds I, *Amer. J. Math.*, **90** (1968), 568–626.

[Gr4] P. A. Griffiths, Periods of integrals on algebraic manifolds II, *Amer. J. Math.*, **90** (1968), 805–865.

[Gr5] P. A. Griffiths, Periods of integrals on algebraic manifolds: Summary of main results and discussion of open problems, *Bull. Amer. Math. Soc.*, **76** (1970), 228–296.

[Gr6] P. A. Griffiths, Hodge theory and geometry, *Bull. London Math. Soc.*, **36** (2004), 721–757.

[GrS] P. A. Griffiths and W. Schmid, Locally homogeneous complex manifolds, *Acta Math.*, **123** (1969), 253–302.

[GuS] V. Guillemin and M. Stenzel, Grauert tubes and the homogeneous Monge–Ampere equation I, II, *J. Differential Geom.*, **34** (1991), 561–570, **35** (1992), 627–641.

[Gu] R. C. Gunning, *Introduction to Holomorphic Functions of Several Variables: Vol.* I: *Function Theory, Vol.* II: *Local Theory, Vol.* III: *Homological Theory*, Wadsworth/Brooks–Cole, Stamford, CT, 1990.

[GuR] R. C. Gunning and H. Rossi, *Analytic Functions of Several Complex Variables*, Prentice–Hall, Engelwood Cliffs, NJ, 1965.

[Ha] S. Halverscheid, *Maximal Domains of Definition of Adapted Complex Structures for Symmetric Spaces of Non-Compact Type*, Schriftenreihe des Graduiertenkollegs

"Geometrie und mathematische Physik" 39, Ruhr-Universität Bochum, Bochum, Germany, 2001.

[HC0] Harish-Chandra, On a lemma of F. Bruhat, *J. Maths. Pures Appl.*, **35** (1956), 203–210.

[HC1] Harish-Chandra, Representations of semisimple Lie groups IV, *Amer. J. Math.*, **77** (1955), 743–777.

[HSc] H. Hecht and W. Schmid, On integrable representations of a semisimple Lie groups, *Math. Ann.*, **220** (1976), 147–150.

[HH] P. Heinzner and A. T. Huckleberry, *Complex Analytic Invariant Theory from the Hamiltonian Viewpoint*, in preparation (book).

[HeSch] A. Helminck and G. Schwarz, Orbits and invariants associated with a pair of commuting involutions, *Duke Math. J.*, **106**-2 (2001), 237–279.

[Hel] S. Helgason, *Differential Geometry, Lie Groups, and Symmetric Spaces*, Pure and Applied Mathematics 80, Academic Press, New York, 1978.

[HO] J. Hilgert and G. Ólafsson, *Causal Symmetric Spaces: Geometry and Harmonic Analysis*, Perspectives in Mathematics 18, Academic Press, New York, 1997.

[Hoch] G. Hochschild, *The Structure of Lie Groups*, Holden–Day, San Francisco, 1965.

[HoH] J. Hong and A. T. Huckleberry, On closures of cycle spaces of flag domains, to appear.

[H] A. T. Huckleberry, On certain domains in cycle spaces of flag manifolds, *Math. Ann.*, **323** (2002), 797–810.

[HL] A. T. Huckleberry and L. Livorni, A classification of homogeneous surfaces, *Canadian J. Math.*, **33** (1981), 1097–1110.

[HN] A. T. Huckleberry and B. Ntatin, Cycle spaces of G-orbits in $G^{\mathbb{C}}$-manifolds, *Manuscripta Math.*, **112** (2003), 433–440.

[HS] A. T. Huckleberry and A. Simon, On cycle spaces of flag domains of $SL_n(\mathbb{R})$, *J. Reine Angew. Math.*, **541** (2001), 171–208.

[HW1] A. T. Huckleberry and J. A. Wolf, Flag duality, *Ann. Global Anal. Geom.*, **18** (2000), 331–340.

[HW2] A. T. Huckleberry and J. A. Wolf, On cycle spaces of real forms of $SL_n(\mathbb{C})$, in *Complex Geometry: A Collection of Papers Dedicated to Hans Grauert*, Springer-Verlag, New York, 2002, 111–133.

[HW3] A. T. Huckleberry and J. A. Wolf, Schubert varieties and cycle spaces, *Duke Math. J.*, **120** (2003), 229–249.

[HW4] A. T. Huckleberry and J. A. Wolf, Injectivity of the double fibration transform for cycle spaces of flag domains, *J. Lie Theory*, **14** (2004), 509–522.

[Hu1] J. E. Humphreys, *Introduction to Lie Algebras and Representation Theory*, Graduate Texts in Mathematics 9, Springer-Verlag, Berlin, New York, Heidelberg, 1972.

[Hu2] J. E. Humphreys, *Linear Algebraic Groups*, Graduate Texts in Mathematics 21, Springer-Verlag, Berlin, New York, Heidelberg, 1975.

[KZ] W. Kaup and D. Zaitsev, On the CR structure of compact group orbits associated with bounded symmetric domains, *Invent. Math.*, **153** (2003), 45–104.

[K] S. Kobayashi, *Hyperbolic Complex Spaces*, Grundlehren der mathematischen Wissenschaften 318, Springer-Verlag, New York, 1998.

[KMS] I. Kolář, P. W. Michor, and J. Slovák, *Natural Operations in Differential Geometry*, Springer-Verlag, New York, 1993.

[Kna] A. W. Knapp, *Lie Groups: Beyond an Introduction*, 2nd ed., Progress in Mathematics 140, Birkhäuser Boston, Cambridge, MA, 2002.

[KW] A. Korányi and J. A. Wolf, Realization of hermitian symmetric spaces as generalized half-planes, *Ann. Math.*, **81** (1965), 265–288.

[Kos1] B. Kostant, On the conjugacy of real Cartan subalgebras I, *Proc. Nat. Acad. Sci. USA*, **41** (1955), 967–970.

[Kos2] B. Kostant, Lie algebra cohomology and the generalized Borel–Weil theorem, *Ann. Math.*, **74** (1961), 329–387.

[KR] B. Kostant and S. Rallis, Orbits and representations associated with symmetric spaces, *Amer. J. Math.*, **93** (1971), 753–809.

[Kow] O. Kowalski, *Generalized Symmetric Spaces*, Lecture Notes in Mathematics 805, Springer-Verlag, New York, 1980.

[Kra] H.-P. Kraft, *Geometrische Methoden in der Invariantentheorie*, Vieweg Verlag, Brauschweig, Germany, 1984.

[KS] B. Krötz and R. Stanton, Holomorphic extension of representations II: Geometry and harmonic analysis, preprint.

[LS] L. Lempert and R. Szőke, Global solutions of the homogeneous complex Monge-Ampère equation and complex structures on the tangent bundles of Riemannian manifolds, *Math. Ann.*, **290** (1991), 409–428.

[Loo] O. Loos, *Symmetric Spaces* I *and* II, W. A. Benjamin, New York, 1969.

[M] Y. I. Manin, *Gauge Field Theory and Complex Geometry*, Springer-Verlag, New York, 1988.

[M1] T. Matsuki, The orbits of affine symmetric spaces under the action of minimal parabolic subgroups, *J. Math. Soc. Japan*, **31** (1979), 331–357.

[M2] T. Matsuki, Orbits on affine symmetric spaces under the action of parabolic subgroups, *Hiroshima Math. J.*, **12** (1982), 307–320.

[M3] T. Matsuki, Double coset decompositions of reductive Lie groups arising from two involutions, *J. Algebra*, **197** (1997), 49–91.

[M4] T. Matsuki, Stein extensions of Riemann symmetric spaces and some generalizations, *J. Lie Theory*, **13** (2003), 565–572.

[MUV] I. Mirkovič, K. Uzawa, and K. Vilonen, Matsuki correspondence for sheaves, *Invent. Math.*, **109** (1992), 231–245.

[Moh] S. Mohrdieck, Conjugacy classes of nonconnected semisimple algebraic groups, *Transform. Groups*, **8**-4 (2003), 377–395.

[Mo] C. C. Moore, Compactifications of symmetric spaces II (the Cartan domains), *Amer. J. Math.*, **86** (1964), 358–378.

[Mos] G. D. Mostow, Some new decomposition theorems for semisimple Lie groups, *Mem. Amer. Math. Soc.*, **14** (1955), 31–54.

[NS] F. Norguet and Y.-T. Siu, Holomorphic convexity of spaces of cycles, *Bull. Soc. Math. France*, **105** (1977), 193–223.

[N] B. Ntatin, *On the Cycle Spaces Associated to Orbits of a Semisimple Lie Group*, doctoral dissertation, Ruhr-Universität Bochum, Bochum, Germany, 2004.

[NO] M. S. Narasimhan and K. Okamoto, An analogue of the Borel–Weil–Bott theorem for hermitian symmetric pairs of noncompact type, *Ann. Math.*, **91** (1970), 486–511.

[N1] J. D. Novak, Parameterizing maximal compact subvarieties, *Proc. Amer. Math. Soc.*, **124** (1996), 969–975.

[N2] J. D. Novak, *Explicit Realizations of Certain Representations of* $Sp(n;\mathbb{R})$ *via the Penrose Transform*, Ph.D. thesis, Oklahoma State University, Stillwater, OK, 1996.

[O1] A. L. Onishchik, Inclusion relations among transitive compact transformation groups, *Trudy Moskov. Mat. Obšč.*, **11** (1962), 199–142.

[O2] A. L. Onishchik, *Topology of Transitive Transformation Groups*, Johann Ambrosius Barth, Leipzig, Berlin, Heidelberg, 1994.

[PR1] C. M. Patton and H. Rossi, Unitary structures on cohomology, *Trans. Amer. Math. Soc.*, **290** (1985), 235–258.

[PR2] C. M. Patton and H. Rossi, Cohomology on complex homogeneous manifolds with compact subvarieties, *Contemp. Math.*, **58** (1986), 199–211.

[RSW] J. Rawnsley, W. Schmid, and J. A. Wolf, Singular unitary representations and indefinite harmonic theory, *J. Functional Anal.*, **51** (1983), 1–114.

[S1] W. Schmid, *Homogeneous Complex Manifolds and Representations of Semisimple Lie Groups*, Ph.D. thesis, University of California at Berkeley, Berkely, CA, 1967.

[S2] W. Schmid, On a conjecture of Langlands, *Ann. Math.*, **93** (1971), 1–42.

[S3] W. Schmid, Variation of Hodge structure: The singularities of the period mapping, *Invent. Math.*, **22** (1973), 211–319.

[S4] W. Schmid, Some properties of square integrable representations of semisimple Lie groups, *Ann. Math.*, **102** (1975), 535–564.

[S5] W. Schmid, L^2 cohomology and the discrete series, *Ann. Math.*, **103** (1976), 375–394.

[SW] W. Schmid and J. A. Wolf, A vanishing theorem for open orbits on complex flag manifolds, *Proc. Amer. Math. Soc.*, **92** (1984), 461–464.

[Sek] J. Sekiguchi, Remarks on real nilpotent orbits of a symmetric pair, *J. Math. Soc. Japan*, **39**-1 (1987), 127–138.

[Ser] J.-P. Serre, Représentations linéaires et espaces homogènes Kählerians des groups de Lie compacts, in *Séminaire Bourbaki* 1954 Secrétariat Mathématique, Université Paris, Paris, 1954.

[Si] Y.-T. Siu, Every K3 surface is Kähler, *Invent. Math.*, **73** (1983), 139–150.

[Su] M. Sugiura, Conjugate classes of Cartan subalgebras in real semi-simple Lie algebras, *J. Math. Soc. Japan*, **11** (1959), 374–434.

[Sz] R. Szőke, Complex structures on tangent bundles of Riemannian manifolds, *Math. Ann.*, **291** (1991), 409–428.

[TW] J. A. Tirao and J. A. Wolf, Homogeneous holomorphic vector bundles, *J. Math. Mech. (Indiana Univ. Math. J.)*, **20** (1970), 15–31.

[T0] J. Tits, *Sur certains classes d'espaces homogènes de groupes de Lie*, memoir, Belgian Academy of Sciences, Brussels, 1955.

[T1] J. Tits, Espaces homogènes complexes compacts, *Comm. Math. Helv.*, **37** (1962), 111–120.

[V] V. S. Varadarajan, *Lie Groups, Lie Algebras, and Their Representations*, Graduate Texts in Mathematics 102, Springer-Verlag, Berlin, New York, Heidelberg, 1984.

[V1] C. Voisin, *Hodge Theory and Complex Algebraic Geometry* I, Cambridge University Press, Cambridge, UK, 2002.

[V2] C. Voisin, *Hodge Theory and Complex Algebraic Geometry* II, Cambridge Studies in Advanced Mathematics 77, Cambridge University Press, Cambridge, UK, 2003.

[WaW] N. R. Wallach and J. A. Wolf, Completeness of Poincaré series for automorphic forms associated to the integrable discrete series, in *Representation Theory of Reductive Groups*, Progress in Mathematics 40, Birkhäuser Boston, Cambridge, MA, 1983, 265–281.

[War] G. Warner, *Harmonic Analysis on Semisimple Lie Groups* I, Grundlehren der mathematischen Wissenschaften 188, Springer-Verlag, Berlin, New York, Heidelberg, 1972.

[We1] R. O. Wells, Jr., Parameterizing the compact submanifolds of a period matrix domain by a Stein manifold, in *Symposium on Several Complex Variables*, Lecture Notes in Mathematics 104, Springer-Verlag, Berlin, New York, Heidelberg, 1971, 121–150.

[We2] R. O. Wells, Jr., *Differential Analysis on Complex Manifolds*, Graduate Texts in Mathematics 65, Springer-Verlag, Berlin, New York, Heidelberg, 1980.

[WeW] R. O. Wells, Jr., and J. A. Wolf, Poincaré series and automorphic cohomology on flag domains, *Ann. Math.*, **105** (1977), 397–448.

[Wi1] F. L. Williams, On the finiteness of the L^2 automorphic cohomology of a flag domain, *J. Functional Anal.*, **72** (1987) 33–43.

[Wi2] F. L. Williams, On the dimension of spaces of automorphic cohomology, in *Algebraic Groups and Related Topics*, Advanced Studies in Pure Mathematics 6, Kinokuniya–North-Holland, Amsterdam, 1985, 1–15.

[Wi3] F. L. Williams, Finite spaces of non-classical Poincaré theta series, *Contemp. Math.*, **53** (1986), 543–554.

[W0] J. A. Wolf, Isotropic manifolds of indefinite metric, *Comm. Math. Helv.*, **39** (1964), 21–64.

[W1] J. A. Wolf, Discrete groups, symmetric spaces and global holonomy, *Amer. J. Math.*, **84** (1962), 527–542.

[W2] J. A. Wolf, The action of a real semisimple Lie group on a complex manifold I: Orbit structure and holomorphic arc components, *Bull. Amer. Math. Soc.*, **75** (1969), 1121–1237.

[W3] J. A. Wolf, The action of a real semisimple Lie group on a complex manifold II: Unitary representations on partially holomorphic cohomology spaces, *Mem. Amer. Math. Soc.*, **138** 1974.

[W4] J. A. Wolf, Fine structure of hermitian symmetric spaces, in W. M. Boothby and G. L. Weiss, eds., *Symmetric Spaces*, Marcel Dekker, New York, 1972, 271–357.

[W5] J. A. Wolf, Completeness of Poincaré series for automorphic cohomology, *Ann. Math.*, **109** (1979), 545–567.

[W6] J. A. Wolf, *Spaces of Constant Curvature*, 5th ed., Publish or Perish, Houston, 1984.

[W7] J. A. Wolf, Geometric realizations of discrete series representations in a nonconvex holomorphic setting, *Bull. Soc. Math. Belgique*, **42** (1990), 797–812.

[W8] J. A. Wolf, The Stein condition for cycle spaces of open orbits on complex flag manifolds, *Ann. Math.*, **136** (1992), 541–555.

[W9] J. A. Wolf, Exhaustion functions and cohomology vanishing theorems for open orbits on complex flag manifolds, *Math. Res. Lett.*, **2** (1995), 179–191.

[W10] J. A. Wolf, Flag manifolds and representation theory, in *Geometry and Representation Theory of Real and p-Adic Groups*, Progress in Mathematics 158, Birkhäuser Boston, Cambridge, MA, 1998, 273–323.

[W11] J. A. Wolf, Real groups transitive on complex flag manifolds, *Proc. Amer. Math. Soc.*, **129** (2001), 2483–2487.

[W12] J. A. Wolf, Hermitian symmetric spaces, cycle spaces, and the Barlet–Koziarz intersection method for construction of holomorphic functions, *Math. Res. Lett.*, **7** (2000), 551–564.

[WG1] J. A. Wolf and A. Gray, Homogeneous spaces defined by Lie group automorphisms I, *J. Differential Geom.*, **2** (1968), 77–114.

[WG2] J. A. Wolf and A. Gray, Homogeneous spaces defined by Lie group automorphisms II, *J. Differential Geom.*, **2** (1968), 115–159.

[WK] J. A. Wolf and A. Korányi, Generalized Cayley transformations of bounded symmetric domains, *Amer. J. Math.*, **87** (1965), 899–939.

[WZ0] J. A. Wolf and R. Zierau, Cayley transforms and orbit structure in complex flag manifolds, *Transform. Groups*, **2** (1997), 391–405.

[WZ1] J. A. Wolf and R. Zierau, Linear cycle spaces in flag domains, *Math. Ann.*, **316** (2000), 529–545.

[WZ2] J. A. Wolf and R. Zierau, Holomorphic double fibration transforms, in R. S. Doran and V. S. Varadarajan, eds., *The Mathematical Legacy of Harish-Chandra: A Celebration of Representation Theory and Harmonic Analysis*, Proceedings of Symposia in Pure Mathematics 68, American Mathematical Society, Providence, RI, 2000, 527–551.

[WZ3] J. A. Wolf and R. Zierau, A note on the linear cycle space for groups of hermitian type, *J. Lie Theory*, **13** (2003), 189–191.

[Y] S.-T. Yau, On the Ricci curvature of a complex Kähler manifold and the complex Monge–Ampère equation I, *Comm. Pure Appl. Math.*, **21** (1978), 339–411.

[Z] O. Zariski, *Algebraic Surfaces*, Ergebnisse der Mathematik und ihrer Grenzgebiete, Springer-Verlag, Heidelberg, 1935; reprinted with new appendices in 1971.

Index

Symbol Index